Y0-BTW-260

INTRODUCTORY STATISTICS AND PROBABILITY FOR ENGINEERING, SCIENCE, AND TECHNOLOGY

PRENTICE-HALL INTERNATIONAL SERIES
IN INDUSTRIAL AND SYSTEMS ENGINEERING

W. J. Fabrycky and J. H. Mize, Editors

BLANCHARD *Logistics: Engineering and Management*
CINLAR *Introduction to Stochastic Processes*
FABRYCKY, GHARE, AND TORGERSEN *Industrial Operations Research*
FRANCIS AND WHITE *Facility Layout and Location: An Analytical Approach*
GOTTFRIED AND WEISMAN *Introduction to Optimization Theory*
KIRKPATRICK *Introductory Statistics and Probability for Engineering, Science, and Technology*
MIZE, WHITE, AND BROOKS *Operations Planning and Control*
OSTWALD *Cost Estimating for Engineering and Management*
SIVAZLIAN AND STANFEL *Analysis of Systems in Operations Research*
SIVAZLIAN AND STANFEL *Optimization Techniques in Operations Research*
WHITEHOUSE *Systems Analysis and Design Using Network Techniques*

INTRODUCTORY STATISTICS AND PROBABILITY FOR ENGINEERING, SCIENCE, AND TECHNOLOGY

ELWOOD G. KIRKPATRICK

School of Industrial Engineering
Purdue University

PRENTICE-HALL, INC., *Englewood Cliffs, New Jersey*

Library of Congress Cataloging in Publication Data

KIRKPATRICK, ELWOOD G
Introductory statistics and probability for engineering, science, and technology.

1. Mathematical statistics. 2. Probabilities.
I. Title.
QA276.K49 519.5 73-23040
ISBN 0-13-501627-4

10 9 8 7 6 5 4 3 2 1

Printed in the United States of America

PRENTICE-HALL INTERNATIONAL, INC., *London*
PRENTICE-HALL OF AUSTRALIA, PTY. LTD., *Sydney*
PRENTICE-HALL OF CANADA, LTD., *Toronto*
PRENTICE-HALL OF INDIA PRIVATE LIMITED, *New Delhi*
PRENTICE-HALL OF JAPAN, INC., *Tokyo*

To Eileen and Tom

CONTENTS

PREFACE

This textbook is an introduction to some of the basic concepts and techniques of statistical analysis. It is directed primarily at the undergraduate technology and engineering student who has completed one introductory course in calculus. No previous background in either probability or statistics is assumed.

The book can be used for a one or two semester course depending on the emphasis and time devoted to end-chapter applications problems. For a two semester sequence, it is presumed that the instructor will complement this book with a formal design of experiments text of his choice. This book has been designed to satisfy the need for an introductory statistics text oriented to engineering applications, a text that is concise, readable by the average student, of moderate length, and not too mathematical. Two basic objectives of the text are: (1) to select and construct subject matter requirements that are attainable by the average student, and (2) to identify and structure topics to introduce the subject to a student who is meeting statistics for the first time.

To satisfy these objectives at this level, it is necessary not only to acquaint the reader with statistical methods but also to indicate the mathematics required to establish the methods on a sound basis. Thus, the presentation is essentially a compromise between a book that gives only statistical techniques and a book that attempts complete mathematical rigor. Some mathematical foundations are given to structure the subject and to stimulate the interested student to pursue specialized and more advanced courses in statistics. A nominal sacrifice of simplicity has been made at the beginning in order to introduce concepts that are fundamental to the development of the subject. For example, the distribution function and characteristic function are introduced in elementary fashion early in Chaps. 3 and 4. However, emphasis throughout the text is primarily on language, method, and applications problems.

An observation made by the author on many occasions is that junior and senior engineering students, who have had one or two statistics courses, frequently demonstrate a remarkable lack of statistical facility in other courses that use statistical tools for quantitative evaluations. One conjecture as to this puzzling end result is that the mathematics of applied statistics is not too difficult, but the amount of time that is required to pursue topics in depth leaves little time for applications and analysis practice. Operating from this assumption, this text is designed to save time wherever possible. Simple functions are used to illustrate concepts. For example, the uniform distribution function is used frequently in Chaps. 3 and 4 to illustrate obtaining distribution moments using either expected value or characteristic function methods. Other simple linear functions are used in examples, exercises, and problems. This is also the reason for an early presentation of the discrete random variable models.

Simple and direct exercises are given regularly throughout each chapter (more rigorous and applied problems are separated into the end-chapter problem assignments). Extensive use is made of graphical explanations and summaries, particularly for the problems of Chaps. 5, 7, 8, and 10. The purpose throughout is to take cognizance of the fact that technology and engineering students simply do not have the time to study statistics in a career fashion.

The author also feels that another important item related to the aforementioned student difficulty is the verbal factor. A first chore for the average reader is to sort the verbiage and condense it to a meaningful structure for himself—a task usually requiring extensive help by the instructor. To simplify this chore, use is made of discussion–paragraph asides blocked out separately from the main paragraphs. Generally, the main paragraphs read continuously and the discussion paragraphs contribute support material—intuitive explanations, of–interest–only derivations, additional argument, detail, exceptions, special cases, and so forth. Student difficulties seem to stem from too much material presented too rapidly and coincidentally. The pedagogical approach throughout this text is that of separation of material. The discussion paragraph asides separate out two types of supplementary material: (1) additional intuitive explanations and aids for the below–average student, and (2) mathematical foundations to interest the above–average student.

In an introductory textbook, the selection of topics, their order of presentation, and extent of coverage are to some degree subjective decisions. There is no one correct sequence and there are a variety of effective teaching combinations. To meet objectives, frequent tradeoffs were made in developing the text, and sophisticated detail was necessarily omitted in some cases. Only elementary probability concepts are presented for the limited purpose of introducing the general statistical topics. A combinatorial rather than a set–theory approach is used. It is felt that the latter view is too abstract for an introductory text, and that the combinatorial approach is more adaptable to an intuitive counting of events and occurrences. Discrete distribution models are introduced early (Chap. 4) to afford simple examples of obtaining distribution moments and also as a means of introducing probabilities of events of the form $X \leq b$ and thereby gradually leading to the use of tables for making probability statements. The topic of sampling inspection (without analysis) is introduced early to obtain practical applications and examples for the binomial and Poisson models. The topics receiving the most coverage are descriptive statistics and hypothesis testing. The latter topic is perhaps the most important single topic in an introductory statistics textbook as it is fundamental to all advanced statistical topics.

The principal quality control topics (Chap. 8) follow immediately the presentation of hypothesis testing (Chap. 7) to further illustrate tests of hypotheses. Since engineering problems often require a presentation of data showing the observed relationship between variables, a separate chapter on curve fitting (Chap. 9) has been included. This treatment, without any of the assumptions required for regression, makes for a less abrupt entry into the regression topic. The chapter on regression that follows also lends itself to a gradual and more meaningful introduction to analysis of variance.

I am grateful to Mrs. Jong-Ping Hsu for her careful review of the manuscript, and to Dr. J. H. Mize, Oklahoma State University, for his review and many helpful suggestions. I am especially grateful to Mr. Kenneth Hopping for his thorough effort on the tedious job of proofreading.

I am indebted to the Literary Executor of the late Sir Ronald A. Fisher, F.R.S., to Dr. Frank Yates, F.R.S., and to Oliver and Boyd, Edinburgh, for permission to reprint Tables of the chi square, *t*–, and *F*–distributions (Appendixes E, F, and G) from their book *Statistical Tables for Biological, Agricultural and Medical Research;* to Dr. Frederick E. Croxton for permission to reprint the normal distribution Table (Appendix D) from his book *Elementary Statistics with Applications in Medicine and the Biological Sciences*, Dover Publications, Inc., 1959; to the *Annals of Mathematical Statistics* and Professors C. L. Ferris, F. E. Grubbs, and C. L. Weaver for permission to reproduce their Operating Characteristic Curves for Significance Tests (Chap. 7); to Dr. George W. Snedecor for permission to reprint a random numbers Table (Appendix I) from his book *Everyday Statistics;* To Professors A. H. Bowker and G. J. Lieberman and the Stanford University Press for permission to reproduce the hypergeometric table (Appendix C) from their book *Tables of the Hypergeometric Probability Distribution*; and to the American Society for Testing Materials for permission to reproduce Table B.2 (Appendix H) from their *A. S. T. M. Manual on Quality Control of Materials.*

West Lafayette, Indiana ELWOOD G. KIRKPATRICK

INTRODUCTORY STATISTICS AND PROBABILITY FOR ENGINEERING, SCIENCE, AND TECHNOLOGY

INTRODUCTION

Statistics is concerned with methods for the collection, analysis, and interpretation of quantitative data in such a way that conclusions based on the data can be objectively evaluated in terms of probability statements. The collection of data and computation of various indices such as averages and percentages is called descriptive statistics. *Systematic drawing of conclusions from the data is called* statistical inference.

1.1 Statistical Inference

The deductive method (or proof) makes inferences or conclusions from accepted principles. On the other hand, the inductive method draws conclusions from several known cases, reasoning from the particular to the general. The technology of deduction is mathematics, while that of induction is statistics. An old science, which has already developed a number of basic principles, can use mathematics to deduce more information from these principles. A young science depends on statistics to develop the basic principles. All sciences use statistical tools for reaching conclusions from experimental data.

1

Although inductive reasoning—going from the particular to the general—can lead to numerous difficulties, it is still possible to be precise about it in terms of the risk involved in making such an inference. The topic of statistical inference may be regarded in this light. The effort is not aimed at making statements each of which is guaranteed to be 100% correct, but rather at making inferences that have a certain probability of being correct.

1.2 Statistical Investigation

The important phases of a statistical investigation are:

1. Definition of the problem.
2. Design of the experiment.
3. Data collection.
4. Data reduction and computations.
5. Analysis.

A formulation or *definition* of the problem includes: (1) statement of the problem, (2) choice of dependent variable (or variables) to be studied, and (3) identification of the independent variables which may affect the dependent or response variable. A number of basic questions are resolved at this stage. Are the variables measurable, and to what degree of accuracy? Are the independent variables to be held constant, or to vary over specified levels, or to vary freely and be averaged out by a process of randomization? Are levels of the independent variables to be set at certain fixed values (e.g., temperature at 80, 90, and 100°F), or are such levels to be set at random among all possible levels? And so forth.

The *design* of the experiment involves decisions regarding: (1) number of observations to be taken, (2) order of experimentation, (3) method of randomization, and (4) mathematical model used to describe the experiment. The number of observations is not arbitrary. One objective of the design is to enable valid conclusions to be drawn from a minimum number of observations. The order in which the observations are made is also important. Some variables are controlled at specified levels during the experiment. However, in any experiment, there are usually a number of variables that cannot be controlled. Randomization of the order of experimentation tends to average out the effects of the uncontrolled variables. Finally, an essential step in the experimental design is to obtain an appropriate mathematical model. The model defines the response variable as a function of the independent variables being studied and also reflects any restrictions imposed on the experiment due to the method of randomization.

The concluding stage of a statistical investigation is the *analysis*. This stage involves analytical explanation of the data information—identifying significant causes and effects, decision making (e.g., accepting or rejecting a hypothesis), estimating, predicting, etc. The analysis may also involve feedback procedures to design further experiments, once certain hypotheses are tenable.

Much of the statistician's technical vocabulary is made up of common words given special meanings. Some of these words have been used in the preceding general discussion. The balance of this chapter is concerned with: (1) an introductory development of the nomenclature of statistics, and (2) a definition of common measures used in descriptive statistics.

1.3 Population and Sample

An *observation* is a recording of information such as a measurement. For example, if a measurement of a product characteristic is made on three product items from the output of a manufacturing process, there are three observations. A mathematical notation for a set of observations is

$$x_1, x_2, \ldots, x_i, \ldots, x_n$$

where the subscript i identifies the observation (e.g., x_1 is the first observation, x_2 the second observation, and so forth). Any arbitrary observation is designated by x_i where the subscript i is variable in the sense that it denotes any of the observations and need only be replaced by the proper number in order to specify a particular observation.

EXAMPLE 1.3.1

Ten observations of lip-opening pressure (p.s.i.) on an automotive seal are

$$x_i\text{: } 11.0, 9.5, 10.5, 10.5, 12.0, 11.0, 11.5, 10.0, 11.5, 11.0.$$

The first observation is $x_1 = 11.0$, the second observation is $x_2 = 9.5$, and so forth. There are $n = 10$ observations, and the last observation is $x_{10} = 11.0$.

A *population* is a conceptual term meaning the totality of items under consideration. A population may be finite or infinite. The size of a population is denoted by N (finite case) or ∞ (infinite case). Infinite populations occur in connection with repetitive events. Given an experiment that may be repeated, each repetition resulting in an outcome, an infinite number of replications of the experiment is possible. In this sense, the population of outcomes is infinite.

EXAMPLE 1.3.2

A manufacturing lot (batch) of 100 automotive seals has the following lip-opening pressure characteristics:

9.0—111
9.5—1111
10.0—~~1111~~ 11
10.5—~~1111~~ ~~1111~~ ~~1111~~ 11
11.0—~~1111~~ ~~1111~~ ~~1111~~ ~~1111~~ ~~1111~~ 111
11.5—~~1111~~ ~~1111~~ ~~1111~~ 111
12.0—~~1111~~ ~~1111~~ 11
12.5—~~1111~~ 111
13.0—11
13.5—1

Three seals have a lip-opening pressure (L.O.P.) of 9.0 p.s.i., four seals have an L.O.P. of 9.5 p.s.i., and so forth. This is the total of the seals under consideration and thus this is a finite population of L.O.P.s of size $N = 100$.

EXAMPLE 1.3.3

An experiment E involving one of the automotive seals referred to in Example 1.3.2 results in an L.O.P. measurement of 10.512 p.s.i. (i.e., an outcome of one trial of E).

A replication of E will result in another L.O.P. measurement, and because of measurement error it is very possible that this outcome will differ from the preceding outcome of 10.512 p.s.i.

In theory, an infinite number of replications of E is possible and thus the population of all possible outcomes of E is infinite.

A *sample* is a part of a population. The sample size is denoted by n. A *random sample* is a sample selected in such a way that each population item has an equal chance of being selected. The 10 observations in Example 1.3.1, if randomly selected, can be viewed as a random sample from the manufacturing lot in Example 1.3.2. The statistical objective is to obtain a sample that is representative of the population. Suppose that an unknown population consists of the numbers 1, 1, 1, 2, 2, 2, 3, 3, 3. An ideal representative sample is 1, 2, 3. If a biased sample is obtained, for example 2, 2, 3, a false inference may be made regarding the population. Based on the information supplied by the biased sample of 2, 2, 3, an inference may be made that the population average is $\frac{7}{3}$ when, in fact, it is 2.

Statistics is concerned with: (1) using sample information to make inferences regarding an unknown population from which the sample is taken, and (2) occasionally, if the character of a population is known, making inferences regarding the nature of samples that are taken from this population.

EXAMPLE 1.3.4

A manufacturer is concerned with the average life span of his product. Destructive testing is necessary to determine average product life. The entire production output cannot be destroyed to obtain an answer.

A small, but carefully tested, part of the production output (sample) makes possible an inference that the average life of the entire output (population) is 10,000 hr.

EXAMPLE 1.3.5

The product specification for a given process operation is 0.500 ± 0.002 in. The process for this operation is an automatic machine that generates output so rapidly that individual output items cannot be measured consecutively as they come from the process.

The process is monitored by measuring samples of the output (population). Each set of sample measurements yields an average measurement value, which is a basis for making an inference that the process is properly centered at 0.500 in.

EXAMPLE 1.3.6

Product is shipped, in lots of 10,000 items each, from a vendor supplier to a purchaser company. It is not economic for the purchaser company's receiving inspection to measure every product item in each lot.

Inspection is performed by taking a random sample from each lot, and either accepting or rejecting the lot on the basis of measurement information yielded by the sample. Each sample makes possible an inference regarding the nature of the lot (population) from which the sample was taken.

1.4 Frequency Distribution

Consider a sample $x_1, x_2, \ldots, x_i, \ldots, x_n$* where some of the x_i may have the same numerical value. It is convenient to represent the sample by a *frequency distribution*

$$\begin{matrix} x_1, x_2, \ldots, x_j, \ldots, x_k & \\ f_1, f_2, \ldots, f_j, \ldots, f_k & \sum_{j=1}^{k} f_j = n \end{matrix} \tag{1.1}$$

where f_j is the *frequency* or number of times x_j occurs in the sample. The observations $x_1, x_2, \ldots, x_k$ are usually arranged in numerical order.

EXAMPLE 1.4.1

The lip-opening pressure (L.O.P.) observations of Example 1.3.1 are ordered into a frequency distribution as given by Fig. 1.1. There are $n = 10$ observations (some having identical numerical values) and $k = 6$ different numerical values comprising this frequency distribution.

The term "frequency distribution" refers to either the frequency table or the frequency distribution graph. A frequency distribution graph is a point plot of observations x_j versus corresponding frequencies f_j.

Also shown in Fig. 1.1 are a frequency polygon and a histogram representation of the L.O.P. observations. A *frequency polygon* is obtained by joining the points of a frequency distribution with straight-line segments. A *histogram* (or bar chart) is generated by constructing rectangles whose heights are proportional to the respective frequencies f_j. The base of each rectangle is bisected by the corresponding x_j-value.

When sample size n is very large, it is convenient to group observations into classes. Each *class* is defined by a class interval, each interval having boundaries called *class limits*. The midpoint of each interval is called the

---

*The subscript i identifies the observation (e.g., x_1 is the first observation, x_2 the second observation, and so forth). Any arbitrary observation is represented by x_i, where the subscript i is variable in the sense that it denotes any of the observations and need only be replaced by the proper number in order to specify a particular observation.

The subscript j identifies the ordered observations of the frequency distribution (e.g., if the x_j are arranged in numerical order, x_1 is the smallest of the x_j, x_2 is the next to the smallest, and so forth.) Again, the subscript j is variable in the sense that it represents any of the ordered observations of the frequency distribution.

The symbol $\sum$ (upper-case Greek sigma) is the summation operator and j is the index of summation. The values j and k that appear, below and above the operator, respectively, designate the range of summation. Thus, the combined expression $\sum_{j=1}^{k} f_j$ reads, "sum all of the frequencies f_j for $j = 1, 2, \ldots, k$."

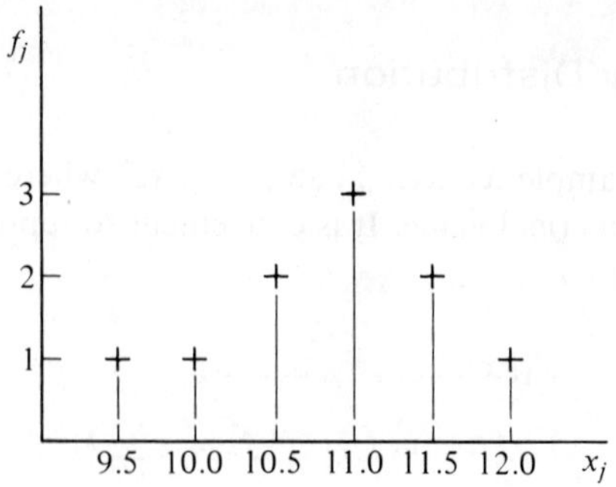

x_j	9.5	10.0	10.5	11.0	11.5	12.0	
f_j	1	1	2	3	2	1	$n = 10$

Frequency distribution.

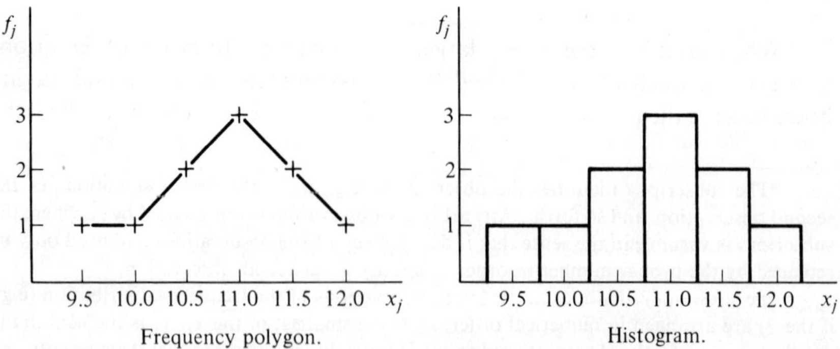

Figure 1.1. Example 1.4.1—frequency distribution table and graph, frequency polygon and histogram.

mid-x value. After the grouping of observations has been effected, the resulting mid-x_j values and corresponding frequencies f_j constitute a frequency distribution.

EXAMPLE 1.4.2

Twenty observations of lip-opening pressure (L.O.P.), each measured to an accuracy of 0.1 p.s.i., are grouped into 0.4 p.s.i. class intervals as shown in Fig. 1.2.

Class Interval	Class Limits	Mid x_j	f_j
	8.75		
8.8– 9.2		9.0	1
	9.25		
9.3– 9.7		9.5	2
	9.75		
9.8–10.2		10.0	3
	10.25		
10.3–10.7		10.5	5
	10.75		
10.8–11.2		11.0	4
	11.25		
11.3–11.7		11.5	3
	11.75		
11.8–12.2		12.0	2
	12.25		
			20

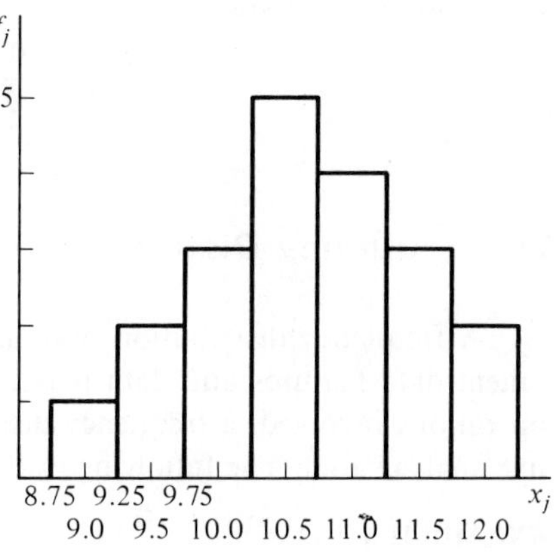

Figure 1.2. Example 1.4.2—grouping observations of L.O.P. into 0.4 p.s.i. class intervals.

Class limits are chosen half-way between successive class intervals. With this choice, class limits have one more significant decimal place than do the original observations x_j.*

*Use of office calculating machines and computers for computations based on frequency distributions makes it possible to easily and rapidly handle a large number of observations. For this reason, grouping data into classes is no longer a common practice.

The grouping procedure introduces systematic error into certain computations involving frequency distributions (ie., actual data is replaced by somewhat fictitious data assigned arbitrarily at the central values of the class intervals). A method of eliminating systematic error due to grouping is given by Shepherd's corrections. See Burington, R. S. and May, D. C. (1953), *Handbook of Probability and Statistics*, Handbook Publishers, Inc., Sandusky, Ohio, pp. 59–60.

EXERCISES

1. Prepare a frequency distribution table and graph for the following x_i observations: 4, 4, 1, 3, 2, 7, 2, 4, 6, 4, 3, 5, 5, 3, 4, 6, 7, 4, 3, 5. Regarding the x_i and x_j notation, what is the numerical value of: (a) n, (b) k, (c) x_i, $i = 5$, and (d) x_j, $j = 5$ for this Exercise? *Ans.* (a) $n = 20$, (b) $k = 7$, (c) $x_5 = 2$, (d) $x_5 = 5$.

2. Construct a frequency distribution table and graph for the following times (hours) between equipment failures:

x_j (time)	0–49	50–99	100–149	150–199	200–249	250–299	300–349	350–399	400–449	
f_j (no. of failures)	2	6	14	18	23	21	12	8	1	105

1.5 Frequency Distribution—An Analysis Tool

A frequency distribution organizes and condenses data to disclose the salient data features and data pattern. With no sophisticated mathematical operations involved, a frequency distribution is by itself a simple, yet effective analysis tool. The following examples illustrate this fact.

EXAMPLE 1.5.1

A customer returns a shipment of seals to the manufacturer claiming defective lip opening pressure. As the frequency distribution, Fig. 1.3, indicates, the percentage of defective seals is too high to warrant salvage operations. The decision to return the entire lot of seals was based on information yielded by a random sample of 80 seals.

EXAMPLE 1.5.2

Product design specifications for two mating components A and B (Fig. 1.4) are: (a) 0.5000/0.5005 in., and (b) 0.4998/0.4993 in. Part B assembles into the slot in part A and clearance* between A and B is restricted to $+0.0002$ in. (minimum) to $+0.0012$ in. (maximum), with average clearance of $+0.0007$ in. considered to be optimum.

A job order specifies 100 assemblies of A and B to be manufactured. Inspection data for production outputs of A and B disclose frequency distributions for dimen-

*Clearance refers to the difference between the specification limits of two mating quality characteristics. A plus clearance value indicates actual physical clearance, whereas a minus clearance value denotes interference (i.e., no clearance). Although this example may be somewhat premature, the interested student can refer to Sec. 5.8 for a more detailed explanation of clearance and fit in connection with specification procedures.

FREQUENCY CHART
QUALITY CONTROL DEPARTMENT

PLANT *N.S.D. Frankfort* JOB TITLE: *L.O.P. Customer return*

INSPECTOR: *G. Carmony* DATE: *12-23-69* PART NO: *8660 × 64*

JOB DESCRIPTION *Sealector 80 pce sample by frequency.*

MACHINE DESCRIPTION: *Semi-Auto sealector*

SPECIFICATIONS: *10.0/16.0* SAMPLE SIZE *80*

CONCLUSION:

9.0	II
9.5	III
10.0	卌
10.5	卌 卌 IIII
11.0	卌 卌 卌 卌 IIII
11.5	卌 卌 卌
12.0	卌 IIII
12.5	卌
13.0	II
13.5	I

Plant *N.S.D. Frankfort* Date *12-23-69*

Part No. *8660 × 64* Quantity *8,570*

Customer *Chevrolet* Location *Detroit*

Reason for return *Lip Opening pressure (sealector rejects)*

N.S.D. Results

Sample *80 pcs*

RGRR *0627*

O.D.
I.D.
Height
L.O.P.
Torque

Comments: *See Attached frequency chart*

Conclusion: *Because of percentage, issue defective material report to sealetor 100% save all rejects, issue work order to assemble corrective springs (High Tension-Low I.D.) to correct Low L.O.P.*

Inspected By: *T. Leazenby.*

Figure 1.3. Example 1.5.1—analysis by means of a sample frequency distribution. *(Courtesy of Federal-Mogul Corp., Frankfort, Indiana.)*

sions a and b as shown in Fig. 1.4. The distributions indicate the following conditions: (1) the A-output is 15% defective, and the B-output is 11% defective, (2) average size of the effective A-output (85 pieces) is 0.5001 in., and average size of the effective B-output (89 pieces) is 0.4996 in.

Condition 2 indicates that, if components A and B are assembled randomly, average clearance is likely to be approximately +0.0005 in. A decision is made to selectively assemble components A and B to obtain an average clearance closer to the optimum average clearance. This procedure also makes it possible to utilize some of the otherwise defective A and B pieces.

Condition 2 is disclosed by the frequency distributions. Even without computation of averages, the distributions indicate a high percentage of a-dimensions close to 0.5000 in. and b-dimensions close to 0.4998 in., which will result in a large number of fits being tighter than optimum (i.e., clearances less than +0.0007 in.).

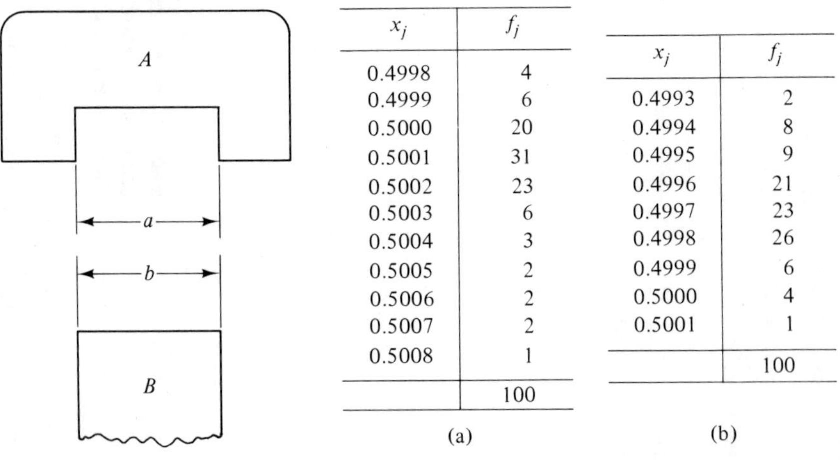

x_j	f_j
0.4998	4
0.4999	6
0.5000	20
0.5001	31
0.5002	23
0.5003	6
0.5004	3
0.5005	2
0.5006	2
0.5007	2
0.5008	1
	100

(a)

x_j	f_j
0.4993	2
0.4994	8
0.4995	9
0.4996	21
0.4997	23
0.4998	26
0.4999	6
0.5000	4
0.5001	1
	100

(b)

Assembly of A and B — Frequency distributions of a and b-dimensions.

Figure 1.4. Example 1.5.2—analysis of clearances by means of a sample frequency distribution.

1.6 Other Frequency Distribution Representation

In certain applications, interest is in how many observations are less than or equal to a specified x-value. Such information is given by a *cumulative frequency distribution*

$$\begin{array}{c} x_1, x_2, \ldots, x_j, \ldots, x_k \\ \sum_{j=1}^{1} f_j, \sum_{j=1}^{2} f_j, \ldots, \sum_{j=1}^{k} f_j \end{array} \tag{1.2}$$

Also, for some purposes it is more convenient to express frequencies in relative units as proportions f_j/n. Thus, Eqs. (1.1) and (1.2) are expressible as relative and cumulative relative frequency distributions, respectively. A *relative frequency distribution* is defined by

$$\begin{aligned} &x_1, x_2, \ldots, x_j, \ldots, x_k \\ &\frac{f_1}{n}, \frac{f_2}{n}, \ldots, \frac{f_j}{n}, \ldots, \frac{f_k}{n} \qquad \sum_{j=1}^{k} \frac{f_j}{n} = 1 \end{aligned} \tag{1.3}$$

A *cumulative relative frequency distribution* is described by

$$\begin{aligned} &x_1, x_2, \ldots, x_j, \ldots, x_k \\ &\sum_{j=1}^{1} \frac{f_j}{n}, \sum_{j=1}^{2} \frac{f_j}{n}, \ldots, \sum_{j=1}^{k} \frac{f_j}{n} \end{aligned} \tag{1.4}$$

EXAMPLE 1.6.1

The lip-opening pressure (L.O.P.) observations of Example 1.3.1 are ordered into a frequency distribution as described by Fig. 1.5. Also shown are cumulative, relative, and cumulative relative frequency distributions for the same set of L.O.P. observations.

Either the relative frequency distribution or the cumulative relative frequency distribution discloses specific frequency proportions. For example, 0.3 of the observed L.O.P.s are 11.0 p.s.i. (from the relative frequency distribution). Also, the cumulative relative frequency distribution indicates that the fraction or proportion of L.O.P.s: (1) less than or equal to 11.0 p.s.i. is 0.7 (2) less than or equal to 10.5 p.s.i. is 0.4, and (3) equal to 11.0 p.s.i. is $0.7 - 0.4 = 0.3$.

The implications of Example 1.6.1 are important in a preparatory sense. If the relative frequency distribution is designated by $f_n(x)$ to indicate a function of the sample observations and the cumulative relative frequency distribution by $F_n(x)$, it can be said that $f_n(x)$ and $F_n(x)$ characterize the distribution of sample observations. In Chap. 3, mathematical models $f(X)$ and $F(X)$, analagous to $f_n(x)$ and $F_n(x)$, are introduced to describe the distribution of population X-values of which the sample observations are possibly a part. That is, if $F_n(x)$ and $F(X)$ agree closely, then the "model" $F(X)$ can be substituted for $F_n(x)$, and $F(X)$ can be utilized to make probability statements regarding various X-values—in the case of Example 1.6.1, probabilities regarding various population L.O.P.s.

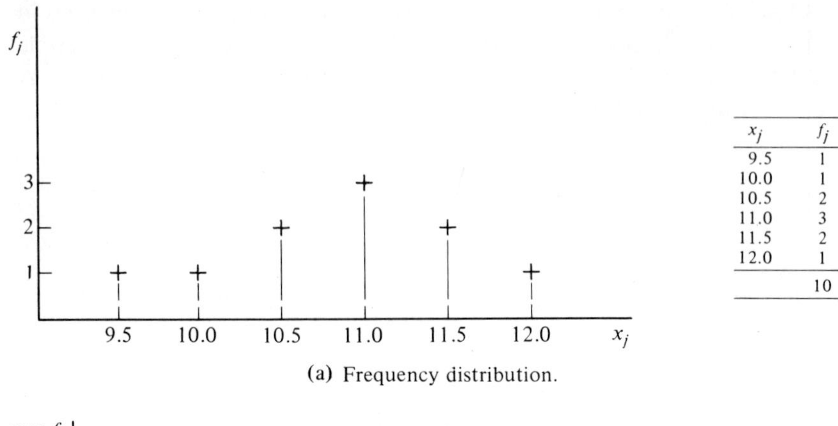

x_j	f_j
9.5	1
10.0	1
10.5	2
11.0	3
11.5	2
12.0	1
	10

(a) Frequency distribution.

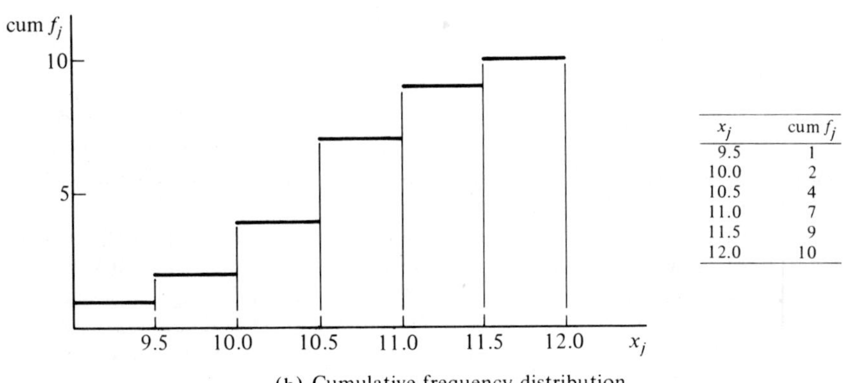

x_j	cum f_j
9.5	1
10.0	2
10.5	4
11.0	7
11.5	9
12.0	10

(b) Cumulative frequency distribution.

Figure 1.5. Example 1.6.1—frequency distribution, cumulative frequency distribution, relative frequency distribution $f_n(x)$, and cumulative relative frequency distribution $F_n(x)$.

EXERCISES

1. (a) Prepare tables and graphs of $f_n(x)$ and $F_n(x)$ for the a-dimension characteristic of Example 1.5.2.

(b) Determine from $F_n(x)$ the fraction or proportion of the a-dimensions corresponding to $0.4999 < x_j \leq 0.5001$ in. *Ans.* $0.61 - 0.10 = 0.51$.

2. (a) Prepare tables and graphs of $f_n(x)$ and $F_n(x)$ for Exercise 2, Sec. 1.4. The $\sum_j f_j/n$ values for each of the k class intervals correspond to x_j less than or equal to the upper class limit of each interval.

(b) Determine from $F_n(x)$ the fraction or proportion of failures corresponding to the times: $149 < x_j \leq 199$ and $149 < x_j \leq 299$.

Ans. $0.38 - 0.21 = 0.17$, $0.80 - 0.21 = 0.59$.

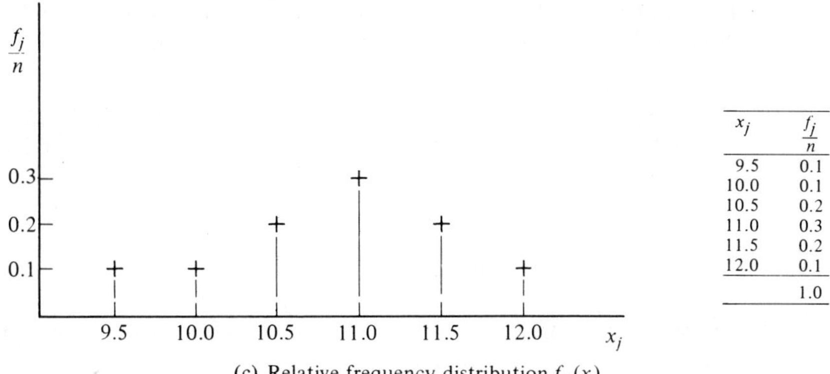

x_j	$\frac{f_j}{n}$
9.5	0.1
10.0	0.1
10.5	0.2
11.0	0.3
11.5	0.2
12.0	0.1
	1.0

(c) Relative frequency distribution $f_n(x)$.

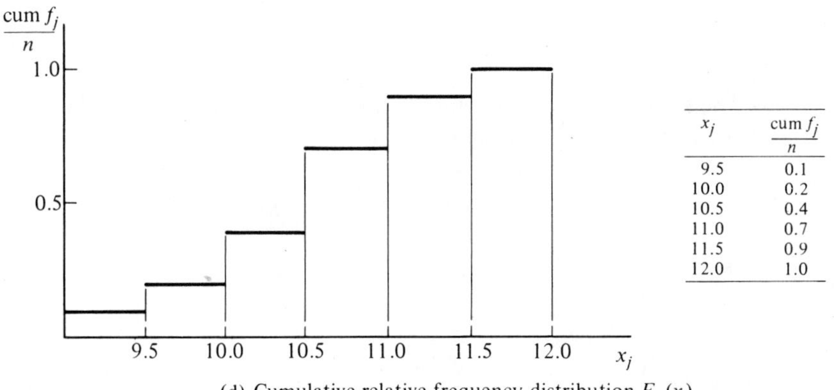

x_j	$\frac{\text{cum } f_j}{n}$
9.5	0.1
10.0	0.2
10.5	0.4
11.0	0.7
11.5	0.9
12.0	1.0

(d) Cumulative relative frequency distribution $F_n(x)$.

Figure 1.5 (cont.)

1.7 Summation Operations

Sums and averages are fundamental elements of all statistical operations. The summation symbol $\sum$ and summation indices were introduced in Sec 1.4. It is presumed that the student is now familiar with the i and j notation for identifying observed values x_i and ordered values x_j. Common summation operations are:

1. The summation of a sum (or difference) is the sum (or difference) of the summations.

$$\sum_{i=1}^{n} (x_i + y_i - z_i) = \sum_{i=1}^{n} x_i + \sum_{i=1}^{n} y_i - \sum_{i=1}^{n} z_i \tag{1.5}$$

2. The summation of the product of a variable and a constant is the product of the constant and the summation of the variable.

$$\sum_{i=1}^{n} cx_i = cx_1 + cx_2 + \ldots + cx_n$$
$$= c(x_1 + x_2 + \ldots + x_n)$$

or

$$\sum_{i=1}^{n} cx_i = c \sum_{i=1}^{n} x_i \tag{1.6}$$

3. The summation of x_i (for $i = 1, 2, \ldots, n$) is equal to the summation of the products $f_j x_j$ (for $j = 1, 2, \ldots, k$).

$$\sum_{i=1}^{n} x_i = \sum_{j=1}^{k} f_j x_j \tag{1.7}$$

4. The summation of a constant is the constant multiplied by the number of terms in the summation.

$$\sum_{i=1}^{n} c = c + c + \ldots + c \quad \text{for } n \text{ terms}$$

or

$$\sum_{i=1}^{n} c = nc \tag{1.8}$$

EXAMPLE 1.7.1

This example illustrates operations 1, 2, and 4 above. Summing the squared deviations $(x_i - c)^2$, where c is a constant,

$$\sum_{i=1}^{n} (x_i - c)^2 = \sum_{i=1}^{n} (x_i^2 - 2cx_i + c^2)$$
$$= \sum_{i=1}^{n} x_i^2 - \sum_{i=1}^{n} 2cx_i + \sum_{i=1}^{n} c^2$$
$$= \sum_{i=1}^{n} x_i^2 - 2c \sum_{i=1}^{n} x_i + nc^2$$

For common statistical operations, partial sums are not encountered frequently. Thus, where the context is clear, the summation notation is simplified by omitting the summation indices. For example, $\sum x_i$ is understood to be

$$x_1 + x_2 + \ldots + x_n$$

where

$$i = 1, 2, \ldots, n.$$

EXERCISES

1. Sample data values x_i (in the order of observation) are 2, 2, 4, 6, 8, 6, 2, 4, 4, 2. (a) Evaluate $\sum_{i=1}^{5} x_i$. (b) Evaluate $\sum_{i=1}^{n} x_i$. (c) Prepare a frequency distribution of x_j-values and verify Eq. (1.7) by computing $\sum_{j=1}^{k} f_j x_j$ and comparing it with the result for (b) above. *Ans.* (a) 22, (b) 40.

2. Suppose that, for the data in Exercise 1, interest is on measuring the data variation from $x = 5$. A possible measure is $\sum_{i=1}^{n} (x_i - 5)$. Compute this sum. Also, compute the same sum by performing the operation $\sum_{j=1}^{k} f_j(x_j - 5)$. *Ans.* -10

3. Other measures for evaluating the data variation from $x = 5$ (Exercise 2) are

$$\sum_{i=1}^{n} |x_i - 5| \quad \text{and} \quad \sqrt{\sum_{i=1}^{n} (x_i - 5)^2}$$

Compute these sums. *Ans.* 20 and 7.07.

4. For the data of Exercise 1, determine a number c such that $\sum_{i=1}^{n} (x_i - c) = 0$.
Ans. $c = 4$

5. Show that

$$\sum_{i=1}^{n} \left[\frac{x_i - c}{\sum_{i=1}^{n} (x_1 - c)} \right] = 1$$

where c is a constant and $\sum_{i=1}^{n} (x_i - c) \neq 0$. Verify this for Exercise 1 using $c = 5$.

1.8 Descriptive Measures

Previous sections of this chapter indicated how tabular and graphic forms of presentation can be used to summarize and, in a limited sense, describe sample observations. A more concise description is obtained by performing arithmetic operations on the data to generate values for one or more descriptive measures called "statistics."

Any function $g(x_1, x_2, \ldots, x_n)$ of the n observations is called a sample *statistic*. A sample is described by a frequency distribution and certain statistics. These statistics are usually denoted by Latin letters.

In a similar manner, a population is defined by a mathematical model $f(X, \theta_1, \theta_2, \ldots, \theta_k)$, a function of an independent variable X and one or more parameters θ_k. Frequently, a parameter may be the population correspondent of the sample statistic (e.g., a sample average and the population average). The parameters of the model are commonly represented by Greek letters. Other notations for distinguishing between sample statistics and corresponding population values are: (1) unprimed and primed letters, and (2) lower case and capital letters. In some instances, these methods will be used for simplicity of expression.

Frequency distribution characteristics commonly measured by statistics are average and variation. Measures of these characteristics are defined in the following sections. Emphasis at this point is on computation methods for obtaining the values of the respective sample statistics. Corresponding population parameters are also given, mainly to develop notation for use in subsequent chapters.

1.9 Measures of Average

Frequency distributions of data usually reveal a tendency for the observations to group about some interior value. This phenomenon is called *central tendency*. A measure of central tendency is a single number that is "representative of" or "typical of" the sample data. This number is descriptive of the sample in the sense that it locates the middle of the distribution. Such a measure is usually an average. Thus, in the following sections, measures of central tendency, location, and average are considered to be synonymous.

The most important measure of central tendency is the arithmetic mean average. Other measures of occasional interest are the median, mode, geometric mean, and harmonic mean.

1.10 Arithmetic Mean

The *arithmetic mean* $\bar{x}$ of the n observations $x_1, x_2, \ldots, x_i, \ldots, x_n$ is the sum of the observations x_i divided by n, that is,

$$\bar{x} = \frac{1}{n}(x_1 + x_2 + \ldots + x_n)$$

or, using the more compact summation notation,

$$\bar{x} = \frac{1}{n}\sum_{i=1}^{n} x_i \tag{1.9}$$

The population mean* μ for a finite population is

$$\mu = \frac{1}{N}\sum_{i=1}^{N} X_i \tag{1.10}$$

From Eq. (1.7), for ordered observations $x_1, x_2, \ldots, x_j, \ldots, x_k$, the mean of n sample observations is

$$\bar{x} = \frac{1}{n}\sum_{j=1}^{k} f_j x_j \tag{1.11}$$

Considering the n observations of x as specifying the positions on a straight line (the x-axis) of a system of n particles of equal weight, the mean $\bar{x}$ corresponds to the center of gravity of the system.

EXAMPLE 1.10.1

The mean of the L.O.P. observations of Example 1.4.1 is computed as follows:

x_j	9.5	10.0	10.5	11.0	11.5	12.0	
f_j	1	1	2	3	2	1	10
$f_j x_j$	9.5	10.0	21.0	33.0	23.0	12.0	108.5

From Eq. (1.11),

$$\bar{x} = \tfrac{1}{10}\sum_{j=1}^{6} f_j x_j$$
$$= \tfrac{1}{10}(108.5) = 10.85$$

EXERCISES

1. Using Eq. (1.11), compute the mean a-dimension of Example 1.5.2.
Ans. 0.50014.

x_j	0.4998	0.4999	0.5000	0.5001	0.5002	0.5003
f_j	4	6	20	31	23	6

0.5004	0.5005	0.5006	0.5007	0.5008	
3	2	2	2	1	100

2. Compute the mean L.O.P. of Example 1.4.2. *Ans.* 10.65.

x_j	9.0	9.5	10.0	10.5	11.0	11.5	12.0	
f_j	1	2	3	5	4	3	2	20

*When there is no ambiguity, the arithmetic mean is referred to simply as the mean.

3. Show by algebraic operations on the sum $(1/n) \sum (x_i \pm c)$ that if a constant c is added to or subtracted from each observation, the mean of the $x_i \pm c$ is $\bar{x} \pm c$. Similarly, show that if each observation is multiplied by a constant c, the mean of the cx_i is $c\bar{x}$. Verify these two relationships for the observations of Example 1.10.1, using $c = 2$.

4. Show by algebraic operations that $\sum (x_i - \bar{x}) = 0$.

1.11 Mean of Means

In certain statistical operations, useful comparisons are made between means of subgroups of observations and the mean of the entire group of observations. For k subgroups of observations, the mean of the means $\bar{\bar{x}}$ is the mean average of $\bar{x}_1, \bar{x}_2, \ldots, \bar{x}_k$ where $\bar{x}_1$ is the mean of the first subgroup, $\bar{x}_2$ the mean of the second subgroup, and so forth.*

For k subgroups of observations, k arithmetic mean values can be generated:

$$
\begin{array}{ll}
x_{11}, x_{12}, \ldots, x_{1i}, \ldots, x_{n_1} & \text{mean: } \bar{x}_1 \\
x_{21}, x_{22}, \ldots, x_{2i}, \ldots, x_{n_2} & \text{mean: } \bar{x}_2 \\
\ldots & \\
x_{j1}, x_{j2}, \ldots, x_{ji}, \ldots, x_{n_j} & \text{mean: } \bar{x}_j \\
\ldots & \\
x_{k1}, x_{k2}, \ldots, x_{ki}, \ldots, x_{n_k} & \text{mean: } \bar{x}_k
\end{array}
$$

The mean of means (or grand mean) of all of the observations is

$$\bar{\bar{x}} = \frac{1}{m} \sum_{j=1}^{k} n_j \bar{x}_j \quad \text{where} \quad m = \sum_{j=1}^{k} n_j \tag{1.12}$$

---

*Subscripts have been used in preceding sections to differentiate between individual observations within a group of observations. The variable is x and the individual observations are

$$x_1, x_2, \ldots, x_i, \ldots, x_n$$

If two subgroups are being considered, the letters x and y can be used to identify the variables. Thus, individual observations are $x_1, x_2, \ldots, x_{n_1}$ and $y_1, y_2, \ldots, y_{n_2}$, where n_1 and n_2 are the respective subgroup sizes.

A more efficient notation, however, is x_1 and x_2 to denote the subgroup variable and a second subscript to identify individual observations. Then, individual observations are

$$x_{11}, x_{12}, \ldots, x_{1i}, \ldots, x_{1n_1}$$

$$x_{21}, x_{22}, \ldots, x_{2i}, \ldots, x_{2n_2}$$

with means $\bar{x}_1$ and $\bar{x}_2$, respectively. This notation is easily extended to any number of subgroups.

and, if

$$n_1 = n_2 = \ldots = n_j = \ldots n_k$$

then

$$\bar{\bar{x}} = \frac{1}{k} \sum_{j=1}^{k} \bar{x}_j \qquad (1.13)$$

EXAMPLE 1.11.1

Three sets of observations and their respective means are: 1, 2, 3 ($\bar{x}_1 = 2$); 6, 4, 5 ($\bar{x}_2 = 5$); and 9, 6, 9 ($\bar{x}_3 = 8$).

$$\bar{\bar{x}} = \frac{1}{m} \sum_{j=1}^{3} n_j \bar{x}_j$$

$$= \tfrac{1}{9}(3 \cdot 2 + 3 \cdot 5 + 3 \cdot 8) = 5$$

or

$$\bar{\bar{x}} = \frac{1}{k} \sum_{j=1}^{3} \bar{x}_j$$

$$= \tfrac{1}{3}(2 + 5 + 8) = 5$$

EXERCISES

1. Compute the separate means $\bar{x}_1$, $\bar{x}_2$, and $\bar{x}_3$ of three subgroups of observations: $x_{1i} = 1, 2, 3$, $x_{2i} = 2, 3, 4, 5, 6$, and $x_{3i} = 3, 4, 5, 6, 7, 8, 9$.
   Ans. $\bar{x}_1 = 2$, $\bar{x}_2 = 4$, $\bar{x}_3 = 6$.
2. Using Eq. (1.12), compute the mean of means $\bar{\bar{x}}$ for the three subgroups of Exercise 1. Verify the result by combining the 15 observations of x_1, x_2, and x_3 and computing $\bar{\bar{x}}$ directly as a mean of one group of 15 observations.
   Ans. 4.53.

1.12 Other Measures of Average

The most commonly used measure of average is the arithmetic mean average. The mean is a clearly defined measure, easily computed, and meaningful even to the nonstatistician. Further, it is more amenable to mathematical treatment than is any other type of average.

There are other average measures, however, that are of occasional interest. Concern here is with the simple case of a statistical decision depending only on a terminal operation of computing an average. It is desirable to select the "best" type of average for the data situation at hand, best in the sense of generating an average value that is most representative of the data. For example, if the sample observations are 0, 0, 0, 0, 15, the mean average is not very informative.*

Two common, though somewhat rough, measures of location are the mode and the median. The *mode* is the x-value associated with the largest frequency. The mode is the most likely or probable x-value. For the data of Fig. 1.1, the mode is 11.0 p.s.i. If the observations are grouped by classes, as in Fig. 1.2, the mode is the mid-x value for the class having the largest frequency. In this case, the mode is 10.5 p.s.i.

The *median* is the middle x-value of an ordered set of observations. Let $x_1, x_2, \ldots, x_n$ be a set of real numbers (some may be identical) arranged in order of magnitude so that

$$x_1 \leq x_2 \leq \ldots \leq x_k \leq \ldots \leq x_n$$

When n is odd, and $n = 2k - 1$, the median is x_k. If n is even, and $n = 2k$, the median is uniquely defined only when $x_k = x_{k+1}$, the common value being the median. If $x_k \neq x_{k+1}$, any value of x in the interval

$$x_k \leq x \leq x_{k+1}$$

satisfies the definitions for the median. In this case, the median is usually taken to be

$$\tfrac{1}{2}(x_k + x_{k+1})$$

EXAMPLE 1.12.1

An ordered set of nine observations is 1, 1, 3, 4, 4, 6, 7, 9, 9. The median is $x_5 = 4$.

If one more observation, $x = 10$, is added to the data set, $k = 5$ and the median is

$$\tfrac{1}{2}(x_5 + x_6) = \tfrac{1}{2}(4 + 6) = 5$$

If the observations are grouped by classes, the preceding definition of the median does not guarantee a unique value. The ambiguity can be eliminated by the following procedure: (1) Define the median as being the x-value for which cumulative frequency is $n/2$, (2) assume a uniform distribution of x-values throughout each class, and (3) construct a cumulative frequency

---

*Many of the illustrative examples in this textbook involve a small number of observations and the observations are small numbers, usually integers. The intent is to free the student (and the instructor) from arithmetic chores and hopefully to focus attention on statistical concepts rather than statistical computations.

However, in practice a considerable number of observations may be involved, and the student should remember that the purpose of an average is to serve as a single number replacement of the data. It is a summary number which should be typical or representative of the data.

table and use straight-line interpolation in the class-limits column to determine the median.

EXAMPLE 1.12.2

For the data of Example 1.4.2 (Fig. 1.2) $n/2$ is 10 and the pertinent cumulative frequencies are:

	Class Limit	cum f
	10.25	6
Median ⟶		10
	10.75	11

The median is

$$10.25 + \frac{10 - 6}{11 - 6}(10.75 - 10.25) = 10.65$$

If a distribution could be represented by a smooth curve,* then the mode is the abscissa of the highest point on the curve. Figure 1.6 shows the location of the mean, median, and mode of a moderately skewed distribution. If the distribution is symmetrical, all three averages coincide.

When the median differs considerably from the mean, it is likely that the median is more representative of the observations. The advantage of the

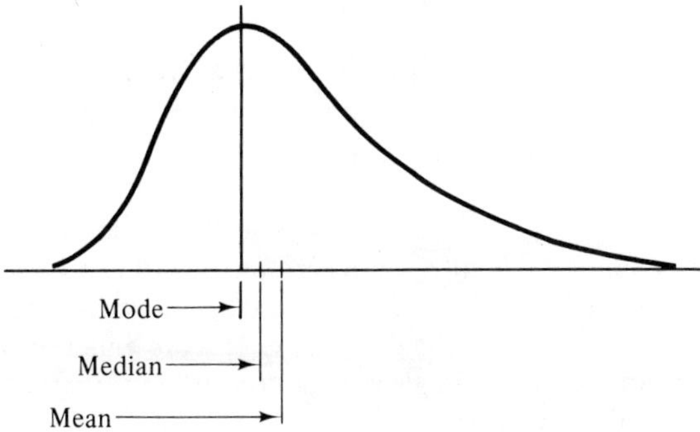

Figure 1.6. Averages for a skewed distribution.

*In subsequent chapters, populations are defined by mathematical models. For a continuous function, the graphic model representation is a continuous curve. Until a formal presentation of the model concept can be made, the student can view the representation (Fig. 1.6) as being the limiting case of a frequency polygon (see Fig. 1.1) where sample size $n \longrightarrow \infty$, or simply as a curve fitted to a large number of points of a frequency distribution plot.

For graphic comparison purposes, the continuous curve representation is used for the balance of this chapter.

median over the mean, as a measure of average, prevails in at least three situations: (1) when occasional and erratic values occur at the ends of a distribution, (2) when the data are presented in a table left open at one or both ends, and (3) when the observations cannot be quantitatively expressed but they can be ordered.

EXAMPLE 1.12.3

An ordered set of seven observations is 1, 2, 2, 4, 4, 7, 8. The mean is 4, and the median is 4. If the number 44 is added to this set (i.e., 44 is the eighth observation), the mean becomes 9, a shift of 5 units, but the median remains unchanged.

Erratic end values (such as 44) occur frequently in distributions of examination grades. In cases of this sort, the median generates a more representative data value than does the mean.

EXAMPLE 1.12.4

The L.O.P. observations of Example 1.4.2 are altered slightly and presented as follows:

x_j	Below 9.3	9.3–9.7	9.8–10.2	10.3–10.7	10.8–11.2	11.3–11.7	11.8–12.2
f_j	1	2	3	5	4	3	2

The first class is called an *open class* and the distribution is referred to as an *open-ended distribution.* A cumulative frequency distribution can be constructed and the median determined without any more information about the x-values less than 9.3 p.s.i.

Averages can easily be misleading. For an average to be meaningful, it should be a number that is representative of a *homogeneous* group. The following example illustrates this point.

EXAMPLE 1.12.5

The annual incomes of a family group are: father, \$10,000; wife and two children, unemployed; maid, \$3,000; gardener, \$2,000. The mean annual income of this group is \$2,500. This is a meaningless figure. It represents not a single person in this group. One cannot make a single reliable conclusion from this mean average value in the absence of the original data.

For certain types of data, a value representative of the data is more readily obtained using either a geometric mean or an harmonic mean average.

The *geometric mean g.m.* of n observations (positive numbers) is the nth root of their product, or

$$\text{g.m.} = (x_1 \cdot x_2 \cdot \ldots \cdot x_n)^{1/n} \tag{1.14}$$

The geometric mean is appropriate when the data exhibit a constant rate of change (i.e., x-values form a geometric progression $x = ar^t$). The growth of many quantities tends to follow this simple exponential function. Thus, with t referring to time, x may represent the population growth of a city, the weight of a quantity, the number of bacteria in a culture, and so forth.

The geometric mean can be used in economics to average "index numbers" which are essentially the ratios of commodity prices at one date to prices at another date. When emphasis is on the rate or percentage of change (rather than the amount of change), the geometric mean is an appropriate average.

EXAMPLE 1.12.6

An ordered set of five numbers is 2, 4, 8, 16, 32. The geometric mean of this set of numbers is

$$\text{g.m.} = (2 \cdot 4 \cdot 8 \cdot 16 \cdot 32)^{1/5}$$
$$= (2^{15})^{1/5} = 8$$

EXAMPLE 1.12.7

An ordered set of five observations is 2.12, 4.08, 8.06, 16.12, 32.31. From Eq. (1.14),

$$\log \text{g.m.} = \frac{1}{n}(\log x_1 + \log x_2 + \ldots + \log x_n)$$

$$\begin{aligned} \log 2.12 &= 0.32634 \\ \log 4.08 &= 0.61066 \\ \log 8.06 &= 0.90634 \\ \log 16.12 &= 1.20737 \\ \log 32.31 &= \underline{1.50934} \\ & 4.56005 \end{aligned}$$

$$\log \text{g.m.} = \tfrac{1}{5}(4.56005)$$
$$= 0.91201$$

and

$$\text{g.m.} = 8.166$$

The geometric mean of the five observations is 8.166.

EXAMPLE 1.12.8

The population of a city was 3,000 in 1960 and 6,100 in 1970. Assuming an approximately constant rate of population change, an estimate of the population at the mid-date (1965) is given by

$$\text{g.m.} = (x_1 \cdot x_2)^{1/2}$$
$$= (3{,}000 \cdot 6{,}100)^{1/2}$$
$$\cong 4{,}278$$

The average annual rate of population increase is obtained by solving for r as follows:

$$x = ar^t$$

$$6{,}100 = 3{,}000r^{10}$$

$$2.03 = r^{10}$$

and

$$r = \sqrt[10]{2.03} = 1.0734$$

The average annual rate of increase is 0.0734 or 7.34%. The population for any intermediate year can now be estimated. For example, if $t = 5$

$$x = (3{,}000)(1.0734)^5 = 4{,}278$$

When ratios such as rates and prices are to be averaged, the harmonic mean average can be used to generate a value that is representative of the data.

The *harmonic mean h.m.* of n observations (positive numbers) is the reciprocal of the arithmetic mean of the reciprocals of the observations, or

$$\text{h.m.} = \frac{1}{\dfrac{1}{n}\left(\dfrac{1}{x_1} + \dfrac{1}{x_2} + \cdots + \dfrac{1}{x_n}\right)}$$

or, more compactly,

$$\text{h.m.} = \frac{1}{\dfrac{1}{n}\sum \dfrac{1}{x_i}} = \frac{n}{\sum \dfrac{1}{x_i}} \tag{1.15}$$

In the case of time rates, the ratio is between two quantities, one of which is in units t of time. The other quantity is in units d of some element such as distance, temperature, work accomplished, and so forth.

A time rate can be expressed either as $x = d/t$ or $x = t/d$. For example, a car that travels at a rate of 30 mph can also be said to travel at a rate of 2 min per mile. Which form is correct d/t or t/d, is indeterminate until it be agreed whether t or d is the fixed element. The choice of fixed element and rate form determines the appropriate average to be used, either arithmetic or harmonic mean.

The following example illustrates some consequences of combinations of fixed element, rate form, and average type. Following the example, operations rules are presented to assure the proper combination.

EXAMPLE 1.12.9

An airplane flies 100 mi at 100 mph, 100 mi at 200 mph., 100 mi at 300 mph., and 100 mi at 400 mph. The distance d is the fixed element and the rate form is

$x = d/t$. The arithmetic mean velocity is

$$\bar{x} = \tfrac{1}{4}(100 + 200 + 300 + 400)$$
$$= 250 \text{ mph}$$

Clearly, this is not a representative velocity. Dividing total distance traveled by total elapsed time, the average velocity is

$$\frac{400 \text{ mi}}{(1 + \frac{1}{2} + \frac{1}{3} + \frac{1}{4}) \text{ hr}} = 192 \text{ mph}$$

But this result is identical to that obtained by computing the harmonic mean:

$$\text{h.m.} = \frac{1}{\frac{1}{4}(\frac{1}{100} + \frac{1}{200} + \frac{1}{300} + \frac{1}{400})}$$
$$= 192 \text{ mph.}$$

where $x_i = d_i/t_i$ denotes the ith rate, $i = 1, 2, 3, 4$.

Operating rules for averaging rates are summarized as follows:*

Fixed Element	*Average Type*	
	h.m.	a.m.
d	$\frac{d}{t}$	$\frac{t}{d}$
t	$\frac{t}{d}$	$\frac{d}{t}$

For example, the left-hand, first-row alternative refers to Example 1.12.9, that is, d is the fixed element and d/t the rate form and hence the harmonic mean is appropriate.

In the case of price rates, a similar discussion holds except that now the unit t of time is replaced by a unit of money. Thus, prices are ratios between two quantities, one of which is in units of money and the other in units of some commodity or service. The correct average to be used depends on how the prices are stated and whether the commodity (or service) unit or the money unit is the fixed element.

*The rate forms given for the arithmetic mean follow from the fact that rates in one form are merely the reciprocals of the same rates in the other form and, from Eq. (1.15),

$$\frac{1}{\text{h.m.}} = \frac{1}{n} \sum \frac{1}{x_i}, \qquad x_i = \frac{d_i}{t_i}$$

EXERCISES

1. Compute the median average for the observations of Example 1.12.4.
Ans. 10.65.
2. Compute the geometric mean average of the five observations: 3.02, 6.12, 12.19, 24.08, 48.05. *Ans.* 12.11.
3. A sum of $1,000 invested at compound interest increases to $1,791 in 10 yr. Use the geometric mean to estimate the value of the investment at the end of 5 yr. *Ans.* $1,338.
4. A future sum of money S corresponding to a present sum P compounded for n years under an interest rate i is

$$S = P(1 + i)^n$$

which is a geometric progression of the form

$$x = ar^t \quad \text{(see Example 1.12.8)}$$

If $1,000 on deposit increases to $2,000 in ten years, what is the interest rate i?
Ans. $i = 0.06$.
5. Three production operators A, B, and C work at the following rates: One unit of work is completed by A in 4 min, by B in 5 min, and by C in 6 min. Compute their average work rate. *Ans.* 4.86 min/1 unit.
At this rate, how many units are completed in a 4-hr shift? *Ans.* 148.
6. Purchases of gasoline are made at four service stations A, B, C, and D. The number of gallons obtained for $1.00 are: A, 3.1; B, 2.9; C, 2.8; and D, 2.6. Let d represent gallons and t dollars. Compute the average price rate.
Ans. $0.35/gal.

1.13 Measures of Variation

The measures of average described in Secs. 1.9–1.12 serve to identify a central or typical value of a set of observations. Such a value locates the frequency distribution along the x-axis.

It is possible to differentiate between data sets (either samples or populations) on the basis of average alone. However, such a differentiation is complete only if the data sets exhibit the same variation. Two or more sets of observations can have identical averages and yet differ considerably in the extent to which the individual observations vary in magnitude. For example, two very different sets of observations are 3, 4, 5, 6, 7 and 0, 0, 5, 10, 10, yet they have the same mean average. Clearly, a more complete data description results from an evaluation that involves not only a measure of average but also a measure of variation.*

Three basic variation measures are range, mean deviation, and standard deviation. The most important measure is the standard deviation. The range has limited statistical utility, and the mean deviation is the least tractable measure for further mathematical treatment. In the following definitive discussion, the mean deviation is presented primarily for the purpose of intuitively developing the standard deviation as a rational measure of variation.

The most simple measure of variation is the *range* r, where r is given by the algebraic difference

$$r = (\text{largest } x_i) - (\text{smallest } x_i) \tag{1.16}$$

EXAMPLE 1.13.1

Two sets of observations are $x_{1i} = 3, 4, 5, 6, 7$ and $x_{2i} = 0, 0, 5, 10, 10$. The means are identical ($\bar{x}_1 = \bar{x}_2 = 5$).

Individual observations comprising the first data set can be described as "varying from 3 to 7." that is, the range r_1 is 4. In like manner, the variation for the second data set is measured by $r_2 = 10$.

The range is easy to compute and it is commonly used in engineering to describe variation. It is a good measure for quickly assessing sample variation. Use of the range as a variation measure is statistically justified for small samples, particularly where repetitive samples are obtained from the same source and an average of the separate ranges is utilized.

A number of serious limitations restrict the usefulness of the range as a general measure of variation: (1) The range is based on only two observations and thus utilizes only a fraction of the information supplied by the sample; (2) the range is unduly influenced by one unusual observation; (3) the range does not measure the variation exhibited by the interior observations (i.e., those between the largest and smallest x_i-values); (4) the range tends to increase with increasing sample size (i.e., an increase in the number of observations cannot possibly decrease the range already determined by a lesser number of observations); and (5) the range is the least stable measure of variation for all but the smallest sample sizes (i.e., for repeated samples

*The study of variation is the primary concern of modern statistics. Historically, early statistics was greatly concerned with large masses of data. Variation itself was not an object of study, but was viewed rather as a troublesome circumstance which detracted from the value of the average.

Modern statisticians consider the study of any phenomena as first and foremost an examination of the variation that is demonstrated. Furthermore, although averages are still very useful measures, emphasis is no longer on averages of large aggregates of data. Interest is primarily on decision making from limited information (i.e., from small economical samples).

from the same source, the ranges thus obtained exhibit more variation from sample to sample than do other available variation measures).

The sample range measures variation in terms of the difference between the largest and the smallest of the sample observations. Thus, the range is dependent on only two observations. Intuitively, a superior measure of variation is one that depends on all of the observations comprising the sample. To construct such a measure, the mean is taken to be the reference and variation is viewed as being the amount that each observation differs from the mean. The quantity $x_i - \bar{x}$ measures the variation of one observation, and it is natural to sum the quantities $x_i - \bar{x}$ for $i = 1, 2, \ldots, n$ to obtain a measure of sample variation.

Since the sum $\sum (x_i - \bar{x})$ is equal to zero, it is necessary to sum either absolute values or squares of the deviations $x_i - \bar{x}$ to obtain a meaningful measure of sample variation. Either sum, $\sum |x_i - \bar{x}|$ or $\sum (x_i - \bar{x})^2$, measures total sample variation, and either sum can be divided by n to produce a measure of "average variation." This leads to the following definitions.

The *mean deviation m.d.* of the n observations $x_1, x_2, \ldots, x_i, \ldots, x_n$ is the mean of the absolute deviations $|x_i - \bar{x}|$, that is,

$$\text{m.d.} = \frac{1}{n} \sum_{i=1}^{n} |x_i - \bar{x}| \tag{1.17}$$

The *variance* s_x^2 of the n observations $x_1, x_2, \ldots, x_i, \ldots, x_n$ is the mean of the squared deviations $(x_i - \bar{x})^2$, that is,

$$s_x^2 = \frac{1}{n} \sum_{i=1}^{n} (x_i - \bar{x})^2 \tag{1.18}$$

The absolute value of a variable is not very tractable in mathematical operations. Therefore, the mean deviation is not favored by statisticians since it is unwieldy in theoretical and mathematical discussions. In applications problems, the mean deviation is useful in dealing with small samples when no elaborate analysis is required. For certain engineering and economic applications,* the mean deviation is favored because of its simplicity and ease of computation.

EXAMPLE 1.13.2

The two sets of observations given in Example 1.13.1 are repeated here: x_{1i}: 3, 4, 5, 6, 7 and x_{2i}: 0, 0, 5, 10, 10 (identical means $\bar{x}_1 = \bar{x}_2 = 5$). The respective mean deviations are

$$\begin{aligned} \text{m.d.}_1 &= \tfrac{1}{5} \sum_{i=1}^{5} |x_{1i} - 5| \\ &= \tfrac{1}{5}(6) = 1.2 \end{aligned}$$

---

*For example, the National Bureau of Economic Research favors the mean deviation as a variation measure in forecasting business cycles.

and

$$\text{m.d.}_2 = \tfrac{1}{5} \sum_{i=1}^{5} |x_{2i} - 5|$$

$$= \tfrac{1}{5}(20) = 4$$

The variances are

$$s_{x_1}^2 = \tfrac{1}{5} \sum_{i=1}^{5} (x_{1i} - 5)^2$$

$$= \tfrac{1}{5}(10) = 2$$

and

$$s_{x_2}^2 = \tfrac{1}{5} \sum_{i=1}^{5} (x_{2i} - 5)^2$$

$$= \tfrac{1}{5}(100) = 20$$

Both measures, the variance and the mean deviation, are useful in comparing the variations of different samples. Clearly, the variance measure exaggerates the numerical differences. The order of magnitude of s_x^2 is large, obviously because of the squaring operation. To reduce the magnitude of the values obtained by this measure, it is reasonable to alter the measure slightly by taking the square root of the resulting s_x^2-value. The square root of the variance is called the *standard deviation.* This is the most commonly used measure of variation in statistics.*

A refinement of Eq. (1.18) is obtained by dividing by $n - 1$ instead of by n. The resulting measure is denoted here by

$$\hat{s}_x^2 = \frac{1}{n-1} \sum_{i=1}^{n} (x_i - \bar{x})^2 \tag{1.19}$$

This definition has the advantage that the sample variance is an unbiased† estimate of the population variance. A mathematical explanation of "unbiasedness" is premature at this point. The student is asked to accept the fact that s_x^2 is a biased estimate of the population variance (i.e., s_x^2 from sample

---

*Although this measure has been arbitrarily constructed, it should be clear that the sum of the squared deviations $(x_i - \bar{x})^2$, called the "sum of squares", actually measures variation. If the observations are identical, the sum of squares is zero. If the observations differ only slightly, the sum of squares is small. Conversely, if the observations differ considerably, the sum of squares is large.

In the language of mechanics, if the n observations of x specify the positions on a straight line of a system of n particles of equal weight, the mean $\bar{x}$ corresponds to the center of gravity of the system and the standard deviation s_x corresponds to the radius of gyration measured from the center of gravity.

†The concept of unbiased estimators is discussed in Chap. 7. In subsequent chapters, the notational distinction between Eqs. (1.17) and (1.18) is unnecessary and s_x^2 is understood to also pertain to Eq. (1.19).

to sample tends to be smaller than the population variance whereas the corrected sample variance $\hat{s}_x^2$ is an unbiased estimate.)
A definition of the population variance (finite population) corresponding to Eq. (1.18) is

$$\sigma_x^2 = \frac{1}{N}\sum_{i=1}^{N}(X_i - \mu)^2 \tag{1.20}$$

This definition is given here to develop notation to be used in subsequent chapters.

EXERCISES

1. It has been stated that the range does not measure variation of the intermediate observations of a sample. Verify this statement by computing the range and mean deviation for each of the following sets of observations: $x_{1i} = 2, 6, 6, 6, 10$ and $x_{2i} = 2, 4, 6, 8, 10$. *Ans.* $r_1 = r_2 = 8$, m.d.$_1 = 1.6$, m.d.$_2 = 2.4$.

2. Using summation operations, algebraically show that if each observation is multiplied by a constant c, the variance of the cx_i is $c^2 s_x^2$. Verify this for the x_{1i} observations of Exercise 1. Also, algebraically show and numerically verify that addition of a constant c to each observation leaves the variance s_x^2 unaffected.

3. Using summation operations, algebraically show that for n identical observations $x_i = c$ the mean is $\bar{x} = c$ and the variance s_x^2 is zero.

1.14 Computation of Variance

There are several computation methods for obtaining the variance. A choice of method depends on the circumstances: (1) whether computation is longhand or by use of a calculator, and (2) the number and the nature of the observations. If the observations are integers and n is small, it is reasonable to use the definitive form of Eq. (1.19), that is,

$$s_x^2 = \frac{1}{n-1}\sum_{i=1}^{n}(x_i - \bar{x})^2$$

If computation proceeds from a frequency distribution table with ordered observations x_j and corresponding frequencies f_j, then from Eq. (1.7) the above expression for the variance becomes

$$s_x^2 = \frac{1}{n-1}\sum_{j=1}^{k} f_j(x_j - \bar{x})^2 \tag{1.21}$$

A variance computation form,* which eliminates the chore of determining the separate $x_i - \bar{x}$ deviations, is

$$s_x^2 = \frac{1}{n-1}\left[\sum_{i=1}^{n} x_i^2 - n\bar{x}^2\right] \tag{1.22}$$

and this form, when used with ordered x_j and corresponding frequencies f_j of a frequency distribution table, becomes

$$s_x^2 = \frac{1}{n-1}\left[\sum_{j=1}^{k} f_j x_j^2 - n\bar{x}^2\right] \tag{1.23}$$

EXAMPLE 1.14.1

The L.O.P. observations of Example 1.10.1 are repeated here and the computation is extended to obtain the variance value.

x_j	f_j	$f_j x_j$	$x_j - \bar{x}$	$(x_j - \bar{x})^2$	$f_j(x_j - \bar{x})^2$
9.5	1	9.5	−1.35	1.8225	1.8225
10.0	1	10.0	−0.85	0.7225	0.7225
10.5	2	21.0	−0.35	0.1225	0.2450
11.0	3	33.0	0.15	0.0225	0.0675
11.5	2	23.0	0.65	0.4225	0.8450
12.0	1	12.0	1.15	1.3225	1.3225
	10	108.5			5.0250

The mean average is

$$\bar{x} = \frac{1}{n}\sum f_j x_j$$

$$= \tfrac{1}{10}(108.5) = 10.85$$

and thus the three right-hand columns are developed from the deviations

$$x_j - \bar{x} = x_j - 10.85$$

*Eq. (1.22) follows from an equivalent expression that can be derived for the sum of the squared deviations, that is,

$$\begin{aligned}\sum (x_i - \bar{x})^2 &= \sum (x_i^2 - 2x_i\bar{x} + \bar{x}^2)\\ &= \sum x_i^2 - 2\bar{x}\sum x_i + \sum \bar{x}^2\\ &= \sum x_i^2 - 2n\bar{x}^2 + n\bar{x}^2\end{aligned}$$

or

$$\sum (x_i - \bar{x})^2 = \sum x_i^2 - n\bar{x}^2 \tag{1.24}$$

The variance is

$$s_x^2 = \frac{1}{n-1} \sum f_j(x_j - \bar{x})^2$$
$$= \tfrac{1}{9}(5.0250) = 0.5583$$

EXAMPLE 1.14.2

Eq. (1.23) is used to compute the variance of the L.O.P. observations of Example 1.14.1.

x_j	f_j	x_j^2	$f_j x_j^2$
9.5	1	90.25	90.25
10.0	1	100.00	100.00
10.5	2	110.25	220.50
11.0	3	121.00	363.00
11.5	2	132.25	264.50
12.0	1	144.00	144.00
	10		1182.25

The variance is

$$s_x^2 = \frac{1}{n-1}[\sum f_j x_j^2 - n\bar{x}^2]$$
$$= \tfrac{1}{9}[1182.25 - 10(10.85)^2]$$
$$= 0.5583$$

EXERCISES

1. Sample observations x_i are 2, 2, 4, 6, 8, 6, 2, 4, 4, 2. Compute the mean and variance, the variance being obtained by the method of Eq. (1.21).
   Ans. $\bar{x} = 4$, $s_x^2 = \frac{40}{9}$.
2. Repeat Exercise 1, obtaining the variance by the method of Eq. (1.23).
3. Compute the mean and variance for the a-dimension observations of Example 1.5.2.
   Ans. $\bar{x} = 0.50014$, $s_x^2 = 3.46(10^{-8})$.

1.15 Change of Scale (Data Coding)

Computation of statistical measures (such as the mean and variance) may involve a considerable amount of arithmetic, particularly if sample size n is large and the observations x_i are decimal numbers. A substantial reduction of arithmetic operations can be accomplished by *coding* the original data. Coding is simply a linear transformation of the form

$$u_j = \frac{x_j - x_o}{c} \tag{1.25}$$

where x_o is an arbitrarily chosen new origin (selected in a way to make the data values smaller numbers) and the divisor c changes the scale of the abscissa axis. The choice of c accomplishes either or both of the objectives: (1) make the data values smaller numbers, and (2) convert decimal values to integers.

If the x_j-values of the frequency distribution are equally spaced, then c is equal to the difference between successive x_j-values, that is,

$$c = x_{j+1} - x_j \qquad j = 1, 2, \ldots, k$$

If the x_j-values are not equally spaced, then c may be taken equal to a common divisor of the x_j (see Example 1.15.3). In the case of the x_j-values being unequally spaced and no common divisor existing, the arithmetic advantage of converting x_j-values to small integer u-values must be foregone. However, the magnitude of the x_j-values can be reduced using a transformation $u_j = x_j - x_0$ where x_0 is either the largest of the x_j or x_0 is an x_j-value near the middle of the ordered x_j of the frequency distribution.

EXAMPLE 1.15.1

Using a linear transformation $u_j = (x_j - 11.0)/0.5$, Example 1.14.1 is repeated here.

x_j	f_j	u_j	f_ju_j	$u_j - \bar{u}$	$(u_j - \bar{u})^2$	$f_j(u_j - \bar{u})^2$
9.5	1	−3	−3	−2.7	7.29	7.29
10.0	1	−2	−2	−1.7	2.89	2.89
10.5	2	−1	−2	−0.7	0.49	0.98
11.0	3	0	0	0.3	0.09	0.27
11.5	2	1	2	1.3	1.69	3.38
12.0	1	2	2	2.3	5.29	5.29
	10		−3			20.10

The mean average in u-units is

$$\bar{u} = \frac{1}{n} \sum f_j u_j$$
$$= \tfrac{1}{10}(-3) = -0.3$$

and thus the three right-hand columns of the computation table are developed from the deviations

$$u_j - \bar{u} = u_j - (-0.3)$$

The variance in u-units is

$$s_u^2 = \frac{1}{n-1} \sum f_j(u_j - \bar{u})^2$$
$$= \tfrac{1}{9}(20.10) = 2.2333$$

Transforming $\bar{u}$ and s_u^2 to $\bar{x}$ and s_x^2 respectively is done using the following relationships.*

$$\bar{x} = c\bar{u} + x_o$$
$$= 0.5(-0.3) + 11.0$$
$$= 10.85$$

and

$$s_x^2 = c^2 s_u^2$$
$$= (0.5)^2 2.2333$$
$$= 0.5583$$

The choice of x_o near the middle of the frequency distribution ($x_o = 11.0$ in Example 1.15.1) generates the smallest possible integer u-values. If computation is performed on a calculator, a more convenient choice is x_o being

---

*These relationships are easily obtained by summation operations. From the u-transformation,

$$u_j = \frac{x_j - x_o}{c}$$

it follows that

$$x_j = cu_j + x_o$$

Then, the mean average is

$$\bar{x} = \frac{1}{n} \sum_{j=1}^{k} f_j x_j = \frac{1}{n} \sum_{j=1}^{k} f_j(cu_j + x_o)$$
$$= \frac{1}{n} \sum_{j=1}^{k} f_j cu_j + \frac{1}{n} \sum_{j=1}^{k} f_j x_o$$

or

$$\bar{x} = c\bar{u} + x_o \tag{1.26}$$

The variance is

$$s_x^2 = \frac{1}{n} \sum_{j=1}^{k} f_j(x_j - \bar{x})^2$$
$$= \frac{1}{n} \sum_{j=1}^{k} f_j(cu_j + x_o - c\bar{u} - x_o)^2$$
$$= \frac{1}{n} \sum_{j=1}^{k} f_j[c(u_j - \bar{u})]^2$$
$$= c^2 s_u^2$$

or

$$s_x = cs_u \tag{1.27}$$

equal to the smallest of the x_j (i.e., in Example 1.15.1, x_o would be 9.5) thus making all u_j-values positive numbers. Clearly, any linear transformation of the form given by Eq. (1.25) merely shifts the origin from zero to x_o and alters the units on the abscissa axis. This is illustrated for Example 1.15.1 by the frequency distribution shown in Fig. 1.7.

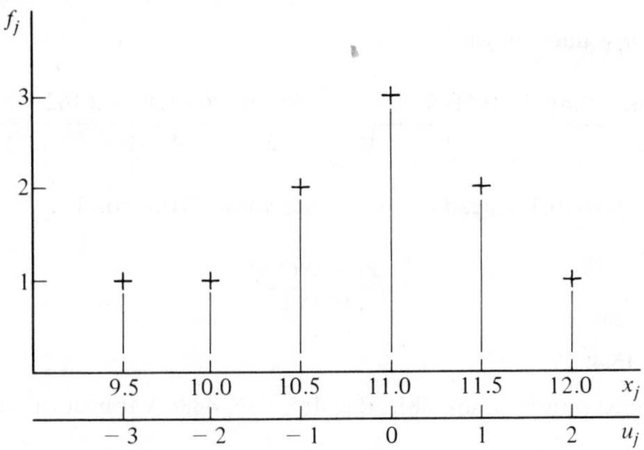

Figure 1.7. Example 1.15.1—change of scale.

EXAMPLE 1.15.2

Example 1.15.1 is repeated here, using Eq. (1.23) to obtain the variance of the L.O.P. observations.

x_j	f_j	u_j	$f_j u_j$	u_j^2	$f_j u_j^2$
9.5	1	−3	−3	9	9
10.0	1	−2	−2	4	4
10.5	2	−1	−2	1	2
11.0	3	0	0	0	0
11.5	2	1	2	1	2
12.0	1	2	2	4	4
	10		−3		21

The variance is

$$\begin{aligned} s_u^2 &= \frac{1}{n-1}[\sum f_j u_j^2 - n\bar{u}^2] \\ &= \tfrac{1}{9}[21 - 10(-0.3)^2] \\ &= \tfrac{1}{9}(20.10) = 2.2333 \end{aligned}$$

and

$$\begin{aligned} s_x^2 &= c^2 s_u^2 \\ &= (0.5)^2 2.2333 \\ &= 0.5583 \end{aligned}$$

EXAMPLE 1.15.3

The following x_j observations are unequally spaced, and the largest common divisor for the fourth decimal place is 0.0002. The u-transformation is

$$u_j = \frac{x_j - 0.6152}{0.0002}$$

producing the u_j-values shown.

x_j	0.6142	0.6144	0.6146	0.6152	0.6156	0.6160	0.6162	0.6164
u_j	−5	−4	−3	0	2	4	5	6

Note that if x_7 is 0.6163 instead of 0.6162, the u-transformation is

$$u_j = \frac{x_j - 0.6152}{0.0001}$$

EXAMPLE 1.15.4

Sample observations x_i are 481, 482, 485, 486, 488. A transformation

$$u_j = x_j - 481$$

generates u_j-values of 0, 1, 4, 5, 7.

EXERCISES

1. Sample observations x_i are 2, 2, 4, 6, 8, 6, 2, 4, 4, 2. Compute the mean and variance, the variance being obtained by the method of Eq. (1.21). Perform the computation in u-units, assuming an origin $x_o = 4$. *Ans.* $\bar{x} = 4$, $s_x^2 = \frac{40}{9}$.
2. Repeat Exercise 1, obtaining the variance by the method of Eq. (1.23).
3. Compute the mean and variance for the a-dimension observations of Example 1.5.2. Perform the computation in u-units, assuming an origin $x_o = 4$. *Ans.* $\bar{x} = 0.50014$, $s_x^2 = 3.46(10^{-8})$.

1.16 Z-Transformation

The most important coding device utilized in statistics is the *Z-transformation.* The Z-transformation is defined by

$$Z = \frac{\text{variable} - \text{mean of the variable}}{\text{standard deviation of the variable}} \tag{1.28}$$

The numerator part of the transformation shifts the origin to the mean, and the denominator part changes the abscissa scale by calibrating it in multiples of the standard deviation. The resulting units for the independent variable are thus freed from the original units of measure in which the observations are expressed. A graphic illustration of a Z-transformation is given by Fig. 1.8.

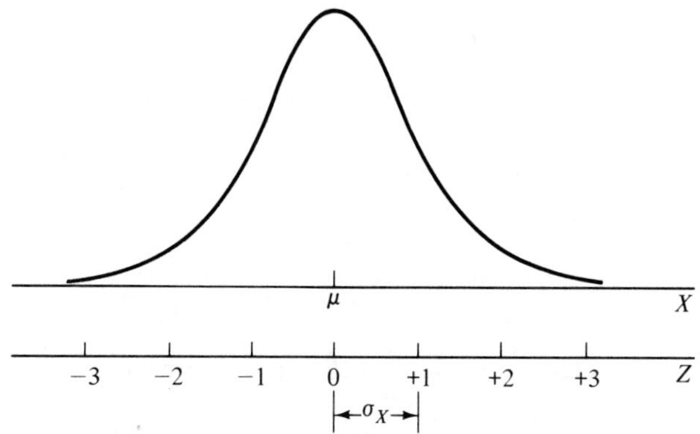

Figure 1.8. Z-transformation of X.

If the independent variable is X, a Z-transformation on the sample x-values is

$$z_j = \frac{x_j - \bar{x}}{s_x} \tag{1.29}$$

and for the corresponding population is

$$Z_j = \frac{X_j - \mu}{\sigma_X} \tag{1.30}$$

In either case, sample or population, the Z-variable is called the standardized form of the independent variable.

1.17 Moments

Sections 1.9–1.16 deal with two types of descriptive statistics: measures of average and variation. By computing a mean and variance, one can identify the nature of a sample set of data and, to some extent, the population represented by the data. This procedure is a part of a general scheme of distribution identification and comparison, a description by means of moments.

A moment in statistics is analogous to a moment in mechanics. A summary of the moment concept in mechanics is shown in Fig. 1.9. The moment with respect to the origin O is the product of the force F applied in a direction PP' and the distance D perpendicular to the line PP', or simply $F \times D$.

The moment concept in statistics is obtained by substituting frequency f_j for force F, and distance x_j from the origin for the distance D. The statistical moment with respect to the origin zero is the product $f_j x_j$.

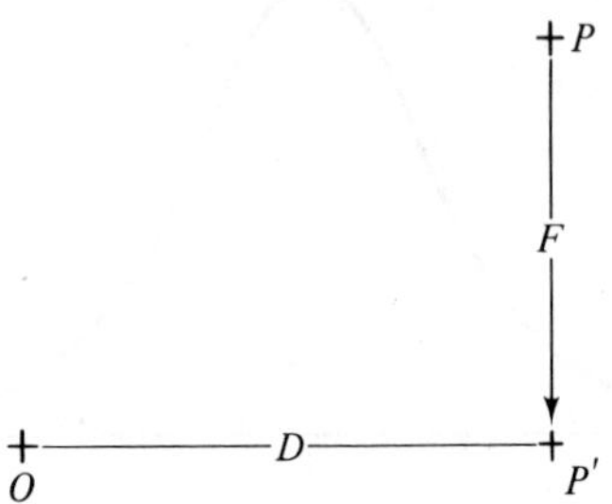

Figure 1.9. Moment concept in mechanics.

The rth moment about the origin $x = 0$ of the observations $x_1, x_2, \ldots, x_j, \ldots, x_k$ is defined by

$$m_{r:0} = \frac{1}{n} \sum_{j=1}^{k} f_j x_j^r \tag{1.31}$$

The rth moment about the mean $x = \bar{x}$ is

$$m_{r:\bar{x}} = \frac{1}{n} \sum_{j=1}^{k} f_j (x_j - \bar{x})^r \tag{1.32}$$

In the population case, the corresponding moments about zero and the mean, respectively, are denoted by $M_{r:0}$ and $M_{r:\mu}$. Clearly, the first moment about zero is the mean, and the second moment about the mean is the variance, for both the sample and population cases. Generally, in applied statistics, only these moments are used to characterize distributions.*

*In early statistical studies, measures of skewness and kurtosis were also used to identify and compare distributions. A departure from symmetry, or skewness, was measured by the third moment about the mean in standard units, or

$$a_3 = \frac{1}{n} \sum_{j=1}^{k} f_j \left(\frac{x_j - \bar{x}}{s_x} \right)^3 = \frac{1}{n} \sum_{j=1}^{k} f_j z_j^3$$

A measure of peakedness, or kurtosis, was measured by the fourth moment about the mean in standard units, or

$$a_4 = \frac{1}{n} \sum_{j=1}^{k} f_j \left(\frac{x_j - \bar{x}}{s_x} \right)^4 = \frac{1}{n} \sum_{j=1}^{k} f_j z_j^4$$

EXERCISES

1. Determine the respective z-values for the a-dimension observations of Example 1.5.2. (Use the mean and variance values obtained in Exercise 3, Sec. 1.15).

Ans. $-1.61, -1.07, \ldots, +3.76$.

2. Given three different frequency distributions A, B, and C already coded in u-units:

	A	B	C
u_j	f_j	f_j	f_j
−3	0	1	0
−2	3	1	1
−1	6	5	10
0	7	11	6
1	6	5	5
2	3	1	2
3	0	1	1

(a) Verify that these distributions have identical means and variances.
(b) Plot a histogram for each distribution and thus demonstrate that the distributions differ from the standpoints of skewness and kurtosis.

1.18 Significant Digits and Data Accuracy

All statistical conclusions from data are based on numbers observed or calculated. The number of places of figures to be retained in the presentation of results depends on the use to be made of the results. Thus, no general rule can be safely laid down. However, some rules are necessary to give a systematic procedure to the calculations. The rules stated in this section are viewed as a guide and not a dictum, and their purpose is twofold—to assure that: (1) numbers reported are significant, and (2) no significant numbers are lost in calculations.

A *significant digit* is defined as any digit that contributes to the specification of the magnitude of the number aside from serving to locate the decimal point. The number 132 has three significant figures, the number 0.0132 has three significant figures, and 132,000 has three or more significant figures depending on whether the zeros are actual measurements or not. For example, if the unit of measurement is 100, the number 132,000 has four significant figures.

The *accuracy* of a reported number is assumed to be to plus or minus half of the smallest unit of measurement. Thus, for each of the preceding numbers, the respective accuracies are 132 ± 0.5, 0.0132 ± 0.00005, and $132{,}000 \pm 50$ (assuming 100 as the unit of measurement).

The accuracy of a reported number cannot be increased by a conversion factor. For example, a measurement reported as $2\frac{3}{32}$ and converted to 2.09375 has an accuracy of $\pm\frac{1}{64}$. Neither can the accuracy of a number be increased by an arithmetic operation. If the number 132 is multiplied by 2.4 to obtain 316.8, the accuracy is still ± 0.5.

The following operational rules are based on the preceding definition of a significant digit and assume that the smallest "significant number" represents the smallest unit of measurement of the reported data (e.g., the number 132 has three "significant digits," the "smallest significant number" is 1, and the "accuracy" is assumed to be ± 0.5).

Rule 1. Report all data to the smallest significant digit.

Rule 2. Arithmetic operations on the data are carried out to at least two more significant figures than the original data. No rounding off is done until the final result is obtained.

Rule 3. In addition and subtraction, report the answer to the largest least significant unit of measurement. Thus, $2.1 + 3.23 = 5.3$.

Rule 4. In multiplication and division, report the result to the same number of significant digits as the measurement with the largest significant number that is involved in the calculation. Thus, $132(2.1) = 277$.

Rule 5. In the rounding off of reported results, the following procedures are suggested:

(*a*) If the first digit to be dropped is less than 5, the last digit retained is left unchanged.

(*b*) If the first digit to be dropped is greater than 5, or is 5 followed by digits greater than zero, the last digit retained is increased by one.

(*c*) If the first digit to be dropped is 5 followed by zeros, the last digit retained is left unchanged if it is even and is increased by one if it is odd.

An example of (*a*) and (*b*) is the number 2.141653, which when reported to two places is 2.14, to three places is 2.142, and to four places is 2.1417. Examples of (*c*) are 2.14500 reported as 2.14, and 2.13500 reported as 2.14.

Rules regarding the number of significant digits when reporting a mean and variance, taken from the *ASTM Manual on Quality Control of Materials** are:

Rule 6. Report the standard deviation to three significant digits. This is an arbitrary rule, and it is clearly affected and altered by Rules 1-5.

Rule 7. Use the following schedule to report a mean average:

**ASTM Manual on Quality Control of Materials* (Jan. 1951), American Society for Testing Materials, Philadelphia, Pa., p. 48.

Observations and the	*Number of Observed Values is*		
0.1, 1, 10, etc., units		2–20	21–200
0.2, 2, 20, etc., units	less than 4	4–40	41–400
0.5, 5, 50, etc., units	less than 10	10–100	101–1000
Retain the Following Number of Places of Figures in the Average	Same number of places as in single values	1 more place than in single values	2 more places than in single values

For example, if $n = 10$, and if observations are to the nearest 1 lb, then the mean average is expressed to the nearest 0.1 lb.

GLOSSARY OF TERMS

Descriptive statistics
Statistical inference
Observation
Population
Sample
Random sample
Frequency
Frequency distribution
Class
Class limits
Mid-x
Histogram
Relative frequency distribution
Cumulative frequency distribution
Cumulative relative frequency distribution
Frequency polygon
Statistic
Central tendency
Arithmetic mean
Mean of means
Mode
Median
Open class
Open-ended distribution
Geometric mean
Harmonic mean
Variation
Range
Mean deviation
Variance
Standard deviation
Sum of squares
Coding
Z-transformation
Standardized variable
Moment
Skewness
Kurtosis

GLOSSARY OF FORMULAS

Page

(1.1) Frequency distribution. 7

$$x_1, x_2, \ldots, x_j, \ldots, x_k$$
$$f_1, f_2, \ldots, f_j, \ldots, f_k \qquad \sum f_j = n$$

(1.2) Cumulative frequency distribution. 12

$$x_1, x_2, \ldots, x_j, \ldots, x_k$$
$$\sum_{j=1}^{1} f_j, \sum_{j=1}^{2} f_j, \ldots, \sum_{j=1}^{k} f_j$$

(1.3) Relative frequency distribution. 13

$x_1, x_2, \ldots, x_j, \ldots, x_k$

$\frac{f_1}{n}, \frac{f_2}{n}, \ldots, \frac{f_j}{n}, \ldots, \frac{f_k}{n} \qquad \sum \frac{f_j}{n} = 1$

(1.4) Cumulative relative frequency distribution. 13

$x_1, x_2, \ldots, x_j, \ldots, x_k$

$\sum_{j=1}^{1} \frac{f_j}{n}, \sum_{j=1}^{2} \frac{f_j}{n}, \ldots, \sum_{j=1}^{k} \frac{f_j}{n}$

(1.5) Summation of a sum or difference. 15

$\sum (x_i + y_i - z_i) = \sum x_i + \sum y_i - \sum z_i$

(1.6) Summation of cx_i where c is a constant. 16

$\sum cx_i = c \sum x_i$

(1.7) Summations of x_i and x_j. 16

$\sum_{i=1}^{n} x_i = \sum_{j=1}^{k} f_j x_j$

(1.8) Summation of a constant. 16

$\sum_{i=1}^{n} c = nc$

(1.9) Definition of sample mean. 18

$\bar{x} = \frac{1}{n} \sum x_i$

(1.10) Definition of population mean. 19

$\mu = \frac{1}{N} \sum X_i$

(1.11) Mean computed from a frequency distribution. 19

$\bar{x} = \frac{1}{n} \sum f_j x_j$

(1.12) Mean of means, unequal subgroup sizes. 20

$\bar{\bar{x}} = \frac{1}{m} \sum n_j \bar{x}_j \quad m = \sum n_j$

(1.13) Mean of means, equal subgroup sizes. 21

$\bar{\bar{x}} = \frac{1}{k} \sum \bar{x}_j \quad n_1 = n_2 = \ldots = n_k$

(1.14) Geometric mean. 24

g.m. $= (x_1 \cdot x_2 \cdot \ldots \cdot x_n)^{1/n}$

(1.15) Harmonic mean. 26

h.m. $= \dfrac{1}{\frac{1}{n} \sum \frac{1}{x_i}}$

(1.16) Range. 29

$r =$ (largest x_i) $-$ (smallest x_i)

(1.17) Mean deviation. 30

$$\text{m.d.} = \frac{1}{n} \sum |x_i - \bar{x}|$$

(1.18) Definition of sample variance. 30

$$s_x^2 = \frac{1}{n} \sum (x_i - \bar{x})^2$$

(1.19) Unbiased estimate of the population variance. 31

$$\hat{s}_x^2 = \frac{1}{n-1} \sum (x_i - \bar{x})^2$$

(1.20) Definition of population variance. 32

$$\sigma_X^2 = \frac{1}{N} \sum (X_i - \mu)^2$$

(1.21) Variance computed from a frequency distribution. 32

$$s_x^2 = \frac{1}{n-1} \sum f_j(x_j - \bar{x})^2$$

(1.22) Variance computation formula. 33

$$s_x^2 = \frac{1}{n-1} [\sum x_i^2 - n\bar{x}^2]$$

(1.23) Variance computation formula. 33

$$s_x^2 = \frac{1}{n-1} [\sum f_j x_j^2 - n\bar{x}^2]$$

(1.24) Summation of squared deviations equivalent. 33

$$\sum (x_i - \bar{x})^2 = \sum x_i^2 - n\bar{x}^2$$

(1.25) Transformation, x-units to u-units. 35

$$u_j = \frac{x_j - x_o}{c}$$

(1.26) Transformation, $\bar{u}$ to $\bar{x}$. 36

$$\bar{x} = c\bar{u} + x_o$$

(1.27) Transformation, s_u to s_x. 36

$$s_x = cs_u$$

(1.28) Z-transformation. 38

$$Z = \frac{\text{variable} - \text{mean of the variable}}{\text{standard deviation of the variable}}$$

(1.29) Sample z-transformation of x. 39

$$z_j = \frac{x_j - \bar{x}}{s_x}$$

(1.30) Population Z-transformation of X. 39

$$Z_j = \frac{X_j - \mu}{\sigma_X}$$

(1.31) The r th moment about the origin $x = 0$. 40

$$m_{r:0} = \frac{1}{n} \sum f_j x_j^r$$

(1.32) The rth moment about the origin $x = \bar{x}$. 40

$$m_{r:\bar{x}} = \frac{1}{n} \sum f_j(x_j - \bar{x})^r$$

REFERENCES

BURINGTON, R. S. AND MAY, D. C. (1953), *Handbook of Probability and Statistics*, Handbook Publishers, Inc., Sandusky, Ohio, pp. 59–60.

BURR, I. W. (1953), *Engineering Statistics and Quality Control*, McGraw-Hill Book Company, New York, Chaps. 1, 2.

DIXON, W. J. AND MASSEY, F. J. (1969), *Introduction to Statistical Analysis*, 3rd ed., McGraw-Hill Book Company, New York, Chaps. 1, 2, 3.

HUFF, D. (1954), *How To Lie With Statistics*, W. W. Norton & Company, Inc., New York, Chaps. 2, 3, 4.

MORONEY, M. J. (1962), *Facts From Figures*, Penguin Books, Inc., Baltimore, Md., Chaps. 3, 4, 5, 6.

PROBLEMS

1. Specifications for lip opening pressure (L.O.P.) on an automotive seal are 10.0 to 16.0 psi. (L.O.P. is the amount of air pressure in psi. to force 10 Mcc of air between the seal I.D. and a shaft size plug.) Inspection of a random selection of 80 seals yields the following observations. Prepare a frequency distribution and a relative frequency distribution for this data. What conclusion can you make regarding the possibility of meeting specifications on a production basis?

11.0	13.0	11.0	11.5	9.5	12.0	11.5	11.0
11.5	10.0	12.0	10.5	10.0	11.0	13.0	11.0
11.0	12.0	11.5	11.0	10.5	10.5	11.0	11.5
12.0	11.5	11.0	10.5	11.5	12.0	10.5	10.5
10.5	9.5	11.5	11.0	11.0	9.0	12.5	10.0
10.5	11.0	12.5	10.5	11.0	10.5	11.0	11.0
12.0	12.5	11.0	10.0	11.0	11.5	11.5	11.5
12.0	10.5	9.5	10.5	12.5	9.0	11.0	11.0
12.5	10.5	10.5	11.0	11.5	10.0	11.0	11.0
11.5	12.0	12.0	11.0	11.0	13.5	11.5	11.5

2. Prepare a cumulative frequency and a cumulative relative frequency distribution for Problem 1.

3. Prepare a histogram for the data of Problem 1, grouping the observations into classes 9.0 to 9.5, 10.0 to 10.5, and so forth. Identify class limits and mid-x values.

4. Compute the mean $\bar{x}$ of the 80 observations in Problem 1.

5. Compute the separate means $\bar{x}_1$, $\bar{x}_2$, and $\bar{x}_3$ of the first three groups of observations in Problem 1:

1	2	3
11.0	13.0	11.0
11.5	10.0	12.0
11.0	12.0	11.5
12.0	11.5	11.0
10.5	9.5	11.5

Using either Eq. (1.12) or (1.13), compute the mean of means $\bar{\bar{x}}$ of these three sets of observations.

6. Assume the addition of one more observation $x = 12.6$ to the third set of observations in Problem 5. Compute $\bar{\bar{x}}$ from the three set means $\bar{x}_1$, $\bar{x}_2$, and $\bar{x}_3$.

7. Suppose that the 80 observations in Problem 1 represent a population (i.e., all of the seals under consideration) and the 15 observations in Problem 5 constitute a sample from this population. Identify numerically each element of the following expressions: $\mu = \frac{1}{N}\sum_{i=1}^{N} X_i$ and $\bar{x} = \frac{1}{n}\sum_{i=1}^{n} x_i$.

8. Define the terms population *parameter* and sample *statistic*.

9. Suppose that the following numbers represent observations on three different variables x_i, y_i, and z_i.

x_i	y_i	z_i
11.0	13.0	11.0
11.5	10.0	12.0
11.0	12.0	11.5
12.0	11.5	11.0
10.5	9.5	11.5

Numerically demonstrate or verify Eqs. (1.5) and (1.7).

10. If $c = 3$, demonstrate numerically Eqs. (1.6) and (1.8) for the data of Problem 9.

11. Using summation operations, perform an algebraic evaluation of:
(a) $\sum_{i=1}^{n} (x_i - \bar{x})$, and (b) $\sum_{i=1}^{n} x_i$.

12. Prove the algebraic identity

$$\sum_{i=1}^{n} (x_i - \bar{x})^2 = \sum_{i=1}^{n} x_i^2 - n\bar{x}^2.$$

13. Obtain an expression for the mean $\bar{y}$ where

$$y_i = x_i + (x_i + 2)^2 + 3\bar{x}.$$

Reduce this expression for $\bar{y}$ to its most simple algebraic form.

14. In one city, approximately 75% of the nitrogen dioxide emissions into the urban air are emitted by motor vehicles. The acceptable maximum nitrogen dioxide level has been set at 2 parts per million (ppm). Twenty-five observations x of nitrogen dioxide air content at one high density traffic location yield the following data:

x_j	0.4	0.6	0.8	1.0	1.2	1.4	1.6	1.8	2.0	
f_j	1	3	5	6	4	2	2	1	1	25

Using either Eq. (1.22) or Eq. (1.23), compute the variance s_x^2 directly, working in x-units.

15. Repeat Problem 14 using a u-transformation to obtain integer u-units. Compute $\bar{u}$ and s_u^2, and convert these statistics to $\bar{x}$ and s_x^2.

16. An important linear transformation is $z_j = (x_j - \bar{x})/s_x$. Convert the x_j of Problem 14 to standard z-units.

17. In certain problems, it is advantageous to convert deviations $x_i - \bar{x}$ to integer u-units. Consider a set of observations 3, 6, 9, 12, 15, 18. What linear transformation $u_i = (x_i - \bar{x})/c$ will convert $x_i - \bar{x}$ deviations to integers?

18. Delete the observation $x = 18$ from the data of Problem 17. What is the linear u-transformation that will convert $x_i - \bar{x}$ deviations to integers?

19. Ten observations of the transverse strength X of bricks in psi units are:

x_j	568	570	572	574	
f_j	4	3	2	1	10

Compute the mean and variance following the procedure of Example 1.14.1.

20. Repeat Problem 19 following the procedure of Example 1.15.1.

21. Repeat Problem 19 following the procedure of Example 1.14.2 and choosing $x_0 = 568$ for an origin.

22. Repeat Problem 19 following the procedure of Example 1.15.2 and choosing $x_0 = 568$ for an origin.

23. The data of Problem 19 is a sample from a population of tranverse strengths of bricks with mean $\mu = 570.9$ and variance $\sigma_X^2 = 2.3$. Why do sample and population means differ (and also sample and population variances)? Identify numerically each element of Eqs. (1.9) and (1.10) and Eqs. (1.19) and (1.20) for the data stated here and the results of Problem 19.

24. Determine the median and mode for the data of Problem 14.

25. The number of bacteria in a culture increases at a constant rate over a 24 hour period from $3 \cdot 10^6$ to $9 \cdot 10^6$. Using the appropriate average, estimate the number of bacteria present 12 hours after time zero.

26. Two production workers A and B work at the following rates: A: 2 units per hour, and B: 3 units per hour. Or, expressed in a different form, the rates are: A: 30 minutes per output unit, and B: 20 minutes per unit. Using the appropriate average, compute average output per hour and average time required to complete one unit.

27. Evaluate the first and second moments about $x = 0$ of the observations $x_1, x_2, \ldots, x_k$ whose respective frequencies are $f_1, f_2, \ldots, f_k$.

28. Evaluate the first and second moments about $x = \bar{x}$ of the observations x_j whose frequencies are f_j.

29. Ten samples of five observations each from a process operation yield the following measurements of a product characteristic. Prepare a frequency distribution. Work in u-units, choosing an origin $x_0 = 0.503$ in. Compute the mean and variance.

Sample No.	x_1	x_2	x_3	x_4	x_5
1	0.505	0.503	0.501	0.502	0.500
2	.501	.499	.504	.503	.505
3	.501	.503	.505	.502	.502
4	.499	.503	.503	.502	.503
5	.504	.505	.505	.504	.503
6	.503	.503	.504	.503	.503
7	.505	.504	.504	.505	.505
8	.504	.505	.501	.504	.507
9	.503	.505	.501	.500	.502
10	.499	.505	.503	.504	.501

30. Compute the separate means $\bar{x}_1, \bar{x}_2, \ldots, \bar{x}_{10}$ of the 10 samples in Problem 29. Compute the grand mean $\bar{\bar{x}}$ from these 10 means.

31. Twenty observations of breaking strength X of 0.104 in. hard-drawn copper wire give the following values. Choose as an origin $x_0 = 572$ lb for a u-transformation that produces integers. Obtain sums $\sum u$ and $\sum u^2$ and compute the mean and variance.

x_j	562	564	566	572	576	580	582	
f_j	1	3	4	5	3	2	2	20

32. If $x_0 = 582$ lb is replaced by $x_0 = 583$ lb in Problem 31, what is the best u-transformation that is possible to simplify the u-values for subsequent computations?

33. Add a constant $c = 2$ to each observation in Problem 31 and then compute the mean and variance. Verify that the new mean is equal to the old mean plus 2, and that the variance is unchanged.

34. Multiply each observation in Problem 31 by a constant $c = 3$ and then compute the mean and variance. Verify that the new mean is equal to 3 (old mean) and the new variance is 9 (old variance).

35. Given that the mean of the x_i is 16 and the variance is 9. What are the mean and variance of: (a) $y_i = x_i + 5$, (b) $y_i = 3x_i$, and (c) $y_i = 3x_i + 5$?

36. Determine the mean, median, mode, variance, and standard deviation for the following distribution of nitrogen 14 levels X (in ppm) of a lake water over a period of 3 months:

x_j	9–12	13–16	17–20	21–24	25–28	29–32	33–36	
f_j	1	6	22	32	26	8	5	100

ELEMENTARY PROBABILITY

The purpose of Chap. 1 is the introduction of methods of summarizing sample data and the computation of certain descriptive statistics that characterize the data. Some statistical nomenclature is presented, and the concept of population and random sample has been introduced. For the most part, this chapter is concerned with that area of statistics called descriptive statistics.

The main purpose of the textbook, however, is the development of systematic means of drawing conclusions about the population from limited information supplied by the sample data. This operation is called statistical inference and is concerned with making inferences that have a certain probability of being correct. Obtaining appropriate mathematical models, to be used as tools for making inferences, involves elementary probability concepts. The following sections deal with these concepts.

2.1 Fundamental Principle for Counting

Many technical and engineering problems are concerned with counting. Frequently, it is necessary to count the number of ways in which various objects can be selected. Formulas have been developed to simplify the prob-

2

lem of counting. Most of the formulas are based on the following fundamental counting principle.

Let a capital letter denote an event, a characteristic, or an attribute of interest and the corresponding lower-case letter represent the number of ways the event can occur. Then, if A can occur in a different ways, if after the occurrence of A, event B can occur in b ways, and so on, the number of ways in which events $A, B, C, \ldots$, etc., can occur in that order is

$$a \cdot b \cdot c \ldots \quad \text{ways} \tag{2.1}$$

EXAMPLE 2.1.1

How many even two-digit numbers can be constructed using the digits 2, 3, 4, 5, 6, and 7 (repetitions such as 22 being permitted)?

To obtain even numbers, the last digit (event A) must be a 2, 4, or a 6 (i.e., $a = 3$ ways). After this selection for the last digit, the remaining digit (event B) can be any one of the six given digits. Thus, the number of two-digit numbers that can be constructed is

$$a \cdot b = 3 \cdot 6 = 18$$

2.2 Factorials

If n is a positive integer, then the symbol $n!$ (read n factorial) is defined by

$$n! = n(n-1)(n-2)\ldots(3)(2)(1)$$

and for convenience $0!$ is defined to be 1.*

Table 2.1. FACTORIALS

n	$n!$	$\log_{10} n!$	n	$n!$	$\log_{10} n!$
1	1	0.00000	27	$1.0888869450 \times 10^{28}$	28.03698
2	2	0.30103	28	$3.0488834461 \times 10^{29}$	29.48414
3	6	0.77815	29	$8.8417619937 \times 10^{30}$	30.94654
4	2.4×10	1.38021	30	$2.6525285981 \times 10^{32}$	32.42366
5	1.20×10^{2}	2.07918	31	$8.2228386541 \times 10^{33}$	33.91502
6	7.20×10^{2}	2.85733	32	$2.6313083693 \times 10^{35}$	35.42017
7	5.040×10^{3}	3.70243	33	$8.6833176188 \times 10^{36}$	36.93869
8	4.0320×10^{4}	4.60552	34	$2.9523279903 \times 10^{38}$	38.47016
9	3.62880×10^{5}	5.55976	35	$1.0333147966 \times 10^{40}$	40.01423
10	3.628800×10^{6}	6.55976	36	$3.7199332679 \times 10^{41}$	41.57054
11	3.9916800×10^{7}	7.60116	37	$1.3763753091 \times 10^{43}$	43.13874
12	4.79001600×10^{8}	8.68034	38	$5.2302261746 \times 10^{44}$	44.71852
13	$6.227020800 \times 10^{9}$	9.79428	39	$2.0397882081 \times 10^{46}$	46.30959
14	$8.7178291200 \times 10^{10}$	10.94041	40	$8.1591528324 \times 10^{47}$	47.91165
15	$1.3076743680 \times 10^{12}$	12.11650	41	$3.3452526613 \times 10^{49}$	49.52443
16	$2.0922789888 \times 10^{13}$	13.32062	42	$1.4050061178 \times 10^{51}$	51.14768
17	$3.5568742809 \times 10^{14}$	14.55107	43	$6.0415263063 \times 10^{52}$	52.78115
18	$6.4023737057 \times 10^{15}$	15.80634	44	$2.6582715748 \times 10^{54}$	54.42460
19	$1.2164510041 \times 10^{17}$	17.08509	45	$1.1962222086 \times 10^{56}$	56.07781
20	$2.4329020082 \times 10^{18}$	18.38612	46	$5.5026221598 \times 10^{57}$	57.74057
21	$5.1090942172 \times 10^{19}$	19.70834	47	$2.5862324151 \times 10^{59}$	59.41267
22	$1.1240007278 \times 10^{21}$	21.05077	48	$1.2413915592 \times 10^{61}$	61.09391
23	$2.5852016739 \times 10^{22}$	22.41249	49	$6.0828186403 \times 10^{62}$	62.78410
24	$6.2044840173 \times 10^{23}$	23.79271	50	$3.0414093201 \times 10^{64}$	64.48307
25	$1.5511210043 \times 10^{25}$	25.19065	51	$1.5511187533 \times 10^{66}$	66.19064
26	$4.0329146113 \times 10^{26}$	26.60562	52	$8.0658175171 \times 10^{67}$	67.90665

From Volk, W. (1969), *Applied Statistics for Engineers*, McGraw-Hill Book Co., N. Y., p. 21.

*It can be argued that since

$$n! = n(n-1)!$$

then n can be set equal to 1 and

$$1! = 1 \cdot 0!$$

The left-hand member of the equality is equal to 1. Therefore, if the equality is to hold, $0!$ must be equal to 1.

For small values of n, the evaluation of factorials can be performed on an office calculating machine or obtained by reference to a table.* In statistical applications, computation usually involves the quotient of at least two factorials. Thus, computation is simplified by cancellation of coefficient terms common to the numerator and denominator.

EXERCISES

1. Two operations are required to produce a product item. The first operation can be performed on any one of 4 machines, and the second operation on any one of 3 machines. How many different ways can the part be manufactured?
Ans. 12 ways.

2. A work sampling study is to be performed randomly on 6 technicians. This involves one visit to each technician. How many different sequences of visits are possible? *Ans.* 720.

3. Compute the following factorial expressions: (a) $25!/20!$, (b) $5!$ using Table 2.1, and (c) $5!\,6!$ using Table 2.1.
Ans. (a) 6,375,600, (b) $1.2(10^2)$, (c) $1.2(7.2)(10^4)$.

4. A useful approximation to $n!$, when n is large, is Stirling's approximation given by $\sqrt{2\pi}n^{n+1/2}e^{-n}$ where the value of e is 2.718282.... Determine $10!$ using this approximation method. *Ans.* 3,598,600.

2.3 Combinations

A *combination* is any set of elements (events, or characteristics, or attributes of interest) where order is disregarded. Thus, the combination A, B, C is identical to B, A, C, since each contains the same elements. The number of combinations of n elements taken k at a time is

$$\binom{n}{k} = \frac{n!}{k!(n-k)!} \tag{2.2}$$

---

*Table 2.1 gives values of $n!$ and $\log n!$ for values of n up to 52. Tables of $n!$ and $\log n!$ are available in engineering handbooks. *The Handbook of Chemistry and Physics* (1955), 36th ed., Chemical Rubber Publishing Co., Cleveland, Ohio gives values of $n!$ and $\log n!$ for n-values up to 100. The same tables for n-values up to 1,000 are given by Grant, E. L. and Leavenworth, R. S. (1972), *Statistical Quality Control*, 4th ed., McGraw-Hill Book Company, New York.

An extensive table for n-values up to 1,000 is given in Salzer, H. E. (1951), *Tables of $n!$ and $\Gamma(n+\frac{1}{2})$ for the First 1,000 Values of n*, Nat. Bur. Standards, Appl. Math. Ser. 16, Supt. of Documents, U. S. Gov't Printing Office, Wash. 25, D. C.

where by definition,

$$\binom{n}{k} = 0 \quad k > n, \qquad \binom{n}{k} = 1 \quad k = n, \qquad \binom{n}{0} = 1 \; *$$

EXAMPLE 2.3.1

A sample of five parts is selected from a manufacturing lot of 50 parts. How many different samples can be selected? From Eq. (2.2), the number of different samples is

$$\binom{50}{5} = \frac{50!}{5!45!} = \frac{50 \cdot 49 \cdot 48 \cdot 47 \cdot 46}{5 \cdot 4 \cdot 3 \cdot 2 \cdot 1}$$
$$= 2{,}118{,}760$$

EXERCISES

1. A product item is composed of 2 components from 5 possible configurations A, B, C, D, and E. Duplicate configurations $AA, BB, \ldots$ are not permitted. If arrangement is irrelevant (i.e., A, B is the same as B, A), how many different product items are possible? *Ans.* 10.
2. If arrangement is relevant (i.e., A, B is a different product item than B, A, but repetition $A, A; B, B; \ldots$ is not possible), how many different product items are possible in Exercise 1? *Ans.* 20.
3. Permitting repetition $A, A; B, B; \ldots$ in Exercise 2, how many different product items are possible? *Ans.* 25.
4. Another product item is composed of 5 components using one each of A, B, C, D, and E (Exercise 1) but A is available in three sizes, B in three sizes, C in two sizes, and D and E in one size each. How many different product items are possible? *Ans.* 18.

---

*Order or arrangement of the elements is not usually of interest in subsequent chapters. However, if order is a condition, then a formula for the number of arrangements of the elements is necessary. A *permutation* is a set of elements arranged in some order. From Eq. (2.1), the number of permutations of n elements taken k at a time is given by

$$n(n-1)(n-2)\ldots(n-k+1)$$

Multiplication and division of this expression by $(n-k)!$ yields a formula which facilitates computation:

$$P(n, k) = \frac{n!}{(n-k)!} \tag{2.3}$$

where $P(n, k)$ is the symbol for the number of permutations of n elements taken k at a time.

If it is agreed to call the $k!$ different arrangements of the same k elements the same combination, it follows that

$$\binom{n}{k} = \frac{P(n, k)}{k!} = \frac{n!}{k!(n-k)!}$$

which is Eq. (2.2).

5. A small plant has 10 machine operators, 4 set-up men, and 3 supervisors. A shift requires 5 operators, 2 set-up men, and 2 supervisors. How many ways can the shift be manned? *Ans.* 4,536.

2.4 Probability

A *random process* or *random experiment* is a repetitive process or operation that, in a single trial, may result in any one of a number of possible outcomes, such that the particular outcome is determined by chance and is impossible to predict.

Under a given set of conditions, a random experiment E has N exhaustive, mutually exclusive, and equally likely outcomes $A_1, A_2, \ldots, A_N$. If M of the outcomes are associated with the occurrence of an event A and $N - M$ outcomes with the nonoccurrence of A, the probability of the occurrence of A is the ratio M/N.*

Symbolically, the probability is written $P(A \mid E) = M/N$ and is read the "probability of A given the experiment E." In general, E is understood and the probability is designated

$$P(A) = \frac{M}{N} \tag{2.4}$$

The definition of probability, Eq. (2.4), is called *a priori probability* and it is concerned with that class of problems in which a full knowledge of the conditions affecting the events in question is known beforehand. That is, the M and N outcomes of E can be identified and counted. If A is certain to occur, then $P(A) = 1$, and if A is certain not to occur, then $P(\text{not } A) = 1$ or $P(A) = 0$. Thus, the range of probability values is

$$0 \leq P(A) \leq 1 \tag{2.5}$$

Also, if $P(A)$ is denoted by P and $P(\text{not } A)$ by Q, then

$$P + Q = 1 \quad \text{and} \quad P = 1 - Q \tag{2.6}$$

EXAMPLE 2.4.1

A manufacturing lot of 50 parts contains four defective parts. What is the probability that two parts randomly selected from the lot will both be defective?

---

*Two outcomes are *mutually exclusive*, if when one outcome occurs, the other cannot occur. A mutually exclusive and *exhaustive* set of outcomes is such that one of the set must occur, but only one can occur in a single trial of E.

The problem of determining $P(A)$ under certain circumstances is a problem of enumerating the equally likely cases favorable to the occurrence of A and those unfavorable to the occurrence of A. In a given application, it is really a matter of mutual agreement as to what constitutes the N exhaustive, mutually exclusive, and equally likely outcomes.

This is a random experiment E with $N = \binom{50}{2}$ exhaustive, mutually exclusive, and equally likely outcomes with $M = \binom{4}{2}$ outcomes resulting in two defective parts being selected. Thus, the desired probability is

$$P = \frac{\binom{4}{2}}{\binom{50}{2}} = \frac{\frac{4!}{2!2!}}{\frac{50!}{2!48!}}$$

$$= \frac{6}{\frac{50 \cdot 49}{2}} = \frac{12}{50 \cdot 49}$$

$$= 0.005^*$$

It should be noted that the term probability refers to a population of outcomes. Probability is a unique property of a population of outcomes and is not associated with any particular outcome. Anyone who has participated in a crap game appreciates the fact that a 7 may occur more often than is accounted for by its probability of occurrence. In fact, runs of consecutive 7's are not uncommon. Randomness produces occurrence irregularities instead of uniformity. Over the long run, however, it produces stability in relative frequencies. Randomness makes possible the prediction of mass behavior (i.e., the population outcomes), but the behavior of a single outcome is irregular, unpredictable, and, in fact, not associated with probability.

When a die is tossed, and the statement is made that the "probability of a 4 occurring is $\frac{1}{6}$", the value $\frac{1}{6}$ refers to the relative frequency of occurrence of a 4 over the entire conceptual population of an infinitely large number of tosses. With respect to a single toss, the outcome is either a 4 or not a 4. The probability of a 4 is either 1 or 0, although we can never know which before the toss is made.

EXERCISES

1. A box of 6 components contains 2 defectives. (a) How many different samples of size 2 are possible? (b) What is the probability that 2 components randomly selected from the box will both be defective? Verify (a) and (b) by preparing a list of the possibilities. *Ans.* (a) 15, (b) $\frac{1}{15}$.

2. 3 items are randomly selected from a batch containing 5 type A and 4 type B items. What is the probability that all 3 items selected are of type A? *Ans.* $\frac{5}{42}$.

---

*The examples in this chapter are simple, some perhaps even trivial, so that the student can intuitively arrive at a solution. The purpose, in this chapter, is not to develop skill in classical type probability problems, but to develop understanding of elementary probability definitions and operations that are required in statistical applications discussed in subsequent chapters.

3. Prove algebraically and verify numerically that $\binom{n}{k} = \binom{n}{n-k}$.
4. Show that if order or arrangement is relevant when selecting k elements from n given elements that the number of possible selections is $k!\binom{n}{k}$.

2.5 Operations with Events and Probabilities

The symbol $A + B$ denotes the event that at least one of the two events A and B occurs. The symbol A, B represents the joint or simultaneous occurrence of A and B.

A random experiment E has N possible outcomes $A_1, A_2, \ldots, A_N$. Of the N outcomes, M are associated with the occurrence of event A, Q are associated with the occurrence of event B, and R are associated with the joint occurrence of A and B. Then,

$$P(A + B) = \frac{M}{N} + \frac{Q}{N} - \frac{R}{N}$$

or

$$P(A + B) = P(A) + P(B) - P(A, B)^* \tag{2.7}$$

EXAMPLE 2.5.1

A manufacturing lot of 100 parts contains 15 parts possessing a type A defect, five parts with type B defects, and five parts with both type A and B defects. One part is randomly selected from the lot. What is the probability that this part is either type A or type B defective? From Eq. (2.7),

---

*The generalization of Eq. (2.7) to more than two categories of events is obtained by adding all of the odd-number combinations and subtracting all even-number combinations of the events in question. Thus, for events A, B, and C,

$$P(A + B + C) = P(A) + P(B) + P(C) - P(A, B) - P(B, C) - P(A, C) + P(A, B, C) \tag{2.8}$$

An experiment E with 15 possible outcomes is summarized in Fig. 2.1. Each point

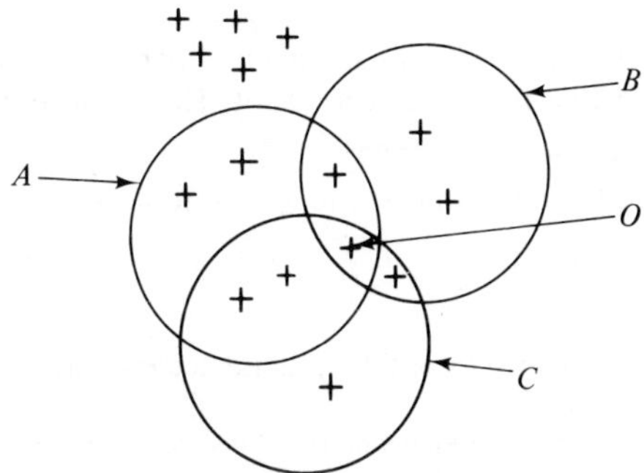

Figure 2.1. Probability of A or B or C.

$$\begin{aligned} P(A + B) &= P(A) + P(B) - P(A, B) \\ &= \tfrac{20}{100} + \tfrac{10}{100} - \tfrac{5}{100} \\ &= \tfrac{25}{100} \end{aligned}$$

If events A and B are mutually exclusive, then the probability $P(A, B) = 0$ and Eq. (2.7) reduces to

$$P(A + B) = P(A) + P(B) \quad P(A, B) = 0 \tag{2.9}$$

EXAMPLE 2.5.2

A manufacturing lot of 100 parts contains 15 parts possessing a type A defect and 10 parts with a type B defect. One part is randomly selected from the lot. From Eq. (2.9), the probability of either a type A or a type B defective part occurring on this random selection is

$$\begin{aligned} P(A + B) &= P(A) + P(B) \\ &= \tfrac{15}{100} + \tfrac{10}{100} \\ &= \tfrac{25}{100} \end{aligned}$$

Frequently, interest is in the probability of a composite event, or related events (e.g., the two consecutive random selections of Example 2.4.1). If A and B are *independent** simple events, the probability of the composite event A, B(i.e., A and also B occurring) is given by

$$P(A, B) = P(A)P(B) \tag{2.10}$$

and this relationship can be generalized to the composite of any number of independent simple events $A, B, C, \ldots$.

EXAMPLE 2.5.3

Two manufacturing lots of 100 parts each are 15% and 10% defective, respectively. One part is randomly selected from each lot. What is the probability of obtaining two defective parts?

Let A denote the occurrence of a defective part from the first lot and B the occurrence of a defective from the second lot. Events A and B are independent.

---

denotes an outcome. The probability of A or B or C is given by

$$\begin{aligned} P(A + B + C) &= \tfrac{6}{15} + \tfrac{5}{15} + \tfrac{5}{15} - \tfrac{2}{15} - \tfrac{2}{15} - \tfrac{3}{15} + \tfrac{1}{15} \\ &= \tfrac{10}{15} \end{aligned}$$

Note that outcome O is counted three times, subtracted out three times, and added in one time.

*Two events A and B are "independent" if the occurrence of A in no way affects the probability of occurrence of B, and conversely.

Thus, from Eq. (2.10),

$$
\begin{aligned}
P(A, B) &= P(A)P(B) \\
&= \frac{15}{100}\left(\frac{10}{100}\right) = \frac{15}{1,000}
\end{aligned}
$$

The following three examples (three different solutions of the same problem) illustrate the importance of carefully defining outcomes and noting whether or not the outcomes (or events) are mutually exclusive and independent.

EXAMPLE 2.5.4

An experiment E has $N = 4$ possible outcomes yielding any one of the numbers 1, 2, 3, and 4. The experiment is repeated once. In the two trials of E, what is the probability of obtaining a 4 at least once?

For the composite event, two trials of E, there are $4 \cdot 4 = 16$ possible outcomes. One definition of the outcomes of the two trials of E associated with the event of interest (a 4 at least once) is

A: a 4 on trial 1, no 4 on trial 2

B: no 4 on trial 1, a 4 on trial 2

C: a 4 on trial 1, a 4 on trial 2

Since the composite events comprising each outcome A, B, and C are independent, from Eq. (2.10), the respective probabilities of A, B, and C are:

$$
\begin{aligned}
P(A) &= \tfrac{1}{4}\cdot\tfrac{3}{4} = \tfrac{3}{16} \\
P(B) &= \tfrac{3}{4}\cdot\tfrac{1}{4} = \tfrac{3}{16} \\
P(C) &= \tfrac{1}{4}\cdot\tfrac{1}{4} = \tfrac{1}{16}
\end{aligned}
$$

and since A, B, and C are mutually exclusive, from Eq. (2.9), the probability of $A + B + C$ is

$$
\begin{aligned}
P(A + B + C) &= P(A) + P(B) + P(C) \\
&= \tfrac{3}{16} + \tfrac{3}{16} + \tfrac{1}{16} \\
&= \tfrac{7}{16}
\end{aligned}
$$

EXAMPLE 2.5.5

Another definition of the outcomes of the two trials of E in Example 2.5.4 is

A: a 4 on trial 1

B: a 4 on trial 2

The respective probabilities are $P(A) = \frac{1}{4}$ and $P(B) = \frac{1}{4}$. However, A and B are not mutually exclusive. Thus, from Eq. (2.7), the probability of $A + B$ is

$$\begin{aligned} P(A + B) &= P(A) + P(B) - P(A, B) \\ &= \tfrac{1}{4} + \tfrac{1}{4} - (\tfrac{1}{4} \cdot \tfrac{1}{4}) \\ &= \tfrac{7}{16} \end{aligned}$$

EXAMPLE 2.5.6

Another definition of the outcomes of the two trials of E in Example 2.5.4 is

A: at least one 4 occurs in two trials

B: no 4 occurs in two trials

From Eq. (2.10),

$$P(B) = \tfrac{3}{4} \cdot \tfrac{3}{4} = \tfrac{9}{16}$$

and from Eq. (2.6),

$$\begin{aligned} P(A) &= 1 - P(B) \\ &= 1 - \tfrac{9}{16} \\ &= \tfrac{7}{16} \end{aligned}$$

2.6 Conditional Probability

The symbol $P(B \mid A)$ is read "the conditional probability of B, given that A has occurred." The implication is that the occurrence of A may alter the probability of the occurrence of B. Thus, if A and B are dependent, the probability of the composite event A, B is given by

$$P(A, B) = P(A)P(B \mid A)^* \tag{2.11}$$

EXAMPLE 2.6.1

Consider the composite event of the two consecutive random selections in Example 2.4.1. From Eq. (2.11), the probability of obtaining two defective parts is

$$P(A, B) = P(A)P(B \mid A)$$

where A is the event, obtaining a defective on the first random selection, and B is the event, obtaining a defective on the second random selection, or

$$\begin{aligned} P(A, B) &= \tfrac{4}{50}(\tfrac{3}{49}) \\ &= \frac{12}{50 \cdot 49} \\ &= 0.005 \end{aligned}$$

---

*Another definition of independence is now possible using this new notation. Events A and B are independent if

$$P(A \mid B) = P(A) \quad \text{or} \quad P(B \mid A) = P(B)$$

EXERCISES

1. A box contains 25 parts, of which 10 are defective. Two parts are selected at random and without replacement from the box. What is the probability that: (a) both are effective, (b) both are defective, and (c) one is effective and one is defective? *Ans.* (a) 0.35, (b) 0.15, (c) 0.5.

2. An experiment E is the random toss of a die and the outcomes for the upturned face are any one of the numbers 1, 2, . . . , 6. The experiment is repeated twice. In the 3 trials of E, what is the probability of obtaining a 1 at least once? *Ans.* $\frac{91}{216}$

3. A box of 6 components contains 2 defectives. If the 6 components are randomly selected for testing, what is the probability of discovering the second defective on the third test? *Ans.* $\frac{2}{15}$.

4. What is the probability that the first defective component is detected on the third test in Exercise 3? *Ans.* $\frac{1}{5}$.

2.7 Baye's Theorem of Inverse Probability

An extension of the principle of conditional probability, Eq. (2.11), can be applied to establish the probability that an event, which has already occurred, might have occurred in a particular way. This is called *posterior probability*. To compute posterior probabilities, use is made of Baye's theorem:

A random process E has N exhaustive and mutually exclusive outcomes $A_1, A_2, \ldots, A_N$, and B is a chance event such that $P(B) \neq 0$. Then,

$$P(A_i \mid B) = \frac{P(B \mid A_i)P(A_i)}{\sum_{j=1}^{N} P(B \mid A_j)P(A_j)} \tag{2.12}$$

EXAMPLE 2.7.1

A product part is being produced on three machines A, B, C. Machine A produces 45% of the total output, machine B 20%, and machine C 35%. Per cent defective output, on the average, for each machine is: A: 10%, B: 5%, and C: 1%.

If one product part is randomly selected from the total output of the three machines and is found to be defective, what is the probability that it was produced by machine A?

Let D denote a defective part, then

$$P(A) = 0.45, \qquad P(D \mid A) = 0.10$$
$$P(B) = 0.20, \qquad P(D \mid B) = 0.05$$
$$P(C) = 0.35, \qquad P(D \mid C) = 0.01$$

and

$$P(A \mid D) = \frac{P(D \mid A)P(A)}{P(D \mid A)P(A) + P(D \mid B)P(B) + P(D \mid C)P(C)}$$

$$= \frac{(0.10)(0.45)}{(0.10)(0.45) + (0.05)(0.20) + (0.01)(0.35)}$$

$$= 0.77$$

2.8 Relative Frequency Probability

One limitation to a priori probability (Sec. 2.4) is that it is not applicable to problems where N is very large or infinite and thus it is inconvenient or impossible to identify and count the outcomes. Frequently, in engineering applications an alternative empirical probability definition is required.

Consider a sequence of replications of an experiment E under identical conditions. Let the event or attribute of interest be A. At the first trial of E, event A is observed to occur a_1 times for the n_1 outcomes of E; at the second trial of E, event A is observed to occur a_2 times for the n_2 outcomes of E; and so on. For k replications of E, the *relative frequency probability* of A is defined to be

$$\frac{f}{n} = \frac{a_1 + a_2 + \ldots + a_j + \ldots + a_k}{n_1 + n_2 + \ldots + n_j + \ldots + n_k} = \frac{\sum a_j}{\sum n_j} \qquad (2.13)$$

The individual relative frequencies may vary with replications of E. However, as the number of replications increases, the combined relative frequency ratio f/n soon becomes relatively stable and, in the long run, tends to converge to the exact a priori probability $P(A)$. Also, the relative frequency definition of probability is compatible with the a priori probability definition in two other important respects. The range of probability values is $0 \leq P(A) \leq 1$ [see Eq. (2.5)] and the additivity law [Eq. (2.9)] for the probability of mutually exclusive events also holds.

EXAMPLE 2.8.1

A manufacturing lot of $N = 1{,}000$ parts has 100 defective parts. A defective part is the event A of interest. Thus,

$$P(A) = 0.10$$

For economic reasons, a sampling inspection is performed. A sample of 50 parts randomly selected from the population of 1,000 parts yields four defectives. The relative frequency probability is

$$P(A) = \tfrac{4}{50} = 0.08$$

2.9 Probability Based on Algebra of Sets

The probability concepts of Secs. 2.4–2.8 have been presented in combinatorial terms. This approach is simple and sufficient for elementary probability considerations required in an introductory statistics textbook. Advanced probability treatment is more rigorously and compactly based on the algebra of sets. For those who prefer the set notation and language, a brief restatement of Secs. 2.4–2.8 in set language is given here.

A *set* is an aggregate or collection of objects. Associated with any random experiment E there is a set of all possible outcomes. This set S is called a *sample space*. If s is a member of S, it is called an *element* of S or a *sample point* and is designated by $s \in S$ which reads "s belongs to S." If s is not an element of S, this fact is denoted by $s \notin S$.

Basic operations on sets are possible. The *union* of two sets A and B is the set of elements that belong to either A or B (or both). This union is designated $A \cup B$. The *intersection* of sets A and B is the set of elements that belong to both A and B. This intersection is denoted $A \cap B$. If U is the *universal set* of all objects under consideration, the *complement* of a set (with respect to U) is the set of elements of U that do not belong to A. The complement is denoted by $\bar{A}$.

An event A of interest to the experimenter is associated with a particular sample space S, which is in turn associated with a random experiment E. If the sample space is considered to be the universal set, then event A is simply some subset of S. This is denoted by $A \subset S$ and it means that each element of A is also an element of S. Since an event is a set, the operations defined for sets are also defined for events; laws and properties for sets also hold for events.

A *null* or *empty* set, denoted $\varnothing$, contains no elements whatsoever. If the intersection of two or more events associated with a particular outcome of E is $\varnothing$, the events are said to be *mutually exclusive*.

Corresponding to Eq. (2.4), a definition of *a priori probability* is: If a random experiment E gives rise to a sample space having a finite number N of equally likely outcomes, then the *probability* of A, denoted $P(A)$, is the ratio of the number of outcomes contained in A to the total number of outcomes of E.

The operations of Secs 2.5–2.6 stated in set language are:

$$P(A) = 1 - P(\bar{A}) \qquad \text{[Eq. (2.6)]}$$

$$P(A \cup B) = P(A) + P(B) - P(A \cap B) \qquad \text{[Eq. (2.7)]}$$

$$P(A \cup B) = P(A) + P(B) \qquad \text{[Eq. (2.9)]}$$

$$P(A \cap B) = P(A)P(B) \qquad \text{[Eq. (2.10)]}$$

$$P(A \cap B) = P(A)P(B|A) \qquad \text{[Eq. (2.11)]}$$

and the solutions to section examples proceed in identical manner as previously presented yielding the same numerical solution values.

GLOSSARY OF TERMS

Factorial
Combination
Permutation
Random experiment
Outcome
Event
Exhaustive
Mutually exclusive
Equally likely
Probability
Conditional probability
A priori probability
Posterior probability
Relative frequency probability

GLOSSARY OF FORMULAS

Page

(2.1) Fundamental counting principle. 51

$a \cdot b \cdot c \ldots$ ways

(2.2) Combinations. 53

$$\binom{n}{k} = \frac{n!}{k!(n-k)!}$$

(2.3) Permutations, n items, taken k at a time. 54

$$P(n, k) = \frac{n!}{(n-k)!}$$

(2.4) A priori probability. 55

$$P(A) = \frac{M}{N}$$

(2.5) Range of probability values. 55

$$0 \leq P(A) \leq 1$$

(2.6) $P(A)$ and $Q = P(\text{not } A)$. 55

$$P + Q = 1$$

(2.7) Probability, A and B not mutually exclusive. 57

$$P(A + B) = P(A) + P(B) - P(A, B)$$

(2.8) Equation (2.7) for events A, B, and C. 57

$$P(A + B + C) = P(A) + P(B) + P(C) - P(A, B) - P(B, C) - P(A, C) + P(A, B, C)$$

(2.9) Equation (2.7) but A and B mutually exclusive. 58

$$P(A + B) = P(A) + P(B)$$

(2.10) Probability, A and B independent. 58

$$P(A, B) = P(A)P(B)$$

(2.11) Conditional probability, A and B dependent. 60

$$P(A, B) = P(A)P(B \mid A)$$

(2.12) Baye's theorem. 61

$$P(A_i \mid B) = \frac{P(B \mid A_i)P(A_i)}{\sum_{j=1}^{N} P(B \mid A_j)P(A_j)}$$

(2.13) Relative frequency probability. 62

$$P(A) = \frac{f}{n} = \frac{\sum a_j}{\sum n_j}$$

REFERENCES

CHOU, YA-LUN (1969), *Statistical Analysis*, Holt, Rinehart & Winston, Inc., New York, Chap. 5.

FELLER, W. (1957), *An Introduction to Probability Theory and its Applications*, Vol. 1, 2nd ed., John Wiley & Sons, Inc., New York, Chaps. 1–5.

HUFF, D. AND GEIS, I. (1959), *How To Take A Chance*, W. W. Norton & Company, Inc., New York.

MORONEY, M. J. (1962), *Facts From Figures*, Penguin Books, Inc., Baltimore, Md., Chaps. 1–3.

PROBLEMS

1. Assemblies of 4 parts each are made by combining component parts A and B. From 5 parts A and 5 parts B available, how many ways can 4 parts be selected if each assembly is to be comprised of 2 A-parts and 2 B-parts?

2. How many ways can the selection be made in Problem 1 if there are no restrictions on the numbers of A-and B-parts in a 4-part assembly?

3. How many ways can the selection be made in Problem 1 if: (a) there must be at least one A-part in an assembly, and (b) if there must be at least one A and also at least one B-part in an assembly?

4. How many different license plates are possible if each license number is to consist of: (a) a four-digit number, (b) a letter followed by a four-digit number, and (c) two letters followed by a four-digit number? (All digits cannot be zero.)

5. A sample of 10 parts is selected from a manufacturing lot of 50 parts. How many different samples can be selected? (use Table 2.1).

6. Define a *random process* or *random experiment*. What is meant by "mutually exclusive" outcomes?

7. What is the principal characteristic of the class of applications problems to which a priori probability applies?

8. A manufacturing lot of 50 parts is 10% defective. What is the probability that 3 parts randomly selected from the lot will all be defective?

9. What is the probability in Problem 8 that 2 of the 3 parts selected will be defective?

10. What is the probability in Problem 8 that the number of defectives in the selection is less than or equal to 2?

11. A manufacturing lot of 100 parts is 10% defective because of a type A defect, 5% defective due to a type B defect, and 5% defective for type C defect reasons. Three parts in the lot possess both A and C defects, and two parts have all three type defects. What is the probability $P(A + B + C)$?

12. A lot of 100 parts contains 12 parts with a type A defect and 8 parts with a type B defect. One part is randomly selected from the lot. What is the probability $P(A + B)$?

13. Assume two consecutive random selections in Problem 12. What is the probability of obtaining in the two selections: (a) one A-defective, (b) no A-defective, (c) at least one A-defective, (d) at most one A-defective, and (e) two A-defectives?

14. A lot of 1,000 parts is 5% defective. A random sample of 25 parts is selected. Set up the expression (but do not compute) for the probability of obtaining exactly 3 defectives.

15. Set up (do not compute) the expression for the probability of 3 or less defectives in Problem 14.

16. A random experiment E has possible independent outcomes 1, 2, and 3 with respective probabilities 0.1, 0.2, and 0.7. The experiment is repeated once. Interest is on the possibility of outcome 2 in the two independent trials of E. One definition of the outcomes of two trials of E is: A: a 2 on trial 1, and B: a 2 on trial 2. Are A and B mutually exclusive? Define the outcomes of two trials of E (involving outcome 2) differently such that the outcomes are mutually exclusive.

17. What is the probability in Problem 16 (two trials of E) of: (a) a 2 exactly once, (b) a 2 at least once, and (c) a 2 at most once?

18. Referring again to Problem 16, with three trials of E what is the probability of: (a) no 2, and (b) a 2 at least twice?

19. Four radio signals are emitted successively. The reception of any one signal is independent of the reception of another. The respective probabilities of reception of the signals are 0.1, 0.2, 0.3, and 0.4. What is the probability of: (a) no signals received, (b) four signals received, and (c) two signals received?

20. Two machines A and B are required to fill a production order. Machine A produces 75% and machine B 25% of the total output. The percent defective output for A is 5% and for B is 2%. One product part is randomly selected from the total output and is observed to be defective. What is the probability that it was produced by machine B?

21. Three vendors A, B, and C supply 25%, 50%, and 25% respectively of the total quantity of a component used in a certain product assembly. The respective probabilities of failure of this component are 0.1, 0.2, and 0.4 for the three manufacturers. Determine the probability that a randomly selected component will not fail.

22. Define *posterior* probability and *relative frequency* probability.

23. The probability of failure of a tool is 0.2. There are two mutually exclusive causes of failure, A and B, with failure due to A occurring three times as often as failure due to B. What are the separate probabilities of the two failure types?

24. Suppose in Problem 23 that failure causes are not mutually exclusive, and the probability of tool failure due to the joint occurrence of A and B is 0.04. What are the separate probabilities of the two failure types?

25. Suppose in Problem 23 that tool failure occurs only when there is a joint occurrence of causes A and B. Assuming that A and B are independent here, what are the separate probabilities of the two failure types?
*Problems 26–29 are recommended.

26. An experiment E has possible independent and equally likely outcomes 1, 2, 3, and 4. The outcome 2 is considered to be a success and other outcomes are failures. What is the probability of: (a) a success, and (b) a failure?

27. The experiment E in Problem 26 is repeated twice. Thus, there are 3 trials of E. List the different ways that 2 successes can occur in 3 trials of E. What is the combinatorial expression [Eq. (2.2)] for the number of ways 2 successes can occur in 3 trials?

28. What is the probability, in Problem 27, of exactly 2 successes in 3 trials of E?

29. Denote (for Problems 26–28) the number of trials of E by n, the required number of successes by X, the probability of a success at a single trial of E by p, and the probability of a failure at a single trial of E by q. Write an expression for the probability of X successes in n trials of E. (Ans. $\binom{n}{X} p^X q^{n-X}$. This is a probability model that is studied in Chap. 4.)

30. Suppose that there are n possible trials of E in Problem 29, where $n = 1, 2, \ldots, k, \ldots$. What is the probability $P(k = n)$, the number of trials to the first success? In other words, specifying k, the desired probability is the probability of the first success occurring on the $k = n$th trial. (Cue: the first success occurs on the nth trial if and only if the first $n - 1$ trials are failures, each with probability $q = 1 - p$. The expression that is derived here for $P(k = n)$ is another probability model that is studied in Chap. 4.)

31. Another question related to Problem 30 is "at which trial w does the kth success occur?" In other words, specifying both k and w, the desired probability is the probability of the kth success occurring on the wth trial. (Cue: the kth success occurs on the wth trial only if there are exactly $k - 1$ successes in the preceding $w - 1$ trials and a success occurs on the wth trial. What is the probability of exactly $k - 1$ successes in $w - 1$ trials (see Problem 29)? The expression that is derived here for the desired probability is another probability model that is studied in Chap. 4.)

PROBABILITY DISTRIBUTIONS

In the preceding chapter, the probability of an event A *is defined as the ratio* M/N, *where* M *is the number of outcomes associated with the occurrence of* A *and* N *is the total number of outcomes of a random experiment or process* E.

A limitation of this definition is that it is not applicable if the outcomes of E *cannot be enumerated. Thus, it is necessary to extend the definition of probability to cover problems that are not of a combinatorial nature. The definition will now be extended to cover probabilities of events of the form* $X \leq b$, *where* X *is a random variable and* b *is a constant. This extension of the definition of probability should agree with the definitions* [*Eqs.* (*2.4*) *and* (*2.13*)] *and conditions* [*Eqs.* (*2.5*) *and* (*2.9*)].

3.1 Random Variable

Consider a random process E whose outcome is a real number X. Assume that E is capable of being repeated indefinitely. The variable X can assume specific values, each of which has a certain probability of occurrence. Then, X is called a *random variable* associated with E. If E is repeated n times,

3

the successive values assumed by X are denoted by $x_1, x_2, \ldots, x_n$ and these numbers are called a *random sample* of the random variable X.

Broadly speaking, a variable which eludes predictability in assuming its different possible values is called a random variable. Specifically, it must have a defined range of possible values and a probability associated with each value. Some examples of experiments and random variables are listed as follows:

E	X
1. Random selection and measurement of a product part characteristic, the part being manufactured in quantity on a process machine.	1. Measurement of the characteristic.
2. Random selection and measurement of a machine operator.	2. Time required to do a particular operation.
3. Random selection of 50 product parts just manufactured.	3. The proportion of parts that are defective.

There are two broad types of error, *systematic* and *random*. For example, a rule calibrated at one temperature reads systematically incorrectly at another temperature. However, even when all systematic variation is eliminated, there remains random error comprised of a large number of small error effects, combining in a random manner to generate random variable values.

Another example is a production operation on a process machine. Systematic error causes are tooling, setup, process adjustments, process drift, stock, operator, etc. Assume that these error causes are eliminated or at least minimized. There still remain random error causes. The process machine is itself an assembly of many component parts. There are fits, clearances, end-plays, etc., involving the machine's component parts. These factors, many in number, each small in magnitude, combine in a random manner. This total or combined random variation effect is reflected in the random variation of the product output being manufactured by the machine.

3.2 Probability Function

A function $f(X)$ of a discrete random variable X is a *probability function* if it satisfies the conditions

$$f(X) \geq 0, \qquad \sum_X f(X) = 1 \tag{3.1}$$

EXAMPLE 3.2.1

A probability function for a discrete random variable X is

X	1	2	3	4	5	
$f(X)$	0.1	0.2	0.4	0.2	0.1	1.0

and is graphically represented in (a) of Fig. 3.1. The function $f(X)$ is a probability function since all $f(x)$ values are positive and the sum of the $f(X)$ values is unity.

A probability function is used as a model to characterize a distribution of probabilities of X for a population of X-values. Thus, the probability $P(X = 1)$ is 0.1, the probability $P(X = 2)$ is 0.2, and so forth.

Clearly, $f(X)$ is an analogous representation to the relative frequency distribution $f_n(x)$ of Example 1.6.1. Whereas $f_n(x)$ describes relative frequencies (or proportions) for the sample observations, the function $f(X)$ gives the respective probabilities of X for a population of X-values.

X	$f(X)$
1	0.1
2	0.2
3	0.4
4	0.2
5	0.1
	1.0

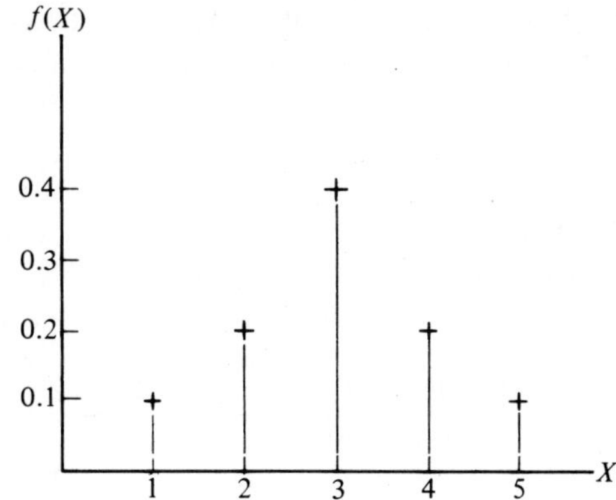

(a) Probability function

$X \leqslant b$	$F(X)$
$X \leqslant 1$	0.1
$X \leqslant 2$	0.3
$X \leqslant 3$	0.7
$X \leqslant 4$	0.9
$X \leqslant 5$	1.0

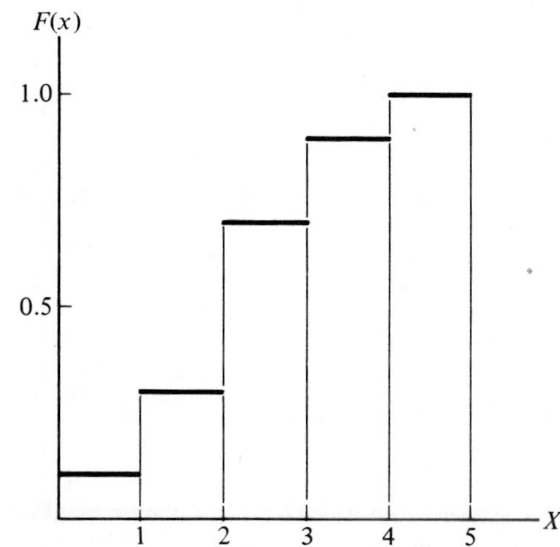

(b) Cumulative distribution function

Figure 3.1. Example 3.2.1—probability and cumulative distribution functions of a discrete random variable X.

3.3 Cumulative Distribution Function

A function $F(X)$ of a real variable X is a *cumulative distribution function* if

$$\left.\begin{array}{l} F(-\infty) = 0, \qquad F(+\infty) = 1 \\ F(X) \text{ is a nondecreasing function, continuous to the right} \end{array}\right\} \qquad (3.2)$$

The cumulative distribution function describes the probability of $X \leq b$ where X is a random variable and b is a constant, that is,

$$P(X \leq b) = F(b) = \sum_{X \leq b} f(X)$$

EXAMPLE 3.3.1

The cumulative distribution function associated with the probability function $f(X)$ given in Example 3.2.1 is

$X \leq b$	$X \leq 1$	$X \leq 2$	$X \leq 3$	$X \leq 4$	$X \leq 5$
$F(b)$	0.1	0.3	0.7	0.9	1.0

and is graphically represented in (b) of Fig. 3.1. Either $f(X)$ or $F(X)$ yields required probabilities. For example, from $f(X)$,

$$P(X = 2) = f(2) = 0.2$$

and, from $F(X)$,

$$\begin{aligned} P(X = 2) &= F(2) - F(1) \\ &= 0.3 - 0.1 = 0.2 \end{aligned}$$

Similarly, from $f(X)$,

$$\begin{aligned} P(2 < X \leq 4) &= f(3) + f(4) \\ &= 0.4 + 0.2 = 0.6 \end{aligned}$$

and, from $F(X)$,

$$\begin{aligned} P(2 < X \leq 4) &= F(4) - F(2) \\ &= 0.9 - 0.3 = 0.6 \end{aligned}$$

Although probabilities of X can be obtained from either $f(X)$ or $F(X)$, the method of using the cumulative distribution function $F(X)$ to obtain a probability value is more important since tabled probability values for various probability distribution models are usually in the $F(X)$ form.*

---

*Let a and b be any real numbers such that $a < b$. The events $X \leq a$ and $a < X \leq b$ are mutually exclusive and their sum is $X \leq b$. By the additivity law of probability, Eq. (2.9),

$$P(X \leq b) = P(X \leq a) + P(a < X \leq b)$$

or

$$P(a < X \leq b) = P(X \leq b) - P(X \leq a)$$

and

$$P(a < X \leq b) = F(b) - F(a). \tag{3.3}$$

Thus, Eq. (3.3) makes it possible to obtain a probability of X occurring in a specified interval by reading $F(b)$ and $F(a)$ from a table and performing the indicated subtraction.

EXERCISES

1. A random variable X can assume only the values 1, 2, and 3 with respective probabilities $\frac{1}{5}$, $\frac{3}{10}$, and $\frac{1}{2}$. Prepare tables and graphs of the probability function $f(X)$ and the cumulative distribution function $F(X)$.
2. Verify numerically the respective conditions, Eqs. (3.1) and (3.2), for $f(X)$ and $F(X)$ of Exercise 1.
3. Demonstrate Eq. (3.3) relative to Exercise 2 by determining: (a) $P(X = 2)$, and (b) $P(1 < X \leq 3)$.
4. Prepare tables and graphs of $f(X)$ and $F(X)$ for a random variable X which is the value of the upturned face of a randomly tossed die.
5. Two dice are randomly tossed. The random variable X is the total number of ones and sixes that appear. Determine the probability function of X.
Ans. $X = 0, 1, 2$ and $f(X) = \frac{4}{9}, \frac{4}{9}, \frac{1}{9}$.
6. Two dice are randomly tossed. The random variable is the sum of the values of the upturned faces. Prepare $f(X)$ and $F(X)$ tables and graphs. What is the probability $P(3 \leq X \leq 7)$? *Ans.* 0.55.

3.4 Continuous Random Variables

A discrete variable is a variable that takes on only certain disconnected values such as the integers $X = 1, 2, \ldots, 5$ in Example 3.2.1. A continuous variable is a variable whose values can change by an amount that can be made arbitrarily small (e.g., $X = 2, 1.9, 1.99$, and so forth). Many important probability distribution models are functions of continuous random variables.

In Example 3.3.1, the desired probability $P(2 < X \leq 4)$ obtained from the cumulative distribution function is

$$P(2 < X \leq 4) = F(4) - F(2)$$

and generalizing for the probability of X being in any interval,

$$\begin{aligned} P(a < X \leq b) &= F(b) - F(a) \\ &= \sum_{X \leq b} f(X) - \sum_{X \leq a} f(X) \end{aligned}$$

where the summations are straightforward arithmetic operations.

A function $f(X)$ of a continuous random variable X is called a *density function* if it satisfies the conditions

$$f(X) \geq 0, \qquad \int_{-\infty}^{\infty} f(X)\,dX = 1 \tag{3.4}$$

and the associated function $F(X)$ is a cumulative distribution function if it satisfies the conditions of Eq. (3.2). The cumulative distribution function describes the probability of $X \leq b$ where X is a random variable and b is a constant, that is,

$$P(X \leq b) = F(b) = \int_{-\infty}^{b} f(x)\,dX.$$

Further, the probability of X being in any interval (a, b) is given by

$$\begin{aligned} P(a \leq X \leq b) &= F(b) - F(a)^{*} \\ &= \int_{-\infty}^{b} f(x)\,dX - \int_{-\infty}^{a} f(X)\,dX = \int_{a}^{b} f(X)\,dX \end{aligned}$$

The names "probability function" and "density function" are standard terminology. Clearly, both terms refer to the same function $f(X)$ but distinguish between discrete and continuous random variables X. To simplify notation and description in subsequent chapters, the notation *D.F.* is used to refer to either the probability or the density function (i.e., *D.F.* for distribution function). Similarly, the notation *C.D.F.* is used to refer to the cumulative distribution function $F(X)$.

EXAMPLE 3.4.1

The Rockwell hardness for a particular type of steel varies from 40 to 60 on a Rockwell B scale. Let X denote Rockwell hardness and assume that X is uniformly distributed as shown in Fig. 3.2. What is the probability of a Rockwell hardness value between 50 and 55?

---

*It follows from the definition of $f(X)$ that only a probability statement regarding a continuous random variable X being in an interval is possible (i.e., the probability of X taking on a specific value is always zero, using $f(X)$ as the probability distribution model). Thus,

$$P(a \leq X \leq b) = P(a < X \leq b) = P(a \leq X < b) = P(a < X < b)$$

The argument is familiar to the student of calculus. For example, consider a probability $P(X = c)$. Specializing the interval (a, b) to be an interval ΔX symmetrically placed about $X = c$,

$$P\left(X - \frac{\Delta X}{2} < X = c \leq X + \frac{\Delta X}{2}\right) = \int_{X-(\Delta X/2)}^{X+(\Delta X/2)} f(X)\,dX$$

or, as an approximation,

$$= f(X)\,\Delta X$$

If $\Delta X \longrightarrow 0$, converging on $X = c$, then $f(X)\,\Delta X \longrightarrow 0$, and $P(X = c) = 0$.

For the uniform distribution model to be a *D.F.*, the function $f(X)$ must equal $\frac{1}{20}$ (i.e., so that the total area beneath the $f(X)$ curve between $X = 40$ and $X = 60$ is equal to unity). Thus, the probability $P(50 \leq X \leq 55)$ is given by

$$
\begin{aligned}
P(50 \leq X \leq 55) &= F(55) - F(50) \\
&= \int_{40}^{55} \tfrac{1}{20}\, dX - \int_{40}^{50} \tfrac{1}{20}\, dX \\
&= \frac{X}{20}\bigg|_{40}^{55} - \frac{X}{20}\bigg|_{40}^{50} = \frac{1}{4}
\end{aligned}
$$

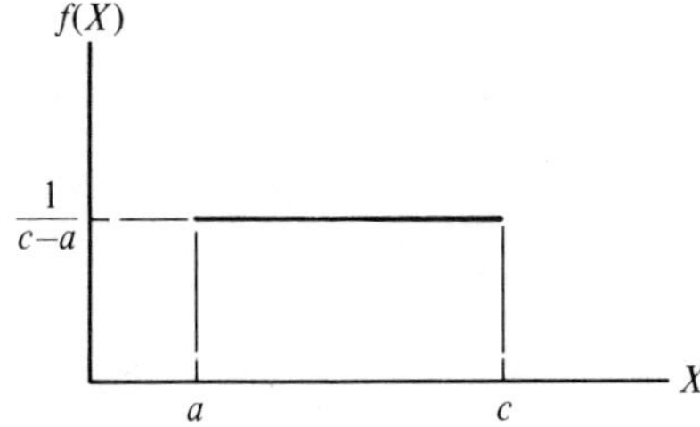

$$
f(X) = \begin{cases} 0 & X < a \\ \dfrac{1}{c-a} & a \leqslant X \leqslant c \\ 0 & X > c \end{cases}
$$

Uniform distribution model.

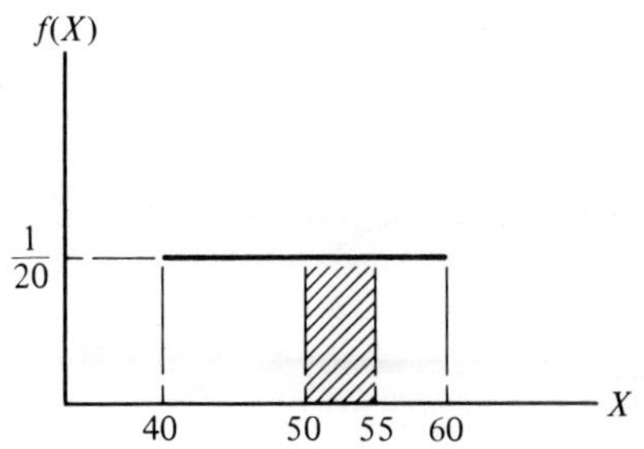

$$
f(X) = \begin{cases} 0 & X < 40 \\ \dfrac{1}{20} & 40 \leqslant X \leqslant 60 \\ 0 & X > 60 \end{cases}
$$

Model for this application problem.

Figure 3.2. Example 3.4.1—distribution of Rockwell hardness values.

EXAMPLE 3.4.2

A continuous random variable X has a *D.F.*

$$
f(X) = \tfrac{1}{2} - aX, \qquad 0 \leq X \leq 4
$$

What is the probability $P(1 \leq X \leq 2)$?

The function $f(X)$ is graphically represented as shown in Fig. 3.3. For $f(X)$ to be a *D.F.*, the integral of $f(X)$ between the limits 0 and 4 must be equal to unity. Constructing this equality and solving for the constant a,

$$\int_0^4 (\tfrac{1}{2} - aX)\, dX = 1$$

or

$$\frac{X}{2} - \frac{aX^2}{2}\bigg|_0^4 = 1$$

and

$$2 - 8a = 1$$
$$a = \tfrac{1}{8}$$

The required probability $P(1 \leq X \leq 2)$ is given by

$$P(1 \leq X \leq 2) = F(2) - F(1)$$
$$= \int_1^2 \left(\frac{1}{2} - \frac{X}{8}\right) dX = \frac{5}{16}$$

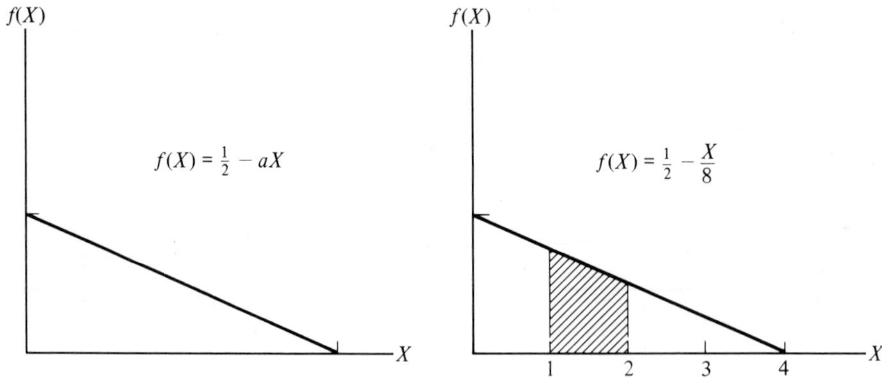

Triangular distribution model. Model for this application problem.

Figure 3.3. Example 3.4.2—probability distribution model.

3.5 Probability Distribution Table

Tables of probability values for commonly used statistical models are based on the cumulative distribution function $F(X)$. Table formats vary from one textbook to another. Some common presentations are examined here. Usually *C.D.F.* values are tabled for models involving a discrete random variable X, and required probabilities are obtained from differences $F(b) - F(a)$ as illustrated in Sec. 3.2. If the probability distribution model involves a continuous random variable X, tables usually give one or both "tail areas" developed from the *C.D.F.* The following example illustrates this practice.

EXAMPLE 3.5.1

A continuous random variable X has a *D.F.* given by

$$f(X) = \begin{cases} 1 - X & 0 \leq X \leq 1 \\ 1 + X & -1 \leq X \leq 0 \end{cases}$$

A graph of $f(X)$ is given in (a) of Fig. 3.4 along with a table of *C.D.F.* values formed from successive X-values that differ by 0.1 over the range $-1 \leq X \leq 0$. Each *C.D.F.* value is designated by α and is obtained from

$$\alpha = \int_{-1}^{X_\alpha} f(X)\, dX$$

Since $f(X)$ is symmetrical, left-hand tail areas α are identical to right-hand tail areas for all absolute values of X. Thus, for convenience only right-hand tail areas,

$$\alpha = 0.5 - \int_{0}^{X_\alpha} f(X)\, dX,$$

are tabulated as shown in (c) of Fig. 3.4. This is the usual table format for symmetrical probability distribution models that occur in subsequent chapters.

EXAMPLE 3.5.2

Using the probability distribution model and the X_α and α table of Fig. 3.4, the following probabilities are determined (and graphically shown in Fig. 3.5):

(a) $P(X \leq -0.7) = 0.045$, $P(X \geq 0.7) = 0.045$
(b) $P(X \geq -0.7) = 1 - 0.045 = 0.955$
(c) $P(X \leq 0.7) = 1 - 0.045 = 0.955$
(d) $P(-1 \leq X \leq 0.2) = 1 - 0.320 = 0.680$
(e) $P(-0.3 \leq X \leq 0.7) = 1 - (0.245 + 0.045) = 0.710$
(f) $P(0.3 \leq X \leq 0.7) = 0.245 - 0.045 = 0.200$

EXAMPLE 3.5.3

In a certain application, a random variable W, distributed over a range (2, 12), tends to follow the probability distribution model of Example 3.5.1. What is the probability $P(W \geq 5)$?

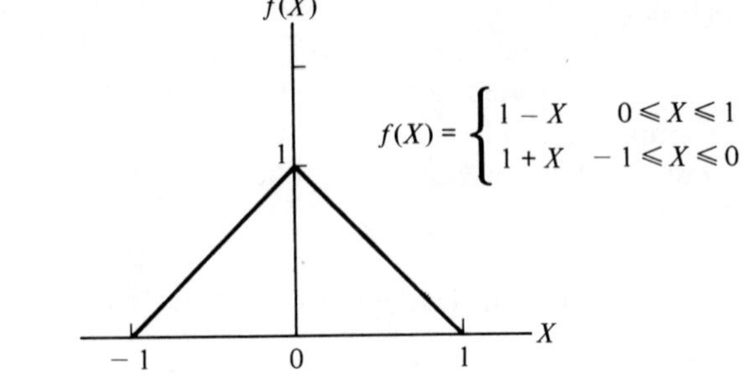

(a) Probability distribution model.

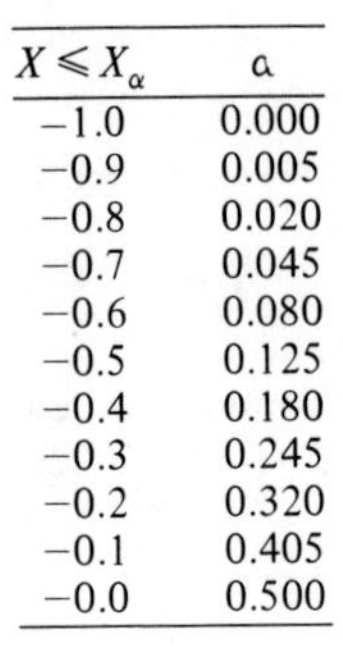

$X \leqslant X_\alpha$	α
−1.0	0.000
−0.9	0.005
−0.8	0.020
−0.7	0.045
−0.6	0.080
−0.5	0.125
−0.4	0.180
−0.3	0.245
−0.2	0.320
−0.1	0.405
−0.0	0.500

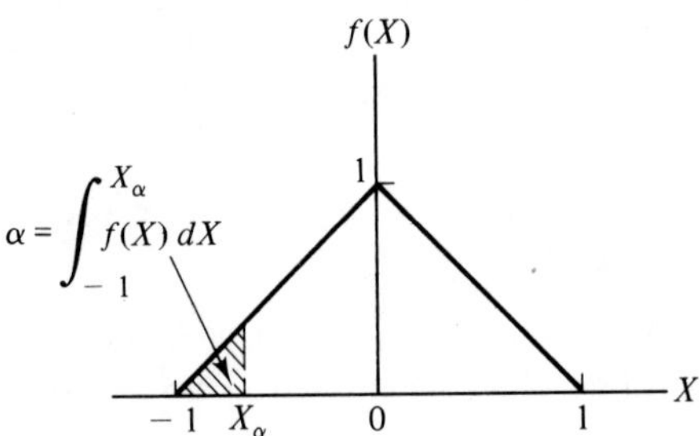

(b) *C.D.F.* values for $-1 \leqslant X \leqslant 0$.

X_α	α
1.0	0.000
0.9	0.005
0.8	0.020
0.7	0.045
0.6	0.080
0.5	0.125
0.4	0.180
0.3	0.245
0.2	0.320
0.1	0.405
0.0	0.500

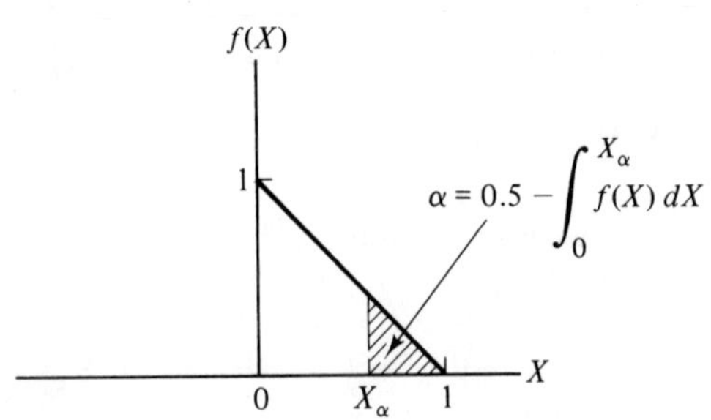

(c) Tail areas α for $-1 \leqslant X \leqslant 1$.

Figure 3.4. Example 3.5.1—tail areas α, standard probability distribution table.

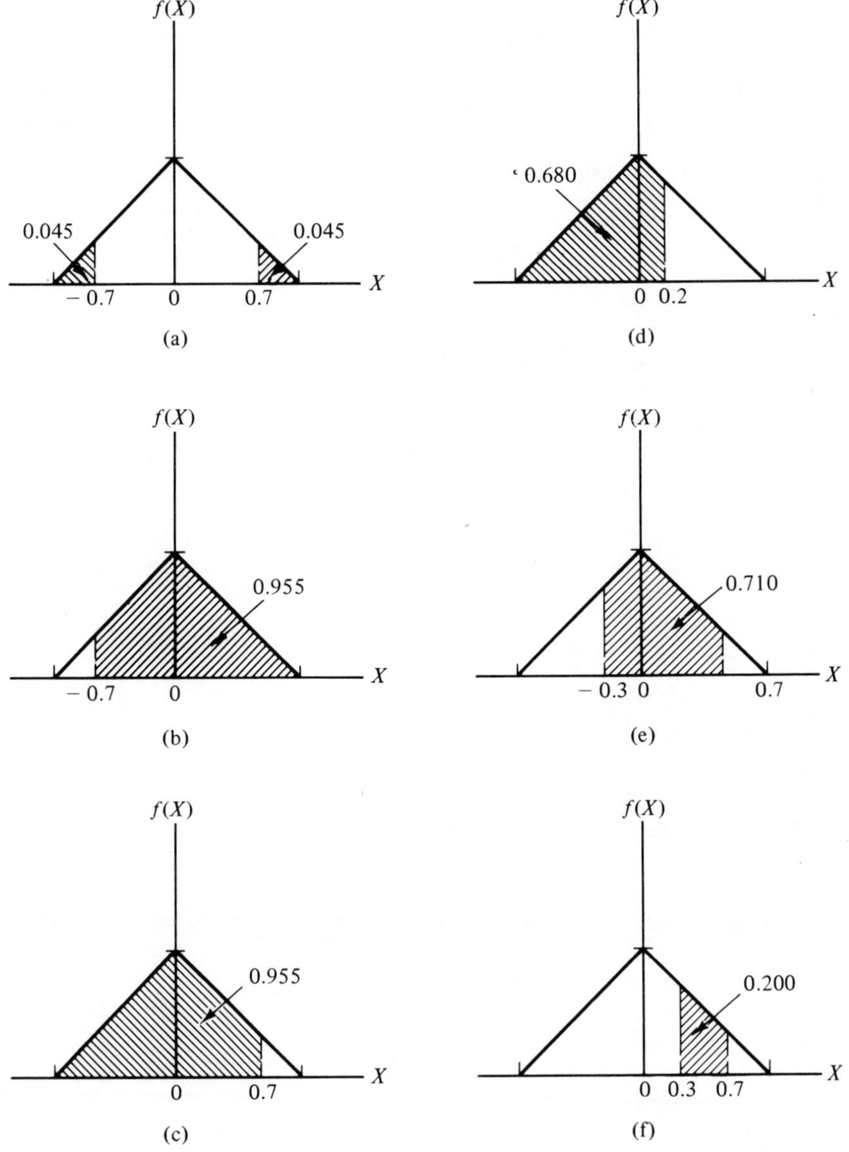

Figure 3.5. Example 3.5.2—probabilities obtained from the X_α and α table of Figure 3.4.

In W units, the function $f(W)$ shown in Fig. 3.6 is not a density function since the total area beneath the $f(W)$ curve and between $W = 2$ and $W = 12$ is not equal to unity. However, a transformation

$$X = \frac{W - 7}{5}$$

yields the identical density function of Example 3.5.1. Further, the X-value corresponding to $W = 5$ is

$$X_5 = \frac{5 - 7}{5} = -0.4$$

and

$$\alpha = \int_{2}^{5} f(W)\,dW = \int_{-1}^{-0.4} f(X)\,dX = 0.180$$

and

$$\begin{aligned} P(W \geq 5) &= P(X \geq -0.4) = 1 - \alpha \\ &= 1 - 0.180 = 0.820 \end{aligned}$$

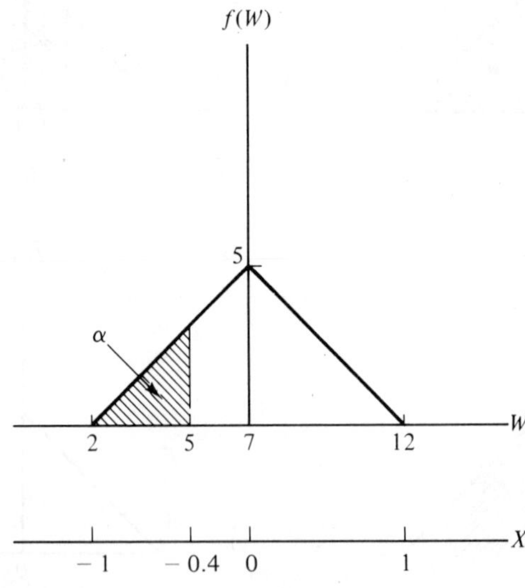

Figure 3.6 Example 3.5.3—transformation of W units to X units.

EXERCISES

1. Construct an X_α and α table for the following density function and determine the probability $P(-0.2 \leq X \leq 0.3)$. *Ans.* 0.74.

$$f(X) = \begin{cases} 2 - 4X & 0 \leq X \leq 0.5 \\ 2 + 4X & -0.5 \leq X \leq 0 \end{cases}$$

2. A random variable W, distributed over a range (2, 6), tends to follow the probability distribution model of Exercise 1. Change the W units to X units to use the model of Exercise 1, and compute the probability $P(W \leq 3)$. *Ans.* 0.125.

GLOSSARY OF TERMS

Random variable
Random sample
Systematic error
Random error
Cumulative distribution function, *C.D.F.*
Probability and density functions, *D.F.*
Discrete random variable
Continuous random variable

GLOSSARY OF FORMULAS

		Page
(3.1)	Conditions for a *D.F.* (discrete X). $f(X) \geq 0, \quad \sum_X f(X) = 1$	71
(3.2)	Conditions for a *C.D.F.* $F(-\infty) = 0, \quad F(+\infty) = 1$; $F(X)$ is a nondecreasing function, continuous to the right	71
(3.3)	Obtaining probability values. $P(a < X \leq b) = F(b) - F(a)$	72
(3.4)	Conditions for a *D.F.* (continuous X). $f(X) \geq 0, \quad \int_{-\infty}^{\infty} f(X)\,dX = 1$	74

REFERENCES

Bowker, A. H. and Lieberman, G. J. (1972), *Engineering Statistics*, Prentice-Hall, Inc., Englewood Cliffs, N. J., Chap. 2.

CHOU, YA-LUN (1969), *Statistical Analysis*, Holt, Rinehart & Winston, Inc., New York, Chap. 6.

SPIEGEL, M. R. (1961), *Schaum's Outline of Statistics*, Schaum Publishing Co., New York, Chap. 6.

WINE, R. L. (1964), *Statistics for Scientists and Engineers*, Prentice-Hall, Inc., Englewood Cliffs, N. J., Chap. 3.

PROBLEMS

1. State the two principal characteristics of a *random variable*.

2. Give a measurement application or illustration of *systematic* and *random* error.

3. A random variable X can assume only the values 1, 2, 3, and 4 with respective probabilities 0.1, 0.2, 0.4, and 0.3. Prepare tables and graphs of the probability and cumulative distribution functions.

4. For a discrete random variable, a probability statement regarding X assuming a specific value is possible and this probability is obtainable either from $f(X)$ or $F(X)$. Demonstrate this procedure for $P(X = 2)$ in Problem 3.

5. The distribution of a uniformly distributed random variable X is described by $f(X) = \frac{1}{3}$ where $1 \leq X \leq 4$. Demonstrate the procedure $F(2.5) - F(2)$ for obtaining the probability $P(2 \leq X \leq 2.5) = \frac{1}{6}$. Using this probability distribution model, what is the probability $P(X = 2)$?

6. Given a random experiment E capable of being repeated indefinitely. The outcomes of E are $X = 1, 2, 3$, and 4 with respective probabilities 0.1, 0.2, 0.4, and 0.3 (see Problem 3). A random sample of the random variable X is obtained by repeating E such that there are 100 trials of E yielding the frequency distribution:

x_j	1	2	3	4	
f_j	8	22	42	28	100

(a) Prepare tables and graphs for $f_n(X)$ and $F_n(X)$, (b) Why do $F(X)$ and $F_n(X)$ differ?, (c) What conclusions are possible dependent on the degree of difference?

7. State the conditions for a cumulative distribution function (C.D.F.).

8. State the conditions for a probability or density function (D.F.). Is the following function a D.F.: $f(X) = \frac{1}{4}, 0 \leq X \leq 5$?

9. A continuous random variable X has a D.F. $f(X) = a(X + 4), 1 \leq X \leq 5$. (a) Evaluate the constant a, (b) Determine the probability $P(X \leq 2)$, (c) Determine the probability $P(2 \leq X \leq 3)$.

10. A D.F. is $f(X) = aX, 0 \leq X \leq 4$. Evaluate a and then determine the probabilities: (a) $P(X \leq 3)$, (b) $P(X \leq 1)$, and (c) $P(1 \leq X \leq 3)$.

11. A D.F. is $f(X) = 1 - aX$, $0 \leq X \leq 2$. Evaluate a and then determine the probabilities: (a) $P(0.5 \leq X \leq 1)$, and (b) $P(1 \leq X \leq 1.5)$.

12. A random variable X is uniformly distributed over the interval (1, 5). Prepare graphs of the density and cumulative distribution functions. What is the probability $P(2 < X \leq 3)$?

13. A random variable X has a density function $f(X) = a(X + 3)$ where $2 \leq X \leq 8$. Evaluate the constant a, and determine the probability $P(3 < X < 5)$.

14. A random variable X has a density function

$$f(X) = \begin{cases} 0 & X < 0 \\ \dfrac{e^{-0.0001X}}{a} & X \geq 0 \end{cases}$$

Evaluate the constant a to assure that $f(X)$ is a *D.F.*

DISCRETE DISTRIBUTION MODELS

In Chap. 3, the definition of probability was extended to cover probabilities of events of the form $X \leq b$. *This extension involved the concepts of random variable and probability distributions. The purpose of this chapter is to apply these concepts to some common distribution models involving a discrete random variable* X.

In each application, the mean and variance are of interest, and the mathematical model refers to a population. A slightly different definition is developed for both the mean and variance—definitions similar to those of Sec. 1.17, where the mean was viewed as a first moment and the variance as a second moment. To do this, it is necessary to first consider the concepts of mathematical expectation and characteristic function.

4.1 Mathematical Expectation

Consider a random experiment E and a function $g(X)$ dependent on the random variable X. The function $g(X)$ is also a random variable. If the following sums or integrals exist, the *mathematical expectation* of the random variable $g(X)$ is defined by

4

$$E[g(X)] = \begin{cases} \sum_X g(X)f(X) & \text{discrete } X \\ \int_{-\infty}^{\infty} g(X)f(X)\,dx & \text{continuous } X \end{cases} \tag{4.1}$$

where $f(X)$ is the *D.F.* for the model under consideration and the symbol $E[g(X)]$ reads *expected value of* $g(X)$. Either brackets or parentheses are accepted notation. Thus, the expected value of X, for example, is denoted either $E[X]$ or $E(X)$.

The expected values of certain functions $g(X)$ are of particular interest in describing the properties of distribution models. For example, if $g(X) = X$, then

$$E(X) = \left\{ \begin{matrix} \sum_X Xf(X) \\ \int_{-\infty}^{\infty} Xf(X)\,dX \end{matrix} \right\} = \mu_X \tag{4.2}$$

and

$$E[(X-\mu)^2] = \left\{ \begin{matrix} \sum_X (X-\mu)^2 f(X) \\ \int_{-\infty}^{\infty} (X-\mu)^2 f(X)\,dX \end{matrix} \right\} = \sigma_X^2 \tag{4.3}$$

The quantity $E[(X - \mu)^r]$ is called the rth *moment about the mean*, and $E(X^r)$ the rth *moment about zero* [corresponding to Eqs. (1.32) and (1.31), respectively]. Thus, the population mean μ is the first moment about zero, and the population variance σ_X^2 is the second moment about the mean.*

---

*It is important that the student appreciate the correspondence between expected value definitions and previous definitions of the mean and variance. A small numbers example is given here to illustrate the correspondence. Figure 4.1 shows a frequency distribution and a relative frequency distribution of six observations on X. The sample mean is obtained from

$$\bar{x} = \frac{1}{n} \sum_j x_j f_j = \frac{1}{6}(3 + 4 + 3) = \frac{10}{6}$$

or by

$$\bar{x} = \sum_j x_j \left(\frac{f_j}{n}\right)$$
$$= 1(\tfrac{3}{6}) + 2(\tfrac{2}{6}) + 3(\tfrac{1}{6}) = \tfrac{10}{6}$$

Assume a population comprised of the same six X-values. Then, the population mean is

$$\mu = E(X) = \sum_X Xf(X)$$
$$= 1(\tfrac{3}{6}) + 2(\tfrac{2}{6}) + 3(\tfrac{1}{6}) = \tfrac{10}{6}$$

x_j	f_j	$f_j x_j$
1	3	3
2	2	4
3	1	3
	6	10

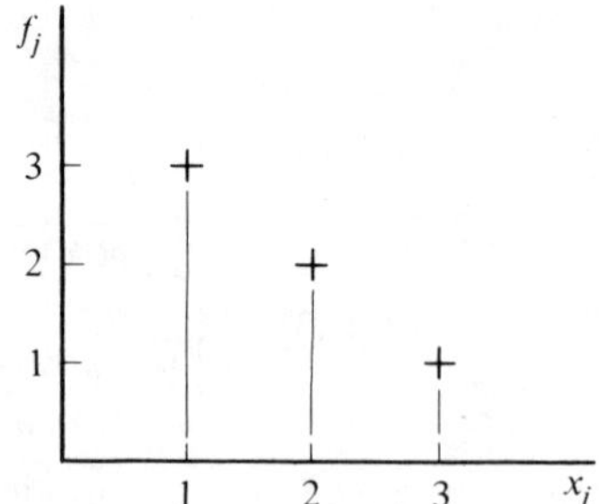

x_j	$\frac{f_j}{n}$
1	$\frac{3}{6}$
2	$\frac{2}{6}$
3	$\frac{1}{6}$
	1.0

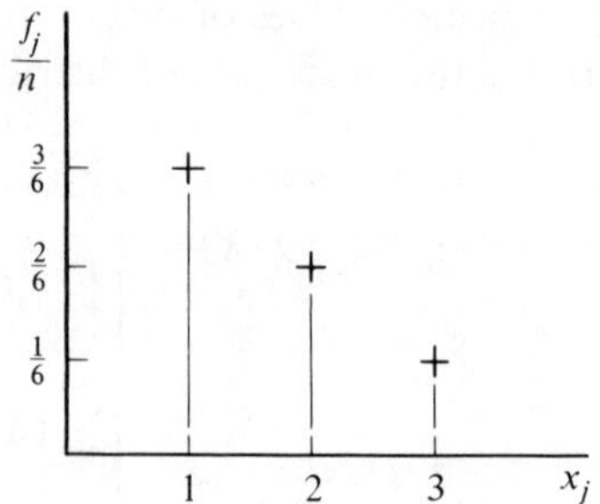

Figure 4.1. Expected value of x.

Expected value operations, since they deal with summations (or limits of sums), are similar to and follow the summation rules, Eqs. (1.5) to (1.8). These operations are summarized here:

$$E(aX) = \begin{Bmatrix} \sum_X aXf(X) \\ \int_{-\infty}^{\infty} aXf(X)\,dX \end{Bmatrix} = aE(X) = a\mu \tag{4.4}$$

$$E\{[a(X-\mu)]^2\} = \begin{Bmatrix} \sum_X [a(X-\mu)]^2 f(X) \\ \int_{-\infty}^{\infty} [a(X-\mu)]^2 f(X)\,dx \end{Bmatrix} = a^2\sigma_X^2 \tag{4.5}$$

$$E(a) = \begin{Bmatrix} \sum_X af(X) \\ \int_{-\infty}^{\infty} af(X)\,dX \end{Bmatrix} = a \tag{4.6}$$

where each summation is for the case of a discrete random variable X, each integral is for the case of a continuous random variable X, and a is a constant.

EXAMPLE 4.1.1

From Eq. (1.20), the population variance for a discrete random variable X is given by

$$\sigma_X^2 = \frac{1}{N}\sum_{j=1}^{k} f_j(X_j - \mu)^2$$

Performing the indicated squaring and summation operations, this expression reduces to

$$\sigma_X^2 = \frac{1}{N}\sum_{j=1}^{k} f_jX_j^2 - N\mu^2 \tag{4.7}$$

To illustrate expected value operations, an analogous form of the population variance is developed here. From Eq. (4.3),

$$\begin{aligned}
\sigma_X^2 = E[(X-\mu)^2] &= \int_{-\infty}^{\infty} (X-\mu)^2 f(X)\,dx \\
&= \int_{-\infty}^{\infty} (X^2 - 2\mu X + \mu^2) f(X)\,dX \\
&= \int_{-\infty}^{\infty} X^2 f(X)\,dX - 2\mu\int_{-\infty}^{\infty} Xf(X)\,dX + \mu^2\int_{-\infty}^{\infty} f(X)\,dX \\
&= \int_{-\infty}^{\infty} X^2 f(X)\,dX - 2\mu^2 + \mu^2
\end{aligned}$$

$$\sigma_X^2 = \int_{-\infty}^{\infty} X^2 f(X)\,dX - \mu^2$$

or

$$= E(X^2) - [E(X)]^2 \tag{4.8}$$

EXERCISES

1. Consider a random toss of one fair die. Let the random variable X be the value of the upturned face of the die. Compute the mean μ, using expected value operations. *Ans.* $\frac{7}{2}$.

2. Compute the variance for Exercise 1 using expected value operations. Also, compute the variance using Eq. (4.8). *Ans.* $\frac{35}{12}$.

3. Two dice are randomly tossed and the random variable X is the sum of the upturned faces. Compute the mean and variance using expected value operations. *Ans.* $\mu = 7, \sigma_X^2 = \frac{35}{6}$.

4. A random variable X assumes values 1, 2, 3, 4, with respective probabilities 0.3, 0.2, 0.1, and 0.4. Compute $E(3X + 2)$. *Ans.* 9.8.

5. For the uniform distribution introduced in Example 3.4.1, compute the mean using Eq. (4.2).

4.2 Chebyshev's Inequality

If a probability distribution is known, all of its moments can be computed. If all of the moments are known, the distribution is characterized. In practice, knowledge of the first two moments is usually sufficient to characterize the distribution such that the application is adequately handled. This raises the question: to what extent do the first two moments, μ and σ_X^2, characterize the probability distribution of the random variable? The answer is given by *Chebyshev's inequality*:*

For any real number k, the probability that the random variable X lies in the inverval $(\mu - k\sigma_X, \mu + k\sigma_X)$ is larger than $1 - (1/k^2)$, that is,

$$P(\mu - k\sigma_X \leq X \leq \mu + k\sigma_X) > 1 - \frac{1}{k^2} \tag{4.9}$$

where X is a random variable having any distribution with mean μ and variance σ_X^2.

EXAMPLE 4.2.1

If $k = 2$, the following probability is obtained:

$$P(\mu - 2\sigma_X \leq X \leq \mu + 2\sigma_X) > 1 - \frac{1}{2^2}$$

*This is sometimes spelled Tchebysheff or Tchebichev.

Clearly, for the distribution models to be considered, it is important to obtain the first two moments:

$$E(X) = \mu, \qquad E[(X-\mu)^2] = \sigma_X^2$$

These moments can be developed either by expected value operations or by use of a characteristic function of the distribution. Frequently, the latter method is more direct and efficient, and, for this reason, it is now presented.

4.3 Characteristic Function of a Distribution

To any random variable X with density or probability function $f(X)$, there can always be associated another function $\phi_X(t)$, called the *characteristic function*, by the relation

$$\phi_X(t) = E(e^{iXt}) = \begin{cases} \sum_X e^{iXt} f(X) \\ \int_{-\infty}^{\infty} e^{iXt} f(X)\, dX \end{cases} \tag{4.10}$$

where the sum refers to the discrete X case and the integral to the continuous X case. The coefficient term e^{iXt} is a complex-valued function given by $e^{iXt} = \cos Xt + i \sin Xt$ and thus $\phi_X(t)$ is also a complex-valued function.

For many distributions, moments are difficult to determine directly by expected value operations as in Sec. 4.1. The characteristic function affords an indirect method of obtaining the moments for a given distribution model. Although this technique has little to do with the applications examples of this text, it is briefly presented here for appreciation by the superior student. Some distribution moments are obtained using this method. For applications purposes, the average student can accept the moments (mean and variance) that are stated and disregard their derivation.

The function $\phi_X(t)$ is a function of t only, the subscript X being used to indicate the variable of the distribution. The moments for a given distribution model can be computed using the relationship

$$E(X^k) = \frac{\phi_X^{(k)}(0)}{i^k} \qquad i^2 = -1* \tag{4.11}$$

---

*Expanding e^{iXt} in a power series and substituting in Eq. (4.10),

$$\begin{aligned} \phi_X(t) &= \sum_X e^{iXt} f(X) \\ &= \sum_X \left[1 + iXt - \frac{X^2t^2}{2!} - \frac{iX^3t^3}{3!} + \frac{X^4t^4}{4!} + \ldots\right] f(X) \end{aligned}$$

Successive differentiations with respect to t yield

EXAMPLE 4.3.1

Using the characteristic function method, develop expressions for the mean and variance of the uniform distribution first introduced in Example 3.4.1 and shown in Fig. 4.2.

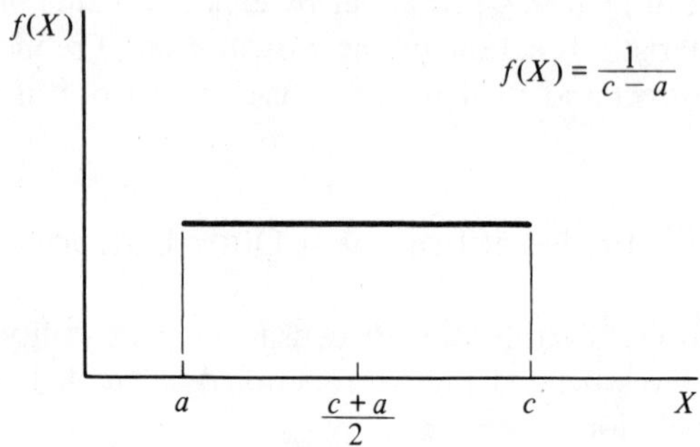

Figure 4.2. Rectangular or uniform distribution.

$$\phi'_X(t) = \sum_X f(X)\left[0 + iX - \frac{2X^2t}{2!} - \frac{3iX^3t^2}{3!} + \ldots\right]$$

$$\phi'_X(0) = i\sum_X Xf(X) = iE(X)$$

$$\phi''_X(t) = \sum_X f(X)\left[0 + 0 - \frac{2X^2}{2!} - \frac{6iX^3t}{3!} + \ldots\right]$$

$$\phi''_X(0) = -\sum_X X^2f(X) = -E(X^2)$$

and so forth, thus giving

$$\begin{aligned} \phi_X(0) &= 1 \\ \phi'_X(0) &= iE(X) \\ \phi''_X(0) &= -E(X^2) \\ \phi'''_X(0) &= -E(X^3) \\ &\vdots \end{aligned}$$

or, more compactly stated as in Eq. (4.11),

$$E(X^k) = \frac{\phi_X^{(k)}(0)}{i^k}$$

In a similar fashion, moments about the mean can be developed using the characteristic function:

$$\phi_{(X-\mu)}(t) = E(e^{i(X-\mu)t}) \tag{4.12}$$

whence

$$E[(X-\mu)^k] = \frac{\phi_{(X-\mu)}^{(k)}(0)}{i^k} \tag{4.13}$$

This is a special case of a more general application where the characteristic function of a random variable $g(X)$ has a characteristic function

$$\phi_{g(X)}(t) = E(e^{ig(X)t}) = \begin{Bmatrix} \sum_X e^{ig(X)t}f(X) \\ \int_{-\infty}^{\infty} e^{ig(X)t}f(X)\,dX \end{Bmatrix} \tag{4.14}$$

From Eq. (4.10), the characteristic function is

$$\phi_X(t) = E(e^{iXt}) = \int_a^c e^{iXt}\frac{1}{c-a}\,dX$$

Successive differentiations yield

$$\phi'_X(t) = \frac{1}{c-a}\int_a^c iXe^{iXt}\,dX$$

and

$$\phi'_X(0) = \frac{i}{c-a}\cdot\frac{X^2}{2}\bigg|_a^c = \frac{i}{c-a}\cdot\frac{c^2-a^2}{2}$$

$$= \frac{i(c+a)}{2}$$

and, from Eq. (4.11),

$$\mu = E(X) = \frac{\phi'_X(0)}{i} = \frac{c+a}{2} \tag{4.15}$$

Also,

$$\phi''_X(t) = \frac{1}{c-a}\int_a^c i^2X^2e^{iXt}\,dX$$

and

$$\phi''_X(0) = \frac{i^2}{c-a}\cdot\frac{X^3}{3}\bigg|_a^c = -\frac{1}{c-a}\cdot\frac{c^3-a^3}{3}$$

$$= -\frac{c^2+ac+a^2}{3}$$

and, from Eq. (4.11),

$$E(X^2) = \frac{\phi''_X(0)}{i^2} = \frac{-\dfrac{c^2+ac+a^2}{3}}{-1} = \frac{c^2+ac+a^2}{3}$$

From Eq. (4.8),

$$\sigma_X^2 = E(X^2) - [E(X)]^2$$

$$= \frac{c^2+ac+a^2}{3} - \left(\frac{c+a}{2}\right)^2$$

$$= \frac{(c-a)^2}{12} \tag{4.16}$$

EXAMPLE 4.3.2

Using only expected value definitions and operations, repeat the development in Example 4.3.1. From Eq. (4.2), the mean μ is given by

$$E(X) = \int_a^c X\left(\frac{1}{c-a}\right)dX = \frac{1}{c-a}\cdot\frac{X^2}{2}\bigg|_a^c$$

$$= \frac{1}{c-a}\cdot\frac{c^2-a^2}{2} = \frac{c+a}{2}$$

Using Eq. (4.8), the variance σ_X^2 is

$$\sigma_X^2 = \int_a^c X^2\left(\frac{1}{c-a}\right)dX - \mu^2 = \frac{1}{c-a}\cdot\frac{X^3}{3}\bigg|_a^c - \mu^2$$

$$= \frac{1}{c-a}\cdot\frac{c^3-a^3}{3} - \left(\frac{c+a}{2}\right)^2$$

$$= \frac{(c-a)^2}{12}$$

EXERCISES

1. In Example 3.4.3, the *D.F.* is

$$f(X) = \frac{1}{2} - \frac{X}{8}, \qquad 0 \leq X \leq 4$$

Using the characteristic function method, develop expressions for the mean and variance for this function. *Ans.* $\mu = \frac{4}{3}, \sigma_X^2 = \frac{8}{9}$.

2. Repeat Exercise 1, using expected value definitions and operations to develop expressions for the mean and variance.

3. A given *D.F.* is

$$f(X) = e^{-X}, \qquad X \geq 0$$

Using the characteristic function method, develop expressions for the mean and variance for this function. *Ans.* $\mu = 1, \sigma_X^2 = 1$.

4. Repeat Exercise 3, using expected value definitions and operations to develop expressions for the mean and variance.

4.4 Bernoulli Distribution Model

A *Bernoulli trial* is a random experiment E that has exactly two outcomes, generally termed success and failure, and denoted 1 (for success) and 0 (for failure), with probabilities p and $q = 1 - p$, respectively. The distribution model refers to a single Bernoulli trial, and the *D.F.* is

$$f(X) = p^X q^{1-X} \qquad X = 0, 1 \tag{4.17}$$

If $p = 0.6$, for example, then

$$f(0) = 0.6^0 0.4^{1-0} = 0.4$$
$$f(1) = 0.6^1 0.4^{1-1} = 0.6$$

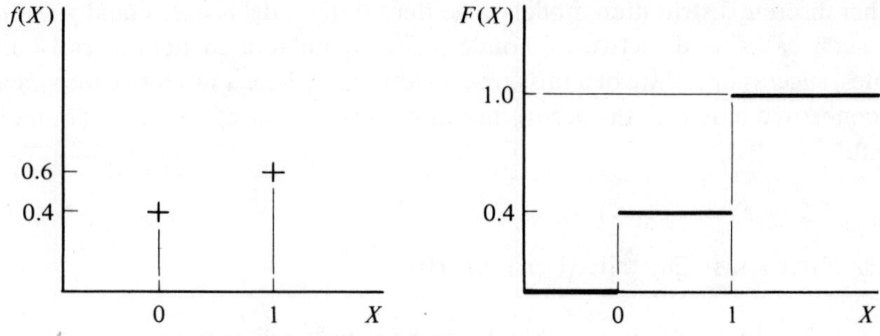

Figure 4.3. Bernoulli distribution, $p = 0.4$.

Thus, $f(0)$ is the probability of a failure and $f(1)$ is the probability of a success, and the *D.F.* and *C.D.F.* are as shown in Fig. 4.3.

The moments for the Bernoulli distribution are easily obtained. The characteristic function is

$$\begin{aligned}\phi_X(t) &= E(e^{iXt}) = \sum_{X=0}^{1} e^{iXt}p^X q^{1-X}\\ &= q + pe^{it}\end{aligned}$$

The first and second derivatives, evaluated at $t = 0$, are

$$\begin{aligned}\phi'_X(t) &= ipe^{it}, \qquad \phi'_X(0) = ip\\ \phi''_X(t) &= i^2pe^{it}, \qquad \phi''_X(0) = i^2p\end{aligned}$$

From Eq. (4.11),

$$\mu = E(X) = \frac{\phi'_X(0)}{i} = \frac{ip}{i} = p \tag{4.18}$$

and

$$E(X^2) = \frac{\phi''_X(0)}{i^2} = \frac{i^2p}{i^2} = p$$

From Eq. (4.8), we have the following;

$$\begin{aligned}\sigma_X^2 &= E(X^2) - [E(X)]^2 = p - p^2\\ &= p(1 - p) = pq^*\end{aligned} \tag{4.19}$$

The Bernoulli distribution model has limited application value. It is presented here primarily because it furnishes a basis for the derivation of

---

*An alternative derivation of the moments for this model, using expected value definitions and operations, is given here. The expected values $E(X)$ and $E(X^2)$ are obtained

other discrete distribution models. The Bernoulli model is occasionally useful in such cases as defective or nondefective manufactured items, yes or no votes, success or failure of a mission—in any case where a random experiment is concerned only with the occurrence or nonoccurrence of an event at a single trial.

4.5 Binomial Distribution Model

A *binomial* random variable X can be considered as the number of successes (or the number of 1's) in n independent Bernoulli trials. The *D.F.* is

$$f(X) = \binom{n}{X} p^X q^{n-X} \quad X = 0, 1, 2, \ldots, n \tag{4.20}$$

where p is the probability of success at each trial.*

The binomial distribution is symmetrical if $p = 0.5$ and is skewed when

---

as follows:

X	$f(X)$	$Xf(X)$	$X^2f(X)$
0	q	0	0
1	p	p	p
$\Sigma \rightarrow$	1	$E(X) = p$	$E(X^2) = p$

Thus,

$$\mu = E(X) = \sum_X Xf(X) = p$$

and

$$\begin{aligned} \sigma_X^2 &= E(X^2) - [E(X)]^2 \\ &= p - p^2 \\ &= p(1 - p) = pq \end{aligned}$$

*This function can be developed intuitively using the probability operations of Chap. 2 by considering a simple example, the probability of obtaining exactly three 7's in five random tosses of a pair of unbiased dice.

Depending on which of the five tosses produces the three 7's, there are $\binom{5}{3} = 10$ mutually exclusive and independent successful outcomes. From Eq. (2.10), each successful outcome has a probability p^3q^2, where the probability p of a 7 is $\frac{1}{6}$ and the probability q of no 7 is $\frac{5}{6}$. Then, from Eq. (2.09),

$$P\text{ (three 7.'s/5 tosses)} = 10(\tfrac{1}{6})^3(\tfrac{5}{6})^2$$

But, this is the expression for the binomial *D.F.*; that is,

$$P(X = 3) = \binom{n}{X} p^X q^{n-X} \quad \text{for} \quad X = 3, n = 5.$$

$p \neq 0.5$. For p less than 0.5, the distribution is skewed to the right, and if p is greater than 0.5 it is skewed to the left. The degree of skewness increases as p approaches either 0 or 1. The skewness, independent of the size of p, diminishes as sample size n increases. Figures 4.4–4.6 illustrate the binomial *D.F.* for $n = 5$ and $p = 0.50, 0.25, 0.10$.

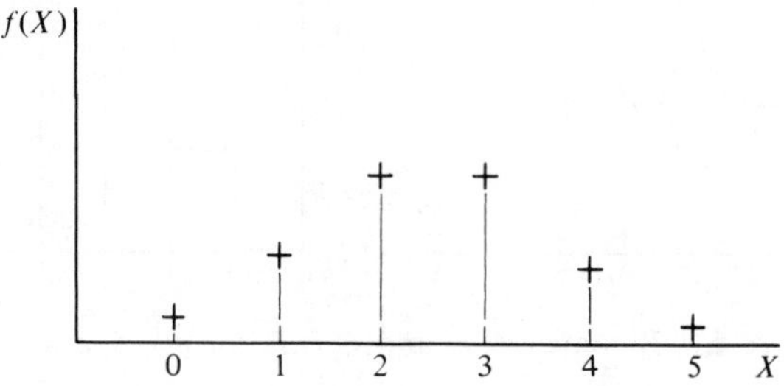

Figure 4.4. Binomial distribution, $n = 5, p = 0.50$.

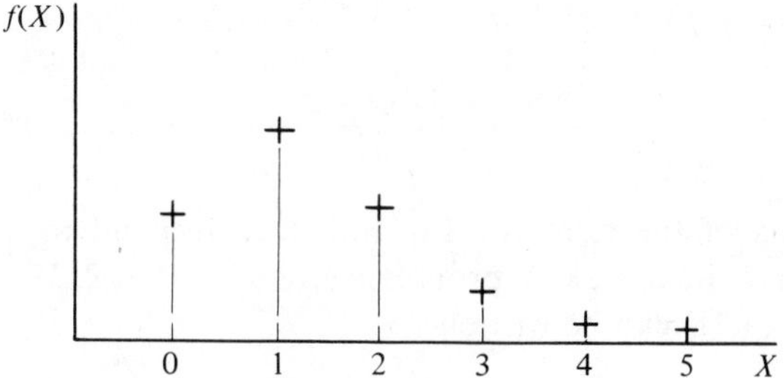

Figure 4.5. Binomial distribution, $n = 5, p = 0.25$.

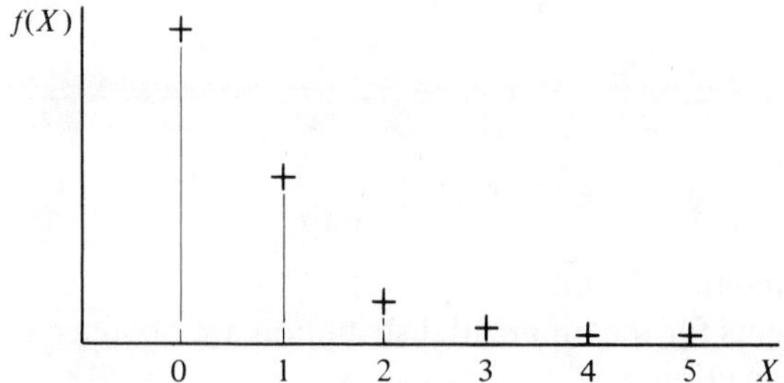

Figure 4.6. Binomial distribution, $n = 5, p = 0.10$.

Figure 4.7 shows the binomial *D.F.* and *C.D.F.* for $n = 5$ and $p = 0.25$.

If $n = 1$, the binomial distribution reduces to the Bernoulli distribution. The binomial distribution model derives its name from the observation that the successive terms in the binomial expression $(q + p)^n$ gives the respective

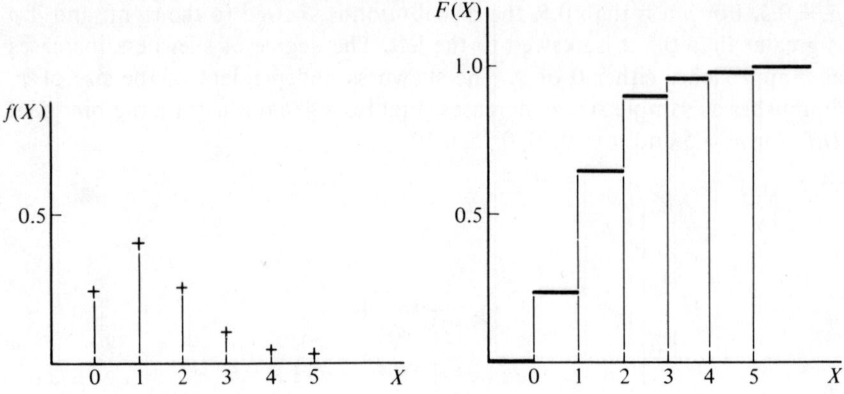

Figure 4.7. Binomial *D.F.* and *C.D.F.* for $n = 5$, $p = 0.25$.

$f(X)$ values. That is, the binomial expansion is

$$(q + p)^n = q^n + npq^{n-1} + \frac{n(n-1)p^2q^{n-2}}{2!} + \frac{n(n-1)(n-2)p^3q^{n-3}}{3!} + \ldots + p^n \tag{4.21}$$

The coefficients of the terms of Eq. (4.21) correspond to a combination expression for n items taken progressively, $X = 0, 1, 2, \ldots, n$ at a time. Therefore Eq. (4.21) can be written

$$(q + p)^n = \binom{n}{0}p^0q^{n-0} + \binom{n}{1}pq^{n-1} + \binom{n}{2}p^2q^{n-2} + \ldots + \binom{n}{n}p^nq^{n-n}$$

or

$$(q + p)^n = \sum_{X=0}^{n}\binom{n}{X}p^Xq^{n-X} \tag{4.22}$$

which is the binomial *C.D.F.*

The moments for the binomial distribution are obtained as follows. The characteristic function is

$$\phi_X(t) = E(e^{iXt}) = \sum_{X=0}^{n} e^{iXt}\binom{n}{X}p^Xq^{n-X}$$

$$= \sum_{X=0}^{n}\binom{n}{X}(pe^{it})^Xq^{n-X}$$

and, from Eq. (4.22),

$$= (pe^{it} + q)^n$$

The first and second derivatives, evaluated at $t = 0$, are

$$\phi'_X(t) = n(pe^{it} + q)^{n-1}pe^{it}i$$
$$= inpe^{it}(pe^{it} + q)^{n-1}$$
$$\phi'_X(0) = inp(p + q)^{n-1} = inp$$

and

$$\phi''_X(t) = i^2npe^{it}(pe^{it} + q)^{n-1} - n(n-1)p^2e^{2it}(pe^{it} + q)^{n-2}$$
$$\phi''_X(0) = -np - n(n-1)p^2$$

From Eq. (4.11),

$$\mu = E(X) = \frac{\phi'_X(0)}{i} = \frac{inp}{i} = np \tag{4.23}$$

and

$$E(X^2) = \frac{\phi''_X(0)}{i^2} = \frac{-np - n(n-1)p^2}{i^2} = np + n(n-1)p^2$$

From Eq. (4.8),

$$\sigma_X^2 = E(X^2) - [E(X)]^2$$
$$= np + n(n-1)p^2 - n^2p^2$$
$$= npq \tag{4.24}$$

EXAMPLE 4.5.1

A binomial model describes the number of successes X in n independent Bernoulli trials. A fair die is tossed five times, and the occurrence of a 7 is considered a success. Each toss is a Bernoulli trial since there are only two outcomes—either a 7 occurs or it does not occur. There are $n = 5$ Bernoulli trials.

The probability of a success at each trial is $\frac{1}{6}$, and the probability of a failure is $\frac{5}{6}$. The respective probabilities of a 7 occurring $X = 0, 1, 2, 3, 4,$ and 5 times are:

$$P(X = 0) = \binom{5}{0}\left(\frac{1}{6}\right)^0\left(\frac{5}{6}\right)^5 = \frac{3{,}125}{7{,}776}$$

$$P(X = 1) = \binom{5}{1}\left(\frac{1}{6}\right)^1\left(\frac{5}{6}\right)^4 = \frac{3{,}125}{7{,}776}$$

$$P(X = 2) = \binom{5}{2}\left(\frac{1}{6}\right)^2\left(\frac{5}{6}\right)^3 = \frac{1{,}250}{7{,}776}$$

$$P(X = 3) = \binom{5}{3}\left(\frac{1}{6}\right)^3\left(\frac{5}{6}\right)^2 = \frac{250}{7{,}776}$$

$$P(X = 4) = \binom{5}{4}\left(\frac{1}{6}\right)^4\left(\frac{5}{6}\right)^1 = \frac{25}{7{,}776}$$

$$P(X = 5) = \binom{5}{5}\left(\frac{1}{6}\right)^5\left(\frac{5}{6}\right)^0 = \frac{1}{7{,}776}$$

The mean and variance for this distribution are

$$\mu = np = 5(\tfrac{1}{6}) = (\tfrac{5}{6})$$
$$\sigma_X^2 = npq = 5(\tfrac{1}{6})(\tfrac{5}{6}) = \tfrac{25}{36}$$

With large sample size n the computation of binomial probabilities is tedious. Extensive tables for the binomial *C.D.F.* are available.* A small table of cumulative binomial probabilities is given in Appendix A. This table can be used to obtain either cumulative or individual binomial probabilities.

EXAMPLE 4.5.2

A manufacturing lot of 250 parts is 10% defective. An experiment E consists of 25 trials, each trial being a random selection of one part from the lot, observing if the part is defective or effective, and then replacing it in the lot before the next selection is made.† What is the probability of obtaining a total of five defective parts for the 25 selections?

Each trial of E is a Bernoulli trial (i.e., there are two outcomes). Let X denote a success (obtaining a defective). The probability of a success at a single trial of E is $p = 0.10$. There are $n = 25$ trials. These are binomial conditions and thus, from Eq. (4.20),

$$\begin{aligned} P(X = 5) &= \binom{25}{5}(0.10)^5(0.90)^{25-5} = F(5) - F(4) \\ &= 0.9666 - 0.9020 \quad \text{(from Appendix A)} \\ &= 0.0646 \end{aligned}$$

EXAMPLE 4.5.3

For the conditions of Example 4.5.2, what is the probability of obtaining two or less defectives? This is the probability of 0, or 1, or 2 defectives, or,

$$P(X = 0) + P(X = 1) + P(X = 2) = \binom{25}{0}(0.10)^0(0.90)^{25}$$
$$+ \binom{25}{1}(0.10)^1(0.90)^{24} + \binom{25}{2}(0.10)^2(0.90)^{23}$$

or, more compactly,

$$= \sum_{X=0}^{2} \binom{25}{X}(0.10)^X(0.90)^{25-X} = F(2)$$
$$= 0.5371 \quad \text{(from Appendix A)}$$

---

*For example, *Tables of the Binomial Probability Distribution* (1950), Nat. Bur. Standards Appl. Math. Ser. 6, Supt. of Documents, U. S. Govt. Printing Office, Wash. 25, D. C.

†Of course, replacing a defective part in the lot is not very practical. However, in this introductory problem this requirement assures exact binomial conditions (constant probability p from trial to trial). This is discussed further in a more practical situation in Example 4.5.6.

EXAMPLE 4.5.4

For the conditions of Example 4.5.2, what is the probability of obtaining four or more defectives? Constructing the probability expression to utilize Appendix A,

$$
\begin{aligned}
P(X \geq 4) &= 1 - P(X \leq 3) \\
&= 1 - \sum_{X=0}^{3} \binom{25}{X} (0.10)^X (0.90)^{25-X} \\
&= 1 - 0.7636 \quad \text{(from Appendix A)} \\
&= 0.2364
\end{aligned}
$$

Binomial probabilities are somewhat tedious to compute. Computation can be simplified by use of a recursion formula. From Eq. (4.20),

$$P(X = 0) = \binom{n}{0} p^0 q^n = q^n$$

$$
\begin{aligned}
P(X = 1) &= \binom{n}{1} pq^{n-1} = np\left(\frac{q^n}{q}\right) \\
&= P(X = 0)(n)\left(\frac{p}{q}\right) = P(X = 0)\left(\frac{n-0}{0+1}\right)\left(\frac{p}{q}\right)
\end{aligned}
$$

$$
\begin{aligned}
P(X = 2) &= \binom{n}{2} p^2 q^{n-2} \\
&= \frac{n!}{2!\,(n-2)!} p^2 q^{n-2} = \frac{n(n-1)}{2}(p)(p)\left(\frac{q^{n-1}}{q}\right) \\
&= P(X = 1)\left(\frac{n-1}{2}\right)\left(\frac{p}{q}\right) = P(X = 1)\left(\frac{n-1}{1+1}\right)\left(\frac{p}{q}\right)
\end{aligned}
$$

and so forth, the general term expression being

$$P(X = k + 1) = P(X = k)\left(\frac{n-k}{k+1}\right)\left(\frac{p}{q}\right) \qquad (4.25)$$

EXAMPLE 4.5.5

Suppose in Example 4.5.1 that $P(X = 3)$ is either given or computed and the $P(X = 4)$ is required. Then, from Eq. (4.25), with $k = 3$,

$$
\begin{aligned}
P(X = 4) &= P(X = 3)\left(\frac{5-3}{3+1}\right)\left(\frac{\frac{1}{6}}{\frac{5}{6}}\right) \\
&= \frac{250}{7{,}776}\left(\frac{1}{2}\right)\left(\frac{1}{5}\right) \\
&= \frac{25}{7{,}776}
\end{aligned}
$$

EXAMPLE 4.5.6

A sampling inspection plan is a rule which states whether to accept or reject a lot (or batch) of manufactured items. The decision to accept or reject is based on information supplied by a small random sample drawn from the lot.

For a specified plan, a lot is accepted if the number of observed defective items X in a sample of $n = 25$ items is less than or equal to 1. What is the probability of acceptance P_a of a 10% defective lot? Using the binomial model,

$$P_a = \sum_{X=0}^{1} \binom{25}{X} (0.1)^X (0.9)^{25-X}$$

$$= (0.9)^{25} + \frac{25!}{1!\,24!}(0.1)(0.9)^{24}$$

$$= 0.2712^*$$

EXERCISES

1. What is the probability of obtaining exactly two 1's in six random tosses of a fair die? *Ans.* 0.2009.

2. A production machine manufactures product items in such a way that there is a probability of 0.01 that a given item is defective. What is the probability that of four items: (a) none is defective, and (b) at most one is defective? (c) What is the average number of defectives in lots of 50 of these items? *Ans.* (a) 0.9606, (b) 0.9994, (c) 0.5.

3. Derive Eqs. (4.23) and (4.24) using expected value definitions and operations.

4. A sampling inspection plan operates as follows. A random sample of 50 items is drawn from a lot of manufactured items. If no more than two defective items are observed, the lot is accepted. Compute the probability of acceptance P_a of a 10% defective lot. Assume a large lot size N so that the binomial model can be used. *Ans.* 0.1117.

---

*For the binomial model to give an exact probability P_a, the probabilities p and q must be constant from trial to trial—in this case, at each draw of a single product item for the purpose of inspection. Thus, the value $P_a = 0.2712$ obtained here is only an approximation.

A good approximation, using the binomial model, is possible if lot size N (the population) is large relative to sample size n so that p and q are approximately constant. A rough rule to assure a good approximation is that the ratio n/N should be less than 0.1.

To generalize, p and q are constant if sampling is: (1) from an infinite population, or (2) with replacement from a finite population. Sampling "with replacement" means that each item drawn randomly from the population is returned to the population prior to the next draw (or trial). Thus, the character of the finite population remains unchanged and the probability p is constant from trial to trial. If sampling is without replacement from a finite population, the conditions stated in the preceding paragraph are necessary in order to obtain a good approximation.

5. Verify the recursion relationship of Eq. (4.25) by numerically demonstrating that $P(X = 3) = P(X = 2)\ \{(20 - 2)/(2 + 1)(0.1/0.9)\}$ for binomial conditions of $n = 20$ and $p = 0.1$. Cue: moving from one computation to the next to obtain the respective probabilities of $X = 0, 1, 2, \ldots$, note that $p^n = pp^{n-1}$ and $q^n = q^{n+1}/q$.

6. One card is drawn from a well shuffled 52-card deck. The card is observed, replaced in the deck, and the deck is shuffled. This procedure is repeated four times. Using the binomial distribution model, the respective probabilities of obtaining $X = 0, 1, 2, 3, 4$, or 5 spades is

X	0	1	2	3	4	5
$f(X)$	$\frac{243}{1{,}024}$	$\frac{405}{1{,}024}$	$\frac{270}{1{,}024}$	$\frac{90}{1{,}024}$	$\frac{15}{1{,}024}$	$\frac{1}{1{,}024}$

Compute $\mu = E(X)$ and $\sigma_X^2 = E(X^2) - [E(X)]^2$. Then, verify the results for μ and σ_X^2 by computation using Eqs. (4.23) and (4.24).

Ans. $\mu = 1.25$, $\sigma_X^2 = 0.9375$.

4.6 Poisson Distribution Model

The *D.F.* for the *Poisson* distribution model is

$$f(X) = \frac{e^{-m}m^X}{X!} \qquad X = 0, 1, 2, \ldots, n \tag{4.26}$$

where m is a positive parameter and e is the base of natural logarithms and equals 2.718282 . . . .

Poisson *D.F.* and *C.D.F.* for $m = 2$ are shown in Fig. 4.8. The *D.F.* is positively skewed, but the skewness decreases as $E(X)$ increases.

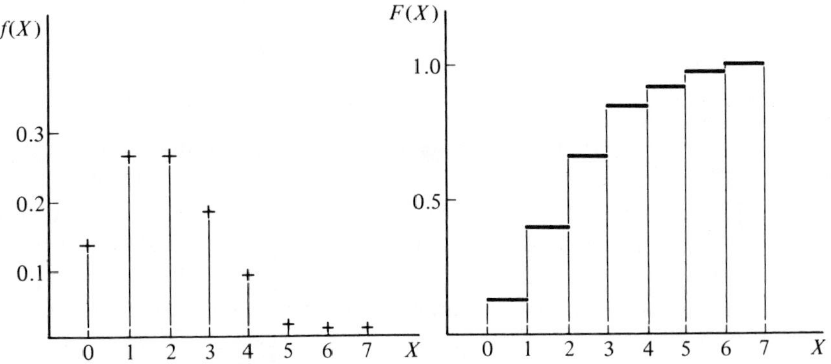

Figure 4.8. Poisson *D.F.* and *C.D.F.* for $m = 2$.

The moments for the Poisson distribution are obtained as follows. The characteristic function is

$$\phi_X(t) = E(e^{iXt}) = \sum_{X=0}^{\infty} e^{iXt} \cdot \frac{e^{-m}m^X}{X!}$$

$$= e^{-m} \sum_{X=0}^{\infty} \frac{(me^{it})^X}{X!} = e^{-m}e^{me^{it}}$$

The first and second derivatives, evaluated at $t = 0$, are

$$\phi'_X(t) = e^{-m}mie^{it}e^{me^{it}}$$

$$\phi'_X(0) = mi$$

and

$$\phi''_X(t) = e^{-m}i^2(m^2e^{2it}e^{me^{it}} + me^{it}e^{me^{it}})$$

$$\phi''_X(0) = -m^2 - m$$

From Eq. (4.11),

$$\mu = E(X) = \frac{\phi'_X(0)}{i} = \frac{mi}{i} = m \tag{4.27}$$

and

$$E(X^2) = \frac{\phi''_X(0)}{i^2} = \frac{-m^2 - m}{-1} = m^2 + m$$

From Eq. (4.8),

$$\sigma_X^2 = E(X^2) - [E(X)]^2$$

$$= m^2 + m - m^2 = m \tag{4.28}$$

Thus, the Poisson *D.F.* can be rewritten as

$$f(X) = \frac{e^{-\mu}\mu^X}{X!} \quad X = 0, 1, 2, \ldots, n \tag{4.29}$$

with mean $\mu = np$ and variance $\sigma_X^2 = np$.

Like the binomial distribution model, the Poisson distribution gives the probability of X number of successes in n independent Bernoulli trials, where p is the probability of a success and q the probability of a failure at each trial. In addition to the constant probability requirement (p and q), p must be sufficiently small and n sufficiently large for the Poisson model to be appropriate. A rough rule is that p should be less than 0.1 and n greater than 16.*

*This can be stated roughly as follows. The Poisson model describes the probability of the frequency of an event wherein the opportunity is large for its occurrence, yet the probability of its occurrence is small.

Tables of the Poisson *C.D.F.* are available.* An abbreviated table of cumulative Poisson probabilities is given in Appendix B.

If tables are not available, computation can be simplified by use of a recursion formula. The Poisson *D.F.* for $X = 0, 1, 2, \ldots, n$ is a series of terms

$$\frac{e^{-\mu}\mu^0}{0!}, \frac{e^{-\mu}\mu^1}{1!}, \frac{e^{-\mu}\mu^2}{2!}, \ldots, \frac{e^{-\mu}\mu^n}{n!}$$

and

$$P(X=0) = \frac{e^{-\mu}\mu^0}{0!}$$

$$P(X=1) = \frac{e^{-\mu}\mu}{1!} = P(X=0)\cdot\frac{\mu}{1}$$

$$P(X=2) = \frac{e^{-\mu}\mu^2}{2!} = P(X=1)\cdot\frac{\mu}{2}$$

$$\cdots$$

$$P(X=k+1) = \frac{e^{-\mu}\mu^{k+1}}{(k+1)!} = P(X=k)\cdot\frac{\mu}{k+1} \tag{4.30}$$

Since $P(0)$ in every case is equal to $e^{-\mu}$, and μ^0 and $0!$ are both equal to 1, it is a simple procedure to evaluate $P(0)$ and multiply successively by $\mu/1$, $\mu/2, \ldots,$ and so forth until the desired value $P(k+1)$ is obtained.

EXAMPLE 4.6.1

The mean number of defects per subassembly of a manufactured product is 2. The opportunity for a defect to occur is large, each subassembly having hundreds of characteristics that could be defective. What is the probability of four or more defects in a given subassembly?

These are Poisson conditions, namely, large opportunity for a defect to occur, yet small frequency of occurrence of a defect. Using the Poisson distribution model and the recursion formula, Eq. (4.30),

$$P(0) = \frac{e^{-2}2^0}{0!} = e^{-2} = 0.13534$$

$$P(1) = P(0)\cdot\tfrac{2}{1} = 0.27068$$

$$P(2) = P(1)\cdot\tfrac{2}{2} = 0.27068$$

$$P(3) = P(2)\cdot\tfrac{2}{3} = 0.18045$$

$$\begin{aligned} P(4 \text{ or more}) &= 1 - [P(0) + P(1) + P(2) + P(3)] \\ &= 1 - 0.85715 \\ &= 0.14285 \end{aligned}$$

*An extensive table with a range of $\mu = np$ values from 0.001 to 100 is in Molina, E. C. (1942), *Poisson's Exponential Binomial Limit*, D. Van Nostrand Co., Inc., Princeton, N. J.

Or, using Appendix B, the probability $P(X \leq 3)$ is read as 0.85715, and

$$P(X \geq 4) = 1 - P(X \leq 3)$$

The Poisson distribution model can be used to describe the frequency of events in a given interval of time, length, area, assembly, and so forth. Some applications are hourly traffic load, telephone circuit traffic, etc., where the total traffic is large but the expected traffic for any single hour is relatively small. Another application is frequency of radioactive disintegration (as measured by Geiger counts per minute) where the number of particles available for disintegration is large, yet the expected number disintegrating per unit time is small.

Another application of the Poisson distribution model is its use as a limiting form of the binomial distribution. Consider a binomial *D.F.* with very large n and very small p. Then, the respective terms of Eq. (4.20) can be approximated by the corresponding terms of the Poisson *D.F.* Eq. (4.29), where $\mu = np$. For fixed X this follows from the following argument: From Eq. (4.20),

$$f(X) = \binom{n}{X} p^X q^{n-X}$$

or

$$\begin{aligned} f(X) &= \frac{n!}{X!(n-X)!} p^X (1-p)^{n-X} \\ &= \frac{n(n-1)(n-2)\ldots(n-X+1)}{X!} p^X (1-p)^{n-X} \end{aligned}$$

Dividing each of the X terms $n, n-1, n-2, \ldots,$ by n, and multiplying the entire expression by n^X,

$$f(X) = \frac{1[1-(1/n)][1-(2/n)]\ldots[1-(X-1/n)]}{X!}(np)^X(1-p)^{n-X}$$

and, for $np = \mu$,

$$f(X) = \frac{1[1-(1/n)][1-(2/n)]\ldots[1-(X-1/n)]}{X!}\mu^X\left(1-\frac{\mu}{n}\right)^n\left(1-\frac{\mu}{n}\right)^{-X}$$

Let $n \longrightarrow \infty$ and $p \longrightarrow 0$ in such a way that np remains fixed at a value μ; then

$$1\left(1-\frac{1}{n}\right)\left(1-\frac{2}{n}\right)\ldots\left(1-\frac{X-1}{n}\right) \longrightarrow 1$$

$$\left(1-\frac{\mu}{n}\right)^n \longrightarrow e^{-\mu}$$

$$\left(1-\frac{\mu}{n}\right)^{-X} \longrightarrow 1$$

and thus

$$\lim_{n \to \infty} f(X) = \frac{e^{-\mu}\mu^X}{X!}$$

EXAMPLE 4.6.2

In Example 4.5.6, the probability of acceptance P_a of a 10% defective lot is 0.2712 (using the binomial model).

Using the Poisson distribution model to approximate the binomial, an approximation to P_a is computed as follows:

$$\mu = np = 25(0.10) = 2.5$$

The probability of $X = 0, 1$ is obtained by

$$\begin{aligned} P(X = 0) &= e^{-2.5} \\ \log P(X = 0) &= -2.5(0.0434294) \\ &= -1.085735 \\ &= 0.914264 - 2 \\ P(X = 0) &= 0.0820855 \end{aligned}$$

and

$$\begin{aligned} P(X = 1) &= 0.0820855(2.5) \\ &= 0.20521 \end{aligned}$$

Therefore,

$$\begin{aligned} P_a &= P(X = 0) + P(X = 1) \\ &= 0.0820855 + 0.20521 \\ &= 0.28730^* \end{aligned}$$

EXERCISES

1. Incoming calls to a telephone exchange occur randomly but at a mean rate of four calls per minute. Using the Poisson distribution model and the recursion formula, Eq. (4.30), compute the probability of: (a) one incoming call during a specified 1-min interval, and (b) at least two calls during the 1-min period.
Ans. (a) 0.073, (b) 0.909.

2. Verify the answers to Exercise 1 by reference to Appendix B.

3. Incoming calls to a telephone exchange occur randomly but at a mean rate of 300 calls per hour. Using the Poisson model and Appendix B, determine the probability of: (a) two incoming calls during a specified 1-min interval, and (b) at least five calls during the 1-min period. *Ans.* (a) 0.084, (b) 0.560.

---

*A better approximation is possible if sample size n is larger. In practice, standard sampling inspection plans specify sample sizes considerably larger than 25.

4. Defects occur randomly in electrical cable but at a mean rate of one defect per each 2,000 ft of cable. Using the Poisson model, compute the probability of no more than one defect in 500 ft of cable. *Ans.* 0.974.

5. The mean number of defects per assembly of a manufactured product is 4. Using the Poisson model, determine the probability of five or more defects in a given assembly. *Ans.* 0.372.

6. Repeat Exercise 4, Sec. 4.5, using the Poisson distribution to approximate the probability of acceptance P_a. *Ans.* 0.124.

7. Derive Eqs. (4.27) and (4.28) using expected value definitions and operations.

4.7 Hypergeometric Distribution Model

The *D.F.* for the *hypergeometric* distribution model is

$$f(X) = \frac{\binom{Np}{X}\binom{Nq}{n-X}}{\binom{N}{n}} \qquad X = 0, 1, 2, \ldots, n \tag{4.31}$$

where p is the probability of success and $q = 1 - p$ is the probability of failure prior to the first trial of $\dot{E}$, N is population size, and n is sample size.

The binomial, Poisson, and hypergeometric models—each gives the probability of X number of successes in n independent Bernoulli trials. Both binomial and Poisson distributions assume a constant probability of success p, from trial to trial. Clearly, p and thus also q are constant only if sampling is: (1) from an infinite population, or (2) with replacement from a finite population. If sampling is without replacement, the question of whether or not the binomial and Poisson models yield adequate approximations depends on how large population size N is, relative to sample size n.

The hypergeometric distribution gives the exact probability of X number of successes in n Bernoulli trials and does not assume a constant probability p at each trial. Hence, the hypergeometric model is appropriate for applications that involve sampling without replacement from finite populations.

The moments for the hypergeometric distribution are obtained in the same manner as that for the other discrete distribution models. It can be shown that

$$E(X) = \mu = np \tag{4.32}$$

and

$$\sigma_X^2 = npq\left(\frac{N-n}{N-1}\right) \tag{4.33}$$

Tables of the hypergeometric *C.D.F.* are available.* An abbreviated table of cumulative hypergeometric probabilities is given in Appendix C.

EXAMPLE 4.7.1

A box contains six black and four white chips. If four chips are randomly drawn from the box, what is the probability that two or fewer black chips are drawn?

Using the hypergeometric model, with $N = 10$, $n = 4$, $p = 0.6$, and $q = 0.4$,

$$P(X \leq 2) = \sum_{X=0}^{2} \frac{\binom{6}{X}\binom{4}{4-X}}{\binom{10}{4}}$$

$$= 0.547619 \quad \text{(from Appendix C)}$$

If a table of hypergeometric probabilities is not available, then the probability is

$$P(X \leq 2) = \frac{\binom{6}{0}\binom{4}{4}}{\binom{10}{4}} + \frac{\binom{6}{1}\binom{4}{3}}{\binom{10}{4}} + \frac{\binom{6}{2}\binom{4}{2}}{\binom{10}{4}}$$

and, even with the aid of a table of factorials, the arithmetic is tedious.

EXAMPLE 4.7.2

In Example 4.5.6, the probability of acceptance P_a of a 10% defective lot is 0.2712 (using the binomial model).

Assume that lot size is 200. Then, using the hypergeometric distribution as a model, with $N = 200$, $n = 25$, $p = 0.1$, and $q = 0.9$, the probability of acceptance can be computed exactly:

$$P_a = \sum_{X=0}^{1} \frac{\binom{20}{X}\binom{180}{25-X}}{\binom{200}{25}}$$

The computation is tedious, even with the aid of a table of factorials. This is the reason for using either a binomial or Poisson model to obtain an approximation to P_a.

---

*An extensive table of hypergeometric probabilities is Lieberman, G. J. and Owens, D. B. (1961), *Table of the Hypergeometric Probability Distribution*, Stanford University Press, Stanford, Calif.

4.8 Summary—Discrete Models

The discrete models introduced thus far are closely related. Table 4.1 summarizes the Bernoulli, binomial, Poisson, and hypergeometric distribution models, defining the random variable in each case, and stating the conditions, *D.F.*, and moments for each distribution.

In each case, the distribution model is a function of the random variable X and certain parameters, that is,

Bernoulli model: $f(X, p)$
Binomial model: $f(X, n, p)$
Poisson model: $f(X, \mu)$
Hypergeometric model: $f(X, p, n, N)$

A good approximation of a hypergeometric probability can be obtained by the more easily computed binomial probability. For fixed n, as N increases, the approximation improves. When N becomes infinite, the hypergeometric and binomial probabilities are identical, that is,

$$\lim_{N \to \infty} f(X, p, n, N) = f(X, n, p)$$

A binomial probability can, in turn, be approximated by a Poisson probability, if n is large and p is small. Then,

$$\binom{n}{X} p^X q^{n-X} \cong e^{-np} \cdot \frac{(np)^X}{X!}$$

where the individual binomial terms are replaced by corresponding Poisson terms with $\mu = np$. The question arises, how large should n be, and likewise how small should p be? If n is greater than 100 and p is less than 0.01, the binomial and Poisson probabilities agree to three decimal places for every value of the random variable X. However, if p is sufficiently small, satisfactory approximations of the binomial probability by a Poisson probability can be made even when n is as small as 10. Given a small p, increasing n improves the approximation. Conversely, for fixed n, decreasing p improves the approximation.

EXERCISES

1. A sampling inspection plan is: $N = 200$, $n = 10$, and accept the lot if the observed number of defectives X is less than or equal to 1. Using the hypergeometric model, compute the probability of acceptance P_a of a 10% defective lot.
Ans. 0.73715.

Table 4.1. SUMMARY OF THE BERNOULLI, BINOMIAL, POISSON, AND HYPERGEOMETRIC DISTRIBUTION MODELS

Distribution	*Random Variable*	*Probability*	*Frequency Function*	*Moments*
Bernoulli	$X = 0, 1$ at each trial.	$p = P(1)$ $q = P(0)$ at each trial. p and q are constants.	$P(X) = p^X q^{1-X}$	$\mu = p, \sigma_X^2 = pq$
Binomial	0, 1 outcomes at each trial.		$P(X) = \binom{n}{X} p^X q^{n-X}$	$\mu = np, \sigma_X^2 = npq$
Poisson	X = number of 1's in n trials.		$P(X) = \frac{e^{-\mu}\mu^X}{X!}$	$\mu = \sigma_X^2 = np$
Hypergeometric	$X = 0, 1, 2, \ldots, n$	p and q are variable.	$P(X) = \frac{\binom{Np}{X}\binom{Nq}{n-X}}{\binom{N}{n}}$	$\mu = np, \sigma_X^2 = npq\left(\frac{N-n}{N-1}\right)$

2. Obtain an approximation to P_a in Exercise 1 by using the Poisson model.

Ans. 0.736.

3. Obtain an approximation to P_a in Exercise 1 by using the binomial model.

Ans. 0.7361.

4.9 Other Discrete Models

The distribution models of Secs. 4.5–4.8 are the most commonly used discrete models. The basic underlying model here is that of Bernoulli trials. Three other discrete models, of occasional use to the engineer, are summarized in the following paragraphs.

Consider a random experiment E which yields k possible mutually exclusive outcomes $E_1, E_2, \ldots, E_k$, with respective probabilities $p_1, p_2, \ldots, p_k$ that $E_1, E_2, \ldots, E_k$ will occur $X_1, X_2, \ldots, X_k$ times, respectively. The experiment is repeated n times independently.

The probability of obtaining exactly X_1 occurrences of E_1, X_2 occurrences of $E_2, \ldots, X_k$ occurrences of E_k is given by the *multinomial distribution*

$$f(X_1, X_2, \ldots, X_k) = \frac{n!}{X_1!X_2!\ldots X_k!} p_1^{X_1} p_2^{X_2} \ldots p_k^{X_k} \tag{4.34}$$

where the X_i are positive integers and $p_i > 0$ for $i = 1, 2, \ldots, n$ and

$$\sum_{i=1}^{k} p_i = 1, \qquad \sum_{i=1}^{k} X_i = n$$

EXAMPLE 4.9.1

Product parts are graded into 4 categories for selective assembly: (1) +0.001 in. oversize, (2) +0.002 in. oversize, (3) −0.001 in. undersize, and (4) −0.002 in. undersize. The respective probabilities of manufactured parts occurring in these categories are: (1) 0.2, (2) 0.4, (3) 0.3, and (4) 0.1. Fifteen product parts are produced. What is the probability that of the 15 parts 3 are #1 rated, 8 are #2 rated, 2 are #3 rated, and 2 are #4 rated?

$$\begin{aligned} P(X_1 &= 3, X_2 = 8, X_3 = 2, X_4 = 2) \\ &= \frac{15!}{3!8!2!2!}(0.2)^3(0.4)^8(0.3)^2(0.1)^2 \\ &= 0.006 \end{aligned}$$

This distribution model, which is a generalization of the binomial model, is called the multinomial distribution since Eq. (4.34) is the general term in the multinomial expansion

$$(p_1 + p_2 + \ldots + p_k)^n$$

In the multinomial model, the number of occurrences of each possible outcome in n independent trials may be considered as an individual variable, and its mean and variance are

$$\mu_i = E(X_i) = np_i, \qquad \sigma^2_{X_i} = np_i(1 - p_i) \tag{4.35}$$

The binomial distribution model describes the number of successes that occur in a fixed number of Bernoulli trials. A possible related question is, "At what trial will the first success occur?" For example, if the proportion defective of a manufacturing lot is p, how many parts will be inspected before a defective part is observed?

Assuming independent trials and a constant p-value, the *D.F.* of the number of trials k to the first success* is given by the *geometric* distribution

$$P(k = n) = f(n) = (1 - p)^{n-1}p \quad n = 1, 2, \ldots, k, \ldots \tag{4.36}$$

The *C.D.F.* is

$$F(n) = 1 - (1 - p)^n \dagger$$

A geometric *D.F.* is shown in Fig. 4.9. The mean and variance for the

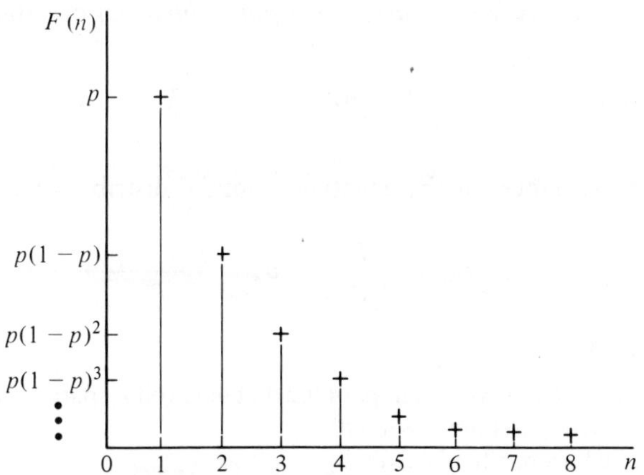

Figure. 4.9. Geometric distribution.

---

*The first success occurs on the nth trial if and only if: (1) the first $n - 1$ trials are failures, which occur with probability $(1 - p)^{n-1}$, and (2) the nth trial is a success, which occurs with probability p.

†The probability that $k \leq n$ is the probability that there is at least one success occurrence in n trials, or

$$1 - P(\text{no successes in } n \text{ trials}) = 1 - (1 - p)^n$$

geometric distribution are

$$\mu = E(k) = \frac{1}{p}, \qquad \sigma_k^2 = \frac{(1-p)}{p^2} \tag{4.37}$$

EXAMPLE 4.9.2

A manufacturing lot is 10% defective. A 100% inspection is required to sort out defective product parts. What is the probability that more than six product parts will be inspected before discovery of the first defective product part?

$$P(n > 6) = 1 - f(6)$$

From Eq. (4.36),

$$\begin{aligned} P(n = 6) &= (1 - 0.10)^5(0.10) \\ &= 0.059 \end{aligned}$$

Thus

$$\begin{aligned} P(n > 6) &= 1 - P(n = 6) \\ &= 1 - 0.059 = 0.94 \end{aligned}$$

The geometric distribution model is concerned with the question, "At which trial does the first success occur?" A more general question is, "At which trial does the kth success occur?" The trial number w, at which the kth success occurs, has a *Pascal* or *negative binomial* distribution:

$$f(w) = \binom{w-1}{k-1}(1-p)^{w-k}p^k \quad w = k, k+1, \ldots * \tag{4.38}$$

The mean and variance for the negative binomial distribution are

$$\mu = E(w) = \frac{k}{p}, \qquad \sigma_w^2 = \frac{k(1-p)}{p^2} \tag{4.39}$$

EXAMPLE 4.9.3

What is the probability in Example 4.9.2 that ten product parts will be inspected before two defective parts are observed?

The required probability is given by

$$\begin{aligned} f(10) &= \binom{10-1}{2-1}(1 - 0.10)^{10-2}(0.10)^2 \\ &= 9(0.9)^8(0.1)^2 = 0.004 \end{aligned}$$

---

*The kth success occurs on the wth trial only if there are exactly $k - 1$ successes in the preceding $w - 1$ trials and a success occurs on the wth trial. The probability of exactly $k - 1$ successes in $w - 1$ trials is given by the binomial distribution model $f(k - 1, w - 1, p)$.

EXERCISES

1. The probability of a product part: (1) conforming to specifications is 0.7, (2) being oversize is 0.2, and (3) being undersize is 0.1. Ten product parts are produced. What is the probability that six conform to specifications, two are oversize, and one is undersize? *Ans.* 0.059.
2. A large lot of manufactured items has a proportion defective $p = 0.05$. What is the probability that, during an inspection operation, the first observed defective will occur on the fourth item inspected? *Ans.* 0.04287.
3. In Exercise 2, what is the probability that the first observed defective will occur in the first four items inspected? *Ans.* 0.18550.

GLOSSARY OF TERMS

Mathematical expectation
Expected value
Chebyshev's inequality
Characteristic function
Bernoulli trial
Binomial variable
Poisson variable
Hypergeometric variable
Sampling with replacement
Sampling without replacement
Multinomial distribution
Geometric distribution
Negative binomial distribution

GLOSSARY OF FORMULAS

Page

(4.1) Expected value of $g(X)$. 85

$$E[g(X)] = \begin{cases} \sum_X g(X)f(X) \\ \int_{-\infty}^{\infty} g(X)f(X)\,dX \end{cases}$$

(4.2) Expected value of X. 85

$$E(X) = \left\{\begin{matrix} \sum_X Xf(X) \\ \int_{-\infty}^{\infty} Xf(X)\,dX \end{matrix}\right\} = \mu_X$$

(4.3) Expected value of $(X - \mu)^2$. 85

$$E[(X-\mu)^2] = \left\{\begin{matrix} \sum_X (X-\mu)^2 f(X) \\ \int_{-\infty}^{\infty} (X-\mu)^2 f(X)\,dX \end{matrix}\right\} = \sigma_X^2$$

(4.4) Expected value of aX. 87

$$E(aX) = \left\{\begin{matrix} \sum_X aXf(X) \\ \int_{-\infty}^{\infty} aXf(X)\,dX \end{matrix}\right\} = aE(X) = a\mu$$

(4.5) Expected value of $[a(X - \mu)]^2$. 87

$$E\{[a(X-\mu)]^2\} = \left\{\begin{matrix} \sum_X [a(X-\mu)]^2 f(X) \\ \int_{-\infty}^{\infty} [a(X-\mu)]^2 f(X)\,dX \end{matrix}\right\} = a^2\sigma_X^2$$

(4.6) Expected value of a. 87

$$E(a) = \left\{\begin{matrix} \sum_X af(X) \\ \int_{-\infty}^{\infty} af(X)\,dX \end{matrix}\right\} = a$$

(4.7) Population variance. 87

$$\sigma_X^2 = \frac{1}{N}\sum_{j=1}^{k} f_j X_j^2 - N\mu^2$$

(4.8) Population variance. 87

$$\sigma_X^2 = \int_{-\infty}^{\infty} X^2 f(X)\,dX - \mu^2 = E(X^2) - [E(X)]^2$$

(4.9) Chebyshev's inequality. 88

$$P(\mu - k\sigma_X \leq X \leq \mu + k\sigma_X) > 1 - \frac{1}{k^2}$$

(4.10) Characteristic function $\phi_X(t)$. 89

$$\phi_X(t) = E(e^{iXt}) = \left\{\begin{matrix} \sum_X e^{iXt} f(X) \\ \int_{-\infty}^{\infty} e^{iXt} f(X)\,dX \end{matrix}\right.$$

(4.11) Moments about zero. 89

$$E(X^k) = \frac{\phi_X^k(0)}{i^k} \qquad i^2 = -1$$

(4.12) Characteristic function $\phi_{(X-\mu)}(t)$. 90

$$\phi_{(X-\mu)}(t) = E(e^{i(X-\mu)t})$$

(4.13) Moments about the mean. 90

$$E[X-\mu)^k] = \frac{\phi_{(X-\mu)}^{(k)}(0)}{i^k}$$

(4.14) Characteristic function $\phi_{g(X)}(t)$. 90

$$\phi_{g(X)}(t) = E(e^{ig(X)t}) = \left\{\begin{matrix} \sum_X e^{ig(X)t} f(X) \\ \int_{-\infty}^{\infty} e^{ig(X)t} f(X)\,dX \end{matrix}\right.$$

(4.15) Uniform distribution mean. 91

$$\mu = \frac{c + a}{2}$$

(4.16) Uniform distribution variance. 91

$$\sigma_X^2 = \frac{(c-a)^2}{12}$$

(4.17) Bernoulli *D.F.* 92

$$f(X) = p^X q^{1-X} \quad X = 0, 1$$

(4.18) Bernoulli distribution mean. 93

$$\mu = p$$

(4.19) Bernoulli distribution variance. 93

$$\sigma_X^2 = pq$$

(4.20) Binomial *D.F.* 94

$$f(X) = \binom{n}{X} p^X q^{n-X} \quad X = 0, 1, 2, \ldots, n$$

(4.21) Binomial expansion. 96

$$(q + p)^n = q^n + npq^{n-1} + \frac{n(n-1)p^2 q^{n-2}}{2!} + \frac{n(n-1)(n-2)p^3 q^{n-3}}{3!} + \cdots + p^n$$

(4.22) Binomial expansion. 96

$$(q + p)^n = \sum_{X=0}^{n} \binom{n}{X} p^X q^{n-X}$$

(4.23) Binomial distribution mean. 97

$$\mu = np$$

(4.24) Binomial distribution variance. 97

$$\sigma_X^2 = npq$$

(4.25) Binomial recursion formula. 99

$$P(X = k + 1) = P(X = k)\left(\frac{n-k}{k+1}\right)\left(\frac{p}{q}\right)$$

(4.26) Poisson *D.F.* 101

$$f(X) = \frac{e^{-m} m^X}{X!} \quad X = 0, 1, 2, \ldots, n$$

(4.27) Poisson distribution mean. 102

$$\mu = m$$

(4.28) Poisson distribution variance. 102

$$\sigma_X^2 = m$$

(4.29) Poisson *D.F.* 102

$$f(X) = \frac{e^{-\mu} \mu^X}{X!} \quad X = 0, 1, 2, \ldots, n$$

(4.30) Poisson recursion formula. 103

$$P(X = k + 1) = P(X = k)\frac{\mu}{k + 1}$$

(4.31) Hypergeometric *D.F.* 106

$$f(X) = \frac{\binom{Np}{X}\binom{Nq}{n-X}}{\binom{N}{n}} \quad X = 0, 1, 2, \ldots, n$$

(4.32) Hypergeometric distribution mean. 106

$$\mu = np$$

(4.33) Hypergeometric distribution variance. 106

$$\sigma_X^2 = npq\left(\frac{N-n}{N-1}\right)$$

(4.34) Multinomial *D.F.* 110

$$f(X_1, X_2, \ldots, X_k) = \frac{n!}{X_1!\,X_2!\ldots X_k!} p_1^{X_1} p_2^{X_2} \ldots p_k^{X_k}$$

X_i positive integers, $p_i > 0$, $\sum_{i=1}^{k} p_i = 1$, $\sum_{i=1}^{k} x_i = n$

(4.35) Multinomial moments. 111

$$\mu_i = E(X_i) = np_i, \qquad \sigma_{X_i}^2 = np_i(1 - p_i)$$

(4.36) Geometric *D.F.* 111

$$P(k = n) = f(n) = (1 - p)^{n-1}p \quad n = 1, 2, \ldots, k, \ldots$$

(4.37) Geometric moments. 112

$$\mu = E(k) = \frac{1}{p}, \qquad \sigma_k^2 = \frac{1-p}{p^2}$$

(4.38) Negative binomial *D.F.* 112

$$f(w) = \binom{w-1}{k-1}(1 - p)^{w-k}p^k \quad w = k, k + 1, \ldots$$

(4.39) Negative binomial moments. 112

$$\mu = E(w) = \frac{k}{p}, \qquad \sigma_w^2 = \frac{k(1-p)}{p^2}$$

REFERENCES

BENJAMIN, J. R. AND CORNELL, C. A. (1970), *Probability, Statistics and Decision for Civil Engineers*, McGraw-Hill Book Company, New York, Chaps. 2, 3.

CHOU, YA-LUN (1969), *Statistical Analysis*, Holt, Rinehart & Winston, Inc., New York, Chaps. 6, 7.

GUTTMAN, I. AND WILKS, S. S. (1971), *Introductory Engineering Statistics*, John Wiley & Sons, Inc., New York, Chap. 3.

OSTLE, BERNARD (1963), *Statistics in Research*, 2nd ed., Iowa State University Press, Ames, Iowa, Chaps. 3, 4.

PROBLEMS

1. Define the *expected value* of a random variable $g(X)$.

2. If X is a random variable, then $g(X) = X + (X + 2)^2 + 3\mu$ is also a random variable. Using expected value definitions and operations, determine the expected value of $g(X)$. (Assume that X is a discrete random variable.)

3. Write the characteristic function associated with the random variable $g(X)$ in Problem 2.

4. The expected value of X corresponds to the mean of a population. Using the expected value definition, Eq. (4.2), determine the expected value of X if X can assume values 1, 2, 3, and 4 with respective probabilities 0.1, 0.2, 0.4, and 0.3.

5. The probabilities in Problem 4 are $f(X)$ values for the *D.F.* describing the population. Suppose that the X-values 1, 2, 3, and 4 are sample observations with respective relative frequencies 0.1, 0.2, 0.4, and 0.3. Determine the mean $\bar{x}$.

6. The mean $\mu = E(X)$ is determined in Problem 4. Write the characteristic function associated with the *D.F.* in Problem 4 and determine μ using Eq. (4.11).

7. Compute the variance for the *D.F.* in Problem 4 using the expected value method.

8. Compute the variance for the *D.F.* in Problem 4 using the characteristic function method.

9. Determine the mean and variance for the following *D.F.* using the expected value method: $f(X) = \frac{1}{3}$, $1 \leq X \leq 4$.

10. Repeat Problem 9 using the characteristic function method.

11. Determine the mean and variance for the following *D.F.* using the expected value method: $f(X) = a(X + 4)$, $1 \leq X \leq 5$.

12. Repeat Problem 11 using the characteristic function method.

13. Determine the mean and variance for the following *D.F.* using the expected value method: $f(X) = aX$, $0 \leq X \leq 4$.

14. Repeat Problem 13 using the characteristic function method.

15. Determine the mean and variance for the following *D.F.* using the expected value method: $f(X) = 1 - aX$, $0 \leq X \leq 2$.

16. Repeat Problem 15 using the characteristic function method.

17. Determine the mean and variance for the following *D.F.* using the expected value method:

$$f(X) = \begin{cases} 1 - X & 0 \leq X \leq 1 \\ 1 + X & -1 \leq X \leq 0 \end{cases}$$

18. Expand $(q + p)^n$ for $n = 3$ and verify that this gives the sum of the binomial probabilities for $X = 0, 1, 2,$ and 3 [see Eq. (4.21)].

19. A proof test of a crane hook involves the application of a load somewhat in excess of the working load but less than any damaging load. Suppose for very old crane hooks that the probability of the hook failing at a single load application is $p = 0.01$. Using Eq. (4.20), compute the binomial probabilities of 0, 1, 2, . . . , 5 failures in $n = 5$ tests of a given crane hook.

20. Using the recursion formula, Eq. (4.25), compute the respective probabilities of $X = 0, 1, 2, 3, 4$ if $n = 4$ and $p = 0.2$.

21. Compute the mean and variance of the binomial *D.F.* in Problem 20.

22. Graph the *D.F.* and *C.D.F.* for Problem 20.

23. Lots of 50 product items each are either accepted or rejected by a crude sampling scheme of randomly inspecting 10 items without replacement from a lot and accepting the lot if there are no observed defectives. Compute to two decimal places the hypergeometric probability of no defectives in a random sample of 10 items if the lot is 10% defective.

24. Compute the mean and variance for the hypergeometric *D.F.* in Problem 23.

25. The conditions of Problem 23 are not good binomial conditions. Why? However, use the binomial model and approximate the hypergeometric probability in Problem 23 (to two decimal places). Also, compute the mean and variance for this binomial *D.F.*

26. An acceptance testing procedure for television tubes is to randomly select ten tubes without replacement from the lot and test them. If no tubes fail the test, the remaining ones in the lot are accepted. Otherwise the lot is rejected. Lot size is 100. (a) Compute the exact probability of acceptance of a 20% defective lot. (b) Approximate this probability using the binomial model.

27. Assume in Problem 26 that lot size is 200, sample size is 20, and lots are accepted whenever 0 or 1 defective is observed in a sample. What is the binomial probability of acceptance of a 10% defective lot? (Use Appendix A table.)

28. What is the probability of acceptance of a 5% defective lot in Problem 27? [Compute $P(X = 0)$ and use the recursion formula, Eq. (4.25), to obtain $P(X = 1)$.]

29. Repeat Problem 28 using a Poisson model to approximate the binomial probability required. Compute $P(X = 0)$ and use the recursion formula, Eq. (4.30), to obtain $P(X = 1)$. Verify your results by reference to Appendix B table.

30. State the conditions underlying the use of a binomial model and the additional conditions required to obtain a good approximation to to the binomial probability by means of a Poisson model.

31. A water processing plant makes daily measurements regarding the number of algae per milliliter of lake water. The probability is $p = 0.15$ that for a given measurement the algae level will exceed 800 per milliliter. What is the probability of exceeding this level 3 times in 10 independent measurement trials?

32. In testing a relay, the probability is $p = 0.05$ that it fails to make satisfactory contact in a single trial, and this probability remains unchanged over a large number of trials. Assuming that outcomes of successive trials are independent, what is the probability of two or fewer contact failures in 20 trials?

33. Suppose in Problem 32 that over a period of time the mean number of contact failures in 25 trials is 4. What conclusion can be made regarding the probability p?

34. A new alloy is being tested for permanent set (i.e., deformation or strain remaining after release of load). In 36 test trials, the mean number of occurrences of permanent set is 2.16. What is the binomial probability of a permanent set occurrence at one trial?

35. Calls arrive at a telephone switchboard such that the number of calls per fifteen-minute period follows a Poisson *D.F.* with mean 20. More than 30 calls per fifteen-minute period constitutes an overload. What is the probability of such an overload occurring?

36. The number of particles emitted from a radioactive source tends to follow a Poisson *D.F.* with mean 1.4 (coded). If the probability of no emissions is 0.25, what is the probability of 3 or more emissions? (Use the recursion formula.)

37. Defects in a manufactured tape occur randomly, but on the average of one per 2000 feet of tape. Assuming that the frequency of occurrence of these defects follows a Poisson distribution, what is the probability that a 5000 foot roll of tape has: (a) no defects, (b) at most 2 defects, and (c) at the least 2 defects?

38. A production machine is inoperative for breakdown and repair reasons on the average of once every 10 hours. Breakdown frequency tends to follow a Poisson distribution. For a given production order, this machine is required for 30 hours of operation. Only 2 breakdowns can be sustained if the production order is to be filled. What is the probability of filling the order?

39. The average number of surface defects for an enameled product part surface is 0.5 and the frequency of these defects follows a Poisson *D.F.* Prepare a frequency distribution for $X = 0, 1, 2, 3,$ and 4 defects for 1,000 product part surfaces that are examined.

40. For a certain fatigue test, the probability is 0.2 that a steel specimen will fail under maximum load. What is the probability that fewer than 5 specimen pieces are tested before the first specimen piece fails?

41. What is the probability in Problem 40 that the first specimen failure will occur on the second specimen that is tested?

42. A fatigue test of a product item cost \$200 for each test-trial preparation and \$100 for the product item that is destroyed by the test. If the product item does not fail under test and thus is not destroyed, the product-item cost is only \$40. The test is to be repeated until a product specimen failure is obtained. The probability of product failure at any test is 0.1, and individual trials are independent. If the expected cost of an event is defined to be the cost associated with the event multiplied by the probability of occurrence of the event, what is the expected cost of the entire test procedure described here?

43. Show that $(1 - p)^{n-1}p$, the expression for a geometric probability, is a *D.F.*

44. If $p = 0.2$ and $n = 5$, plot the *D.F.* and *C.D.F.* for: (a) a binomial model, and (b) a Poisson model.

45. If $p = 0.2$, plot the geometric *D.F.* and *C.D.F.* for $n = 1, 2, \ldots, 6$.

46. If $p = 0.2$ and $k = 2$, plot the negative binomial *D.F.* and *C.D.F.* for $w = 2, 3, \ldots, 7$.

NORMAL DISTRIBUTION MODEL

The distribution models of Chap. 4 are discrete models—each D.F. *is a function of a discrete random variable and certain parameters. In Chaps. 5 and 6, distribution models are presented which involve a continuous random variable. One of these distributions, the normal distribution model, is the most important distribution in statistics. For this reason, an entire chapter is devoted to a consideration of this model.*

5.1 The Normal Distribution

A variable X is called a *normally* distributed random variable if its *D.F.* has the form

$$f(X) = \frac{1}{\sigma_X\sqrt{2\pi}} e^{-(X-\mu)^2/2\sigma_X^2} \tag{5.1}$$

where μ and σ_X are the mean and standard deviation, $e = 2.71828$, and $\pi = 3.1416$, respectively.

5

5.2 Properties of the Normal Distribution

Figure 5.2 shows the *D.F.* and *C.D.F.* for the normal distribution model.* The *D.F.* is unimodal, asymptotic to the X-axis, and symmetric about the mean. The function is a maximum when $X = \mu$ and has inflection points at $X = \mu \pm \sigma_X$.

The *D.F.* is a function of the random variable X and parameters μ and σ_X. Thus, $f(X, \mu, \sigma_X)$ designates a family of normal distributions, each distribution being dependent on specific values assumed by μ and σ_X. This is illustrated by Figure 5.3.

---

*This distribution model can be generated from a differential equation of the form

$$\frac{dY}{dX} = \frac{(m - X)Y}{a}$$

where m and a are constants. This function is *unimodal* (see Fig. 5.1) since X occurs only as a linear factor in the numerator, and a solution of $dY/dX = 0$ for X yields only one root. Since Y is a coefficient in the numerator, as $dY/dX \longrightarrow 0$, then $Y \longrightarrow 0$ and the function is asymptotic to the X-axis. Further, the function is nonnegative, continuous, and defined for all values of X in the interval $(-\infty, +\infty)$.

5.3 Normal Model in Standard Form

From Eq. (1.30), a Z-transformation on the random variable X is

$$Z = \frac{X - \mu_X}{\sigma_X}$$

A separation of variables yields the expression

$$\frac{dY}{Y} = \left(\frac{m - X}{a}\right) dX$$

and integrating,

$$\int \frac{dY}{Y} = \int \left(\frac{m - X}{a}\right) dX$$

$$\log Y + c_0 = -\frac{(m - X)^2}{2a} + c_1$$

or

$$\log Y = -\frac{(m - X)^2}{2a} + k \quad k = c_1 - c_0$$

and

$$Y = ke^{-(m-X)^2/2a}$$

From Eq. (3.4), for the function under consideration to be a *D.F.*,

$$\int_{-\infty}^{\infty} ke^{-(m-X)^2/2a}\, dX = 1$$

Solving for the adjusting constant k, let $t = X - m$, then $dX = dt$ and

$$k \int_{-\infty}^{\infty} e^{-t^2/2a} dt = k\sqrt{2\pi a} = 1 \quad \text{and} \quad k = \frac{1}{\sqrt{2\pi a}}$$

From Eq. (4.2), the first moment is

$$E(X) = \mu = \frac{1}{\sqrt{2\pi a}} \int_{-\infty}^{\infty} Xe^{-(m-X)^2/2a}\, dX$$

Evaluating this expression, let $t = X - m$, then $dX = dt$ and

$$E(X) = \mu = \frac{1}{\sqrt{2\pi a}} \int_{-\infty}^{\infty} (t + m)e^{-t^2/2a}\, dt$$

$$= \frac{1}{\sqrt{2\pi a}} \left[\int_{-\infty}^{\infty} te^{-t^2/2a}\, dt + m \int_{-\infty}^{\infty} e^{-t^2/2a}\, dt\right]$$

The first integral can be evaluated by parts and over the interval $(-\infty, +\infty)$ is equal to zero. The second integral is equal to $\sqrt{2\pi a}$, and thus $m = \mu$. From Eq. (4.3), the variance is

$$\sigma_X^2 = E[(X - \mu)^2] = \frac{1}{\sqrt{2\pi a}} \int_{-\infty}^{\infty} (X - \mu)^2 e^{-(X-\mu)^2/2a}\, dX$$

and letting $t = X - \mu$, then $dt = dX$ and

$$\sigma_X^2 = \frac{1}{\sqrt{2\pi a}} \int_{-\infty}^{\infty} t^2 e^{-t^2/2a}\, dt$$

which can be integrated by parts yielding $a = \sigma_X^2$.

Therefore, since $k = 1/\sqrt{2\pi a}$, $m = \mu$, and $a = \sigma_X^2$, the *D.F.* can be expressed as

$$f(X) = \frac{1}{\sigma_X\sqrt{2\pi}} e^{-(X-\mu)^2/2\sigma_X{}^2}$$

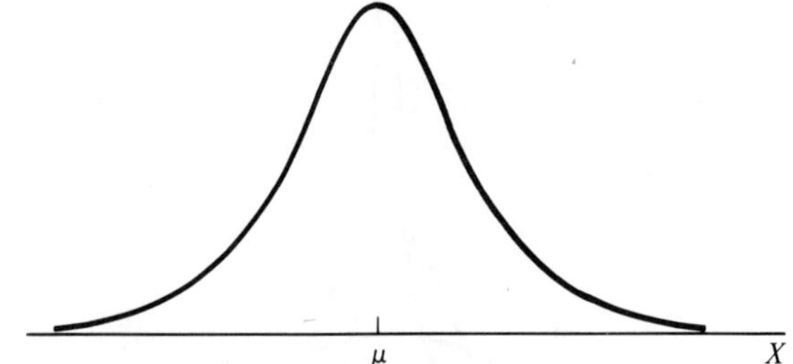

Figure 5.1. Normal distribution model.

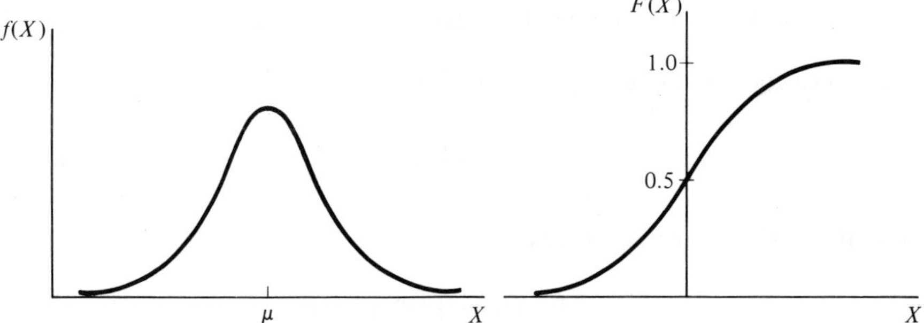

Figure 5.2. Normal distribution function.

This transformation affects both the mean and variance. When $X = \mu$, the corresponding mean μ_Z in Z-units is zero. Also,

$$\sigma_Z^2 = \frac{1}{N}\sum_{i=1}^{N}(Z - \mu_Z)^2 = \frac{1}{N}\sum_{i=1}^{N}\left(\frac{X - \mu_X}{\sigma_X} - 0\right)^2$$

$$= \frac{1}{N\sigma_X^2}\sum_{i=1}^{N}(X - \mu_X)^2 = \frac{\sigma_X^2}{\sigma_X^2} = 1^*$$

Further, the form of the *D.F.* expression is altered. In X-units, the function is

$$f(X) = \frac{1}{\sigma_X\sqrt{2\pi}}e^{-(X-\mu_X)^2/2\sigma_X{}^2}$$

In Z-units, the function is

$$f(Z) = \frac{1}{\sigma_Z\sqrt{2\pi}}e^{-(Z-\mu_Z)^2/2\sigma_Z{}^2}$$

---

*A common notation is $N(\mu, \sigma_X^2)$ which reads "X is normally distributed with mean μ and variance σ_X^2." Thus, the notation $N(0, 1)$ indicates that a Z-transformation has been applied to the random variable X, and "Z is normally distributed with mean zero and variance one."

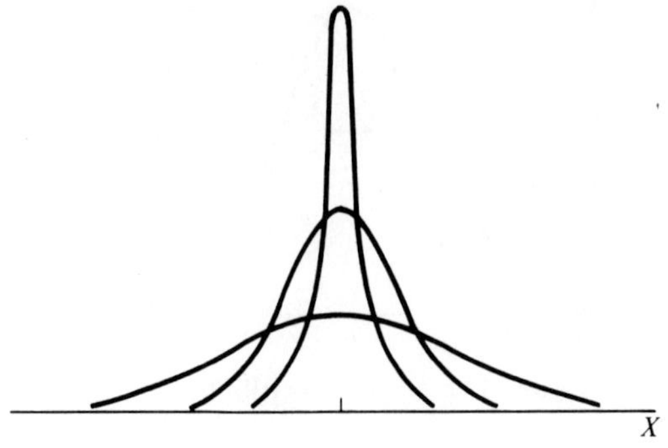

Figure 5.3. Normal distributions, $\sigma = 0.5$, 1.0, and 2.0.

but, since $\mu_Z = 0$ and $\sigma_Z^2 = 1$, the *D.F.* becomes

$$f(Z) = \frac{1}{\sqrt{2\pi}} e^{-Z^2/2} \tag{5.2}$$

and the *C.D.F.* in standard form is

$$F(Z) = \int f(Z)\, dZ \tag{5.3}$$

and this function is tabulated in Appendix *D*.*

The net effect of the *Z*-transformation is to shift the origin from $X = 0$ to $X = \mu$ and calibrate the *Z*-axis in multiples of σ_X as shown in Fig. 5.5. Since certain probabilities occur frequently in applications problems, it is convenient to memorize these values. For example, 68.26% of the total area beneath the *D.F.* curve lies between $\mu \pm 1\sigma$ (shaded area in Fig. 5.5). Table 5.1 summarizes the areas within $\mu \pm 1\sigma$, $\pm 2\sigma$, and $\pm 3\sigma$.

*Tables for $F(Z)$ are constructed differently in different textbooks. Appendix D gives the tail area α beneath the curve, for various *Z*-values, that is,

$$\alpha = 0.5 - \int_0^{Z_\alpha} f(Z)\, dZ$$

as indicated in Fig. 5.4.

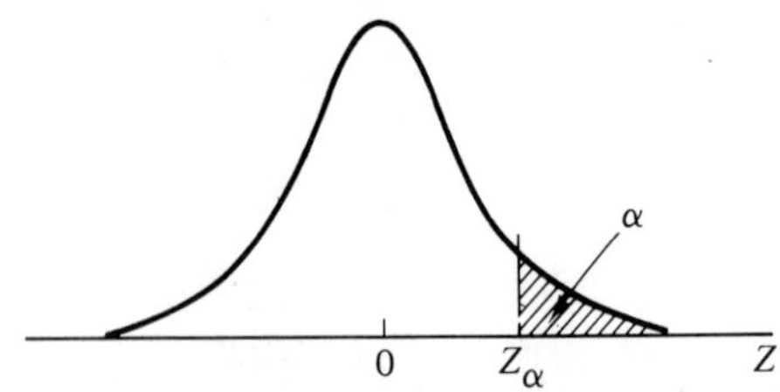

Figure 5.4. Distribution tail area α.

Table 5.1. CENTRAL AREAS, NORMAL DISTRIBUTION

$\mu \pm KZ$	% *area*
$\mu \pm Z$	68.26
$\mu \pm 2Z$	95.44
$\mu \pm 3Z$	99.73

If X is $N(\mu, \sigma_X^2)$, then Z is $N(0, 1)$. The effect of this statement on the computation of probabilities of normally distributed random variables is that it is unnecessary to return to the integral form $F(X)$ or $F(Z)$.* An ine-

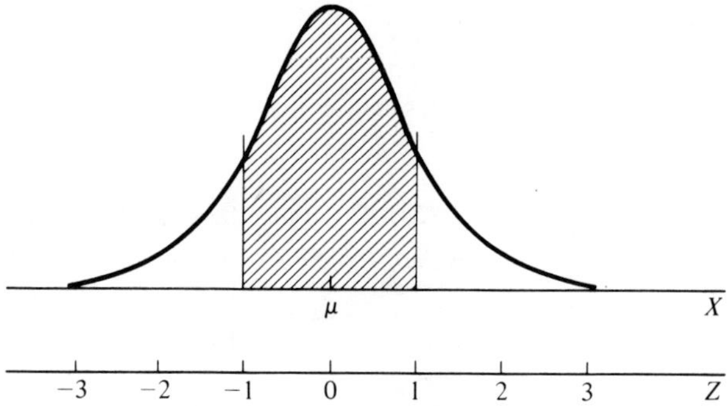

Figure 5.5. Z-transformation.

---

*The normal *D.F.* is not expressible in terms of elementary functions and, therefore, cannot be integrated directly. However, letting $h = 1/\sqrt{2}$,

$$\int_0^Z f(Z)\,dZ = \frac{h}{\sqrt{\pi}} \int_0^Z e^{-h^2Z^2}\,dZ$$

which is a function that can be evaluated by expanding the integrand into a power series and integrating as many terms as may be needed. The result of this approximation is

$$\left.\begin{aligned} &\int_0^Z f(Z)\,dZ = 0.5 - \frac{e^{-Y^2}}{2Y\sqrt{\pi}}\left[1 - \frac{1}{2Y^2} + \frac{3}{4Y^4} - \frac{3\cdot 5}{8Y^6} + \ldots + (-1)^n T_{n+1} + \ldots\right] \\ &\text{where } Y = hZ,\ n+1 \text{ is the number of the term, and} \\ &T_{n+1} = \frac{1\cdot 3\cdot 5\cdot \ldots \cdot (2n-1)}{2^n Y^{2n}} \end{aligned}\right\} \quad (5.4)$$

This series is semi-convergent (i.e., converges until a minimum term is reached, then diverges). The general term T_{n+1} decreases for $n \leq Y^2$. To illustrate this method of evaluation of the *D.F.*, consider the integral

$$\int_0^{Z=3} f(Z)\,dZ$$

Since $h = 1/\sqrt{2}$ and $Z = 3$, the evaluation is for

quality on X is changed to an inequality on Z, and the probability is read directly from the table in Appendix D. For example,

$$P(X \leq b) = \int_{-\infty}^{b} f(X)\,dX = P(Z \leq Z_b) = \int_{-\infty}^{Z_b} f(Z)\,dZ$$

where Z_b is the standard form for $X = b$, that is,

$$Z_b = \frac{b - \mu_X}{\sigma_X}$$

Also,

$$P(X \geq b) = 1 - P(X \leq b)$$

and

$$P(a \leq X \leq b) = \int_{a}^{b} f(X)\,dX = P(Z_a \leq Z \leq Z_b) = \int_{Z_a}^{Z_b} f(Z)\,dZ$$

where Z_a and Z_b are the standard forms of a and b, respectively.

EXAMPLE 5.3.1

A random variable X is $N(50, 25)$. Compute the probability $P(X \geq 60)$. The Z-value corresponding to $X = 60$ is obtained from the Z-transformation:

$$Z_{60} = \frac{60 - 50}{5} = 2.0$$

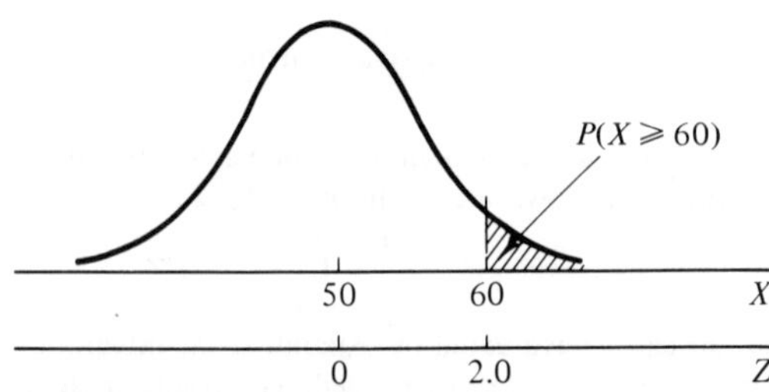

Figure 5.6. Example 5.3.1—normal model.

$$Y = hz = \frac{3}{\sqrt{2}}$$

and

$$n \leq Y^2 = \tfrac{9}{2} = 4\tfrac{1}{2} \quad \text{(use five terms)}$$

Then,

$$\int_{0}^{Z=3} f(Z)\,dZ = 0.5 - \frac{e^{-9/2}}{2\sqrt{\pi}} \cdot \frac{\sqrt{2}}{3}\left[1 - \frac{1}{9} + \frac{3}{81} - \frac{15}{729} + \frac{105}{6{,}561}\right]$$
$$= 0.49864$$

which agrees to four decimal places with the tabled value for $Z = 3$ (Appendix D):

$$0.50000 - 0.00135 = 0.49865$$

The probability of $X \geq 60$ is

$$P(X \geq 60) = P(Z \geq 2.0)$$
$$= 0.0228 \quad \text{(from Appendix D)}$$

Also, compute the probability $P(55 \leq X \leq 60)$.

$$Z_{55} = \frac{55 - 50}{5} = 1.0$$

and

$$P(55 \leq X \leq 60) = P(Z \geq 1.0) - P(Z \geq 2.0)^{*}$$
$$= 0.1587 - 0.0228$$
$$= 0.1359$$

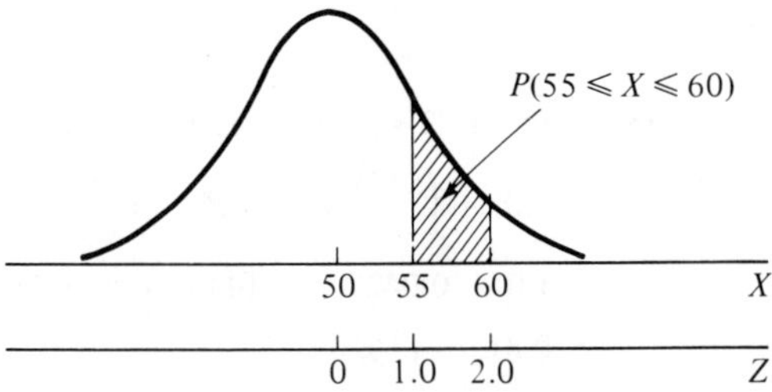

Figure 5.7. Example 5.3.1—normal model.

EXAMPLE 5.3.2

The life X of a certain type of electronic tube is normally distributed with mean $\mu = 200$ hr and standard deviation $\sigma_Y = 22$ hr. What is the expected percentage of tubes requiring replacement at or before 150 hr?

$$Z_{150} = \frac{150 - 200}{22} = -2.27$$

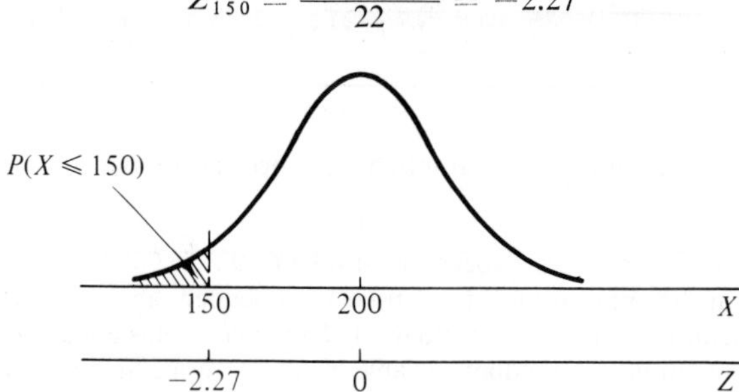

Figure 5.8. Example 5.3.2—normal model.

---

*In applications problems, the student is encouraged to sketch the distribution model and identify the area segments corresponding to the required probabilities. This procedure is helpful in indicating the necessary additions and subtractions of tabled probability values.

and

$$P(X \leq 150) = P(Z \leq -2.27) = \int_{-\infty}^{-2.27} f(Z)\,dZ^*$$

$$= 0.0116 \quad \text{(from Appendix D)}$$

Therefore, 1.16% of the tubes are expected to be replaced at or before 150 hr.

EXAMPLE 5.3.3

The tension X for a certain type of piston ring is normally distributed with mean $\mu = 13.1$ lb. and standard deviation $\sigma_X = 1.8$ lb. Manufacturing specifications for these rings are 10 to 15 lb tension. What is the percentage of manufactured rings that is expected to be within specifications?

$$Z_{10} = \frac{10 - 13.1}{1.8} = -1.72, \qquad Z_{15} = \frac{15 - 13.1}{1.8} = +1.06$$

and

$$P(10 \leq X \leq 15) = P(-1.72 \leq Z \leq +1.06)\dagger$$

$$= 1.0 - \int_{-\infty}^{-1.72} f(Z)\,dZ - \int_{+1.06}^{+\infty} f(Z)\,dZ$$

$$= 1.0 - 0.0427 - 0.1446 \quad \text{(from Appendix D)}$$

$$\cong 0.81$$

Thus, 81% of the rings are expected to meet specifications.

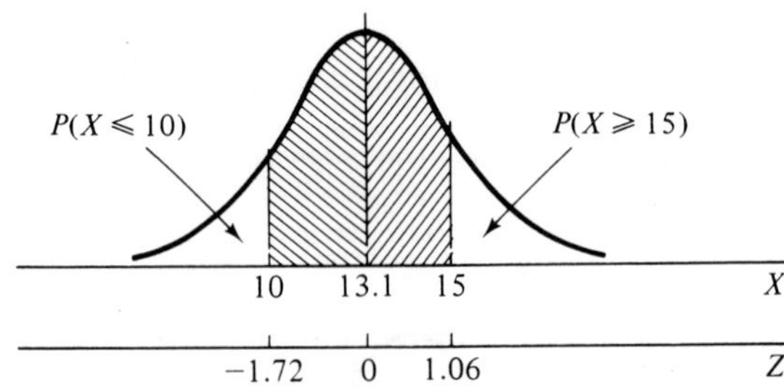

Figure 5.9. Example 5.3.3—normal model.

---

*The range of the X-values associated with the *C.D.F.* is $(-\infty, +\infty)$. Yet the range of the X-values in the application (life of tubes) is bounded, and the X-values are non-negative by the nature of the physical situation. To reconcile this apparent contradiction, note that the assumption of a random variable having a normal distribution is simply an assumption regarding the form of a mathematical model that, at best, is only an approximation to the real physical situation.

†It should be noted (Sec. 3.4) that for a continuous random variable only a probability statement concerning an interval is possible. Thus,

$$P(a \leq X \leq b) = P(a < X \leq b) = P(a \leq X < b) = P(a < X < b)$$

EXERCISES

1. Determine the following probability values $F(Z) = \int_{Z_1}^{Z_2} f(Z)\,dZ$ for the (Z_1, Z_2) values: $(0, \infty)$, $(-1.2, +1.2)$, $(-\infty, +1.2)$, $(+1.2, +\infty)$, and $(-1.2, +\infty)$. In each case, sketch the *D.F.*

2. The finished diameter of a product part is normally distributed with mean 0.700 in. and standard deviation 0.010 in. What is the probability that the diameter will exceed 0.715 in. on a part that is randomly selected? *Ans.* 0.0668.

3. The length of a production part is normally distributed with mean 0.750 in. and standard deviation 0.003 in. The production specifications are 0.750 ± 0.0006 in. What percentage of parts is expected to be defective?

Ans. 84.14%.

4. In Exercise 3, what is the maximum allowable standard deviation that will permit no more than 13.36% defective product? *Ans.* $\sigma_X = 0.0004$ in.

5. Evaluate $F(Z) = \int_0^{2.8} f(Z)\,dZ$ using the approximation method given by Eq. (5.4) and compare the result with the tabled value (Appendix D).

Ans. 0.4975.

6. Gage repeat error is a measure of the variation of gage readings and is measured by $\pm 3\sigma_X$ from a true reading. A product part specification is $0.495^{+0.005}_{-0.000}$ in. Gage repeat error is 0.0009 in. (i.e., $6\sigma_X = 0.0009$). What is the probability of rejection of an acceptable product unit whose true size is 0.4998 in.?

Ans. 0.0918.

5.4 Normal Approximation to the Binomial

With large sample size n, the computation of binomial probabilities becomes tedious. In Sec. 4.6, the binomial distribution is approximated by the Poisson distribution, when n is large and the probability of a success p, at each trial, is small. If neither p nor q is near zero, and n is sufficiently large, the normal distribution model yields good approximations to binomial probabilities.*

*It can be shown that as sample size n increases, the binomial distribution approaches a normal distribution with mean $\mu = np$ and variance $\sigma_X^2 = npq$. The accuracy of this approximation improves as n increases, of course. Also, for fixed n, the accuracy improves as p approaches 0.5.

Certain n and p combinations that result in good approximations are given in Table 5.2. This table is taken from Cochran, W. G. (1953), *Sampling Techniques*, John Wiley & Sons, Inc., New York, p. 41.

Table 5.2. MINIMUM SAMPLE SIZE TO USE NORMAL MODEL TO APPROXIMATE A BINOMIAL POPULATION

P	*n at least equal to*
0.5	30
0.4 or 0.6	50
0.3 or 0.7	80
0.2 or 0.8	200
0.1 or 0.9	600
0.05 or 0.95	1400

EXAMPLE 5.4.1

A random sample of $n = 100$ is taken from a binomial population, where the constant probability of success is $p = 0.4$. What is the probability that the number of successes, X, in the sample will be between 30 and 45 inclusive?

Using the normal distribution model, to approximate the binomial probability,

$$\mu = np = 100(0.4) = 40$$

and

$$\sigma_X^2 = np(1 - p) = 100(0.4)(0.6) = 24$$

$$\sigma_X = \sqrt{24} = 4.9$$

The pertinent Z-values are

$$Z_{30} = \frac{30 - 40}{4.9} = -2.04, \qquad Z_{45} = \frac{45 - 40}{4.9} = +1.02$$

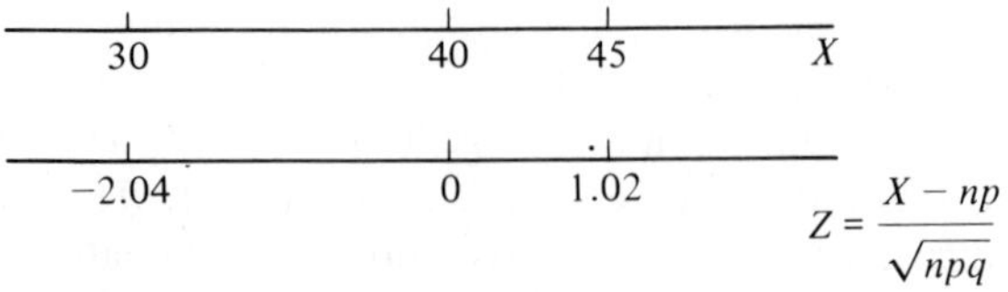

Figure 5.10. Example 5.4.1—normal approximation to the binomial model, probability $P(30 \leq X \leq 45)$.

The desired probability is

$$P(30 \leq X \leq 45) = P(-2.04 \leq Z \leq 1.02)$$

$$= 1.0 - \left[\int_{-\infty}^{-2.04} f(Z)\,dZ + \int_{+1.02}^{+\infty} f(Z)\,dZ\right]$$

$$= 1.0 - (0.0207 + 0.1539)$$
$$= 0.8254$$

The exact probability (from Appendix A) is

$$0.8689 - 0.0148 = 0.8541$$

However, a continuous distribution is being used to approximate the exact probability

$$\sum_{X=30}^{45} \binom{100}{X} (0.4)^X (0.6)^{100-X}$$

which is represented by a histogram as shown in Fig. 5.11.

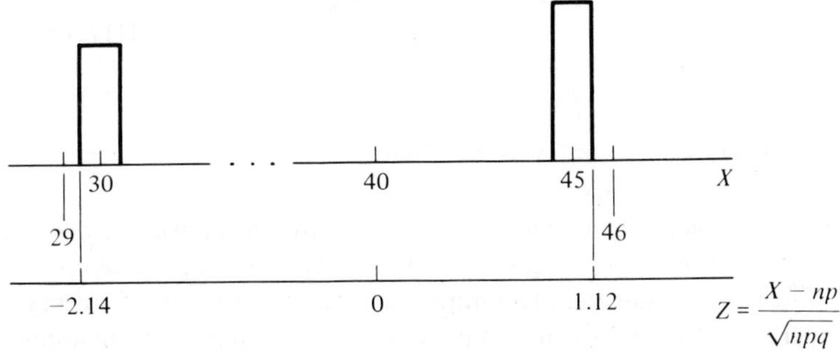

Figure 5.11. Example 5.4.1—normal approximation to the binomial model, probability $P(29.5 \leq X \leq 45.5)$.

In the preceding solution, using the continuous model, the sum of the probabilities

$$P(X \leq 29) + P(30 \leq X \leq 45) + P(X \geq 46) \neq 1$$

Clearly, a better approximation* is obtained if the area under the normal curve corresponding to the desired probability includes the lower boundary of the bar whose mid-X value is 30 and the upper boundary of the bar whose mid-X value is 45. Then

$$Z_{29.5} = \frac{29.5 - 40}{4.9} = -2.14, \qquad Z_{45.5} = \frac{45.5 - 40}{4.9} = +1.12$$

and

$$P(-2.14 \leq Z \leq 1.12) = 0.8524$$

---

*The error in the normal approximation, with this continuity correction, never exceeds

$$\frac{0.140}{\sqrt{np(1-p)}} \tag{5.5}$$

This relationship is from Raff, M. S. (June 1956), "On Approximating the Point Binomial," *American Statistical Association*, Vol. 51, No. 274, pp. 293–303.

EXERCISES

1. A fair coin is randomly tossed 10 times. Compute the probability of obtaining 4, 5, 6, or 7 heads by using the: (a) binomial model, and (b) the normal approximation to the binomial probability. *Ans.* (a) 0.7734, (b) 0.7718.

2. Using Eq. (5.5), compute the maximum error for the procedure in Exercise 1(b). *Ans.* 0.0885.

3. A sampling inspection plan operates as follows. A random sample of $n = 50$ items is drawn from a large lot of manufactured items. If no more than two defective items are observed, the lot is accepted. Compute the probability of acceptance P_a of a 10% defective lot using the: (a) binomial model, and (b) normal approximation to the binomial probability.
Ans. (a) 0.1117, (b) 0.1190.

5.5 Normalizing Transformation

Many physical phenomena produce data that are normally distributed. Also, some physical events result in data that can be transformed to normal data. Since many statistical techniques are based on assumed normality of the underlying distribution, it is important to be able to transform nonnormal to normal data.* The most common transformations are

$$Y = \log_e X \quad \text{and} \quad Y = \sqrt{x} \tag{5.6}$$

To determine if one of these transformations produces normality, use can be made of special graph papers that are commercially available.

When the basic measurements (observations) are easy to obtain in large numbers, an effective normalization of the data can almost always be achieved (except for extremely nonnormal data) by averaging. This assumes that the questions of interest relative to the population concerned can be rephrased in terms of the corresponding sampling distribution of means of sufficiently large random samples.†

---

*This is an extensive topic which is beyond the scope of this textbook. For example, an entire book is devoted to log transformations [see Aitchison, J. and Brown, J. A. (1957), *The Log Normal Distribution*, Cambridge University Press, New York].

Also, an excellent summary of this topic is given in Nat. Bur. Standards (August 1963), *Experimental Statistics*, Handbook 51, Supt. of Documents, U. S. Govt. Printing Office, Wash. 25, D. C., Chap. 20.

†This statement is somewhat premature, but will be meaningful to the student after studying the sampling distribution of the mean in Secs. 5.6 and 5.7.

5.6 Central Limit Theorem

Let $y = x_1 + x_2 + \ldots + x_n$ where $x_1, x_2, \ldots, x_n$ are identically distributed, independent random variables each having mean μ and finite variance σ_X^2. Then, the distribution of

$$Z = \frac{y - n\mu}{\sqrt{n}\,\sigma_X}$$

approaches the normal distribution with mean zero and variance 1 as n increases without bound.

The theorem is also valid when the variables are not identically distributed, if their variances are reasonably homogeneous. Thus, the theorem also implies that the sum of a large number of random variables is approximately normally distributed regardless of the distributions of the individual random variables. This statement has important implications regarding the applications problems in Sec. 5.9. Also note that

$$\frac{y - n\mu}{\sqrt{n}\,\sigma_X} = \left(\frac{\bar{x} - \mu}{\sigma_X}\right)\sqrt{n}$$

This means that the theorem also implies that the mean of n identically distributed random variables is approximately normally distributed regardless of the distribution of the individual variables.*

*The essence of the preceding paragraphs, as far as applications are concerned, is that repeated samples of size n from even a nonnormal population generate an approximately normal distribution of either sums or means as indicated by Fig. 5.12 (i.e., for each sample either a sum or mean is computed; repeated sampling creates a sequence or distribution of either sums or means).

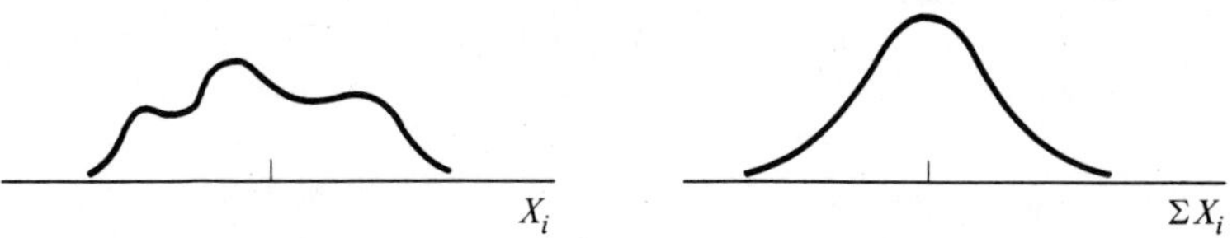

Figure 5.12. All samples $x_1, x_2, \ldots, x_n$ from a non-normal X_i population generate an approximately normal population of $\sum$ sums X_i.

5.7 Sampling Distribution of the Mean

As a consequence of the central limit theorem, if samples of size n are repeatedly* taken from an X-population (and for each sample a mean $\bar{x}$ is computed), the $\bar{x}$-population thus generated possesses the following properties:

1. If the X-population is normal, then the $\bar{x}$-population is normal. If the X-population is not normal, but sample size n is sufficiently large, then the $\bar{x}$-population is approximately normal.
2. If the X-population is infinite (or finite and sampling is with replacement) with mean μ and variance σ_X^2, then the $\bar{x}$-population is distributed with mean μ and variance $\sigma_{\bar{x}}^2$, where

$$\sigma_{\bar{x}}^2 = \frac{\sigma_X^2}{n} \tag{5.7}$$

3. If the X-population is finite with mean μ and variance σ_X^2 and sampling is without replacement, then the $\bar{x}$-population is distributed with mean μ and variance $\sigma_{\bar{x}}^2$, where

$$\sigma_{\bar{x}}^2 = \frac{\sigma_X^2}{n}\left(\frac{N-n}{N-1}\right) \tag{5.8}$$

EXAMPLE 5.7.1

A random variable X is uniformly distributed. All possible samples of size $n = 4$ from this population generate an $\bar{x}$-population that is approximately normal as shown in Fig. 5.13.

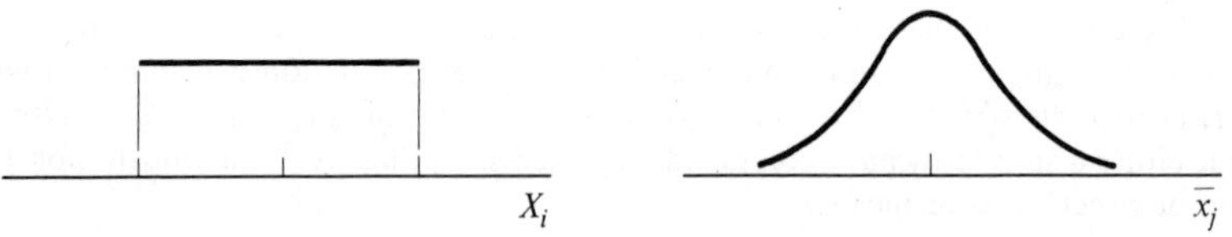

Figure 5.13. Example 5.7.1—all samples x_1, x_2, x_3, x_4 from uniform X_i population generate an approximately normal population of means x_j.

---

*This means that all possible samples of a specified and fixed size n are being considered. Thus, a family of $\bar{x}$-populations dependent on n can be generated. If n is large, the range of the generated $\bar{x}$-population is small. Conversely, if n is small, the range of the $\bar{x}$-population is relatively large. [See Eq. (5.7) and recall from Fig. 5.5 that the range of a distribution can be measured in terms of multiples of σ_X.]

A random variable X has a triangular distribution. All possible samples of size $n = 4$ from this population generate an $\bar{x}$-population that is approximately normal as indicated by Fig. 5.14.

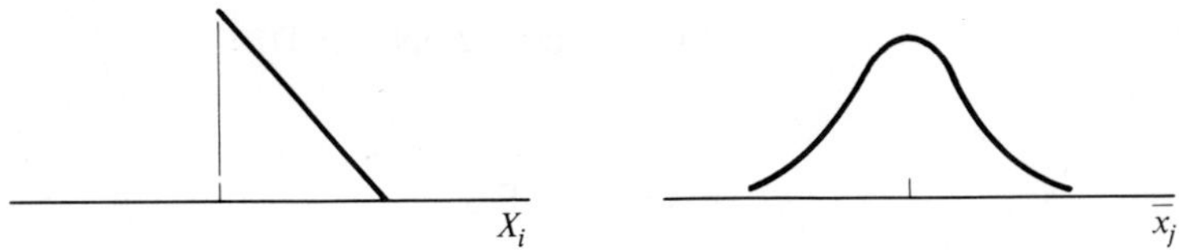

Figure 5.14. Example 5.7.1—all samples x_1, x_2, x_3, x_4 from a triangular X_i population generate an approximately normal population of means $\bar{x}_j$.

EXAMPLE 5.7.2

A random variable X is uniformly distributed over an interval (15, 25). A random sample of size $n = 12$ is taken from this X-population. What is the probability that the mean of the 12 observations is equal to or greater than 21? The distribution models are shown in Fig. 5.15.

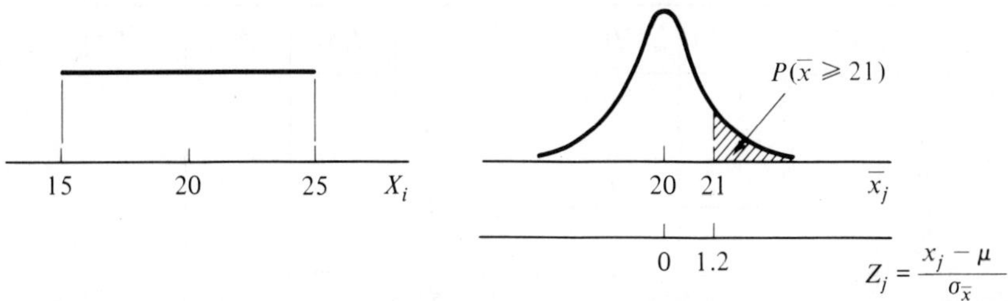

Figure 5.15. Example 5.7.2—normal model.

From Eqs. (4.15) and (4.16), the mean and variance for this X-population are obtained as follows

$$\mu = E(X) = \frac{c + a}{2} = \frac{15 + 25}{2} = 20$$

and

$$\sigma_X^2 = \frac{(c - a)^2}{12} = \frac{(25 - 15)^2}{12} = \frac{100}{12}$$

From Eq. (5.7),

$$\sigma_{\bar{x}}^2 = \frac{\sigma_X^2}{n} = \frac{\frac{100}{12}}{12} = \frac{100}{144}$$

and

$$\sigma_{\bar{x}} = \tfrac{10}{12}$$

Also

$$Z_{21} = \frac{21 - 20}{\frac{10}{12}} = 1.2$$

and thus

$$\begin{aligned} P(\bar{x} \geq 21) &= P(Z \geq 1.2) \\ &= \int_{1.2}^{\infty} f(Z)\,dZ \\ &= 0.1151 \quad \text{(from Appendix D)} \end{aligned}$$

EXERCISES

1. From historical data, it is known that a type of electrical cable has a mean tensile strength of 400 psi and a standard deviation of 20 psi. A random sample of 100 cables is drawn from the production line. What is the probability that the mean tensile strength of this sample is: (a) less than 396 psi, (b) more than 402 psi, and (c) between 395.4 and 404.6 psi inclusive?

Ans. (a) 0.0228, (b) 0.1587, (c) 0.9786.

2. A population consists of three numbers: 2, 4, and 6. Assuming sampling with replacement after each random draw, list the nine possible samples of size $n = 2$. Complete the two frequency distribution tables that follow, compute the mean and variance for each distribution, and thus verify Eq. (5.7).

X	$f(X)$	$Xf(X)$	$X^2f(X)$
2			
4			
6			
Σ			

$\bar{x}$	$f(\bar{x})$	$\bar{x}f(\bar{x})$	$\bar{x}^2f(\bar{x})$
2			
3			
4			
5			
6			
Σ			

3. Repeat Exercise 2 assuming sampling without replacement, and verify Eq. (5.8).

5.8 Specifications

Section 5.9 deals with linear combinations of independent random variables. The applications problems in this section assume a basic knowledge of specifications procedures. To prepare for this section, a brief review summary of specifications in the mechanical industries is presented here.*

A *tolerance* is a specified permissible variation. Tolerance specifications are denoted by two fundamental forms. One form states the design size and

*A comprehensive presentation of this topic is given in Kirkpatrick, E. G. (1970), *Quality Control for Managers and Engineers*, John Wiley & Sons, Inc., New York, Chaps. 4, 5, 7.

tolerance, for example, 0.500 ± 0.002 in., where the *design size* is 0.500 in. and the tolerance is 0.004 in. The other form is a statement of the two limits, that is, 0.502/0.498 in. where 0.502 in. is the *maximum limit*, 0.498 in. is the *minimum limit*, and the tolerance is 0.004 in.

When two or more mating parts are to be assembled, each part being defined by a tolerance specification, the assembly condition of primary interest is that of *clearance*. For example, parts A, B, C, and D are assembled in that order into the assembly space (the slot in E) as shown in Fig. 5.16. Cumulative tolerances (or variation) of the parts determines whether or not part D will assemble.

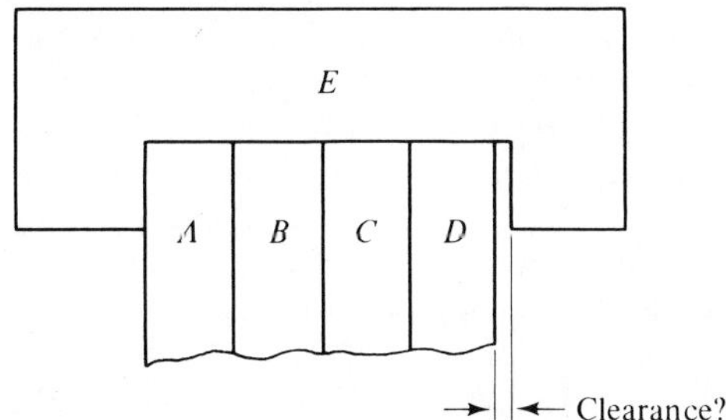

Figure 5.16. Clearance condition.

A strict definition of clearance covers both physical conditions—interference and clearance. A negative clearance denotes interference (i.e., lack of physical clearance), and a positive clearance value signifies actual physical clearance. In the case of negative clearance (press or shrink fits), actual assembly is effected by means of considerable pressure and/or temperature changes to temporarily alter the mating part sizes during assembly.

An example given in Fig. 5.17 summarizes both tolerances and clear-

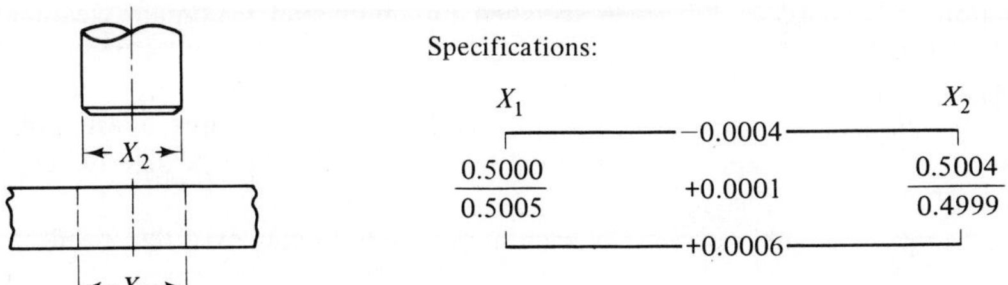

Figure 5.17. Clearance $Y = X_1 - X_2$.*

*Specification limits are commonly expressed in this manner. The *maximum material limit* (*MML*) is written on top (0.5000 for X_1 and 0.5004 for X_2). The *MML* refers to the limit that describes the most material possible remaining on the finished part after processing. Clearly, the *MML* of the shaft A is the maximum limit, and the *MML* of the hole E is the minimum limit.

ances. The *minimum clearance* is -0.0004 in., the *maximum clearance* is $+0.0006$ in., and the *mean clearance* μ_Y is $+0.0001$ in. The simple mathematical model or clearance function is $Y = X_1 - X_2$.

The concepts of basic hole, basic shaft, unilateral, and bilateral tolerance procedures are not discussed here. All examples in Sec. 5.9 follow basic hole and unilateral tolerance procedures as indicated by the following two examples.

EXAMPLE 5.8.1

Tolerance specifications X_1 and X_2 are to be determined for a shaft-hole assembly (Fig. 5.17), given a specified minimum clearance and the tolerance for each part. The design size is $\frac{1}{2}$ in., minimum clearance is $+0.0002$ in., and each component part tolerance is 0.0005 in.

The specifications are shown in Fig. 5.18. The *MML* of X_1 is established first (i.e., it is equal to the decimal equivalent of $\frac{1}{2}$ in.). Application of the minimum clearance establishes the *MML* of X_2. Applying the tolerance of 0.0005 in. in a plus direction from 0.5000 in. determines the maximum limit of X_1 and in a minus direction from 0.4998 in. the minimum limit of X_2.

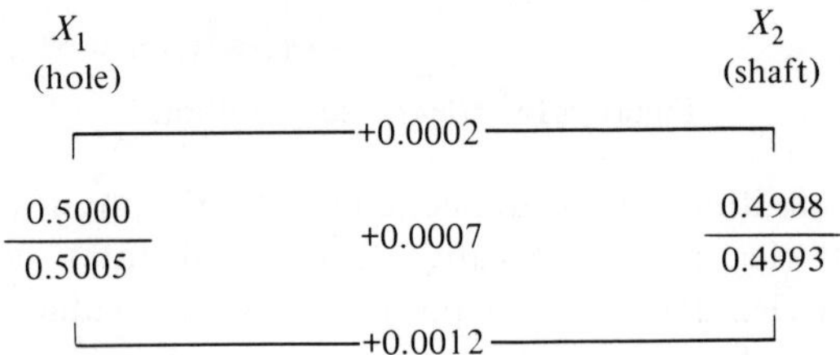

Figure 5.18. Example 5.8.1—specifications for X_1 and X_2.

EXAMPLE 5.8.2

Tolerance specifications X_1 and X_2 are to be determined for the same physical situation of Example 5.8.1, given specified minimum and maximum clearances of $+0.0004$ in. and $+0.0016$ in., respectively. Tolerance is to be allocated equally to X_1 and X_2.

The specifications are shown in Fig. 5.19. Again, the *MML* of X_1 is established first (i.e., 0.5000 in.), and application of the minimum clearance determines the *MML* of X_2 to be 0.4996 in.

An infinite number of pairs of minimum material limits exist that result in a

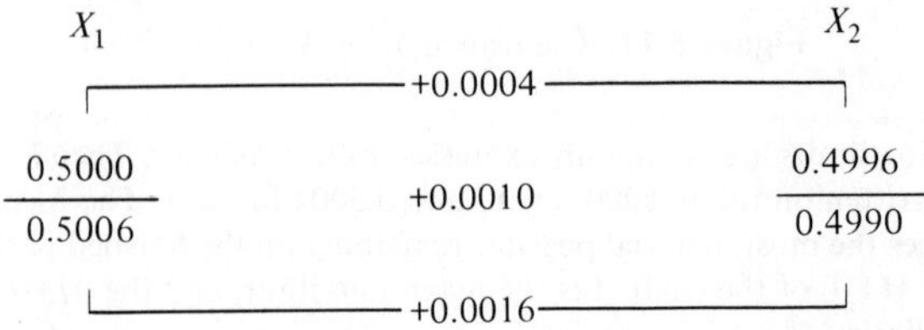

Figure 5.19. Example 5.8.2—equal tolerance allocation.

maximum clearance of +0.0016 in. However, only one pair satisfy the equal tolerance allocation condition.* The tolerance is

$$\tfrac{1}{2}[(+0.0016) - (+0.0004)] = 0.0006 \text{ in.}$$

Applying the 0.0006-in. tolerance in a plus direction from 0.5000 in. determines the maximum limit of X_1 and in a minus direction from 0.4996 in. the minimum limit of X_2.

The applications examples of Sec. 5.9 involve *random assembly* of component parts under a controlled risk of incurring defective assemblies. This is a part of a broader topic called *interchangeability alternatives*, wherein there are alternatives of full interchangeability, modified full interchangeability, the latter alternative consisting of two subalternatives—selective and random assembly. This general treatment is beyond the scope of this textbook. However, a point should be made in order to make the examples of Sec. 5.9 more meaningful. With random assembly, the objective is to increase component tolerances (and hence obtain a considerable decrease in process costs), accept a small percentage of assembly and test difficulties (and a nominal increase in discovery-and-correction-of-defective-assembly costs), and thus minimize total manufacturing costs. This is summarized in Fig. 5.21.

5.9 Combinations of Independent Random Variables

Computation of economic tolerances is based on the following theorem:

If $X_1, X_2, \ldots, X_j, \ldots, X_k$ are independent random variables with respective means,

$$\mu_1, \mu_2, \ldots, \mu_j, \ldots, \mu_k$$

---

*Equal tolerance allocation is usually not practical. Generally, more tolerance is allocated to the interior characteristic (e.g., hole, slot, etc.), which is usually more difficult to process to small tolerances on a production basis. Figure 5.20 shows an unequal specification tolerance solution for the same clearance conditions of this example.

However, in the examples of Sec. 5.9, tolerances are allocated equally to simplify examples and to generate unique solutions.

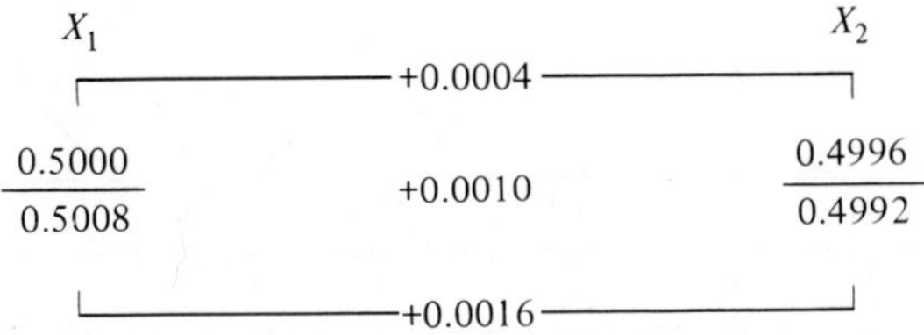

Figure 5.20. Example 5.8.2—unequal tolerance allocation.

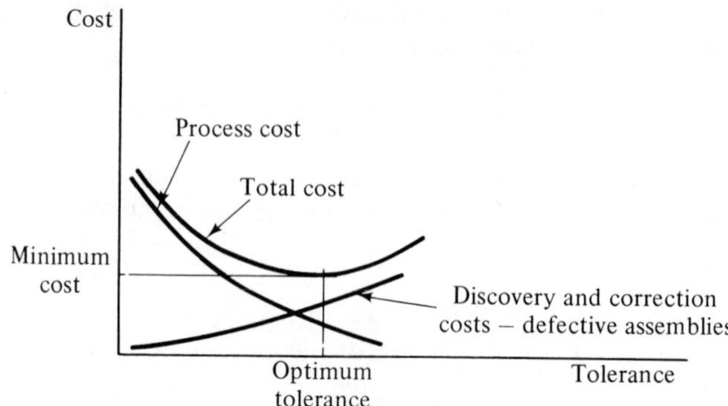

Figure 5.21. Economics of increased tolerances through random assembly.

and variances,

$$\sigma_1^2, \sigma_2^2, \ldots, \sigma_j^2, \ldots, \sigma_k^2$$

and if $a_1, a_2, \ldots, a_j, \ldots, a_k$ are constants and Y is a linear combination of the X_j, that is,

$$Y = a_1X_1 \pm a_2X_2 \pm \ldots \pm a_kX_j$$

then Y is a random variable having the following properties:

1. $\mu_Y = a_1\mu_1 \pm a_2\mu_2 \pm \ldots \pm a_k\mu_k$ (5.9)
2. $\sigma_Y^2 = a_1^2\sigma_1^2 + a_2^2\sigma_2^2 + \ldots + a_k^2\sigma_k^2$ (5.10)
3. If $X_1, X_2, \ldots, X_k$ are each normally distributed, then Y is normally distributed.
4. If the X_j are not all normally distributed and if the variances are approximately homogeneous, then from the central limit theorem (Sec. 5.6) as k increases, the Y-distribution rapidly approaches the normal distribution.

(5.11)

Each X-distribution range measures the variation of the X-variable. The Y-distribution range measures the variation of the linear combination of the X_j. In most product-tolerance cases, the tolerance for the assembly quality characteristic is a linear combination of the tolerances for the respective

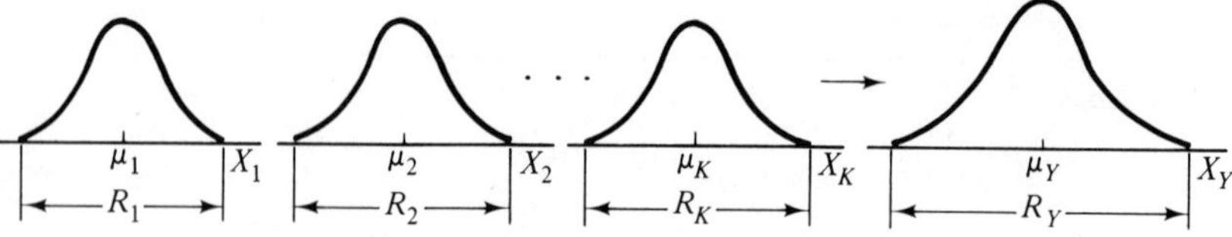

Figure 5.22. Normal models to represent tolerances.

component quality characteristics. Thus, in tolerance applications, each X-distribution range corresponds to the tolerance for that component characteristic. The Y-distribution range corresponds to the tolerance for the assembly characteristic. This correspondence is summarized in Fig. 5.22.

The ranges $R_1, R_2, \ldots, R_k$ represent the respective component characteristic tolerances. The range R_Y represents the assembly characteristic tolerance. Mean values $\mu_1, \mu_2, \ldots, \mu_k$ correspond to the nominal* specification values for the respective components, and μ_Y denotes the nominal specification value for the assembly characteristic.

The standard deviation σ_Y is used as a measure of R_Y (for example, let $R_Y = 6\sigma_Y$). In the same manner, $\sigma_1, \sigma_2, \ldots, \sigma_k$ measure the ranges R_1, $R_2, \ldots, R_k$, respectively. The number of component characteristics generating the assembly characteristic of one product unit is denoted by k. Thus, there are k component distributions involved.

Regarding product-design purposes, applications problems can be separated into two classes: (1) The assembly tolerance is fixed by functional requirements, and the problem is to statistically determine the component tolerances, and (2) component tolerances are fixed, either because of process limitations or the components being standard purchased parts with fixed commercial tolerances, and the problem is to estimate probable cumulative-error effect on the assembly characteristic.

For both design purposes, computation is based on an acceptable risk of incurring defective assemblies. The risk value λ is the probability of occurrence of a defective assembly. The magnitude of λ is dependent on economic considerations and varies from one production situation to another.

The simplified applications examples that follow are restricted to product conditions of size and clearance.†

EXAMPLE 5.9.1

Four component parts assemble adjacent to each other as indicated by Fig. 5.23. Their respective sizes are denoted by X_1, X_2, X_3, and X_4 and the assembly dimension is $Y = 2.460 \pm 0.008$. The Y-value is fixed by functional requirements and the problem is to enlarge the X-specifications tolerances.

The risk of incurring defective assemblies is $\lambda = 0.0027$.‡ To simplify the example, identical component tolerances are assumed.

---

*The term "nominal" in this section refers to the manufacturing interpretation, namely, the mean average of the maximum and minimum specification limits.

†The following examples in this section have been abstracted from Kirkpatrick, E. G. (1970), *Quality Control for Managers and Engineers*, John Wiley & Sons, Inc., New York, Chap. 7. The interested student is encouraged to review the original text of these examples, particularly the implementation responsibilities and the translation of mathematical conditions [Eqs. (5.9)–(5.11)] to their physical and economic equivalents.

‡The λ-value is arbitrary here to construct an example. The choice of λ is discussed at the conclusion of this example.

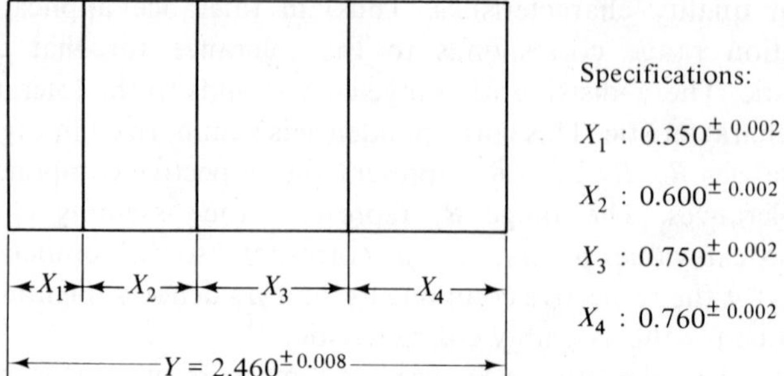

Figure 5.23. Example 5.9.1—specifications, assembly requirement Y fixed.

Figure 5.24 shows the four component distributions. Mean averages correspond to nominal specification values. Distribution ranges are measured by $6\sigma_X$.

The assembly distribution is given in Fig. 5.25. The distribution mean corresponds to the nominal assembly specification value. The fixed assembly tolerance (Fig. 5.23) determines the limit dimensions for Y, 2.452 in. and 2.468 in. The risk value $\lambda/2$ is represented by the tail areas which correspond to the probability of incurring defective assemblies.

Since $\lambda = 0.0027$, the Y-distribution range is $6\sigma_Y$. That is, from Appendix D, corresponding to a Z_α-value of 3.0 the tail area is 0.00135, and thus for $\pm 3Z$ or $\pm 3\sigma_Y$ the sum of the tail areas is 0.0027. From Eq. (5.9),

$$\mu_Y = \mu_1 + \mu_2 + \mu_3 + \mu_4$$
$$= 2.460$$

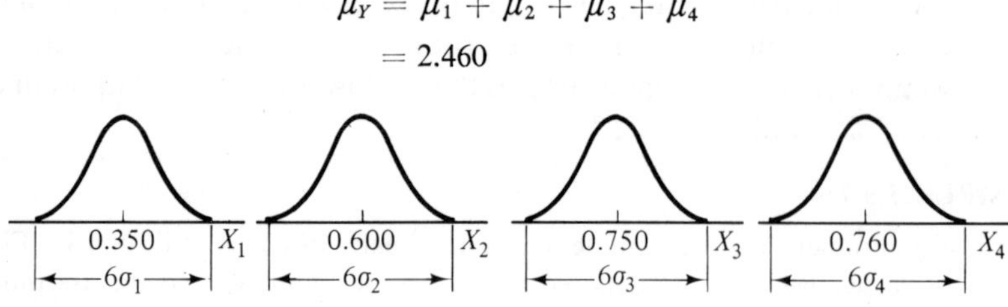

Figure 5.24. Example 5.9.1—components distribution models.

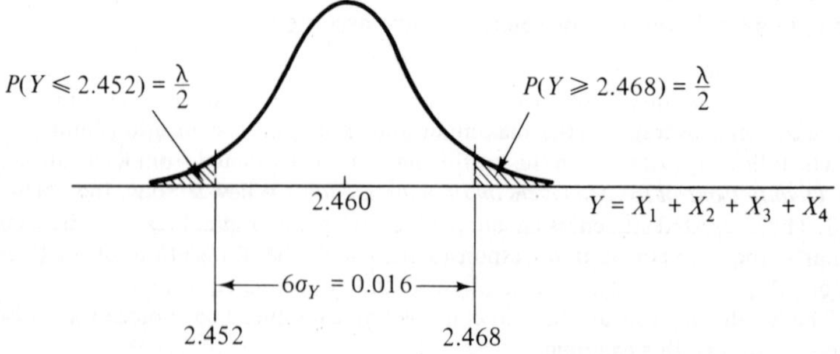

Figure 5.25. Example 5.9.1—assembly distribution model.

From Eq. (5.10),*

$$\sigma_Y^2 = \sigma_1^2 + \sigma_2^2 + \sigma_3^2 + \sigma_4^2$$

and if component tolerances are identical,

$$\sigma_Y^2 = 4\sigma_X^2 \quad \text{where} \quad \sigma_X = \sigma_1 = \sigma_2 = \sigma_3 = \sigma_4$$

Since $6\sigma_Y = 0.016$ (see Fig. 5.25),

$$\left(\frac{0.016}{6}\right)^2 = 4\sigma_X^2$$

and

$$\sigma_X = 0.00133$$

Since the component distribution ranges correspond to the respective component tolerances,

$$\begin{aligned} 6\sigma_X &= 6(0.00133)\dagger \\ &= 0.008 \\ &= \pm 0.004 = \text{component tolerance} \end{aligned}$$

The result or conclusion to Example 5.9.1 is summarized as follows. The component tolerances for X_1, X_2, X_3, and X_4 can be enlarged from ± 0.002 in. to ± 0.004 in. This advantage is gained by relaxing from 100% to 99.73% the percentage of assemblies expected to satisfy the fixed-size requirement of 2.460 ± 0.008 in. This assumes, of course, that the component characteristics X_1, X_2, X_3, and X_4 are independent random variables, each normally distributed with mean average corresponding to the nominal specification value, and that component selection (at assembly work stations) is on a random basis.

The risk of incurring defective assemblies, λ, is arbitrarily taken to be 0.0027 in Example 5.9.1. In practice, λ is economically established from a tradeoff involving two sets of costs: (1) cost of producing components to restrictive small tolerances, and (2) discovery and correction costs at assembly

---

*The tolerance function here is

$$Y = X_1 + X_2 + X_3 + X_4$$

and hence in the general expression, Eq. (5.10),

$$a_1 = a_2 = a_3 = a_4 = 1$$

†Component distribution ranges are always set equal to $6\sigma_X$. This procedure is equivalent to requiring that most of the components (in theory, 99.73%) be manufactured to the statistical tolerance specifications. If this is not the case, the percentage of defective assemblies may actually be greater than 100λ.

On the other hand, the assembly distribution range depends on λ. For example, if λ is equal to 0.0456, then the assembly range (corresponding to the fixed specification tolerance requirement) is $4\sigma_Y$ as indicated by Appendix D.

and test work stations, these costs being due to the occurrence of a small percentage of defective assemblies. Increasing component tolerances usually decreases cost 1 significantly whereas cost 2 increases by only a nominal amount. This advantage is summarized by Fig. 5.21.

EXERCISES

1. Five independent, normally distributed, random variables, $X_1, X_2, \ldots, X_5$, have respective means of 2, 3, 4, 5, and 6, and standard deviations of 0.2, 0.2, 0.3, 0.4, and 0.4. The X_j are related by $Y = X_1 + X_2 + \ldots X_5$. Compute the probability $P(Y \geq 20.06)$. *Ans.* 0.4658.

2. Repeat the solution of Example 5.9.1, assuming that the economics of the situation indicates a risk value $\lambda = 0.0456$. *Ans.* ± 0.006 in.

3. In Example 5.9.1 the assembly tolerance was fixed by functional requirements and the problem was to statistically determine the component tolerances under a risk $\lambda = 0.0027$. Using the same specification values (Fig. 5.23), compute the probable assembly variation assuming now that the component tolerances are fixed. *Ans.* 2.460 ± 0.004 in.

A common assembly condition involving cumulative tolerances is that of clearance between fitting parts. One or more exterior component characteristics match with an interior component characteristic. The resulting assembly condition is either clearance or interference. Examples of clearance are a pin diameter matching with a hole diameter (Fig. 5.17) and a number of mating parts fitting into a given assembly space (Fig. 5.16). The most simple case is that of two matching characteristics X_1 and X_2 where X_1 is the interior and X_2 the exterior characteristic, with interest being on the assembly clearance Y.

EXAMPLE 5.9.2

Component part 2 fits into the slot in part 1 as shown in Fig. 5.26. The mating dimensions are X_1 and X_2.

The characteristic of interest is the clearance between X_1 and X_2, and the tolerance function is $Y = X_1 - X_2$. Current specifications are shown in Fig. 5.27. Minimum clearance is +0.0002 in., maximum clearance is +0.0012 in., and mean clearance is +0.0007 in.

Under a risk $\lambda = 0.0456$, the component tolerances for X_1 and X_2 are to be enlarged. Assuming a normal distribution model, the component and assembly distributions are given by Fig. 5.28. From Eq. (5.10),

$$\sigma_Y^2 = \sigma_1^2 + \sigma_2^2$$

and if component tolerances are identical,

$$\sigma_Y^2 = 2\sigma_X^2 \quad \text{where} \quad \sigma_X = \sigma_1 = \sigma_2$$

Since $4\sigma_Y = 0.0010$,

$$\left(\frac{0.001}{4}\right)^2 = 2\sigma_X^2$$

and

$$\sigma_Y = 0.000177$$

Component tolerance is

$$\begin{aligned} 6\sigma_X &= 6(0.000177) \\ &= 0.001 \end{aligned}$$

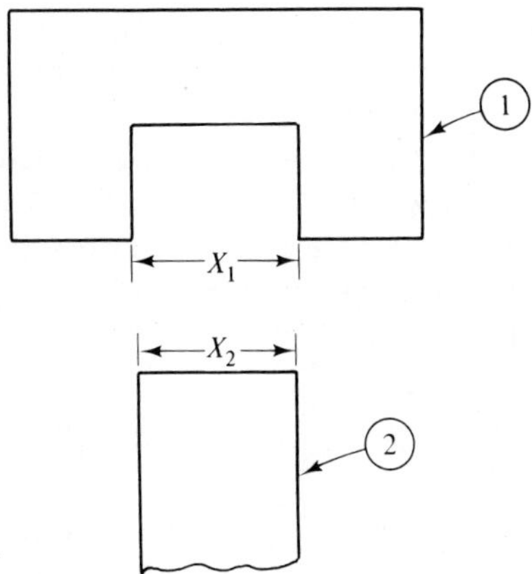

Figure 5.26. Example 5.9.2—clearance condition.

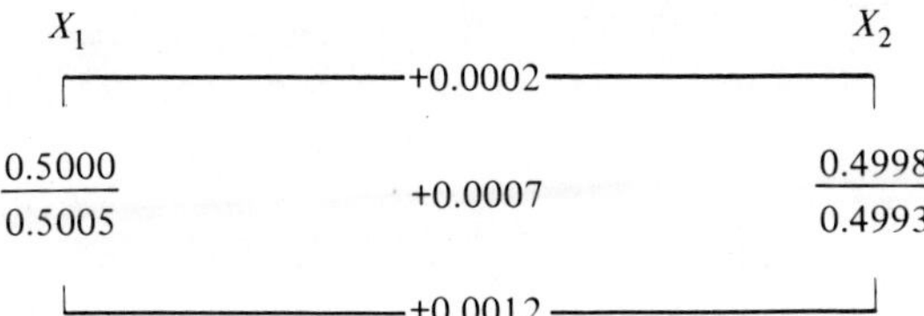

Figure 5.27. Example 5.9.2—specifications for X_1 and X_2.

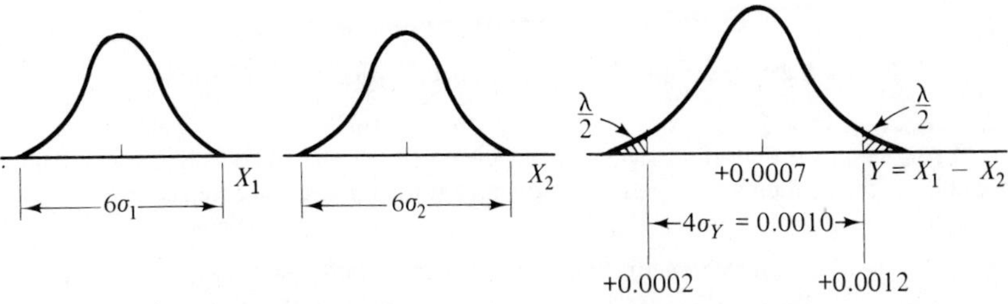

Figure 5.28. Example 5.9.2—distribution models.

The conclusion to Example 5.9.2 is summarized as follows. If X_1 and X_2 are normally distributed, independent random variables, with mean averages corresponding to their respective nominal specification values, the tolerance for X_1 and for X_2 can be increased from 0.0005 in. to 0.001 in. If components are randomly selected for assembly, the expected percentage of assemblies meeting the fixed requirements of clearances from +0.0002 in. to +0.0012 in. is

$$100(1 - 0.0456) = 95.44\%$$

It should be noted that Example 5.9.2 is essentially a two-part problem—the first part being concerned with a determination of the enlarged tolerances for the components, and the second part dealing with a formulation of the new specifications for the components. The following discussion concerns the writing of the new specifications.

The new specifications depend on design and functional requirements. A decision is required as to which is more critical: defective assemblies due to fits that are too tight, or defective assemblies where the fits are too loose. This decision affects the partitioning of the risk λ. For example, if tight fits and loose fits are equally critical, then

$$\frac{\lambda}{2} = P(Y \leq +0.0002) = P(Y \geq +0.0012)$$

and minimum and maximum clearances are, respectively,

$$+0.0007 - 0.0010 = -0.0003$$
$$+0.0007 + 0.0010 = +0.0017^*$$

The new specifications are given by Fig. 5.29. The *MML* of X_1 is estab-

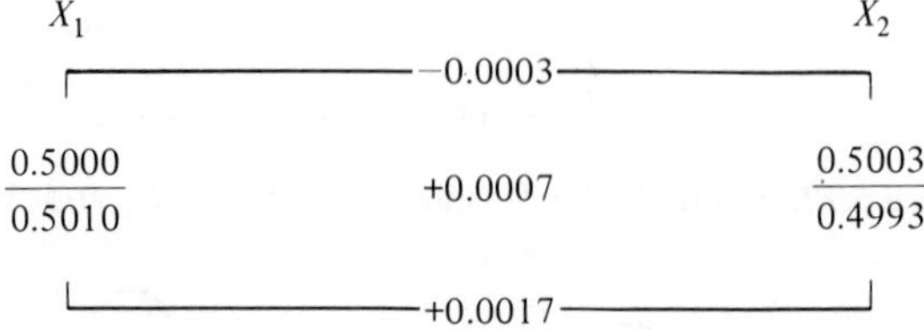

Figure 5.29. Example 5.9.2—new specifications for X_1 and X_2.

*This assumes that the original mean clearance $\mu_Y = +0.0007$ is optimum and should be unchanged under the new specifications. If tolerance is allocated equally to both components, then

$$\text{(mean clearance)} - \text{(tolerance)} = \text{minimum clearance}$$
$$\text{(mean clearance)} + \text{(tolerance)} = \text{maximum clearance}$$

lished first (i.e., 0.5000 in.). Applying the new tolerance (0.001) in a plus direction from 0.5000 determines the maximum limit of X_1. Application of the new minimum and maximum clearances (i.e., -0.0003 in. and $+0.0017$ in.) determines both limits of X_2.

Fig. 5.30 summarizes the solution: 95.44% of the assembly clearances are expected to be between $+0.0002$ and $+0.0012$ in. (the original fixed assembly requirements); 100% of the clearances will be between -0.0003 and $+0.0017$ in. assuming that manufacturing and inspection operations follow the specifications.

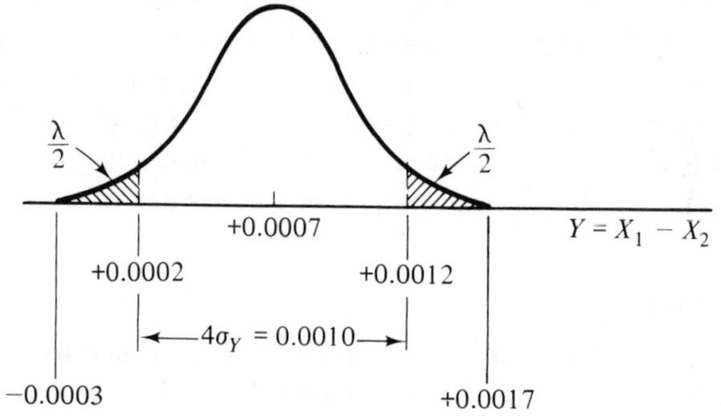

Figure 5.30. Example 5.9.2—summary.

EXERCISES

1. A two-part assembly $Y = X_1 - X_2$ has the following fixed requirements: design size is 0.4000 in., and required clearances are $+0.0004$ to $+0.0016$ in. Assume normally distributed independent random variables X_1 and X_2, a risk $\lambda = 0.0456$, and equal tolerance allocation. Compute the enlarged statistical component tolerances. *Ans.* 0.0012 in.

2. Write the new statistical component specifications for Exercise 1 assuming equal probabilities of tight and loose assembly fits.

Ans. $X_1\colon \frac{0.4000}{0.4012}$, $X_2\colon \frac{0.4002}{0.3990}$.

3. Repeat Example 5.9.2 for the following design and risk conditions. A clearance less than $+0.0002$ in. is considered to be functionally more critical than a clearance greater than $+0.0012$ in. Thus, the following unequal probabilities are adopted: $\lambda = 0.00135 + 0.04365 = 0.045$ where $P(Y \leq +0.0002) = 0.00135$ and $P(Y \geq +0.0012) = 0.04365$. Compute the enlarged statistical component tolerances. *Ans.* 0.0009 in.

4. Determine the new mean clearance for Exercise 3 and write the new statistical component specifications. *Ans.* $\mu_Y = +0.00083$ in., $X_1\colon \frac{0.5000}{0.5009}$, $X_2\colon \frac{0.5001}{0.4992}$.

5. For the assembly given in Fig. 5.16, the specification for the slot width (part E) is 2.002/2.006 in. Clearances are to be from +0.002 to +0.010 in. Compute the enlarged statistical tolerances for the widths of parts *A*, *B*, *C*, and *D* assuming a risk $\lambda = 0.0456$ and equal tolerance allocation. Note that the 2.002/2.006 in. specification is fixed and will not be enlarged (i.e., only the widths of *A*, *B*, *C*, and *D* are treated as being random variables). *Ans.* 0.003 in.

6. Assume in Exercise 5 that the slot width (part E) is also considered to be a random variable. State the variance equation relating the assembly and component variances.

All of the Examples and Exercises to this point have involved mechanical assemblies with the assembly variable Y being a linear combination of the component variables X_j. Linearity is a requirement when combining independent random variables and making use of Eq. (5.10). The following example is presented to illustrate the statistical tolerancing method when it is possible to obtain a linear approximation for Y when Y is a nonlinear tolerance function.

EXAMPLE 5.9.3

Component specifications and the assembly requirement for a voltage-transformer-amplifier combination* are summarized in Fig. 5.31. This is a design situation of working from fixed component specifications and estimating cumulative error effect on the assembly characteristic (output voltage). The risk λ is 0.0027.

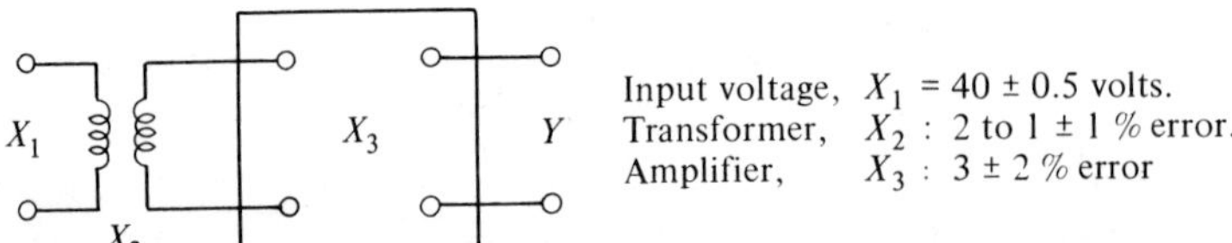

Figure 5.31. Example 5.9.3—voltage, transformer, amplifier combination.

The tolerance distributions are shown in Fig. 5.32. Per cent error has been translated to absolute error for the X_2-and X_3-distributions as follows: The specification limits for the X_2-distribution are

$$\mu_2 \pm 0.01\mu_2$$

and for the X_3-distribution are

$$\mu_3 + 0.02\mu_3$$

*The elements of this example have been taken from Johnson, R. H. (January 1953), "*How To Evaluate Assembly Tolerances*", Product Engineering, Vol. 24, No. 1, pp. 179–181.

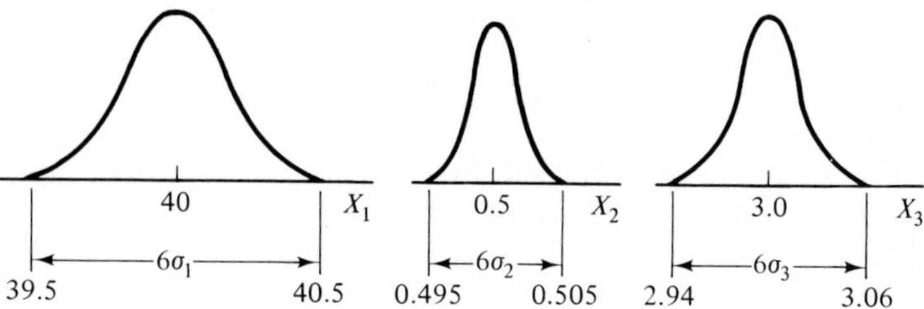

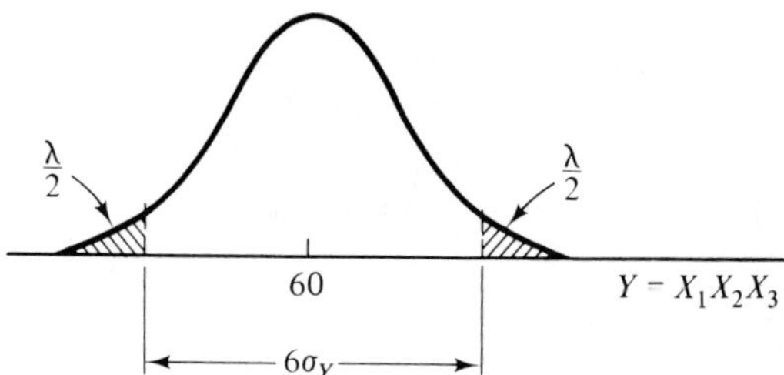

Figure 5.32. Example 5.9.3—distribution models.

The mean output voltage is

$$\begin{aligned}\mu_Y &= \mu_1\mu_2\mu_3 \\ &= 40(0.5)(3.0) \\ &= 60 \text{ V}\end{aligned}$$

The tolerance function $Y = X_1X_2X_3$ is nonlinear. Linearity is a requirement for combining independent random variables [see Eqs. (5.9) and (5.10)]. By the methods of the calculus,* the following linear approximation to Y is obtained:

$$Y \cong \mu_1\mu_2\mu_3 + \mu_2\mu_3(X_1 - \mu_1) + \mu_1\mu_3(X_2 - \mu_2) + \mu_1\mu_2(X_3 - \mu_3)$$

---

*A Taylor's series expansion of the function $Y = X_1X_2X_3$ in powers of $X_1 - \mu_1$, $X_2 - \mu_2$, and $X_3 - \mu_3$, neglecting higher order terms, is

$$\begin{aligned}Y &\cong \mu_1\mu_2\mu_3 + (X_1 - \mu_1)\frac{\partial Y}{\partial X_1}\bigg|_{\mu_1, \mu_2, \mu_3} \\ &\quad + (X_2 - \mu_2)\frac{\partial Y}{\partial X_2}\bigg|_{\mu_1, \mu_2, \mu_3} + (X_3 - \mu_3)\frac{\partial Y}{\partial X_3}\bigg|_{\mu_1, \mu_2, \mu_3} \\ &\cong \mu_1\mu_2\mu_3 + (X_1 - \mu_1)\mu_2\mu_3 + (X_2 - \mu_2)\mu_1\mu_3 + (X_3 - \mu_3)\mu_1\mu_2\end{aligned}$$

For the student who is unfamiliar with the Taylor's series expansion, this example is perhaps too advanced. However, he should appreciate that there are tolerance problems that involve nonlinear functions, and these problems cannot be handled directly using the methods of Sec. 5.9.

or

$$Y - \mu_1\mu_2\mu_3 \cong \mu_2\mu_3(X_1 - \mu_1) + \mu_1\mu_3(X_2 - \mu_2) + \mu_1\mu_2(X_3 - \mu_3)$$

which is linear in the independent variables $X_1 - \mu_1$, $X_2 - \mu_2$, and $X_3 - \mu_3$. The coefficients $\mu_2\mu_3$, $\mu_1\mu_3$, and $\mu_1\mu_2$ correspond to the constants a_1, a_2, and a_3 in the theorem for the combination of independent random variables. From Eq. (5.10),

$$\sigma_Y^2 \cong (\mu_2\mu_3)^2\sigma_1^2 + (\mu_1\mu_3)^2\sigma_2^2 + (\mu_1\mu_2)^2\sigma_3^2$$

$$\cong [0.5(3.0)]^2\left(\frac{1}{6}\right)^2 + [40(3.0)]^2\left(\frac{0.01}{6}\right)^2 + [40(0.5)]^2\left(\frac{0.12}{6}\right)^2$$

$$\sigma_Y \cong \tfrac{1}{6}(3.074)$$

and

$$6\sigma_Y \cong 3.074$$

Thus, probable assembly variation is

$$60 \pm \tfrac{1}{2}(3.074)\ \text{V}$$

or

$$60 \pm 1.537\ \text{V}$$

or

$$60\ \text{V} \pm 2.56\%\ \text{error}$$

The conclusion to Example 5.9.3 is summarized as follows. If component characteristics X_1, X_2, and X_3 are independent random variables, each normally distributed with mean average identical to the nominal specification value, then the probable assembly variation (99.73% of the time) is 60 volts ±2.56% error. Thus, it seems that the assembly requirement (output voltage; 60 V ±3% error) will be satisfied.

GLOSSARY OF TERMS

Unimodal
Normal distribution
Z-transformation
Standardized variable
Normalizing transformation
Central limit theorem
Sampling distribution of the mean
Tolerance
Maximum limit
Minimum limit
Clearance
Nominal specification value
Maximum material limit
Random assembly

GLOSSARY OF FORMULAS

		Page
(5.1)	Normal *D.F.*	120

$$f(X) = \frac{1}{\sigma_X\sqrt{2\pi}}e^{-(X-\mu)^2/2\sigma_X^2}$$

(5.2) Normal *D.F.* in *Z*-units. 124

$$f(Z) = \frac{1}{\sqrt{2\pi}}e^{-Z^2/2}$$

(5.3) Normal *C.D.F.* 124

$$F(Z) = \int f(Z)\,dZ$$

(5.4) Evaluation of *F*(*Z*). 125

$$\int_0^Z f(Z)\,dZ = 0.5 - \frac{e^{-Y^2}}{2Y\sqrt{\pi}}\left[1 - \frac{1}{2Y^2} + \frac{3}{4Y^4} - \frac{3\cdot 5}{8Y^6} + \ldots + (-1)^n T_{n+1} + \ldots\right]$$

$$T_{n+1} = \frac{1\cdot 3\cdot 5\ldots(2n-1)}{2^n Y^{2n}}$$

(5.5) Maximum error using the continuity correction. 131

$$\frac{0.140}{\sqrt{np(1-p)}}$$

(5.6) Normalizing transformations. 132

$$Y = \log_e X \quad \text{and} \quad Y = \sqrt{X}$$

(5.7) Variance, $\bar{x}$-distribution (continuous *X*) 134

$$\sigma_{\bar{x}}^2 = \frac{\sigma_x^2}{n}$$

(5.8) Variance, $\bar{x}$-distribution (discrete *X*). 134

$$\sigma_{\bar{x}}^2 = \frac{\sigma_x^2}{n}\left(\frac{N-n}{N-1}\right)$$

(5.9) Mean of a linear combination of independent random variables. 140

$$\mu_Y = a_1\mu_1 \pm a_2\mu_2 \pm \ldots \pm a_k\mu_k$$

(5.10) Variance of a linear combination of independent random variables. 140

$$\sigma_Y^2 = a_1^2\sigma_1^2 + a_2^2\sigma_2^2 + \ldots + a_k^2\sigma_k^2$$

(5.11) Normality property. 140

If the X_j are each normally distributed, then *Y* is normally distributed. If the X_j are not all normally distributed and if the *X*-variances are approximately homogeneous, then as *k* increases the *Y*-distribution approaches normality.

REFERENCES

BOWKER, A. H. AND LIEBERMAN, G. J. (1972), *Engineering Statistics*, Prentice-Hall, Inc., Englewood Cliffs, N. J., Chap. 3.

COCHRAN, W. G. (1953), *Sampling Techniques*, John Wiley & Sons, Inc., New York, p. 41.

GUTTMAN, I. AND WILKS, S. S. (1971), *Introductory Engineering Statistics*, John Wiley & Sons, Inc., New York, Chap. 7.

JOHNSON, R. H. (January 1953), "How To Evaluate Assembly Tolerances," *Product Engineering*, Vol. 24, No. 1, pp. 179–181.

KIRKPATRICK, E. G. (1970), *Quality Control for Managers and Engineers*, John Wiley & Sons, Inc., New York, Chaps. 4, 5, 7.

NAT. BUR. STANDARDS (August 1963), *Experimental Statistics*, Handbook 51, Supt. of Doc., U. S. Govt. Print. Office, Wash. 25, D. C., Chap. 20.

RAFF, M. S. (June 1956), "On Approximating the Point Binomial," *American Statistical Association*, Vol. 51, No. 274, pp. 293–303.

PROBLEMS

1. A random variable X is $N(6, 16)$. (a) State the value of σ_X. (b) Assume a Z-transformation on X. Give an equivalent statement describing the distribution of Z.

2. Assuming that X is $N(6, 16)$, compute the respective probabilities: (a) $P(X \geq 15.2)$, (b) $P(X \leq 2)$, (c) $P(X \geq 2)$, and (d) $P(2 \leq X \leq 15.2)$. Distinguish between $P(X \leq 2)$ and $P(X < 2)$. (See Sect. 3.4).

3. Breaking strength X of a new fabric is $N(140, 5)$ lb. A customer has an application for this fabric but is concerned with the possibility of breaking strength of some fabric pieces being as low as 132 lb. What is the probability $P(X \leq 132)$?

4. Assume in Problem 3 that the customer's specifications are 134 to 140 lb. What is the expected percentage of rejected fabric that fails to meet his specifications?

5. A size specification for a finished component diameter is 0.500 ± 0.003 in. For the manufacturing operation producing this finished diameter, this size characteristic X is normally distributed with mean 0.500 in. and standard deviation 0.002 in. What is the expected percentage of defective product?

6. Suppose in Problem 5 that the process shifts to a mean setting of 0.497 in. What is the expected percentage of: (a) oversize product, and (b) undersize product? What is the expected percentage of defective product if the mean process setting is 0.498 in.?

7. Process shifts are a concern in Problem 5. Oversize product is salvageable, undersize product is not. If a process control objective is a maximum of 10% undersize product, what is the minimum mean process setting?

8. The range of the individual X-variation in Problem 5 is ± 0.006 in. or 0.012 in., using $6\,\sigma_X$ as a measure of the range. Suppose that the process is monitored by samples of size $n = 4$ and the sample mean is a control criterion. What is the expected range (99.73% of the time) of mean values?

9. If the mean process setting is 0.500 in. in Problem 5, what is the probability $P(\bar{x} \geq 0.501)$ for a random sample of size 16? Determine the probability $P(0.4995 \leq \bar{x} \leq 0.5005)$.

10. Mean average life of an electron tube is $N(1{,}200, 25)$ hours. This information is based on life-test samples of size $n = 100$. What is the probability of an individual tube failing at or before 1,124 hours?

11. A size specification for a product characteristic is 0.250 ± 0.0015 in. The operation that produces this characteristic has been monitored by obtaining mean measurements of samples of 25 product items each. This information has yielded a standard deviation $\sigma_{\bar{x}} = 0.0002$ in. What is the probability of an individual output product item failing to meet specifications if the mean process setting is: (a) 0.249 in., and (b) 0.250 in.?

12. Measurement error for a certain laboratory test is $N(0, 0.0004^2)$ in. How many repeat measurements are required to assure a mean measurement error of 0 ± 0.0002 in. 95.44% of the time?

13. In Example 3.4.2, Rockwell hardness of a certain steel is uniformly distributed over an interval (40, 60) on a B-scale. If samples of 25 specimens each are taken from a shipment of this steel, what is the probability of obtaining a mean hardness value between 48.9 and 51.7?

14. A density function $f(X) = \frac{3}{2} - X,\ 0 \leq X \leq 1$ describes the probability of measurable amounts of a certain chemical in the atmosphere where X is in units of parts per million (ppm). If a sample of 11 random measurements is taken, what is the probability of obtaining a mean X ppm between $\frac{3}{12}$ and $\frac{6}{12}$? Use the expected value method to obtain σ_X, and express results as fractions to simplify computation.

15. Each observation of a work sampling study results in one of two outcomes —either there is, or is not, an avoidable delay. Let X denote the number of avoidable delays recorded for $n = 80$ observations, where the probability of an avoidable delay is $p = 0.3$ and is approximately constant from observation to observation. Using a normal model to approximate the binomial probability, compute the probability $P(19 \leq X \leq 24)$.

16. A sample of 200 parts is randomly selected from a manufacturing lot of 2,500 parts. The lot is 20% defective. Use a normal model to compute the probability of obtaining 30 to 35 defectives.

17. One implication of the central limit theorem is that, in sampling from even a non-normal X-population, distributions of sums or means approach normality. What condition is necessary to assure good normality of the sampling distribution of the mean?

18. What conditions are required for a normal model to yield a good approximation to a binomial probability?

19. Five independent, normally distributed, random variables $X_1, X_2, \ldots, X_5$ have respective means of 1, 2, 3, 4, and 5 and respective standard deviations 0.1, 0.1, 0.1, 0.2, and 0.3. The X_j are related by $Y = X_1 + X_2 + X_3 + X_4 + X_5$. Compute the probability $P(Y \geq 15.06)$.

20. Three components assemble adjacent to each other in an assembly space of fixed specification 0.600 ± 0.003 in. To assure conformance to the assembly requirement, the widths of the components are specified to be 0.200 ± 0.001, 0.185 ± 0.001, and 0.215 ± 0.001 in. Assuming random assembly, a risk $\lambda = 0.01242$, and equal tolerance allocation, determine the enlarged statistical component tolerances.

21. State the three principal conditions that the X_j-component variables must satisfy to assure that $100(1 - \lambda)\%$ of the assemblies in Problem 20 satisfy the assembly requirement.

22. Repeat Problem 20 assuming that it is economic to operate from a risk $\lambda = 0.3174$.

23. Assume a design situation of fixed component tolerances in Problem 20 (i.e., those stated) and compute the probable assembly variation 99.73% of the time.

24. A clearance condition $Y = X_1 - X_2$ is of interest (see Example 5.9.2). Design size is $\frac{5}{8}$ in. and required clearances are $+0.0003$ to $+0.0019$ in. Assume normally distributed independent random variables, random assembly, a risk $\lambda = 0.0456$, and equal tolerance allocation. Compute the enlarged statistical component tolerances.

25. Since $k = 2$ in Problem 24, what important condition must be satisfied by the two X_j-component variables?

26. Write the new statistical component specifications following from the result of Problem 24 assuming equal probabilities of tight and loose assembly fits.

27. A clearance condition $Y = X_1 - X_2$ is of interest. Current component specifications are: X_1: 0.5002 ± 0.0002, and X_2: 0.4993 ± 0.0002 in. Component tolerances are to be increased assuming that the X_j are normally distributed, independent, random variables. Further assumptions are random assembly, equal tolerance allocation, and the following risks: $P(\text{fits too tight}) = 0.0107$, and $P(\text{fits too loose}) = 0.0548$. Compute the enlarged statistical tolerances, and the new mean clearance, and write the new statistical component specifications.*

28. A product feature location is assured by two dimensions X_1 and X_2. Measure-

---

*Another implication of the central limit theorem is that, if normally distributed, independent, random variables X_j are linearly combined, the resulting random variable Y is normally distributed with mean $\mu_Y = \mu_1 \pm \mu_2 \pm \ldots \pm \mu_k$ and variance $\sigma_Y^2 = \sum_{j=1}^{k} \sigma_j^2$. Problems 19–27 are specification tolerance applications of these properties. Another principal application is that of measurement error (see Problems 28-33).

ment error for each dimension is $N(0, 0.0002^2)$ in. The respective measurement errors are considered to be independent random variables. What is the combined measurement error for the product feature location distance 99.73% of the time?

29. In manufacturing, consideration is given to both the manufacturing error from the process operation and the gaging error when inspecting the output from the process. Suppose that for a given product characteristic manufacturing error is $N(0.5000, 0.001^2)$ in. and gaging error is $N(0, 0.0002^2)$ in. What is the range of recorded product measurements 90% of the time?

30. The normal saturation level is approximately 5 parts of oxygen per million parts of water at the bottom of a large lake. During a period of thermal stratification the oxygen level decreases considerably. Assume that this level is $N(2, 0.5^2)$ ppm and that measurement error is $N(0, 0.02^2)$ ppm. What is the probability that a single measurement will exceed 3 ppm?

31. In Problem 30, what is the probability that the mean of 9 measurements will exceed 3 ppm?

32. The acceptable maximum level of nitrogen dioxide in urban air has been set at 2 ppm. This level varies over a year's time. Assume that during one three-month period the nitrogen dioxide level is $N(0.7, 0.02^2)$ ppm and that measurement error is $N(0, 0.001^2)$ ppm. What is the probability that a single measurement will exceed 0.73 ppm?

33. In Problem 32, what is the probability that the mean of 16 measurements will be between 0.69 and 0.73 ppm?

34. A product characteristic is $N(200, 40^2)$ lb. Measurement error is $N(0, 3^2)$ lb. What is the probability that the mean of 16 measurements is less than or equal to 175 lb.?

CONTINUOUS DISTRIBUTION MODELS

In Chap. 5, an important continuous distribution, the normal model, is described. In this chapter, other continuous distribution functions are presented, along with a number of random variables that occur in connection with hypothesis testing (Chap. 7). Generally, the presentation is limited to a definition of the D.F. *and* C.D.F. *with emphasis on how to enter and read tables for the purpose of making probability statements. Since these distributions depend on a parameter called "degrees of freedom," the first section is devoted to this concept.*

6.1 Degrees of Freedom

A number ν called the *number of degrees of freedom* (or *d.f.*) refers to the number of independent variables involved in a mathematical expression. Implied is that there are ν variables that can "freely" assume different values. A number of examples are presented before stating the statistical concept of degrees of freedom.

In the language of physics, a point that can move freely in three space has 3 *d.f.* Thus, a point (x, y, z) has 3 *d.f.* (it can move in three directions in

6

space), a point (x, y) has 2 *d.f.* (it can move freely in a plane), and a point (x) has 1 *d.f.* (it can move freely along a line). For each linear relationship (or constraint) imposed on the variables, one *d.f.* is lost. For example, consider two functions, $f_1(x, y, z)$ and $f_2(x, y, z)$. Each function represents a plane; in each case there are 2 *d.f.* Now, impose the constraint that a specified point must lie on both planes, i.e., it must lie on the line of intersection of the planes. A simultaneous solution of f_1 and f_2 defines the required line and there is $2 - 1 = 1$ *d.f.* One *d.f.* has been lost due to the one constraint that has been imposed.

A few more simple examples follow. Consider a random choice of a pair of numbers (x, y). The choice can be any point in the x, y plane. There are 2 *d.f.* Suppose that one constraint is imposed on the foregoing choice. For example, consider a random choice of a pair of numbers (x, y) whose sum is seven. Only one number can be freely chosen since the second number is fixed as soon as the first number is selected. Thus, there is $2 - 1 = 1$ *d.f.*

A common statistical example is that of the sample variance, which by definition is $s_x^2 = 1/n \sum (x_i - \bar{x})^2$. There are n independent observations x_i, but once the sample is taken, thus fixing $\bar{x}$, there are only $n - 1$ independent deviations $x_i - \bar{x}$ (i.e., $n - 1$ deviations can be independently considered but the nth deviation is fixed. Or, an alternative argument is that there is one linear constraint,

$$\sum_{i=1}^{n} x_i = n\bar{x}$$

and thus one *d.f.* has been lost and there are $n - 1$ *d.f.*

This example leads to an intuitive or computational definition of degrees of freedom which the student can use in all applications. The number of degrees of freedom of a statistic is the number of n independent observations in the sample minus the number m of sample parameters required to compute the statistic. For example, returning again to the sample variance, there is one sample parameter $\bar{x}$ which depends on the sample observations and is required for the computation of the statistic s_x^2. Therefore, there are $n - 1$ *d.f.* associated with the statistic s_x^2.

6.2 Chi-Square Distribution Model

If $X_1, X_2, \ldots, X_\nu$ are independent random variables, each $N(0, 1)$, then the statistic

$$\chi^2 = X_1^2 + X_2^2 + \ldots + X_\nu^2$$

is said to have a χ^2 (*chi-square*) distribution with ν *d.f.* The *D.F.* for the random variable χ^2 is

$$f(\chi^2) = \frac{1}{2^{\nu/2}\Gamma(\nu/2)}(\chi^2)^{(\nu-2)/2}e^{-\chi^2/2} \qquad \chi^2 \geq 0 \tag{6.1}$$

$$= 0 \quad \text{otherwise}$$

where $\mathbf{\Gamma}$ denotes the gamma function*

*To satisfy Eq. (3.4) to obtain a *D.F.*, it is necessary to define $\Gamma(\nu/2)$ such that

$$\int_0^\infty f(\chi^2)\,d\chi^2 = 1$$

Thus,

$$\Gamma\left(\frac{\nu}{2}\right) = \frac{1}{2^{\nu/2}}\int_0^\infty (\chi^2)^{(\nu-2)/2}e^{-\chi^2/2}\,d\chi^2$$

Letting $\chi^2/2 = Y$,

$$\Gamma\left(\frac{\nu}{2}\right) = \int_0^\infty Y^{(\nu-2)/2}e^{-Y}\,dY$$

and integrating by parts,

$$\Gamma\left(\frac{\nu}{2}\right) = \left(\frac{\nu}{2} - 1\right)\Gamma\left(\frac{\nu}{2} - 1\right)$$

and, by repetition,

The *C.D.F.* for the random variable χ^2 is

$$F(\chi^2) = \frac{1}{2^{\nu/2}\Gamma(\nu/2)} \int_0^{\chi^2} (\chi^2)^{(\nu-2)/2} e^{-\chi^2/2}\, d\chi^2 \tag{6.2}$$

where $\Gamma(\nu/2)$ is as defined above. Values of $F(\chi^2)$ are tabulated in Appendix *E*.

The *D.F.* and *C.D.F.* for $\nu = 4$ are shown in Fig. 6.1. The *D.F.* curves for $\nu = 2, 4, 6, 8,$ and 10 are given by Fig. 6.2.

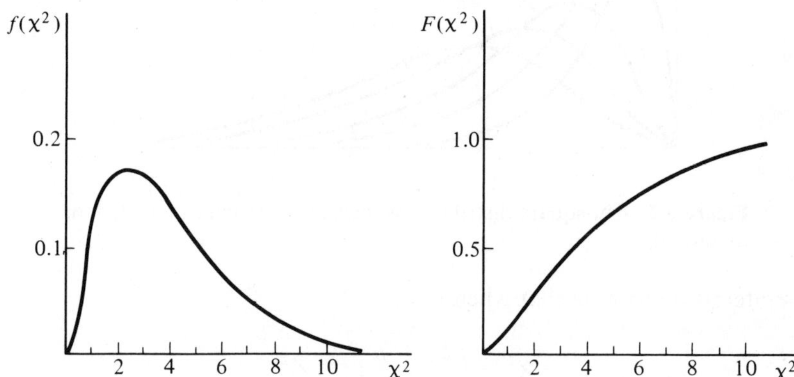

Figure 6.1. Chi-square distribution function for degrees of freedom $\nu = 4$.

$$\Gamma(2) = \begin{cases} \left(\frac{\nu}{2}-1\right)\left(\frac{\nu}{2}-2\right)\left(\frac{\nu}{2}-3\right)\cdots 3\cdot 2\cdot 1\Gamma(1) & \text{if } \nu \text{ is even} \\ \left(\frac{\nu}{2}-1\right)\left(\frac{\nu}{2}-2\right)\left(\frac{\nu}{2}-3\right)\cdots \frac{5}{2}\cdot\frac{3}{2}\cdot\frac{1}{2}\Gamma\left(\frac{1}{2}\right) & \text{if } \nu \text{ is odd} \end{cases}$$

Evaluating the coefficients $\Gamma(1)$ and $\Gamma(\frac{1}{2})$ in this expression,

$$\Gamma(1) = \int_0^\infty e^{-Y}\, dY = 1$$

and

$$\Gamma\left(\frac{1}{2}\right) = \int_0^\infty Y^{-1/2} e^{-Y}\, dY$$

Letting $Y = t^2/2$, $dY = t\, dt$, and

$$\frac{1}{\sqrt{Y}} = \frac{1}{t/\sqrt{2}} = \frac{\sqrt{2}}{t}$$

and thus

$$\Gamma\left(\frac{1}{2}\right) = \int_0^\infty \sqrt{2}\, e^{-t^2/2}\, dt$$

$$= \sqrt{\pi} \int_{-\infty}^\infty \frac{1}{\sqrt{2\pi}} e^{-t^2/2}\, dt = \sqrt{\pi}$$

Summarizing,

$$\nu \text{ even: } \Gamma\left(\frac{\nu}{2}\right) = \left(\frac{\nu}{2}-1\right)!$$

$$\nu \text{ odd: } \Gamma\left(\frac{\nu}{2}\right) = \left(\frac{\nu}{2}-1\right)\left(\frac{\nu}{2}-2\right)\cdots\frac{5}{2}\cdot\frac{3}{2}\cdot\frac{1}{2}\sqrt{\pi}$$

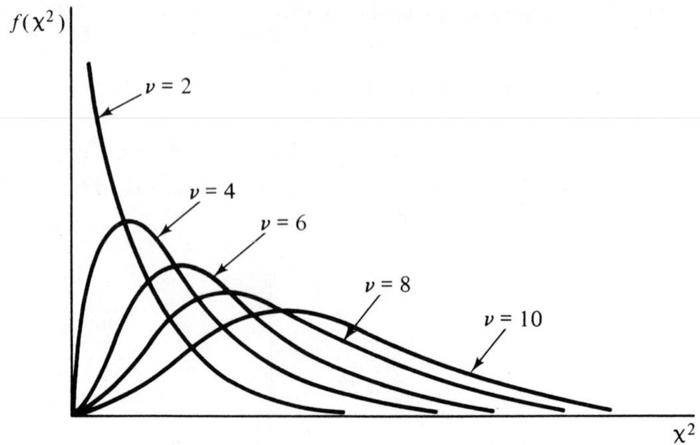

Figure 6.2. Chi-square distributions, degrees of freedom $\nu = 2, 4, 6, 8$, and 10.

It is interesting to note that when $\nu = 1$,

$$\chi^2 = \left(\frac{X_1 - \mu}{\sigma_X}\right)^2 = Z^2$$

where Z is $N(0, 1)$. Figure 6.3 shows the normal distribution in standard form and the χ^2 distribution for $\nu = 1$. The middle portion of the area beneath the normal *D.F.* (between $\pm 1Z$ or $\pm 1\sigma$) becomes the left tail of the χ^2 *D.F.* The tail areas of the normal distribution (outside of $\pm 2Z$) become the right tail of the χ^2 distribution.

Some properties of the chi-square model are summarized as follows:

1. There is an infinite number of χ^2 distributions, each one corresponding to a positive integer value for ν.
2. Since χ^2 is a sum of squared values, the range of the χ^2 distribution is $(0, \infty)$.
3. For $\nu > 30$, the χ^2 random variable is approximately normally distributed.
4. If X is a random variable $N(0, 1)$ and $x_1, x_2, \ldots, x_n$ are n observations of a random sample, then the sum

$$\sum_{i=1}^{n} \left(\frac{x_i - \mu}{\sigma_X}\right)^2$$

   is χ^2 distributed with n *d.f.*
5. If χ_1^2 and χ_2^2 are independent chi-square distributed random variables, with ν_1 and ν_2 *d.f.*, respectively, then the sum

$$\chi^2 = \chi_1^2 + \chi_2^2$$

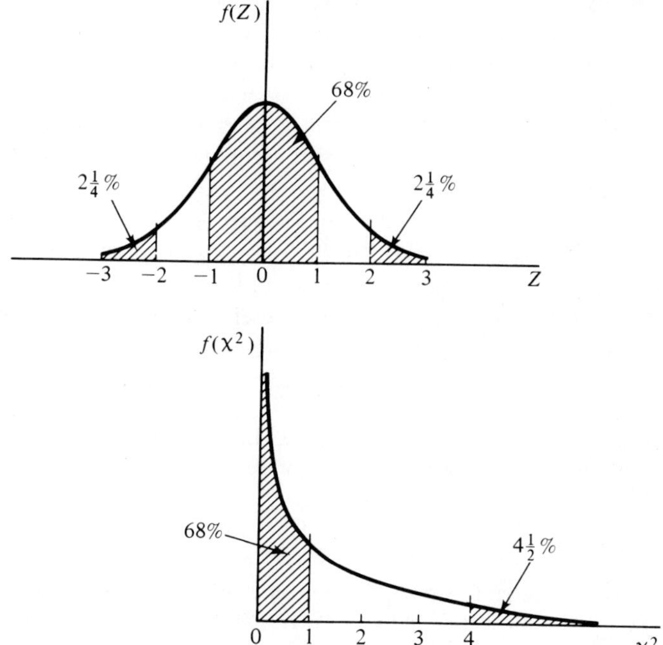

Figure 6.3. Normal distribution model in standard form and chi-square distribution model with degrees of freedom $\nu = 1$.

also has a chi-square distribution with $\nu = \nu_1 + \nu_2$ *d.f.* This is called the additive property of chi-square distributions.

6. The chi-square distribution is completely defined by the number of degrees of freedom. A choice of ν determines the *D.F.* Moreover, the mean is ν and the variance is 2ν*.

---

*Obtaining moments for continuous distribution models usually involves tedious integration and differentiation. One more derivation is given here only to indicate that the procedure is identical to that for discrete distribution models as shown in Chap. 4.

The characteristic function for the chi-square distribution is

$$\phi_{\chi^2}(t) = \int_0^\infty e^{i\chi^2 t} \cdot \frac{1}{2^{\nu/2}\Gamma(\nu/2)}(\chi^2)^{(\nu-2)/2} e^{-\chi^2/2}\, d\chi^2$$

$$= \int_0^\infty \frac{1}{2^{\nu/2}\Gamma(\nu/2)}(\chi^2)^{(\nu-2)/2} e^{-[(1-2it)\chi^2]/2}\, d\chi^2$$

Substitution of $Y = (1 - 2it)\chi^2$ makes possible an evaluation of the integral above. Then,

$$\phi_{\chi^2}(t) = \frac{1}{(1 - 2it)^{\nu/2}}$$

The first and second derivatives evaluated at $t = 0$ are

$$\phi'_{\chi^2}(t) = \frac{2i(\nu/2)}{(1 - 2it)^{(\nu/2)+1}} \quad \text{and} \quad \phi'(0) = i\nu$$

Many physical phenomena exhibit frequencies that follow approximately a chi-square distribution model. Such variables are usually discovered by experience and analysis of frequency distributions. For example, the distribution of life of certain vacuum tubes follows a chi-square distribution. Also, when location of hole patterns is given by x and y coordinates (a frequent practice in manufacturing), measurement errors follow approximately a chi-square distribution.

Other random variables have chi-square distributions because they satisfy the conditions of Eq. (6.1). In the following chapters, concern is more with this case. Certain statistical models are known to have chi-square distributions and for applications problems interest is primarily on the use of Appendix E tables to make probability statements.

EXAMPLE 6.2.1

Returning to the Example of Fig. 6.1, a chi-square distribution for $\nu = 4$ is shown in Fig. 6.4. The table in Appendix E reads

$$P(\chi^2 \geq \chi^2_{\alpha,\nu}) = 1.0 - \int_0^{\chi^2_{\alpha,\nu}} f(\chi^2)\,d\chi^2$$

and thus gives the probability α of obtaining a χ^2-value equal to or greater than a specified $\chi^2_{\alpha,\nu}$. The table is used in two ways: (1) Given α and ν, enter the table and

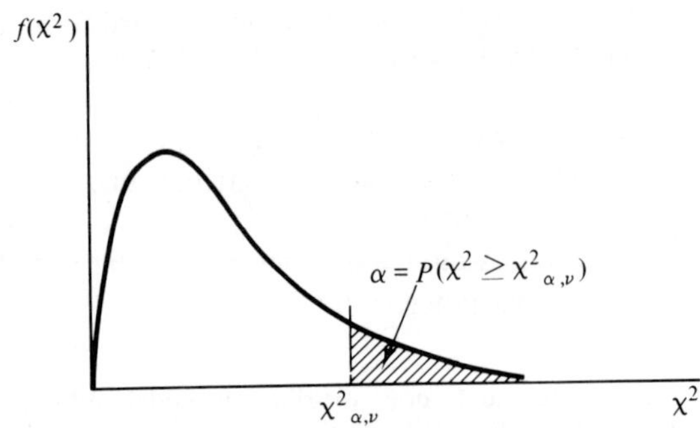

Figure 6.4. Probability from Appendix E table.

---

and

$$\phi''_{\chi^2}(t) = \frac{(2i)^2(\nu/2)[(\nu/2)+1]}{(1-2it)^{(\nu/2)+2}} \quad \text{and} \quad \phi''(0) = -\nu^2 - 2\nu$$

From Eq. (4.11),

$$\mu = E(\chi^2) = \frac{\phi'_{\chi^2}(0)}{i} = \frac{i\nu}{i} = \nu \tag{6.3}$$

and

$$E[(\chi^2)^2] = \frac{\phi''_{\chi^2}(0)}{i^2} = \frac{-\nu^2 - 2\nu}{-1} = \nu^2 + 2\nu$$

From Eq. (4.8),

$$\begin{aligned}\sigma^2_{\chi^2} &= E[(\chi^2)^2] - [E(\chi^2)]^2 \\ &= \nu^2 + 2\nu - \nu^2 = 2\nu\end{aligned} \tag{6.4}$$

read the value $\chi^2_{\alpha,\nu}$ that will be equaled or exceeded by chance alone 100α percent of the time, and (2) given ν and a specified $\chi^2_{\alpha,\nu}$, enter the table and read the probability α that χ^2-values will randomly equal or exceed the specified $\chi^2_{\alpha,\nu}$. Examples follow.

1. Given $\alpha = 0.05$ and $\nu = 4$, enter the $\nu = 4$ row and $\alpha = 0.05$ column of the table and read

$$P(\chi^2 \geq \chi^2_{0.05,4}) = 0.05$$

or

$$P(\chi^2 \geq 9.488) = 0.05$$

2. Given $\nu = 4$ and a χ^2-value of 5.4, explore the $\nu = 4$ row of the table for a value close to 5.4 and read

$$P(\chi^2 \geq 5.385) = 0.25$$

EXERCISES

1. If $\nu = 15$, what is the chi-square value that will be equaled or exceeded 10% of the time? *Ans.* 22.307.

2. Given $\nu = 11$ and $\chi^2 = 12.9$. What is the probability that χ^2-values will randomly equal or exceed 12.9? *Ans.* 0.30.

3. If $d.f. = 9$, a chi-square distribution is defined. Compute the mean and variance of this distribution. *Ans.* $\mu = 9$, $\sigma^2_{\chi^2} = 18$.

6.3 Student's *t*-Distribution Model

If X is a random variable $N(0, 1)$, and χ^2 is a random variable independent of X and chi-square distributed with ν *d.f.*, then the random variable

$$t = \frac{X\sqrt{\nu}}{\sqrt{\chi^2}} \quad \text{or} \quad t = \frac{X}{\sqrt{\chi^2/\nu}}$$

has a *t-distribution* with ν *d.f.**

The *D.F.* for the random variable t is

$$f(t) = \frac{1}{\sqrt{\pi\nu}} \cdot \frac{\Gamma[(\nu + 1)/2]}{\Gamma(\nu/2)} \left(1 + \frac{t^2}{\nu}\right)^{-(\nu+1)/2} \tag{6.5}$$

---

*This distribution was described in *Biometrica*, Vol. 6 (1908) by W. S. Gosset writing under the name of "Student." This model is now commonly called the *t-distribution.*

where

$$t = \frac{X}{\sqrt{\chi^2/\nu}} \qquad -\infty < t < +\infty$$

satisfies the conditions above and ν is an integer.

The *C.D.F.* for the random variable t is

$$F(t) = \frac{1}{\sqrt{\pi\nu}} \cdot \frac{\Gamma[(\nu + 1)/2]}{\Gamma(\nu/2)} \int_{-\infty}^{t} \left(1 + \frac{t^2}{\nu}\right)^{-(\nu+1)/2} dt \tag{6.6}$$

and values of $F(t)$ are tabulated in Appendix F.

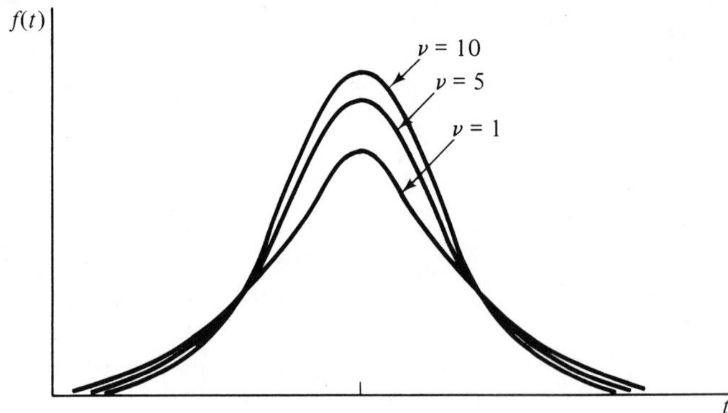

Figure 6.5. t distributions, degrees of freedom $\nu = 1$, 5, and 10.

The *D.F.*'s for $\nu = 1$, 5, and 10 are shown in Fig. 6.5. The t-distribution is symmetric about its mean, and has more area in its tails and is more peaked than the normal distribution. As the degrees of freedom ν increase, the t-distribution tends toward the standardized normal distribution. In fact, except for extreme values of α, values for $t_{\alpha,\nu}$ are approximately equal to Z_α values for the normal distribution whenever $\nu \geq 30$. The mean and variance of the t-distribution are

$$\begin{aligned} \mu &= E(t) = 0 \quad \text{for } \nu > 1 \\ \sigma_t^2 &= \frac{\nu}{\nu - 2} \qquad \text{for } \nu > 2 \end{aligned} \tag{6.7}$$

There are cases of physical phenomena whose frequencies follow approximately the t-distribution. However, the t-probability function is difficult to work with and simpler models are usually used. The most important applications of the t-distribution model occur in connection with random variables that satisfy the conditions of Eq. (6.5), i.e., a random variable that is a ratio, its numerator being a random variable X that is $N(0, 1)$ and its denominator being the root of a chi-square distributed random variable, independent of X, and divided by its *d.f.*

EXAMPLE 6.3.1

The table in Appendix F gives the probability α of obtaining a t-value equal to or greater than a specified $t_{\alpha,\nu}$. The table reads

$$P(t \geq t_{\alpha,\nu}) = 0.5 - \int_0^{t_{\alpha,\nu}} f(t)\,dt$$

as shown in Fig. 6.6. The distribution is symmetric and therefore the left-hand tail areas are identical to the right-hand tail areas designated by the table.

The table is read in the same manner as that for the chi-square distribution. For example, entering the table with a given $\nu = 4$ and $\alpha = 0.05$, the value 2.132 is obtained, and thus

$$P(t \geq t_{0.05,4}) = 0.05$$

or

$$P(t \geq 2.132) = 0.05$$

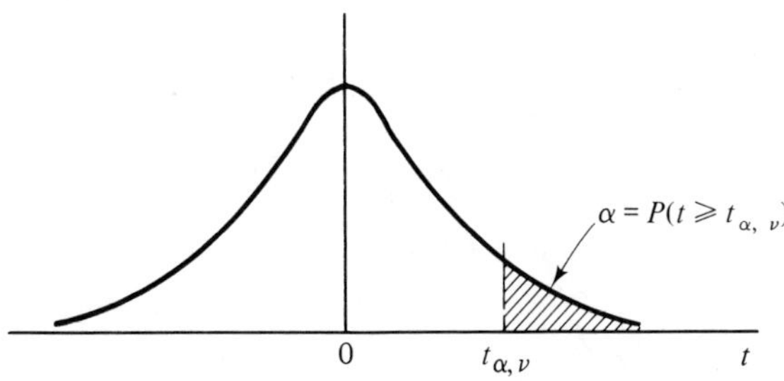

Figure 6.6. Probability from Appendix F table.

EXERCISES

1. If $\nu = 15$, what is the t-value that will be equaled or exceeded 10% of the time? *Ans.* 1.341.

2. Given $\nu = 11$ and $t = 0.54$. What is the probability that t-values will randomly equal or exceed 0.54? *Ans.* 0.30.

6.4 *F*-Distribution Model

If χ_1^2 and χ_2^2 are independent chi-square distributed random variables with ν_1 and ν_2 *d.f.*, respectively, then the random variable

$$F = \frac{\chi_1^2/\nu_1}{\chi_2^2/\nu_2}$$

has an *F-distribution* with ν_1 and ν_2 *d.f.* The *D.F.* for the random variable F is

$$f(F) = \frac{\Gamma[(\nu_1 + \nu_2)/2]}{\Gamma(\nu_1/2)\Gamma(\nu_2/2)}\left(\frac{\nu_1}{\nu_2}\right)^{\nu_1/2} \frac{F^{(\nu_1/2)-1}}{[1 - (\nu_1 F/\nu_2)]^{(\nu_1+\nu_2)/2}} \quad \text{for } F > 0 \qquad (6.8)$$
$$= 0 \quad \text{otherwise}$$

where

$$F = \frac{\chi_1^2/\nu_1}{\chi_2^2/\nu_2}$$

The *C.D.F.* for the random variable F

$$F(F) = \int_0^F f(F)\, dF$$

is tabulated in Appendix G. Density functions for $\nu_1 = 15$, $\nu_2 = 5$, 50, and ∞ are shown in Fig. 6.7.

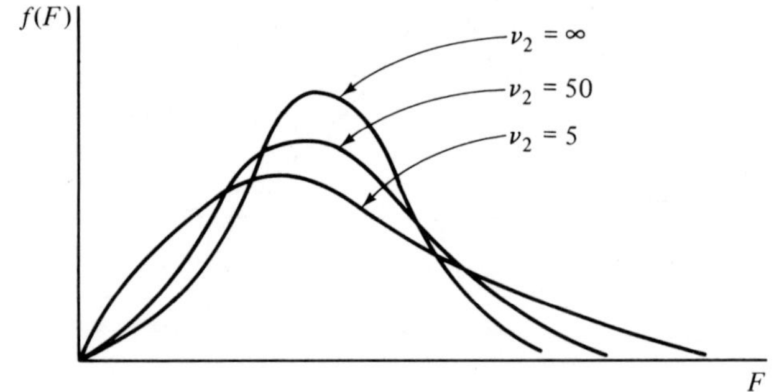

Figure 6.7. F distributions, degrees of freedom $\nu_1 = 15$ and $\nu_2 = 5$, 50, and ∞.

The mean and variance of the F-distribution are given by

$$\mu = E(F) = \frac{\nu_2}{\nu_2 - 2} \quad \text{for } \nu_2 > 2$$
$$\sigma_F^2 = \frac{2\nu_2^2(\nu_1 + \nu_2 - 2)}{\nu_1(\nu_2 - 2)^2(\nu_2 - 4)} \quad \text{for } \nu_2 > 4 \qquad (6.9)$$

As in the case of the t-distribution model, there are physical phenomena whose frequencies follow approximately the F-distribution, but the most important applications of this distribution occur in connection with random variables that satisfy the conditions of Eq. (6.8), i.e., a random variable that is a ratio of two independent chi-square distributed random variables, each divided by its *d.f.*

EXAMPLE 6.4.1

The table in Appendix G gives the probability α of obtaining an F-value equal to or greater than a specified F_{α,ν_1,ν_2}. The table reads

$$P(F \geq F_{\alpha,\nu_1,\nu_2}) = 1.0 - \int_0^{F_{\alpha,\nu_1,\nu_2}} f(F)\,dF$$

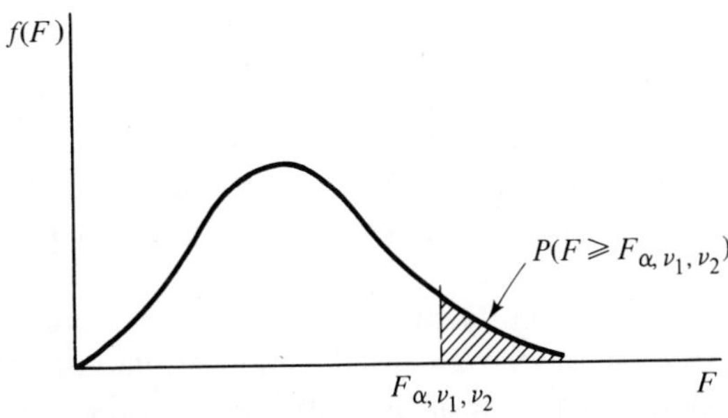

Figure 6.8. Probability from Appendix G table.

as shown in Fig. 6.8. To illustrate reading the table,* for given $\alpha = 0.05$, $\nu_1 = 8$, and $\nu_2 = 6$

$$P(F \geq F_{0.05,8,6}) = 0.05$$

or

$$P(F \geq 4.15) = 0.05$$

F tables usually found in textbooks give values only for $\alpha \leq 0.10$ and this is adequate for hypothesis testing (Chap. 7). If F-values for $1 - \alpha$ are desired, however, use can be made of the relationship

$$F_{1-\alpha\ \nu_1,\nu_2} = \frac{1}{F_{\alpha,\nu_2,\nu_1}} \tag{6.10}$$

EXAMPLE 6.4.2

Given $\nu_1 = 6$, $\nu_2 = 8$, and $\alpha = 0.05$, $F_{0.95\ 6,8}$ is obtained as follows:

$$F_{0.95,6,8} = \frac{1}{F_{0.05,8,6}}$$

$$= \frac{1}{4.15} = 0.241$$

*The essential table differences are as follows. The body of the normal table gives probabilities α, the columns of the chi-square and t-distribution tables are for specified probabilities α, and each page of the F-distribution table is for a specified probability α.

EXERCISES

1. Given $\nu_1 = 10$ and $\nu_2 = 14$, what is the F-value that will be equaled or exceeded 5% of the time? *Ans.* 2.60.

2. If $\nu_1 = 10$ and $\nu_2 = 14$, what is the value of $F_{0.95,10,14}$? *Ans.* 0.350.

6.5 Distribution of $\bar{x}$

Certain random variables are of interest in testing hypotheses (Chap. 7). These variables follow the distributions introduced in Chaps. 5 and 6: normal, chi-square, t-distribution, and F-distribution. The purpose of this and following sections is to identify the appropriate distribution model and to generally relate the random variable to the applications context in which it is to be used. The first case is that of the distribution of the sample mean.

In many applications, it is more efficient to deal with averages than with individual observations. For example, a process machine is monitored to assure that it is properly set to assure output satisfying the specifications. At regular time intervals a sample of the production output is taken and the observations $x_1, x_2, \ldots, x_n$ are averaged to yield a sample mean average $\bar{x}$. Since only individual observations can be compared directly with specifications, the question arises as to what criterion is the $\bar{x}$-value to be measured against. The statistical criterion that is used is the probability of obtaining such an $\bar{x}$-value.

In Sec. 5.7, it was observed that this probability is obtainable from knowledge of the distribution of the mean $\bar{x}$. If all possible samples of a specified size n are taken from a distribution that is $N(\mu, \sigma_X^2)$, a distribution of mean values $\bar{x}_j$ is generated and this distribution is $N(\mu, \sigma_{\bar{x}}^2)$. Equations (5.7) and (5.8) define the relationship between the variances σ_X^2 and $\sigma_{\bar{x}}^2$. Further, if X is not normally distributed, using a sufficiently large sample assures the normality of the $\bar{x}$-distribution.

EXAMPLE 6.5.1

Given that a process is, in fact, centered at a mean of 0.500 in. and its standard deviation is 0.003 in., what is the probability of obtaining a mean average output equal to or greater than 0.5015 in. if sample size is $n = 9$?

A summary of the solution is given in Fig. 6.9. The required Z-value is

$$Z_{0.5015} = \frac{0.5015 - 0.500}{0.003/\sqrt{9}} = 1.5$$

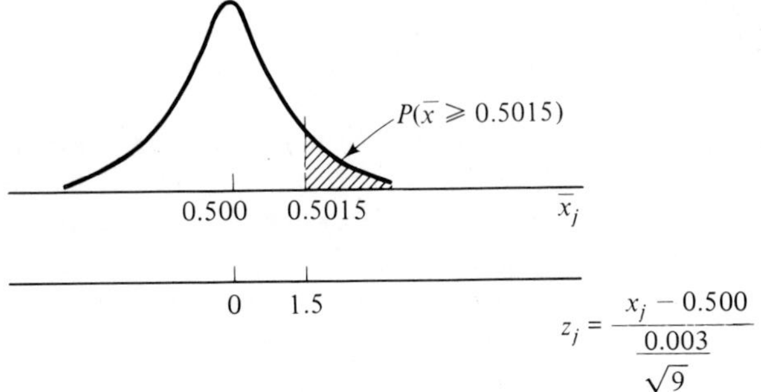

Figure 6.9. Example 6.5.1—normal model.

and the probability is

$$\begin{aligned} P(\bar{x} \geq 0.5015) &= P(Z \geq 1.5) \\ &= 0.0668 \quad \text{(from Appendix D)} \end{aligned}$$

A solution of Example 6.5.1 involves the standardized normal distribution of $\bar{x}$, that is, the distribution of

$$Z_j = \frac{\bar{x}_j - \mu}{\sigma_{\bar{x}}} = \frac{\bar{x}_j - \mu}{\sigma_X/\sqrt{n}}$$

which is a random variable (i.e., X is a random variable and a function of a random variable is itself a random variable). Thus, a first observation of importance for use in Chap. 7 is that the random variable

$$Z_j = \frac{\bar{x}_j - \mu}{\sigma_X/\sqrt{n}} \quad \text{is } N(0, 1) \tag{6.11}$$

In applications, the population standard deviation σ_X is seldom known, and it is natural to replace σ_X by its estimate, s_x, in the expression for the random variable in Eq. (6.11). It can be shown that the resulting random variable

$$t_j = \frac{\bar{x}_j - \mu}{s_x/\sqrt{n}} \quad \text{is } t\text{-distributed with } n - 1 \ d.f.^* \tag{6.12}$$

---

*An intuitive argument for this result is given here to again draw attention to the conditions underlying the t-distribution. From Eq. (6.11),

$$\frac{\bar{x}_j - \mu}{\sigma_X/\sqrt{n}}$$

is $N(0, 1)$, and it can be shown that the random variable

$$\frac{(n-1)s_X^2}{\sigma_X^2}$$

6.6 Distributions Related to s_x^2

Frequently, industrial experimentation is concerned with establishing the variation of physical characteristics such as life and quality of product, performance characteristics, strength of materials, and so forth. Decisions are made with reference to populations, i.e., all of the product to be sold, all of the material received on a shipment, etc. Because of time and money constraints (the expense of inspection and test) or physical limitations (testing may be destructive), the population variation is estimated from information supplied by small random samples. Basically, this involves an estimated comparison of sample variance s_x^2 and population variance σ_X^2 (standard set by specifications).

Comparisons can be made by differences $(s_x^2 - \sigma_X^2)$ or ratios (s_x^2/σ_X^2) and distributions of these measures. A common measure with a known distribution function is the ratio $(n-1)s_x^2/\sigma_X^2$. If $X_1, X_2, \ldots, X_n$ are independent normally distributed random variables, each having mean μ and variance σ_X^2, the random variable

$$\frac{(n-1)s_x^2}{\sigma_X^2} \quad \text{is chi-square distributed with } n-1 \; d.f.^* \qquad (6.13)$$

A more sophisticated comparison is that of the variation of two populations (e.g., two manufacturing processes, two types of material, etc.) where the variances of both populations are unknown and thus necessarily estimated from their respective sample variances. A useful measure for this purpose, with a known distribution function, is the random variable

$$\frac{s_x^2/\sigma_X^2}{s_y^2/\sigma_Y^2} \quad \text{which is } F\text{-distributed with } n_x - 1 \text{ and } n_y - 1 \; d.f. \qquad (6.14)$$

where X and Y refer to the two distinct populations being considered.

---

is chi-square distributed with $n-1$ *d.f.* Then, the ratio

$$\frac{\dfrac{\bar{x}_j - \mu}{\sigma_X/\sqrt{n}}}{\sqrt{\dfrac{(n-1)s_X^2}{\sigma_X^2/(n-1)}}} = \frac{\bar{x}_j - \mu}{s_X\sqrt{n}}$$

is t-distributed with $n-1$ *d.f.* since it satisfies the conditions for a t random variable, i.e., a random variable that is a ratio, its numerator being a random variable that is $N(0, 1)$ and the denominator being a chi-square distributed random variable divided by its d.f.

*This means all possible samples of a specified and fixed size n, from a normal population, are being considered. A variance s_X^2 is computed for each sample. Thus, a distribution of s_X^2-values can be obtained, or a distribution of the random variable $(n-1)s_X^2/\sigma_X^2$ can also be generated. This latter distribution is known to be chi-square distributed.

EXERCISES*

1. Given a process setup such that the mean product output is 0.750 in. and the standard deviation σ_X is known to be 0.002 in. What is the probability of obtaining a mean sample output equal to or greater than 0.752 in. if sample size $n = 4$? *Ans.* 0.0228.

2. Repeat Exercise 1 assuming that σ_X is not known and the sample standard deviation $s_x = 0.002$ in. is used as an estimate. *Ans.* 0.0747.

3. Boxes are automatically filled by a machine and the standard deviation for the operation is known to be $\sigma_X = 0.05$ lb. A control sample of 21 observations from the operation yields a standard deviation $s_x = 0.06$. What is the approximate probability of occurrence of a value for the random variable $(n - 1)s_x^2/\sigma_X^2$ equal to or greater than that obtained? *Ans.* 0.0935.

4. In Exercise 3, two different machines X and Y are being considered for the same job. Random samples of 30 observations from test runs of each machine yield s_x^2- and s_y^2-values. Assuming there is, in fact, no variation difference between the machines ($\sigma_X = \sigma_Y$), how large a ratio s_x^2/s_y^2 would be equaled or exceeded only 10% of the time? *Ans.* 1.625.

6.7 Distributions of $\bar{x} - \bar{y}$

Industrial experimentation often takes the form of comparing two treatments. A sample of n_x items is taken from a population of items receiving treatment X and a sample of n_y items from a population of items having treatment Y. Decisions are made based on the magnitude of the difference $\bar{x} - \bar{y}$ of the sample means. For example, Table 6.1 gives breaking strengths of 0.104-in. hard-drawn copper wire. Sample X is from a population of output from a new and presumably superior process, and sample Y is from a population of output associated with an existing process. If $\bar{x} > \bar{y}$ and the difference $|\bar{x} - \bar{y}|$ is significantly large, it may be concluded that process X is indeed superior. How large a difference is significant depends on the distribution of $\bar{x}_j - \bar{y}_j$.

If X is a random variable $N(\mu_X, \sigma_X^2)$ and Y is a random variable independent of X and $N(\mu_Y, \sigma_Y^2)$, and $\bar{x}_j$ refers to the mean values obtainable from all possible samples of size n_x from the X-population, and $\bar{y}_j$ to the mean values obtainable from all possible samples of size n_y from the Y-population,

*The student should not view these exercises as being decision-making methods. The questions here are academic and the purpose is practice in use of distribution models and associated tables. Decision-making tools, based on these models, are developed in Chap. 7.

Table 6.1. OBSERVATIONS OF BREAKING STRENGTH OF 0.104 IN.-HARD-DRAWN COPPER WIRE. RANDOM SAMPLES OF 20 OBSERVATIONS EACH FROM A NEW PROCESS X AND AN EXISTING PROCESS Y YIELD MEANS AND VARIANCES AS SHOWN.

X	Y
560	559
565	563
.	.
.	.
.	.
575	574
$n_x = 20$	$n_y = 20$
$\bar{x} = 565$	$\bar{y} = 563$
$s_x = 4.9$	$s_y = 5.1$

then the random variable

$$\frac{(\bar{x}_j - \bar{y}_j) - (\mu_X - \mu_Y)}{\sqrt{(\sigma_X^2/n_x) + (\sigma_Y^2/n_y)}} \text{ is } N(0, 1)^* \tag{6.15}$$

*It should be noted that $\bar{x}_j$ and $\bar{y}_j$ distributions are generated as shown in Fig. 6.10 (see Sec. 5.9) and that all possible linear combinations $\bar{x}_j - \bar{y}_j$ generate a distribution of the random variable $\bar{x}_j - \bar{y}_j$ also shown in Fig. 6.10.

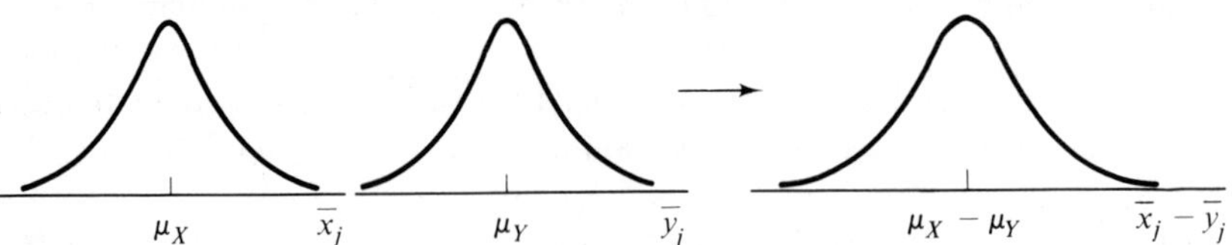

Figure 6.10. Distribution of $\bar{x}_j - \bar{y}_j$, normal model.

The latter distribution is also normal with mean $\mu_X - \mu_Y$ and variance $\sigma^2_{(\bar{x}-\bar{y})}$. To standardize this normal random variable, a Z-transformation is necessary:

$$Z_j = \frac{(\bar{x}_j - \bar{y}_j) - (\mu_X - \mu_Y)}{\sigma_{(\bar{x}-\bar{y})}}$$

However, from Eq. (5.10),

$$\sigma^2_{(\bar{x}-\bar{y})} = \sigma^2_{\bar{x}} + \sigma^2_{\bar{y}}$$

or, from Eq. (5.7),

$$\sigma^2_{(\bar{x}-\bar{y})} = \frac{\sigma_X^2}{n_x} + \frac{\sigma_Y^2}{n_y}$$

and thus the Z-transformation is

$$Z_j = \frac{(\bar{x}_j - \bar{y}_j) - (\mu_X - \mu_Y)}{\sqrt{(\sigma_X^2/n_x) + (\sigma_Y^2/n_y)}}$$

which is the random variable, Eq. (6.15).

If $\sigma_X = \sigma_Y$ in Eq. (6.15) and the variances are factored out of the denominator expression, then the random variable

$$\frac{(\bar{x}_j - \bar{y}_j) - (\mu_X - \mu_Y)}{\sigma\sqrt{(1/n_x) + (1/n_y)}} \text{ is } N(0, 1) \tag{6.16}$$

where σ denotes the identical standard deviations.

The population variances are usually not known in practical applications and, if this is the case, experimentation based on the difference $\bar{x} - \bar{y}$ makes use of the random variable

$$\frac{(\bar{x}_j - \bar{y}_j) - (\mu_X - \mu_Y)}{\sqrt{(1/n_x) + (1/n_y)}\sqrt{[1/(n_x + n_y - 2)][(n_x - 1)s_x^2 + (n_y - 1)s_y^2]}} \tag{6.17}$$

which is *t*-distributed with $n_x + n_y - 2$ *d.f.*

EXERCISES

1. A random variable X is $N(10, 16)$ and another random variable Y is $N(6, 9)$. Samples of nine observations each from these populations yield $\bar{x} = 11$ and $\bar{y} = 5$. What is the probability of obtaining a difference $\bar{x} - \bar{y}$ equal to or larger than the observed difference of 6? *Ans.* 0.1151.

2. Repeat Exercise 1 assuming that the population variances are unknown and their estimates are $s_x^2 = 16$ and $s_y^2 = 9$. *Ans.* 0.129.

GLOSSARY OF TERMS

Degrees of freedom
Chi-square random variable
t random variable
F random variable

GLOSSARY OF FORMULAS

Page

(6.1) Chi-square *D.F.* 158

$$f(\chi^2) = \frac{1}{2^{\nu/2}\Gamma(\nu/2)}(\chi^2)^{(\nu-2)/2}e^{-\chi^2/2} \quad \chi^2 \geq 0$$

(6.2) Chi-square *C.D.F.* 159

$$F(\chi^2) = \frac{1}{2^{\nu/2}\Gamma(\nu/2)}\int_0^{\chi^2}(\chi^2)^{(\nu-2)/2}e^{-\chi^2/2}\,d\chi^2$$

(6.3) Chi-square distribution mean. 162

$\mu = E(\chi^2) = \nu$

(6.4) Chi-square distribution variance. 162

$\sigma^2_{\chi^2} = 2\nu$

(6.5) *t*-density function. 163

$$f(t) = \frac{1}{\sqrt{\pi\nu}} \cdot \frac{\Gamma[(\nu +)1/2]}{\Gamma(\nu/2)} \left(1 + \frac{t^2}{\nu}\right)^{-(\nu+1)/2}$$

(6.6) *t*-cumulative distribution function. 164

$$F(t) = \frac{1}{\sqrt{\pi\nu}} \cdot \frac{\Gamma[(\nu + 1)/2]}{\Gamma(\nu/2)} \int_{-\infty}^{t} \left(1 + \frac{t^2}{\nu}\right)^{-(\nu+1)/2} dt$$

(6.7) *t*-distribution moments. 164

$$\mu = E(t) = 0 \quad \text{for } \nu > 1, \qquad \sigma_t^2 = \frac{\nu}{\nu - 2} \quad \text{for } \nu > 2$$

(6.8) *F*-density function. 166

$$f(F) = \frac{\Gamma[(\nu_1 + \nu_2)/2]}{\Gamma(\nu_1/2)\Gamma(\nu_2/2)} \left(\frac{\nu_1}{\nu_2}\right)^{\nu_1/2} \frac{F^{(\nu_1/2)-1}}{[1 + (\nu_1 F/\nu_2)]^{(\nu_1+\nu_2)/2}}$$

for $F > 0$

(6.9) *F*-distribution moments. 166

$$\mu = E(F) = \frac{\nu_2}{\nu_2 - 2} \quad \text{for } \nu_2 > 2$$

$$\sigma_F^2 = \frac{2\nu_2^2(\nu_1 + \nu_2 - 2)}{\nu_1(\nu_2 - 2)^2(\nu_2 - 4)} \quad \text{for } \nu_2 > 4$$

(6.10) Relationship to obtain $F_{1-\alpha}$ from tabled values of F_α. 167

$$F_{1-\alpha, \nu_1, \nu_2} = \frac{1}{F_{\alpha, \nu_2, \nu_1}}$$

Some random variables and distributions.

(6.11) $Z_j = \dfrac{\bar{x}_j - \mu}{\sigma_x\sqrt{n}}$ is $N(0, 1)$ 169

(6.12) $t_j = \dfrac{\bar{x}_j - \mu}{s_x/\sqrt{n}}$ is *t*-distributed with $n - 1$ *d.f.* 169

(6.13) $\dfrac{(n - 1)s_x^2}{\sigma_X^2}$ is chi-square distributed with $n - 1$ *d.f.* 170

(6.14) $\dfrac{s_x^2/\sigma_X^2}{s_y^2/\sigma_Y^2}$ is *F*-distributed with $n_x - 1$ and $n_y - 1$ *d.f.* 170

(6.15) $\dfrac{(\bar{x}_j - \bar{y}_j) - (\mu_X - \mu_Y)}{\sqrt{(\sigma_X^2/n_x) + (\sigma_Y^2/n_y)}}$ is $N(0, 1)$ 172

(6.16) $\dfrac{(\bar{x}_j - \bar{y}_j) - (\mu_X - \mu_Y)}{\sigma\sqrt{(1/n_x) + (1/n_y)}}$ is $N(0, 1)$ 173

where $\sigma = \sigma_X = \sigma_Y$.

(6.17) $$\frac{(\bar{x}_j - \bar{y}_j) - (\mu_X - \mu_Y)}{\sqrt{(1/n_x) + (1/n_y)}\sqrt{[1/(n_x + n_y - 2)][(n_x - 1)s_x^2 + (n_y - 1)s_y^2]}}$$ 173

is t-distributed with $n_x + n_y - 2$ *d.f.*

REFERENCES

CHOU, YA-LUN (1969), *Statistical Analysis*, Holt, Rinehart & Winston, Inc., New York, Chap. 12.

CRAMER, H. (1946), *Mathematical Methods of Statistics*, Princeton University Press, Princeton, N. J., Chap. 18.

VOLK, W. (1969), *Applied Statistics for Engineers*, 2nd ed., McGraw-Hill Book Company, New York, Chaps. 5, 6.

WINE, R. L. (1964), *Statistics for Scientists and Engineers*, Prentice-Hall, Inc., Englewood Cliffs, N. J., Chaps. 7–9.

PROBLEMS

1. For each linear relationship (or constraint) imposed on the variables in a mathematical expression, one degree of freedom (*d.f.*) is lost. (a) Randomly choose a pair of numbers (X, Y). Specify the associated *d.f.*, (b) suppose that the selection of (X, Y) must satisfy $X + Y = 10$. Specify the *d.f.*, and (c) suppose that in addition to the condition imposed in (b) that the numbers must also satisfy $X^2 + Y^2 = 25$. Specify the *d.f.* Give a geometric explanation of (c).

2. Why are there $n - 1$ *d.f.* associated with the sample variance?

3. A statistic of interest in Chap. 10 is the correlation coefficient r where

$$r = \frac{1/n \sum (x - \bar{x})(y - \bar{y})}{\sqrt{1/(n-1) \sum (x - \bar{x})^2}\sqrt{1/(n-1) \sum (y - \bar{y})^2}}$$

Specify the *d.f.* associated with this statistic.

4. A line $y = a_0 + a_1x + a_2x^2$ is fitted to n observed points (x, y) and thus a_0, a_1, and a_2 depend on the observations. Specify the *d.f.* associated with the statistic y.

5. Certain physical phenomena follow approximately a chi-square distribution. For example, let X_1 and X_2 designate the respective measurement errors associated with the abscissa and ordinate components of the distance between two points P and P' in the XY plane. If X_1 and X_2 are independently and normally distributed as follows: X_1 is $N(0, 0.0002^2)$, and X_2 is $N(0, 0.0001^2)$, then $X_1^2/0.0002^2 + X_2^2/0.0001^2$ is chi-square distributed with 2 *d.f.* What is the probability that the square of the error in the measured distance PP' is greater than 2.408 units, i.e., that $X_1^2/0.0002^2 + X_2^2/0.0001^2 > 2.408$? (The meaning of normally distributed "with mean zero" here is that + errors and − errors are equally likely.)

6. Of more interest for subsequent chapters, however, is the fact that certain statistics have chi-square distributions. For example, a statistic of importance in hypothesis testing (Chap. 7) is $(n-1)s_X^2/\sigma_X^2$, where s_X^2 is sample variance, σ_X^2 is population variance, and n is sample size. A manufacturer is concerned about the variation in weight of 5 lb boxes filled with product by an automatic machine. If the population variance (all of the boxes) is less than or equal to 0.05^2 lb, the filling operation is considered to be satisfactory. A random sample of 20 boxes from the filling operation yields a variance $s_X^2 = 0.07^2$ lb. To determine if the sample s_X^2 value is representative of an unsatisfactory population (variance too large) or is simply not a representative sample variance from a satisfactory population (variance $\leq 0.05^2$), it is necessary to determine the probability of obtaining a sample variance as large as 0.07^2 from a population whose variance is 0.05^2 lb. An efficient indirect method of doing this is to determine the probability of obtaining a value for $(n-1)s_X^2/\sigma_X^2$ as large as the value of this statistic for the random sample in question. Determine this probability.
7. Determine for Problem 1.14 (Chap. 1) the probability of obtaining a value of the statistic $(n-1)s_X^2/\sigma_X^2$ as large as that for the 25 observations of this problem, if the sample is from a population whose variance is 0.103 ppm.
8. For the conditions of Problem 1.14 ($d.f. = 25 - 1$), what is the chisquare value that is randomly equaled or exceeded 10% of the time?
9. For a chi-square distributed random variable with 6 *d.f.*, what is the probability: (a) $P(\chi^2 \geq 16.81)$, (b) $P(\chi^2 \leq 0.87)$, (c) $P(\chi^2 \geq 14)$, and (d) $P(12.59 \leq \chi^2 \leq 16.81)$?
10. If a chi-square distributed random variable has 25 *d.f.*, determine the value c for each of the following: (a) $P(\chi^2 \geq c) = 0.01$, and (b) $P(\chi^2 \leq c) = 0.80$.
11. If χ_1^2 is a chi-square distributed random variable with 5 *d.f.* and χ_2^2 a chi-square distributed random variable with 10 *d.f.*, and χ_1^2 and χ_2^2 are independent, what is the probability that $\chi_1^2 + \chi_2^2$ exceeds 24?
12. An experiment is concerned with the mean breaking strength μ of the manufactured output of 0.104 in. hard-drawn copper wire. A random sample of 20 specimen copper-wire pieces from production yields a mean breaking strength $\bar{x} = 565$ lb. Basic to the decision making procedures of Chap. 7 is the question of how likely or probable is a sample mean of this magnitude, assuming that the mean μ of the population being sampled has a specified value. Compute the probability of a sample mean $\bar{x}$ equaling or exceeding 565 lb. if breaking strength of all possible copper-wire specimens is $N(560, 5^2)$ lb. (see Sec. 6.5).
13. Referring to the conditions of Problem 12, compute the probabilities: (a) $P(\bar{x} \geq 563)$, (b) $P(\bar{x} \geq 562)$, (c) $P(\bar{x} \geq 561)$, and (d) $P(\bar{x} \leq 559)$.
14. The probability distribution model for Problem 12 is the distribution of the random variable

$$z_j = \frac{\bar{x}_j - \mu}{\sigma_X \sqrt{n}}$$

Assume in Problem 12 that the population standard deviation is unknown and reliance is placed on its estimate s_x. What model is appropriate now? Repeat Problem 12 assuming that $s_x = 5$ lb.

15. Repeat Problem 13 assuming that σ_X is unknown and estimated by $s_x = 8.81$ lb.

16. Determine the mean and variance for a *t*-distributed random variable with: (a) 4 *d.f.*, and (b) 20 *d.f.*

17. A consideration of interest related to the conditions of Problem 12 is the difference between two breaking-strength means μ_X and μ_Y, with the first mean referring to a population of copper wire from production treatment X and the second mean referring to a copper wire population receiving production treatment Y. In Chap. 7, concern is with the probability of obtaining a difference $\bar{x} - \bar{y}$ as large as that observed for samples from two specified populations. Compute the probability of obtaining a difference $\bar{x} - \bar{y}$ as large as that observed for the following sample data if $\mu_X = 574$, $\mu_Y = 573$, $\sigma_X = 5.0$, and $\sigma_Y = 4.5$.

x_j	570	572	574	576	578	580	
f_j	1	2	3	2	1	1	10

y_j	568	570	572	576	578	584	
f_j	1	3	3	1	1	1	10

18. Repeat Problem 17 assuming that σ_X and σ_Y are unknown and reliance is placed on the respective estimates s_x and s_y.

19. Suppose that interest is with a comparison of the variations of the two populations of Problem 17. Compute the probability of obtaining a value for the ratio of Eq. (6.14) as large as that observed for the two samples in Problem 17.

20. Determine the value of $F_{0.05,6,8}$ such that $P(F \geq F_{0.05,6,8}) = 0.05$ where F is a random variable with 6 and 8 *d.f.*

21. Determine the value of $F_{0.95,8,6}$ such that $P(F \geq F_{0.95,8,6}) = 0.95$.

22. Determine the value of $F_{0.90,5,10}$ such that $P(F \geq F_{0.90,5,10}) = 0.90$. What is the probability $P(F \leq F_{0.90,5,10})$?

23. If F is a random variable having an F-distribution with 4 and 6 *d.f.*, what are the respective probabilities: (a) $P(F \geq 4.5)$, (b) $P(F \geq 6.2)$, and (c) $P(F \geq 9.1)$?

HYPOTHESIS TESTING AND ESTIMATION

Except for limited applications, the major part of the text to this point has been preparatory for Chaps. 7–12. In Chap. 1, methods of summarizing and describing data are developed, along with the fundamental concept of making comparisons using measures of average and variance. Elementary probability concepts are given in Chap. 2 and extended in Chap. 3 to include probability distributions. Chapters 4–6 summarize the important probability distribution models. Now, in Chap. 7, the general problem of statistics is introduced—that of making inferences about the distribution function f(X) *from the sample* x_1 $x_2, \ldots, x_n$. *This problem can be classified into two closely related categories of problems: (1) testing certain hypotheses regarding the nature of a population, i.e., either the nature of the function representing the population, or the values of the parameters of the function, and (2) specifying* f(X) *when the type of function or family is known but certain parameters are unknown—this is called estimation.*

7

7.1 Rationale of Hypothesis Testing

Statistics is concerned with problems having elements of uncertainty. An experiment involving random variables is performed, and decisions are based on the values assumed by the random variables. The experiment may be a survey, a poll, a sample of product output from a manufacturing process, a formal laboratory experiment, and so forth. The "uncertainty" is reflected in the fact that the same experiment repeated again under the same conditions usually gives different sample results. It is difficult and often impossible to obtain a truly random sample. Consider a simplified population $X = 3, 3, 3, 5, 5, 5, 7, 7, 7$, and assume that each sample of three items that is taken always yields $x = 3, 5, 7$. In this case, there is no uncertainty regarding the nature of the population, and indeed there is no statistical problem. Unfortunately, the chance occurrence of random sample values does not follow such a regular pattern. Thus, statistics necessarily deals with the formulation of optimum procedures or *rules* of actions for making decisions, rules which are specified before the experiment is performed. The objective is not aimed at making statements that are guaranteed to be 100% correct, but rather at making *inferences* that have a certain probability of being correct.

Testing a hypothesis is a formal procedure not too different from that taken by a skilled gambler. For example, a participant in a crap game questions whether the dice and the opponent's method of tossing the dice are truly unbiased and fair. A 7 occurs on the first toss. What decision is possible on the basis of this information? A decision is not easily arrived at since this event (a 7 on the first toss) can readily be explained as a *chance occurrence*, i.e., it is not unlikely that the dice and method of tossing result in random outcomes. Suppose the first five tosses yield consecutive 7's, what is the decision to be? There is uncertainty. After all, five consecutive 7's can occur *by chance alone*. Suppose the first twenty-five tosses yield 7's? Any reasonable player will not accept chance alone as being the reason for this remarkable event. Thus, the dice (and opponent) are *rejected* and the game is terminated.

A formal procedure for testing a hypothesis consists of the following steps:

1. State a hypothesis regarding an event A of interest.
2. Select or construct a mathematical model that describes the probability $P(A)$.
3. Formulate a decision rule, based on the model, that specifies when to accept and when to reject the hypothesis.

The term "test" refers to step 3. A *test* is equivalent to a *rule* for making decisions.

EXAMPLE 7.1.1

A formal statement of a possible test for the preceding dice-game example is:

Hypothesis: The dice are fair.
Decision Rule: Accept the dice if no more than six consecutive 7's occur, otherwise reject the dice.

The decision rule is based on a probability distribution of the occurrence of consecutive 7's [e.g., $P(\text{one } 7) = \frac{1}{6}$, $P(\text{two consecutive 7's}) = \frac{1}{6}\cdot\frac{1}{6}$, and so forth]. If more than six consecutive 7's are observed, the dice are rejected, the argument being that the event—more than six consecutive 7's—can, in theory, be the result of chance but the odds are so heavily against such an occurrence that, in practice, it is concluded that *chance alone* does not account for this event.

The choice of cutoff point (no more than "six" consecutive 7's) in Example 7.1.1 is to some extent subjective, dependent on personal feelings and preferences. Many of us drive an automobile during holiday periods and thus accept a small probability that the decision to drive under high traffic conditions was indeed the wrong decision. Some people accept higher probabilities of making the wrong decision by using other modes of travel. In engineering and technological decision making, the cutoff point is not so arbitrary. If the cost consequences of a wrong decision are substantial, only a small probability of making the wrong decision is acceptable.*

7.2 Nomenclature

There are two general types of statistical hypotheses: (1) a hypothesis that the population has a certain *D.F.*—normal, chi-square, binomial, Poisson, etc., and (2) a hypothesis that a population parameter has a specified value, for example, $\mu_X = 20$, or $\sigma_X = 5$, etc. The major part of Chap. 7 is concerned with hypotheses of the latter type.

Frequently, a primary hypothesis is tested against an alternative hypothesis. For example, a product-characteristic specification is 0.500 ± 0.004 in. A primary hypothesis is that the process setting for this operation is centered at 0.500 in. The alternative hypothesis is that the process setting is so seriously incorrect that it is centered at 0.503 in. Primary or *null* hypotheses are denoted by H_0, for example,

$$H_0\colon \quad \mu = 0.500$$

which reads "hypothesis that the population mean μ is 0.500." If an alternative hypothesis is considered, it is denoted by H_1. Another common notation is $\mu_0 = 0.500$ to relate this value to H_0 and differentiate it from a μ_1-value associated with H_1.

A hypothesis test, by the nature of the decision rule, is an experiment resulting in two decision actions, accepting or rejecting H_0. In terms of decision outcomes, however, there are four, with two being favorable outcomes involving correct decisions. The probabilities of these outcomes are denoted:

$$\alpha = P\text{ (rejecting } H_0\text{, when } H_0 \text{ is true)}$$
$$1 - \alpha = P\text{ (accepting } H_0\text{, when } H_0 \text{ is true)}$$

and

$$\beta = P\text{ (accepting } H_0\text{, when } H_0 \text{ is false)}$$
$$1 - \beta = P\text{ (rejecting } H_0\text{, when } H_0 \text{ is false)}$$

The two possible decision errors are called *Type I* and *Type II* errors.

---

*It is important to note that the probability of making a decision error and the cutoff point used for the decision rule are related. In Example 7.1.1., using ten consecutive 7's as a criterion for rejecting the dice is associated with a smaller probability of "rejecting the dice when they should be accepted," i.e., deciding the dice tosses are not random when, in fact, they are. The higher the number of consecutive 7's required for rejection, the smaller is the probability of making a decision error.

Thus,

$$\alpha = P(\text{Type I error})$$

and

$$\beta = P\ (\text{Type II error})$$

It will be shown later that a quantitative assessment of the probability of a Type II error is possible only if an explicit alternative hypothesis statement is possible. In fact, this probability β is directly related to the magnitude of the H_1-value. For this reason, the following notation is adopted:

$$\beta = P\ (\text{accept } H_0,\ H_0 \text{ false, and } H_1 \text{ true})$$

Summarizing, the decision error probabilities are

$$\begin{aligned} \alpha &= P\ (\text{reject } H_0,\ H_0 \text{ true})^* \\ \beta &= P\ (\text{accept } H_0,\ H_0 \text{ false, and } H_1 \text{ true}) \end{aligned} \tag{7.1}$$

and the probabilities of correct decisions are

$$\begin{aligned} 1 - \alpha &= P\ (\text{accept } H_0,\ H_0 \text{ true}) \\ 1 - \beta &= P\ (\text{reject } H_0,\ H_0 \text{ false, and } H_1 \text{ true}) \end{aligned} \tag{7.2}$$

7.3 Hypothesis Test Concerning the Mean, with Standard Deviation Known

The population mean is an important parameter. In many applications, concern is with whether μ has increased, decreased, or remained unchanged, or interest may be in whether μ is significantly greater than, or less than, an assumed or specified value.

The following example of a hypothesis test concerning the mean is used primarily to introduce hypothesis testing. To separate and simplify an explanation of procedure relative to the probability of Type I and II decision errors, only the Type I error is considered here. This example is repeated in Sec. 7.4, at which time both decision errors are considered.

A manufacturer produces 0.104-in. hard-drawn copper wire with a mean breaking strength of 560 lb. A change in raw material and production method is expected to increase the breaking strength of this product. The standard deviation of breaking strength is known to be 5 lb. and this value is expected to remain unchanged. If there is no change in mean breaking strength, it is desired to have a probability of 0.99 of making this correct decision.

---

*The probability α is commonly called the *significance level.*

It is known that the change of material and production method will not decrease breaking strength. Thus, the test is only for a possible increase in strength. The hypothesis being tested is that mean breaking strength is 560 lb. Rejecting the hypothesis is equivalent to concluding that there is an increase in mean breaking strength. If there is, in fact, no change in mean breaking strength, the probability of reaching this conclusion (accepting H_0, H_0 true) is to be $1 - \alpha = 0.99$, and the probability of not making this conclusion (rejecting H_0, H_0 true) is $\alpha = 0.01$. A summary of the test conditions is:

EXAMPLE 7.3.1

$$X = \text{breaking strength}, \qquad \sigma_X = 5 \text{ lb} \quad \text{(known)}$$

$$H_0: \ \mu = 560 \text{ lb}, \qquad \alpha = 0.01$$

The sample statistic used to test this hypothesis is

$$Z_j = \frac{\bar{x}_j - \mu_0}{\sigma_X/\sqrt{n}}$$

and, from Eq. (6.11), Z_j is $N(0, 1)$. If H_0 is true, all possible samples of size $n = 20$ from the X_i-population of 0.104-in. copper wire produced generates an $\bar{x}_j$-distribution of mean breaking strength as shown in Fig. 7.1.

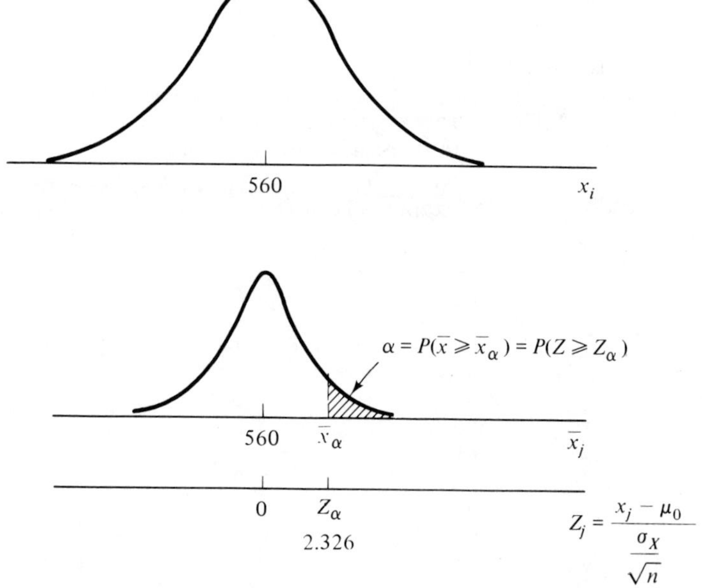

Figure 7.1. $\bar{x}$-distribution model for test of the mean when σX is known.

The $\bar{x}_j$-distribution is the model on which the hypothesis test is based. If X_i is $N(560, \sigma_X^2)$, then $\bar{x}_j$ is $N(560, \sigma_{\bar{x}}^2)$. If X_i is not normally distributed but sample size is sufficiently large, then $\bar{x}_j$ is approximately normally distributed (see Sec. 5.7). In this case, a conservative rule is $n \geq 30$.

A choice of α determines the decision rule. In this case, $\alpha = 0.01$. Thus, there exists an associated mean value $\bar{x}_\alpha$ and a corresponding $Z_\alpha = 2.326$ (see Appendix D) such that

$$\alpha = P(\bar{x} \geq \bar{x}_\alpha) = P(Z \geq Z_\alpha)$$

as represented by Fig. 7.1. Therefore, the decision rule is

$$z_{\text{data}} \geq Z_\alpha = 2.326, \qquad \text{reject } H_0$$
$$\text{otherwise,} \qquad \text{accept } H_0$$

An appropriate model has been selected. A decision rule, based on the model, has been formulated. All that remains is to test the hypothesis. A random sample of 20 copper wire specimens from the production output is tested yielding the breaking strengths given in Table 7.1. The mean of these 20 observations is $\bar{x} = 565$ lb and the corresponding z-value is given by

$$z_{565} = \frac{565 - \mu_0}{\sigma_X/\sqrt{n}} = \frac{565 - 560}{5/\sqrt{20}} = 4.47$$

Therefore, following the decision rule,

$$z_{\text{data}} = 4.47 > Z_\alpha = 2.326$$

and the hypothesis is rejected.

Table 7.1. BREAKING STRENGTH (LB) OF 0.104-IN. HARD-DRAWN COPPER WIRE, 20 OBSERVATIONS FROM PRODUCTION OUTPUT

x_j	f_j
561	1
562	2
563	2
564	3
565	5
566	4
567	0
568	1
569	0
570	2
$\bar{x} = 565$	

Of course, even if H_0 is true in Example 7.3.1, it is possible and indeed likely that a sample mean will not be equal to 560 lb. The criterion underlying the decision rule, however, is the magnitude of $\bar{x} - 560$. Is the observed difference $\bar{x} - 560$ too large to be explained by sampling fluctuations? In this example, the probability is only 0.01 of obtaining a mean value $\bar{x}$ large enough such that the difference $\bar{x} - 560$ in standard z-units is equal to or greater than 2.326. Yet, a mean value has been obtained such that its z-value is 4.47! The conclusion is that the sample is not coming from an X-population with mean $\mu_0 = 560$; it is probably coming from a different population with mean $\mu > \mu_0$.*

EXERCISES

1. State the decision rule for Example 7.3.1 if $\alpha = 0.20$.
Ans. $z_{\text{data}} > Z_\alpha = 0.84$, reject H_0.

2. The standard deviation $\sigma_{\bar{x}}$ depends on sample size. If $n = 4$, what is the test result in Example 7.3.1? *Ans.* $z_{\text{data}} = 2.0$, accept H_0.

3. The standard deviation $\sigma_{\bar{x}}$ also depends on σ_x. If $\sigma_x = 10$, what is the test result in Example 7.3.1? *Ans.* $z_{\text{data}} = 2.236$, accept H_0.

4. Mean tensile strength for a standard steel product is 24,100 psi. A manufacturing improvement is expected to increase mean strength. The standard deviation of tensile strength is 160 psi and this value is not expected to change. Testing a pilot production run of the improved product, 16 random observations of the tensile strength yield a mean strength of 24,280 psi. Test for a possible increase in mean strength at a significance level of $\alpha = 0.05$. *Ans.* $z_{\text{data}} = 4.50$, reject H_0.

7.4 Type II Error and β

A quantitative consideration of the Type II decision error requires a formulation of an alternative hypothesis H_1 which is also of interest to the investigator. The alternative hypothesis is usually critical in the sense that, if H_1 is true and the hypothesis test fails to reach this conclusion, then this decision error generates significant costs.

Each H_1-value is associated with a different β-value, where β is the probability of a Type II decision error. Thus, β is a function of H_1 and this

---

*This argument is analogous to that of the dice-game example. The probability of occurrence of 25 consecutive 7's is so small that no reasonable player will accept *chance alone* as being the reason for this remarkable result. Likewise, the probability is only 0.00000421 (from Appendix D) of obtaining a sample mean with a corresponding z-value as large as 4.47, if H_0 is true. In practice, it is concluded that *chance alone* is not the cause of this unusual result. It is very likely that an increase in mean breaking strength has occurred. Thus, H_0 is rejected.

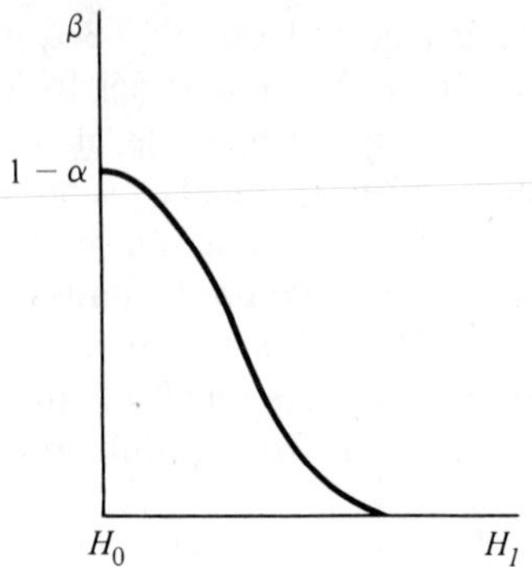

Figure 7.2. Definition form of *OC* curve, α and n fixed.

functional relationship is described by an *operating characteristic* or *OC* curve as shown in Fig. 7.2. For fixed α and n, the ordinate values β give the probabilities P(accepting H_0) for various H_1-values.

In the same manner as the Z-transformation (see Sec. 5.3), an *OC* curve is expressed in standard units by applying the linear transformation

$$d = \frac{\mu_1 - \mu_0}{\sigma_X} \tag{7.3}$$

The effect of this transformation, shown in Fig. 7.3, is to shift the origin to the null hypothesis value and calibrate the abscissa axis in multiples of σ_X.

Figures 7.4–7.6 give *OC* curves in this standardized form for the hypothesis test concerning the mean when the standard deviation is known. Curves are given for $\alpha = 0.01$ and 0.05 and for various n-values.

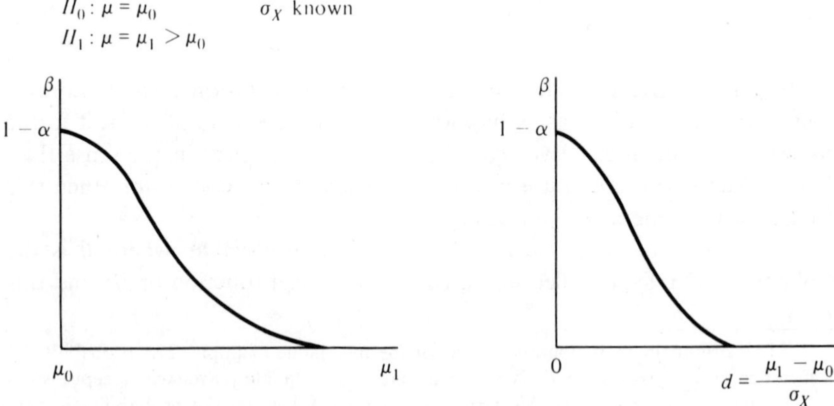

Figure 7.3 *OC* curves, definition and standard form, for test of the mean when σ_X is known.

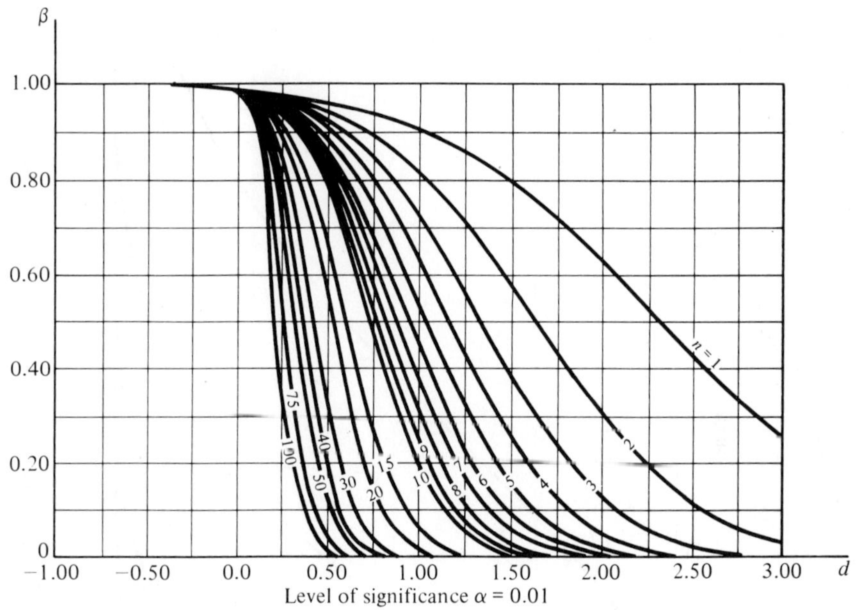

Level of significance α = 0.01

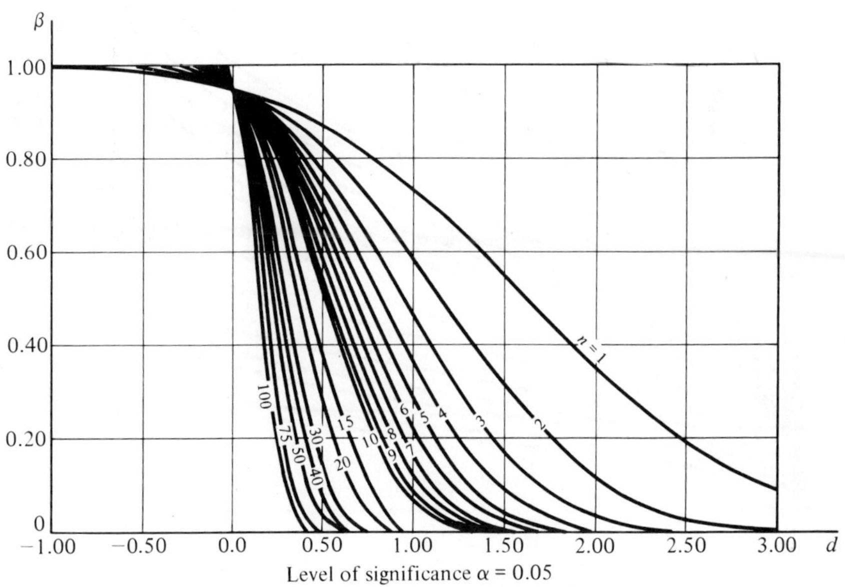

Level of significance α = 0.05

Figure 7.4. *OC* curves for a one-sided normal test.

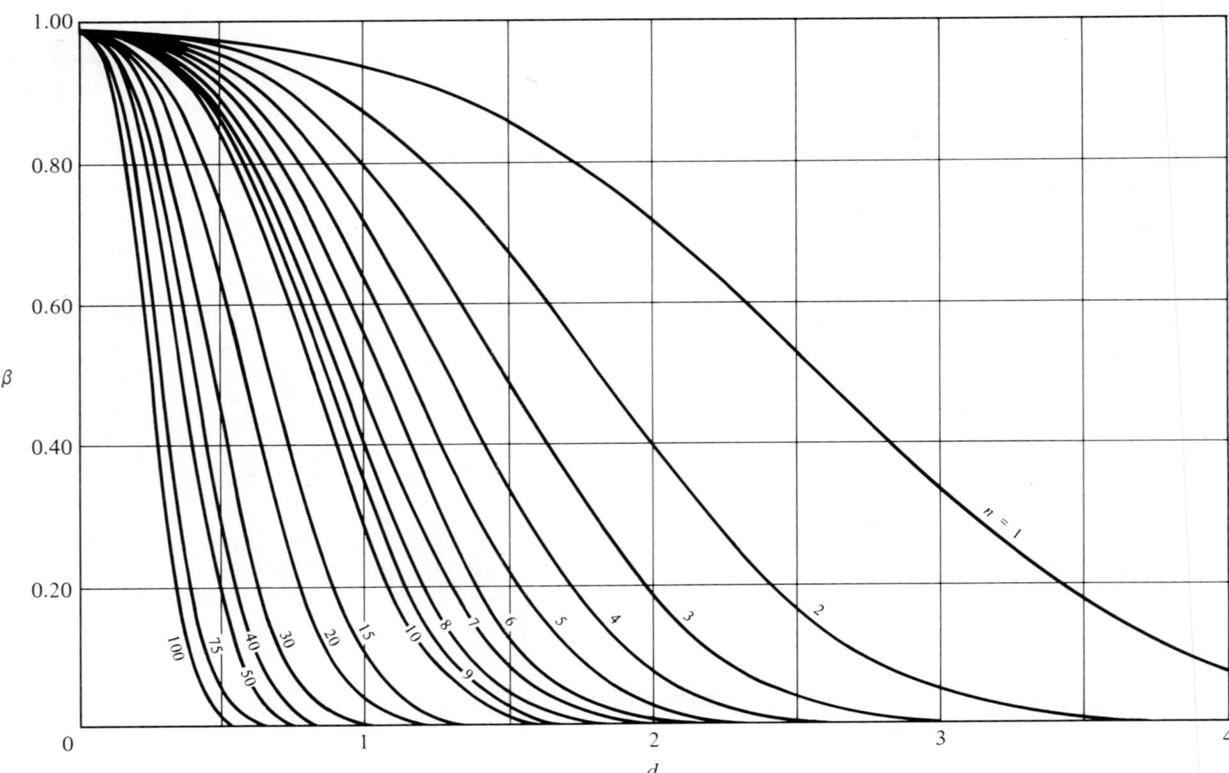

Figure 7.5. *OC* curves for a two-sided normal test, $\alpha = 0.01$.

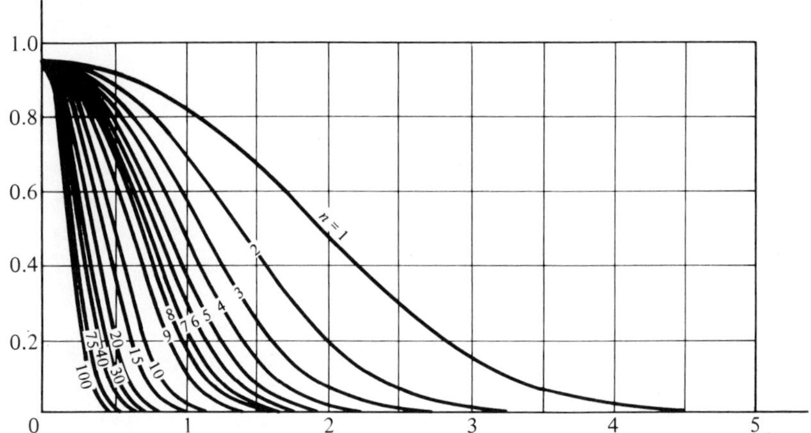

Figure 7.6. *OC* curves for a two-sided normal test, $\alpha = 0.05$.

Sample size is not arbitrary. On the one hand, sample size should be small to minimize the expense of inspection and test required to obtain the n observations. On the other hand, n should be sufficiently large to give a $1 - \beta$ protection against a Type II decision error. To illustrate this latter point, Example 7.3.1 is repeated with consideration now being given to both Type I and II decision errors.

One additional condition is imposed on the original test (Example 7.3.1). If mean breaking strength is actually increased by as much as 3.75 lb, the manufacturer is willing to assume only a 0.15 risk of the hypothesis test failing to detect this fact. This requirement establishes the probabilities:

$$P(\text{reject } H_0, H_0 \text{ false, and } H_1 \text{ true}) = 1 - \beta = 0.85$$

and

$$P(\text{accept } H_0, H_0 \text{ false, and } H_1 \text{ true}) = \beta = 0.15$$

A summary of the test conditions and the test in its entirety is:

EXAMPLE 7.4.1

$$\begin{array}{ll} X = \text{breaking strength}, & \sigma_X = 5 \text{ lb} \quad (\text{known}) \\ H_0: \ \mu = 560 \text{ lb}, & \alpha = 0.01 \\ H_1: \ \mu = 563.75 \text{ lb}, & \beta = 0.15 \end{array}$$

A graphical summary is given in Fig. 7.7. Essentially, there are two models: one under the assumption that H_0 is true, the other assuming that H_1 is true. The probabilities α and β are graphically represented as shown, together with the decision-rule value Z_α and its associated $\bar{x}_\alpha$-value.

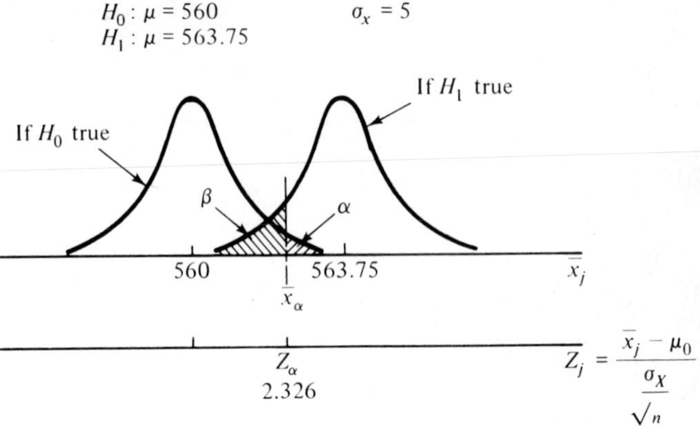

Figure 7.7. Example 7.4.1—$\bar{x}$-distribution models for test of the mean when σ_X is known.

To determine the required sample size, it is necessary to compute d and then enter the family of OC curves with d and β and read the appropriate n-value:

$$d = \frac{\mu_1 - \mu_0}{\sigma_X}$$

$$= \frac{563.75 - 560}{5} = 0.75$$

Entering the set of OC curves, Fig. 7.4, with $d = 0.75$ and $\beta = 0.15$, sample size n is determined to be 20.

The hypothesis test then proceeds as described in Example 7.3.1:

$$z_{565} = \frac{565 - 560}{5/\sqrt{20}} = 4.47$$

and since

$$z_{\text{data}} = 4.47 > Z_\alpha = 2.326$$

the hypothesis H_0 is rejected.*

It should be noted that the hypothesis test (Example 7.4.1) is basically a test of H_0, i.e., whether or not the mean breaking strength is 560 lb. Of course, a rejection of H_0 is equivalent to a conclusion that mean strength has

---

*A more compact argument for the rejection is a simple comparison of z_{data} with the two populations: normal model with mean 560 and normal model with mean 563.75. If H_0 is true, the probability

$$P(\bar{x} \geq \bar{x}_\alpha) = P(Z \geq Z_\alpha = 2.326)$$

is only 0.01. Yet, the observed mean value is 565, and its corresponding z-value is 4.47. Thus, it is unlikely that the sample of 20 observations (with mean 565) is coming from the H_0-population with mean 560. Therefore, H_0 is rejected and the conclusion is that the sample of 20 observations is probably coming from a population with mean $\mu > 560$.

increased. Stated differently, rejecting H_0 is equivalent to detecting an increase in mean strength. The size of the possible increase (i.e., H_1), the risk β, and sample size n are related, as we shall see in Sec. 7.6. A preliminary example is considered here to emphasize the uncertainty that is pervasive to a hypothesis test.

EXAMPLE 7.4.2

To exaggerate, suppose in Example 7.4.1 that sample size is only 4.

Consider the possibility that a sample of four observations yields a mean breaking strength such that, following the decision rule, H_0 is accepted. This result does not assure that a 3.75 lb increase in mean strength has not actually taken place. It simply means that the sample size is not sufficiently large to detect a strength increase of this magnitude.

Referring to Fig. 7.4, when $d = 0.75$ and $n = 4$, the probability β is 0.80. Thus, the probability of rejecting H_0 is only $1 - 0.80 = 0.20$. A sample of size $n = 4$ will detect a 3.75 lb. increase in mean strength only 20% of the time, on the average.

EXERCISES

1. Referring to Example 7.4.1, changing only α and β, determine sample size n for each of the following conditions:
(a) $\alpha = 0.01$, $\beta = 0.05$.
(b) $\alpha = 0.01$, $\beta = 0.26$.
(c) $\alpha = 0.05$, $\beta = 0.10$. *Ans.* (a) 30, (b) 15, (c) 15.

2. Suppose in Example 7.4.1 that sample size is 10. What percentage of the time will the test detect a 3.75 lb. increase in mean breaking strength? *Ans.* 53%.

7.5 One- and Two-Sided Tests

The example traced through Secs. 7.3–7.4 involves a test for an increase in breaking strength. This is called an *upper one-sided test*. Concern is only with a possible increase in the parameter being tested. This situation is summarized in Fig. 7.8, using the test of the mean as an example. The distribution models, probabilities of Type I and II decision errors, and decision-rule value are graphically represented.

Again using the example of Secs. 7.3–7.4, suppose that a temporary increase in customer demand makes it necessary to utilize previously retired production facilities (i.e., retired for reasons of deterioration and obsolescence). A relevant hypothesis test now might concern the question, "is there a decrease in mean breaking strength of the product?" This is an example of a *lower one-sided test*. A summary of this test is given in Fig. 7.9.

$H_0 : \mu = \mu_0$ $\qquad \sigma_X$ known

$H_1 : \mu = \mu_1 > \mu_0$

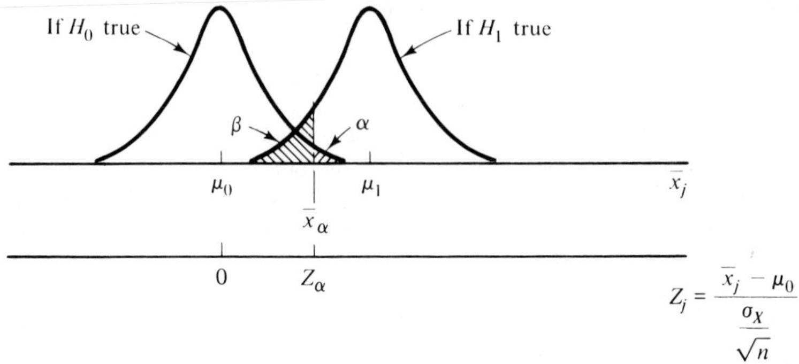

Figure 7.8. Upper one-sided test.

$H_0 : \mu = \mu_0$ $\qquad \sigma_X$ known

$H_1 : \mu = \mu_1 < \mu_0$

If H_1 true

If H_0 true

α $\quad \beta$

μ_1 $\quad \mu_0$ $\quad \bar{x}_j$

$\bar{x}_\alpha$

Z_α $\quad 0$

$$Z_j = \frac{\bar{x}_j - \mu_0}{\dfrac{\sigma_X}{\sqrt{n}}}$$

Figure 7.9. Lower one-sided test.

A *two-sided test* is concerned with variation in both directions from a hypothesized parameter. Using the test of the mean as an example, interest is with both a possible increase or decrease of the mean. For example, consider a product specification 0.500 $\pm$ 0.004 in. A null hypothesis might be that the manufacturing process is centered at a mean of 0.500 in., whereas the alternative hypothesis is that the process centering is seriously incorrect—as bad as $\mu = 0.498$ or 0.502 in.

In this case, the α risk is partitioned (usually equally) to cover decision errors due either to increases or decreases in the population mean. This type of test is summarized in Fig. 7.10.

The graphical summaries and shorthand hypotheses statements are

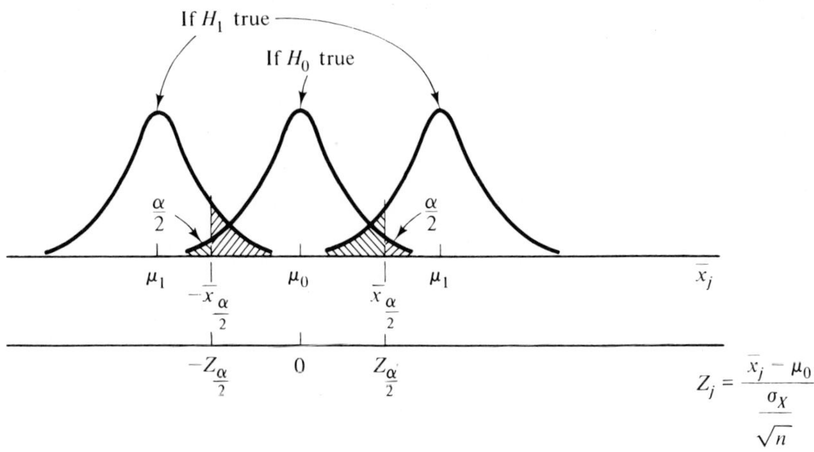

Figure 7.10. Two-sided test.

efficient aids to problem solving. However, as with all symbolism, the detailed conditions that have been shorthanded out tend to be overlooked or compromised. It should be emphasised that a hypothesis test leads only to a decision regarding H_0. For example, a rejection of $H_0\colon \mu = \mu_0$ where $H_1\colon \mu = \mu_1 > \mu_0$ is only a decision that μ is not equal to μ_0 and is probably larger than μ_0. There is no conclusion as to how large is μ. To illustrate, consider the conditions

$$H_0\colon \mu = 10 \qquad \alpha = 0.01$$
$$H_1\colon \mu = 14 \qquad \beta = 0.05$$

This is shorthand notation meaning only that H_0 is to be accepted or rejected, and the value $\mu = 14$ is stated to indicate the basis for establishing sample size n. Thus, n is determined such that, if μ is actually 14, there is a $1 - \beta = 0.95$ probability of detection of the difference $14 - 10 = 4$ (detection in the sense of rejecting H_0 if H_1 is in fact true). Similarly, the shorthand notation

$$H_0\colon \mu = 10 \qquad \alpha = 0.01$$
$$H_1\colon \mu = 6{,}14 \qquad \beta = 0.05$$

means that H_0 is being tested, the difference to be detected is 4 (measured in either direction $\pm$ from μ_0), and this difference coupled with β is the basis for determining sample size n.

EXAMPLE 7.5.1

A test concerns the setting of a manufacturing process. A null hypothesis that the process setting is at $\mu = 0.500$ in. is tested against the alternative that the setting differs from 0.500 by 0.002 in. The test and test conditions are:

$$H_0: \quad \mu = 0.500 \text{ in.}, \qquad \alpha = 0.05$$
$$H_1: \quad \mu = 0.498, 0.502 \text{ in.}, \qquad \beta = 0.10$$
$$\sigma_X = 0.002 \text{ in.} \quad \text{(known)}$$

$H_0: \mu = 0.500$ $\qquad \sigma_x = 0.002$ (known)

$H_1: \mu = 0.498, 0.502$

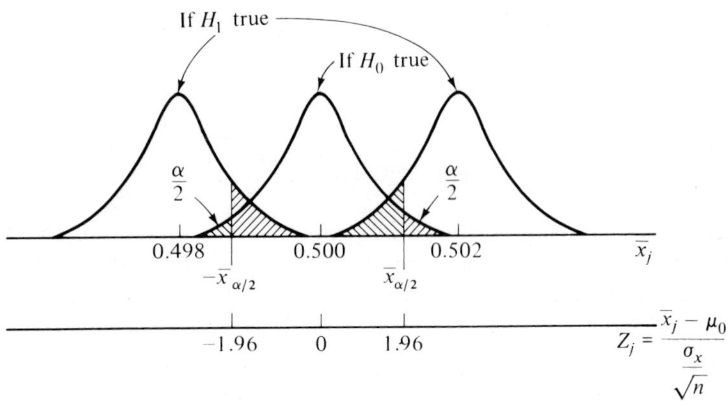

Figure 7.11. Example 7.5.1—distribution models.

Figure 7.11 summarizes the hypothesis test. For $\alpha/2 = 0.025$, the $Z_{\alpha/2}$-value read from Appendix D is 1.96. Thus, the decision rule is

$$|z_{\text{data}}| \geq Z_{\alpha/2} = 1.96, \qquad \text{reject } H_0$$
$$\text{otherwise,} \qquad \text{accept } H_0$$

Determining sample size,

$$d = \frac{|\mu_1 - \mu_0|}{\sigma_X}$$
$$= \frac{0.002}{0.002} = 1.0$$

and entering the *OC* curves, Fig. 7.6, with $d = 1.0$ and $\beta = 0.10$, sample size n is read to be 10.

A random sample of 10 observations yields a mean measurement of 0.5015 in.

The corresponding z-value is

$$z_{0.5015} = \frac{0.5015 - \mu_0}{\sigma_X \sqrt{n}}$$

$$= \frac{0.5015 - 0.500}{0.002\sqrt{10}} = 2.37$$

Therefore,

$$|z_{\text{data}}| = 2.37 > Z_{\alpha/2} = 1.96$$

and the hypothesis is rejected.

EXERCISES

1. Suppose that Example 7.4.1 is a test for a possible decrease in breaking strength of 3.75 lb and

$$H_0: \quad \mu = 560 \text{ lb}, \qquad \alpha = 0.01$$
$$H_1: \quad \mu = 556.25 \text{ lb}, \qquad \beta = 0.15$$

Sketch the distribution models and state the decision rule.

Ans. $z_{\text{data}} \leq Z_\alpha = -2.326$, reject H_0.

2. A test concerns the setting of an automatic machine that fills 5-lb boxes. A null hypothesis that the machine setting is at $\mu = 5$ lb is tested against the alternative that the setting differs from 5 lb by 0.2 lb. The probability of detecting a difference of this magnitude must not be less than 0.90. The standard deviation σ_X is 0.2 lb. Test the hypothesis at a 1% significance level assuming that the sample observations yield a mean weight of 5.11 lb. *Ans.* $|z_{\text{data}}| = 2.13$, accept H_0.

7.6 α, β, n Relationships

Before proceeding with other hypothesis tests, the test of the mean affords a simple example for developing intuition regarding the general concept of testing hypotheses. Because of physical and economic constraints (e.g., destructive testing, time, expense of measurement, etc.), it may not be feasible to examine an entire population. Thus, decision making is based on limited information available from a small random sample. It is desirable to minimize sample size n, and yet also have α and β as small as possible. Unfortunately, all three items, α, β, and n, cannot be minimized. Fixing any two of these three quantities determines the remaining one.

Using the test of the mean to illustrate the relationships between α, β, and n, the sampling situation is graphically described in Fig. 7.12. If H_0 is true, the sample $x_1, x_2, \ldots, x_n$ is coming from a population with mean μ_0.

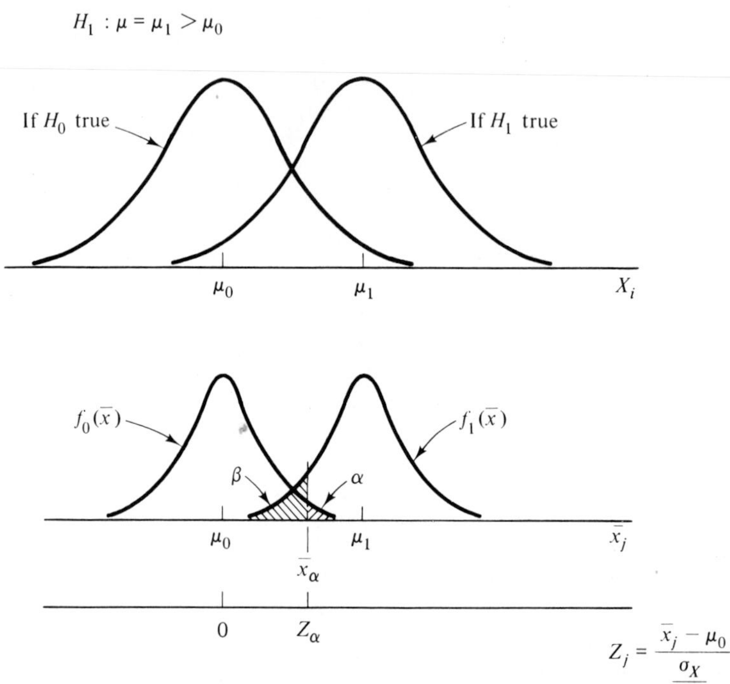

Figure 7.12. X_i-populations and $\bar{x}$-distribution models for upper one-sided test of the mean when σ_X is known.

If H_1 is true, the sample is from a population with mean μ_1. The models underlying the hypothesis test are the $\bar{x}_j$-distributions designated by $f_0(\bar{x})$ and $f_1(\bar{x})$ in Fig. 7.12.

Specifying μ_0 and μ_1 (σ_X being fixed) determines the two X_i-populations (Fig. 7.12) in the sense that their respective locations and ranges are thus established and normality is assumed. Further, if sample size n is fixed, the $\bar{x}_j$-populations are determined with respect to both location and range, the range determination following from the relationship $\sigma_{\bar{x}} = \sigma_X/\sqrt{n}$, with both σ_X and n now being fixed.* Since the $\bar{x}_j$-populations are thus determined, the overlap of the distribution tails is also determined as shown in Fig. 7.12. Now, any choice of the decision rule value Z_α and thus also $\bar{x}_\alpha$ determines both α and β, these two probabilities being given by

$$\alpha = \int_{\bar{x}_\alpha}^{+\infty} f_0(\bar{x})\, d\bar{x} \quad \text{and} \quad \beta = \int_{-\infty}^{\bar{x}_\alpha} f_1(\bar{x})\, d\bar{x}$$

*At this point, it is beneficial for the student to review Sec. 5.7.

From the geometry of Fig. 7.12, four pertinent intuitive observations are possible:

1. Specify μ_0, μ_1, and α.
   Then, increasing n decreases β and conversely.
2. Specify μ_0, μ_1, and n.
   Then, increasing α decreases β and conversely.
3. Specify μ_0, α, and n.
   Then, increasing the magnitude of μ_1 decreases β and conversely.
4. Specify μ_0 and μ_1.
   Then, the only way α and β can both be decreased is by increasing n.

Extending the condition of (3) above affords an intuitive explanation of why the maximum value of β on an *OC* curve is $1 - \alpha$. For fixed α and n, let $\mu_1 = \mu_0$. Then, the $f_1(\bar{x})$ distribution curve shifts to the left and coincides with the $f_0(\bar{x})$ curve and β [represented by the tail area beneath $f_1(\bar{x})$] becomes equal to $1 - \alpha$.

In applications, the usual procedure sequence is: (1) Specify μ_0 and μ_1; (2) in the order of their practical importance, fix two of the three factors, α, β, and n, equal to predetermined small values, and then accept the value thus determined for the remaining factor. Compromises are often required. For example, α and n may be fixed at some desirable values, only to find that the resulting value of β is impractically large. Thus, either α or n or both must necessarily be enlarged slightly.

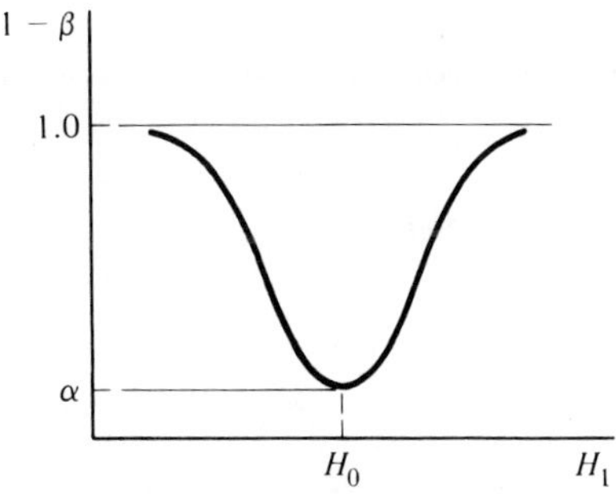

Figure 7.13. Power curve.

Some texts make use of *power curves* instead of *OC* curves to graphically describe α, β, and n relationships. A power curve is defined by Fig. 7.13. The ordinate axis of a power curve gives $1 - \beta$ values. Thus, a power curve describes the probability of not making a Type II decision error for various alternative H_1-values and various α, n combinations. In comparing two tests available for the same purpose, a test is described as having a "high power" or a "low power." A test with high power has a large $1 - \beta$ value, or, conversely, a low β value. Since there are two outcomes, given that H_1 is true,

$$P(\text{reject } H_0, H_1 \text{ true}) + P(\text{accept } H_0, H_1 \text{ true}) = 1$$

where the first probability is $1 - \beta$ and the second is β, conversions can be made between power and OC curve values using the relationship

$$OC \text{ curve ordinate} = 1 - \text{power curve ordinate} \tag{7.4}$$

EXERCISES

1. For a one-sided test of the mean, $\mu_0 = 10$, $\mu_1 = 11.5$, $\sigma_X = 1$, $\alpha = 0.01$, and $\beta = 0.10$. (a) What is the sample size? (b) Increase sample size by 2. What is the new β-value? *Ans.* (a) $n = 6$, (b) $\beta = 0.03$.

2. For a one-sided test of the mean, $\mu_0 = 10$, $\mu_1 = 11$, $\sigma_X = 1$, $\alpha = 0.01$, and $n = 10$. (a) What is the β-value? (b) Increase α to 0.05. What is the new β-value? *Ans.* (a) $\beta = 0.20$, (b) $\beta = 0.07$.

3. Sample size n is increased until it equals the population size N. Then, what is the value of α? of β? *Ans.* $\alpha = \beta = 0$.

7.7 Analytic Solution for β or n

The test of the mean also affords a simple example of indicating how OC curves are derived. Given any two of the factors, α, β, and n, a solution for the third factor is possible from two Z-transformation equations. Two examples are presented here to illustrate the general procedure.

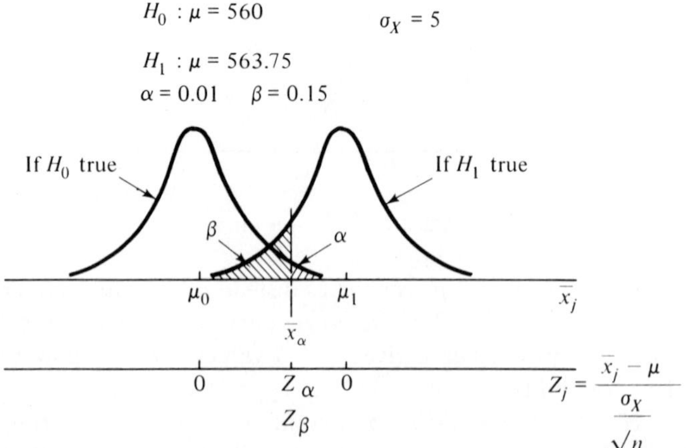

Figure 7.14. Example 7.7.1—summary—analytic solution for n.

EXAMPLE 7.7.1

The hypothesis conditions of Example 7.4.1 are repeated here and summarized in Fig. 7.14. Sample size n is to be determined without reference to an OC curve.

$$H_0: \quad \mu = 560 \text{ lb}, \qquad \alpha = 0.01$$
$$H_1: \quad \mu = 563.75 \text{ lb}, \qquad \beta = 0.15$$

A Z-transformation on $\bar{x}_\alpha$ referred to the origin $\mu_0 = 560$ lb is

$$Z_\alpha = \frac{\bar{x}_\alpha - \mu_0}{\sigma_X/\sqrt{n}}$$

or

$$Z_{0.01} = +2.326 = \frac{\bar{x}_\alpha - 560}{5/\sqrt{n}}$$

A Z-transformation on $\bar{x}_\alpha$ referred to the origin $\mu_1 = 563.75$ lb is

$$Z_\beta = \frac{\bar{x}_\alpha - \mu_1}{\sigma_X \sqrt{n}}$$

or

$$Z_{0.15} = -1.037 = \frac{\bar{x}_\alpha - 563.75}{5/\sqrt{n}}$$

Subtracting, $Z_{0.01} - Z_{0.15}$,

$$3.363 = \frac{(\bar{x}_\alpha - 560) - (\bar{x}_\alpha - 563.75)}{5/\sqrt{n}}$$
$$= 3.75\left(\frac{\sqrt{n}}{5}\right)$$

whence

$$n = 20.11 \quad \text{(thus, use } n = 20\text{)}$$

Summarizing, the solution for n in Example 7.7.1 together with $\alpha = 0.01$ and $\beta = 0.15$ identifies the point (0.75, 0.15) shown on the OC curve in Fig. 7.15. Other points on this curve are obtained by solving for β, given α, n, μ_0, and μ_1. The following example illustrates this procedure.

EXAMPLE 7.7.2

The preceding example is essentially a solution for n, given α and β (and μ_0 and μ_1, of course). In this example, α and n are fixed and the solution is for β.

Again, the hypothesis conditions of Example 7.4.1 are used for illustration:

$$H_0: \quad \mu = 560 \text{ lb}, \qquad \alpha = 0.01$$
$$H_1: \quad \mu = 563.75 \text{ lb}, \qquad n = 20$$

A Z-transformation on $\bar{x}_\alpha$ referred to the origin $\mu_0 = 560$ lb is

$$Z_{0.01} = 2.326 = \frac{\bar{x}_\alpha - 560}{5/\sqrt{20}}$$

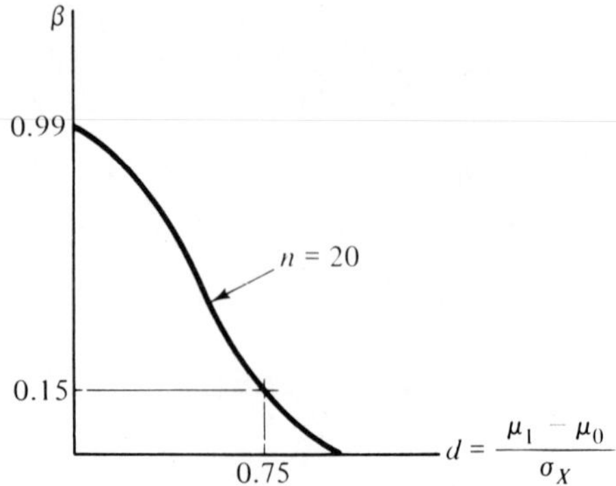

Figure 7.15. Example 7.7.1—identification of the point (0.75, 0.15) on *OC* curve.

whence

$$\bar{x}_\alpha = 562.60$$

A Z-transformation on $\bar{x}_\alpha$ referred to the origin $\mu_1 = 563.75$ lb is

$$Z_\beta = \frac{562.60 - 563.75}{5/\sqrt{20}}$$
$$= -1.03$$

Thus, β is given by the integral

$$\beta = \int_{-\infty}^{-1.03} f_1(Z)\,dZ = 0.1515 \quad \text{(from Appendix D)}$$

One final, albeit incidental, point is worth mentioning here. It should now be clear, from Secs. 7.6–7.7, that stating H_0 and H_1 and choosing α and β is equivalent to determining two points, (H_0-value, $1 - \alpha$) and (H_1-value, β), on an *OC* curve as shown in Fig. 7.16.

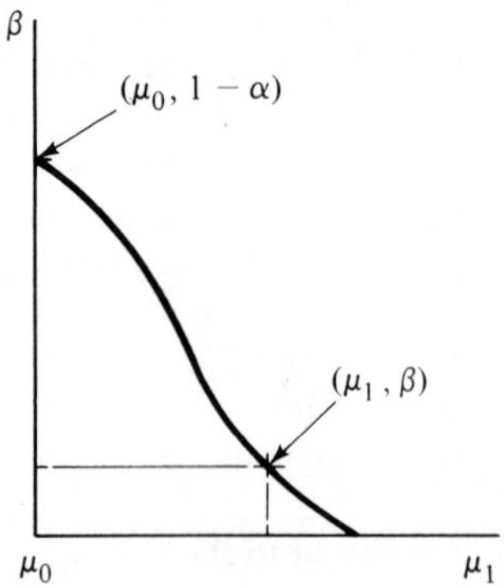

Figure 7.16. *OC* curve (definition form) for a one-sided test of the mean, given α and n.

EXERCISES

1. Determine sample size n without reference to an OC curve using the following information:

$$H_0:\ \mu = 10.0, \quad \alpha = 0.05, \quad \sigma_X = 2$$
$$H_1:\ \mu = 10.5, \quad \beta = 0.10$$

Ans. $n = 137$.

2. Determine β without reference to an OC curve using the following information:

Ans. $\beta = 0.1963$.

$$H_0:\ \mu = 10.0, \quad \alpha = 0.05, \quad \sigma_X = 2$$
$$H_1:\ \mu = 10.5, \quad n = 100$$

7.8 Summary—Test of the Mean With Standard Deviation Known

The test of the mean has been used to introduce hypothesis testing in some detail to develop the rationale of this statistical concept. The general procedure for all tests is the same. From one test to another, the only factors that change are: (1) the statistic used to make the test, e.g., in the test of the mean with σ_X known, the statistic is the random variable Z [see Eq. (6.11)], and (2) the mathematical model employed. In the following sections, much of the previous detail is dropped. For this reason, the student is encouraged to become familiar with the following notations. A summary, such as the following, is given at the end of each subsequent section for review and reference.

Summary—Test of the Mean With Standard Deviation Known

X-distribution normal, or $n \geq 30$.
σ_X known.

$$H_0:\ \mu = \mu_0 \qquad H_1:\ \mu = \mu_1 > \mu_0$$

$$d = \frac{\mu_1 - \mu_0}{\sigma_X} \qquad n:\ \text{enter Fig. 7.4 with } d,\ \beta.$$

Decision Rule:

$$z_{\text{data}} \geq Z_\alpha, \quad \text{reject } H_0$$

where

$$z_{\text{data}} = \frac{\bar{x}_{\text{data}} - \mu_0}{\sigma_X \sqrt{n}}$$

$$H_0:\ \mu = \mu_0 \qquad H_1:\ \mu = \mu_1 < \mu_0$$

$$d = \frac{\mu_0 - \mu_1}{\sigma_X} \qquad n:\ \text{enter Fig. 7.4 with } d,\ \beta.$$

Decision Rule:

$$z_{\text{data}} \leq -Z_\alpha, \quad \text{reject } H_0$$

$$H_0:\ \mu = \mu_0 \qquad H_1:\ \mu = \mu_1 \gtrless \mu_0$$

$$d = \frac{|\mu_1 - \mu_0|}{\sigma_X} \qquad n:\ \text{enter Figs. 7.5, 7.6 with } d,\ \beta.$$

Decision Rule:

$$|z_{\text{data}}| \geq Z_{\alpha/2}, \quad \text{reject } H_0$$

7.9 Test of the Mean With Standard Deviation Unknown

In practice, when testing a mean, the standard deviation σ_X is not usually known. A sample estimate of this parameter is required and s_x is used for this purpose. The statistic used to test H_0: $\mu = \mu_0$ is

$$t_j = \frac{\bar{x}_j - \mu_0}{s_x/\sqrt{n}}$$

If the X_i-population that generates the $\bar{x}_j$-distribution is normal (or, if sample size is sufficiently large thus assuring approximate normality of the $\bar{x}_j$-distribution) and if H_0 is true, this t random variable, which is a function of $\bar{x}$, s_x, and n, is t-distributed with $n - 1$ *d.f.* [see Eq. (6.12)].

EXAMPLE 7.9.1

Returning to Example 7.4.1, assume that σ_X is unknown and let $s_x = 5$ lb. A summary of the test conditions is:

$$\begin{aligned} &X = \text{breaking strength}, && s_x = 5 \text{ lb} \\ &H_0:\ \mu = 560 \text{ lb}, && \alpha = 0.01 \\ &H_1:\ \mu = 563.75 \text{ lb}, && \beta = 0.15 \end{aligned}$$

and a graphical summary is given in Fig. 7.17.

An estimate of σ_X is required for the d-computation to determine sample size. This estimate can be that of s_x, or an estimate based on historical or physical information, and if it is an uncertain estimate it can be established arbitrarily high

$H_0 : \mu = 560$ $\qquad$ $s_x = 5$

$H_1 : \mu = 563.75$

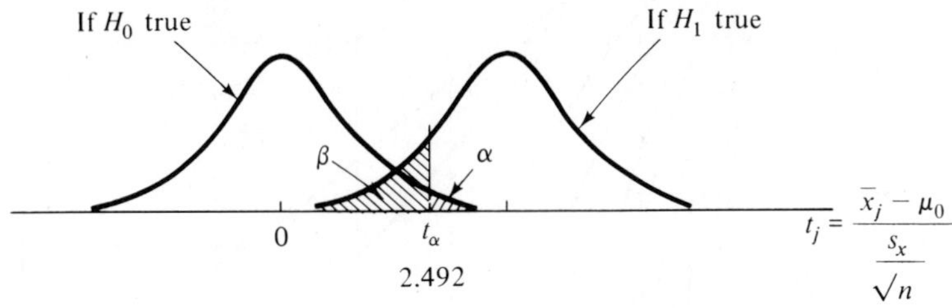

Figure 7.17. Example 7.9.1—*t*-distribution model for an upper one-sided test of the mean when σ_X is unknown.

to obtain sample size sufficiently large to comfortably assure a $1 - \beta$ protection against the Type II decision error. Using $s_x = 5$ lb as an estimate,

$$d = \frac{\mu_1 - \mu_0}{\sigma_X} = \frac{3.75}{5} = 0.75$$

Entering Fig. 7.18 with $d = 0.75$ and $\beta = 0.15$, sample size is read to be 25. The $t_{\alpha,\nu}$-value, for $\alpha = 0.01$ and $d.f. = 25 - 1$, read from Appendix F, is 2.492. Therefore, the decision rule is

$$t_{\text{data}} \geq t_{\alpha,n-1} = 2.492, \qquad \text{reject } H_0$$
$$\text{otherwise,} \qquad \text{accept } H_0$$

A random sample of 25 observations yields a mean breaking strength of 565 lb. The corresponding *t*-value is

$$t_{565} = \frac{565 - \mu_0}{s_x/\sqrt{n}} = \frac{565 - 560}{5/\sqrt{25}} = 5.0$$

and since

$$t_{\text{data}} = 5.0 > t_{\alpha,n-1} = 2.492$$

the hypothesis is rejected.*

*Essentially, the same argument as that of Sec. 7.3 applies here. Only the model is different. If H_0 is true, by chance alone a *t*-value as large as 2.492 is obtained, on the average, only 1% of the time. Yet, the *t*-value computed from the observations is 5.0! Therefore, H_0 is rejected.

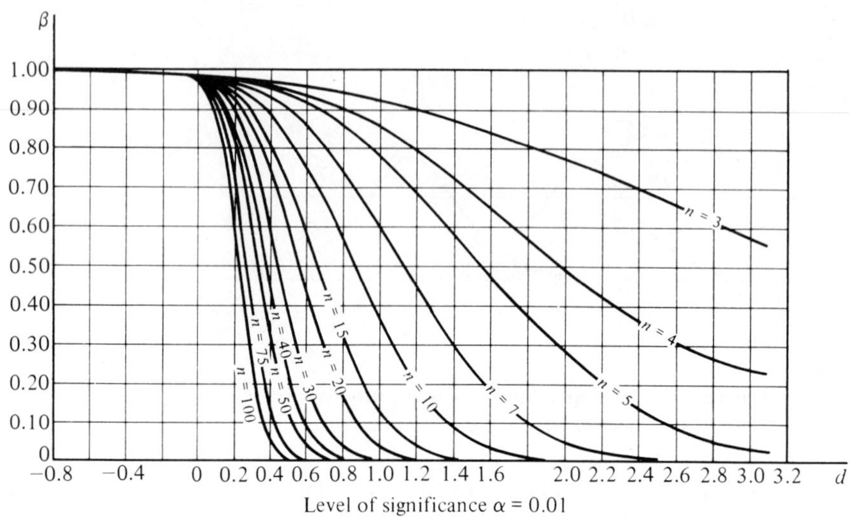

Level of significance $\alpha = 0.01$

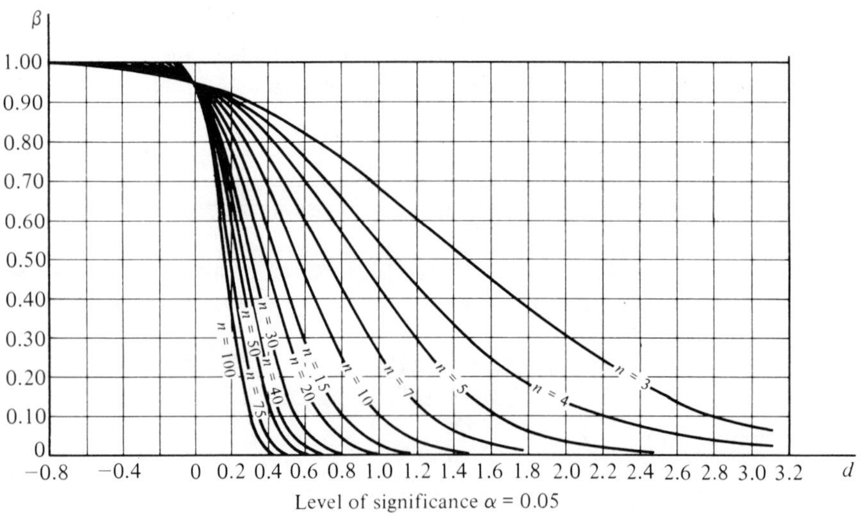

Level of significance $\alpha = 0.05$

Figure 7.18. *OC* curves for a one-sided *t*-test.

Summary—Test of the Mean With Standard Deviation Unknown

X-distribution normal, or $n \geq 30$.
σ_X unknown.

$$H_0: \quad \mu = \mu_0 \qquad H_1: \quad \mu = \mu_1 > \mu_0$$

$$d = \frac{\mu_1 - \mu_0}{\sigma_X} \quad \text{where } \sigma_X \text{ is estimated.}$$

n: enter Fig. 7.18 with d, β.
Decision Rule:

$$t_{\text{data}} \geq t_{\alpha, n-1}, \quad \text{reject } H_0$$

where

$$t_{\text{data}} = \frac{\bar{x}_{\text{data}} - \mu_0}{s_x/\sqrt{n}}$$

$$H_0: \quad \mu = \mu_0 \qquad H_1: \quad \mu = \mu_1 < \mu_0$$

$$d = \frac{\mu_0 - \mu_1}{\sigma_X} \quad \text{where } \sigma_X \text{ is estimated.}$$

n: enter Fig. 7.18 with d, β.
Decision Rule:

$$t_{\text{data}} \leq -t_{\alpha, n-1}, \quad \text{reject } H_0$$

$$H_0: \quad \mu = \mu_0 \qquad H_1: \quad \mu = \mu_1 \gtrless \mu_0$$

$$d = \frac{|\mu_0 - \mu_1|}{\sigma_X} \quad \text{where } \sigma_X \text{ is estimated.}$$

n: enter Figs. 7.19, 7.20 with d, β.
Decision Rule:

$$|t_{\text{data}}| \geq t_{\alpha/2, n-1}, \quad \text{reject } H_0$$

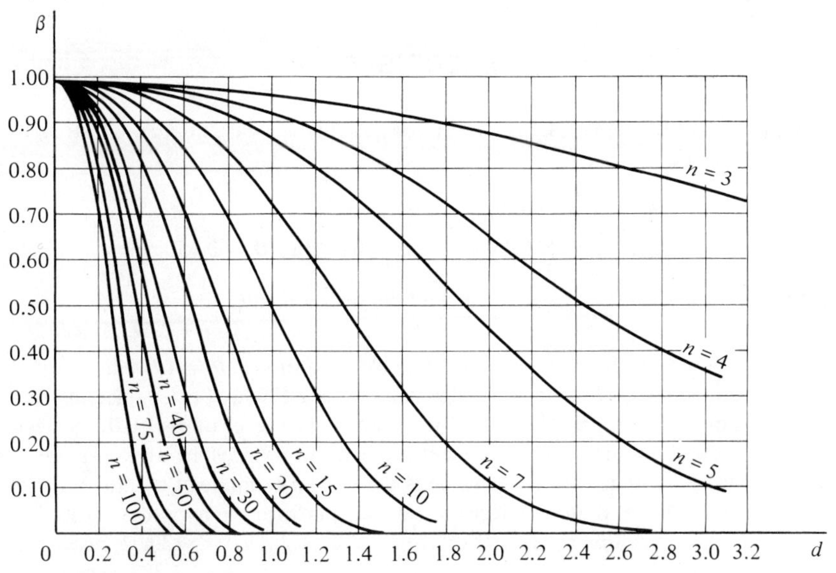

Figure 7.19. *OC* curves for a two-sided t-test, $\alpha = 0.01$.

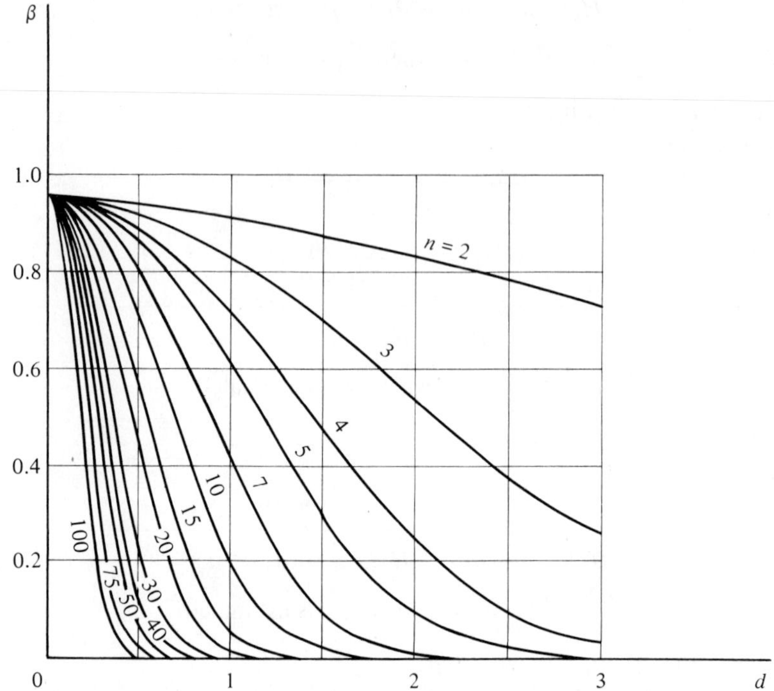

Figure 7.20. *OC* curves for a two-sided *t*-test, $\alpha = 0.05$.

EXERCISES

1. Suppose that Example 7.4.1 is a test for a possible decrease in breaking strength of 3.75 lb and

$$H_0: \quad \mu = 560 \text{ lb}, \qquad s_x = 5 \text{ lb}$$
$$H_1: \quad \mu = 556.25 \text{ lb}, \qquad \alpha = 0.01 \text{ and } \beta = 0.15$$

Sketch the distribution model and state the decision rule.

Ans. $t_{\text{data}} \leq -t_{0.01,24} = -2.492$, reject H_0.

2. A null hypothesis that the mean life of an electron tube is $\mu = 1{,}600$ hr is tested against the alternative that mean life is as low as 1,570 hr. For engineering reasons, sample size is 16. The test is at a significance level of $\alpha = 0.01$. Sixteen random observations of tube life yield a mean life of 1,590 hr and a standard deviation $s_x = 30$ hr. The latter value is used as an estimate of σ_X. (a) What is the value of β? (b) State the decision rule. (c) Perform the hypothesis test.

Ans. (a) $\beta = 0.10$, (b) $t_{\text{data}} \leq -t_{0.01,15} = -2.602$, reject H_0,
(c) $t_{\text{data}} = -1.333$, accept H_0.

7.10 Test That Two Means Are Equal

A frequent hypothesis test in engineering and technology is that of comparing two treatments. A sample of n_x items from a population receiving treatment X and a sample of n_y items from a population receiving treatment Y are compared. Inferences are made based on the magnitude of the difference $\bar{x} - \bar{y}$ of the sample means.

The null hypothesis being tested is H_0: $\mu_X = \mu_Y$. A situation develops here that is similar to the test of the mean. There are three cases, σ_X and σ_Y being: (1) known, (2) unknown but equal, and (3) unknown and not necessarily equal.

In the first case, with σ_X and σ_Y known, the statistic used to test H_0 is

$$Z_j = \frac{(\bar{x}_j - \bar{y}_j) - (\mu_X - \mu_Y)}{\sqrt{(\sigma_X^2/n_x) + (\sigma_Y^2/n_y)}} \tag{7.5}$$

where the underlying random variable is the difference $\bar{x} - \bar{y}$ and this statistic is simply the z-transformation of $\bar{x} - \bar{y}$. From Eq. (6.15), Z_j is $N(0, 1)$. Also if H_0 is true, the difference $\mu_X - \mu_Y$ is zero and the statistic reduces to

$$Z_j = \frac{\bar{x}_j - \bar{y}_j}{\sqrt{(\sigma_X^2/n_x) + (\sigma_Y^2/n_y)}} \tag{7.6}$$

The standard units d-value used to enter the OC curves, Figs. 7.4–7.6, is

$$d = \frac{\mu_X - \mu_Y}{\sqrt{\sigma_X^2 + \sigma_Y^2}}^{*} \qquad n = n_x = n_y \tag{7.7}$$

EXAMPLE 7.10.1

Again referring to the example concerning the breaking strength of 0.104-in. hard-drawn copper wire, Table 6.1 gives 20 observations each of breaking strength of copper wire output from two types of production processes (i.e., "treatments"). The X process is a new and supposedly improved process and the Y process is an existing standard process.

The hypothesis test is H_0: $\mu_X = \mu_Y$ and a rejection of H_0 is equivalent to concluding that process X is indeed superior, using the criterion of mean breaking

---

*In the test of the mean, d is the difference in the hypotheses values $\mu_0 - \mu_1$ divided by the standard deviation of the random variable X. Here, d is also the difference in the hypotheses values $0 - (\mu_X - \mu_Y)$ divided by the standard deviation of the random variable $\bar{x}_j - \bar{y}_j$, which from Eq. (5.10), is

$$\sigma_{X-Y} = \sqrt{\sigma_X^2 - \sigma_Y^2}$$

strength for comparison of X and Y. A summary of the test conditions is:

$$H_0\colon \quad \mu_X = \mu_Y, \qquad \sigma_X = \sigma_Y = 5 \text{ lb} \quad \text{(known)}$$
$$H_1\colon \quad \mu_X > \mu_Y \quad \text{by an amount 5.3 lb}$$
$$\alpha = 0.01 \quad \text{and} \quad \beta = 0.15$$

and a graphical summary is shown in Fig. 7.21.*

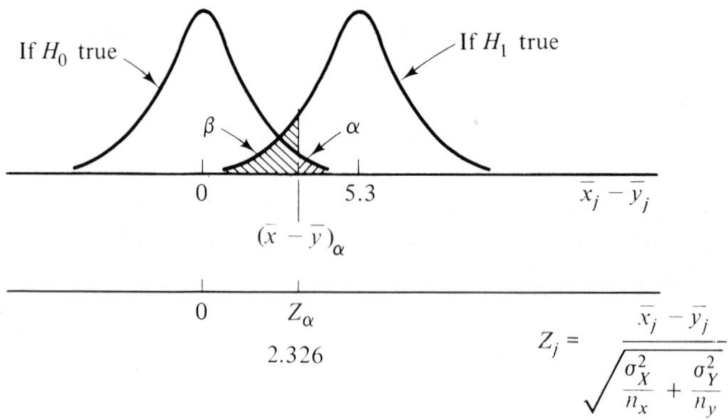

Figure 7.21. Example 7.10.1—$\bar{x}_j - \bar{y}_j$ distribution model for testing the equality of two means.

For $\alpha = 0.01$, the Z-value read from Appendix D is 2.326 and the decision rule is

$$z_{\text{data}} \geq Z_\alpha = 2.326, \qquad \text{reject } H_0$$
$$\text{otherwise, accept } H_0$$

Determining sample size,

$$d = \frac{\mu_X - \mu_Y}{\sqrt{\sigma_X^2 + \sigma_Y^2}}$$
$$= \frac{5.3}{\sqrt{(5)^2 + (5)^2}} = 0.75$$

and entering Fig. 7.4 with $d = 0.75$ and $\beta = 0.15$, sample size is read to be 20.

---

*Two points should be noted: (1) This test differs from the test of the mean (Sec. 7.8) in that there may be little or no knowledge of the separate μ_X- and μ_Y-values; this test is limited to merely whether or not $\mu_X = \mu_Y$, and (2) since an explicit statement of H_1 is required in order to consider β (see Sec. 7.4), a difference magnitude is required (i.e., $\mu_X > \mu_Y$ by 5.3 lb).

A random sample of 20 observations each from the X and Y processes yields mean values $\bar{x} = 565$ lb and $\bar{y} = 563$ lb. The mean difference in standard units is

$$z_{565-563} = \frac{565 - 563}{\sqrt{[(5)^2/20] + [(5)^2/20]}} = 1.27$$

Therefore,

$$z_{\text{data}} = 1.26 < Z_\alpha = 2.326$$

and the hypothesis is accepted.*

Sample sizes n_x and n_y are identical in Example 7.10.1. If for physical reasons sample sizes are fixed and unequal, an OC curve can still be used to determine β by using the relationship

$$n = \frac{\sigma_X^2 + \sigma_Y^2}{(\sigma_X^2/n_x) + (\sigma_Y^2/n_y)} \tag{7.8}$$

and entering the OC curve with d, n and reading β.

The testing procedure for the remaining cases is identical to that of Example 7.10.1 with the exceptions of the statistic used to make the test and the model on which the test is based.

In the second case, σ_X and σ_Y unknown but known to be equal, the statistic is

$$t_j = \frac{\bar{x}_j - \bar{y}_j}{\sqrt{(1/n_x) + (1/n_y)}\sqrt{(1/n_x + n_y - 2)[\sum (x_i - \bar{x})^2 + \sum (y_i - \bar{y})^2]}} \tag{7.9}$$

which is t-distributed with $n_x + n_y - 2$ $d.f.$

For the third case, σ_X and σ_Y unknown and not necessarily equal, the statistic is

$$t_j = \frac{\bar{x}_j - \bar{y}_j}{\sqrt{(s_x^2/n_x) + (s_y^2/n_y)}} \tag{7.10}$$

which is approximately t-distributed when H_0 is true with ν $d.f.$ where ν is

$$\nu = \frac{[(s_x^2/n_x) + (s_y^2/n_y)]^2}{\dfrac{(s_x^2/n_x)^2}{n_x + 1} + \dfrac{(s_y^2/n_y)^2}{n_y + 1}} - 2 \tag{7.11}$$

---

*Again the same argument applies, only the model is different. If H_0 is true, by chance alone a Z-value as large as 2.326 is obtained, on the average, 1% of the time. In this case, the z-value of 1.27 (computed from the observations) can occur 10.4% of the time. Thus, following the decision rule, which is based on an agreed α-value, there is no evidence for rejection and H_0 is accepted.

The Type II decision error cannot be considered since the distribution of the t random variable, Eq. (7.10), is not known when $\mu_X \neq \mu_Y$.

Summary—Testing That Two Means Are Equal

X and Y: independent random variables.
X- and Y-distributions normal, or $n_x \geq 30$ and $n_y \geq 30$, where n_x and n_y are the respective sample sizes.
σ_X and σ_Y known.

$$H_0:\ \mu_X = \mu_Y \qquad H_1:\ \mu_X > \mu_Y$$

$$d = \frac{\mu_X - \mu_Y}{\sqrt{\sigma_X^2 + \sigma_Y^2}}, \qquad n = n_x = n_y$$

n: enter Fig. 7.4 with d, β.

[If $n_x \neq n_y$, use Eq. (7.8).]

Decision Rule:

$$z_{\text{data}} \geq Z_\alpha, \text{ reject } H_0$$

where

$$z_{\text{data}} = \frac{\bar{x}_{\text{data}} - \bar{y}_{\text{data}}}{\sqrt{(\sigma_X^2/n_x) + (\sigma_Y^2/n_y)}}$$

$$H_0:\ \mu_X = \mu_Y \qquad H_1:\ \mu_X < \mu_Y$$

$$d = \frac{\mu_Y - \mu_X}{\sqrt{\sigma_X^2 + \sigma_Y^2}}, \qquad n = n_x = n_y$$

n: enter Fig. 7.4 with d, β.

[If $n_x \neq n_y$, use Eq. (7.8).]

Decision Rule:

$$z_{\text{data}} \leq -Z_\alpha, \qquad \text{reject } H_0$$

$$H_0:\ \mu_X = \mu_Y \qquad H_1:\ \mu_X \gtrless \mu_Y$$

$$d = \frac{|\mu_X - \mu_Y|}{\sqrt{\sigma_X^2 + \sigma_Y^2}}, \qquad n = n_x = n_y$$

n: enter Figs. 7.5, 7.6 with d, β.
Decision Rule:

$$|z_{\text{data}}| \geq Z_{\alpha/2}, \qquad \text{reject } H_0$$

Summary—Testing That Two Means Are Equal

X and Y: independent random variables.
X- and Y-distributions normal, or $n_x \geq 30$ and $n_y \geq 30$, where n_x and n_y are the respective sample sizes.
σ_X and σ_Y unknown, but equal.

$$H_0: \ \mu_X = \mu_Y \qquad H_1: \ \mu_X > \mu_Y$$

$$d = \frac{\mu_X - \mu_Y}{2\sigma}, \qquad n = n_x = n_y, \qquad \sigma = \sigma_X = \sigma_Y,$$

where σ is estimated.
n: enter Fig. 7.18 with d, β and read n, then use $(n + 1)/2$ as sample size.

[If $n_x \neq n_y$, use Eq. (7.8).]

Decision Rule:

$$t_{\text{data}} \geq t_{\alpha, n_x + n_y - 2}, \qquad \text{reject } H_0$$

where

$$t_j = \frac{\bar{x}_j - \bar{y}_j}{\sqrt{(1/n_x) + (1/n_y)}\sqrt{(1/n_x + n_y - 2)[\sum (x_i - \bar{x})^2 + \sum (y_i - \bar{y})^2]}}$$

$$H_0: \ \mu_X = \mu_Y \qquad H_1: \ \mu_X < \mu_Y$$

$$d = \frac{\mu_Y - \mu_X}{2\sigma}, \qquad n = n_x = n_y, \qquad \sigma = \sigma_X = \sigma_Y.$$

where σ is estimated.
n: enter Fig. 7.18 with d, β and read n, then use $(n + 1)/2$ as sample size.

[If $n_x \neq n_y$, use Eq. (7.8).]

Decision Rule:

$$t_{\text{data}} \leq -t_{\alpha, n_x + n_y - 2}, \qquad \text{reject } H_0$$

$$H_0: \ \mu_X = \mu_Y \qquad H_1: \ \mu_X \gtrless \mu_Y$$

$$d = \frac{|\mu_X - \mu_Y|}{2\sigma}, \qquad n = n_x = n_y, \qquad \sigma = \sigma_X = \sigma_Y$$

where σ is estimated.
n: enter Figs. 7.19, 7.20 with d, β and read n, then use $(n + 1)/2$ as sample size.

[If $n_x \neq n_y$, use Eq. (7.8).]

Decision Rule:

$$|t_{\text{data}}| \geq t_{\alpha/2, n_x + n_y - 2}, \qquad \text{reject } H_0$$

Summary—Testing That Two Means Are Equal

X an Y: independent random variables.
X- and Y-distributions normal, or $n_x \geq 30$ and $n_y \geq 30$, where n_x and n_y are respective sample sizes.
μ_X and μ_Y unknown, not necessarily equal.

$$H_0: \ \mu_X = \mu_Y \qquad H_1: \ \mu_X > \mu_Y$$

Type II decision error not considered.
Decision Rule:

$$t_{\text{data}} \geq t_{\alpha,\nu}, \qquad \text{reject } H_0$$

where

$$t_{\text{data}} = \frac{\bar{x}_j - \bar{y}_j}{\sqrt{(s_x^2/n_x) + (s_y^2/n_y)}}$$

and

$$\nu = \frac{[(s_x^2/n_x) + (s_y^2/n_y)]^2}{\dfrac{(s_x^2/n_x)^2}{n_x + 1} + \dfrac{(s_y^2/n_y)^2}{n_y + 1}} - 2$$

$$H_0: \ \mu_X = \mu_Y \qquad H_1: \ \mu_X < \mu_Y$$

Type II decision error not considered.
Decision Rule:

$$t_{\text{data}} \leq -t_{\alpha,\nu}, \qquad \text{reject } H_0$$

$$H_0: \ \mu_X = \mu_Y \qquad H_1: \ \mu_X \gtrless \mu_Y$$

Type II decision error not considered.
Decision Rule:

$$|t_{\text{data}}| \geq t_{\alpha/2,\nu}, \qquad \text{reject } H_0$$

EXERCISES

1. Test at a 1% significance level that $\mu_X = \mu_Y$ against the alternative that $\mu_X \gtrless \mu_Y$ for the conditions: $n_x = n_y = 100$, $\sigma_X = 0.04$, $\sigma_Y = 0.05$, $\bar{x}_{\text{data}} = 6.11$, and $\bar{y}_{\text{data}} = 6.14$. *Ans.* $|z_{\text{data}}| = 4.685$, reject H_0.

2. Repeat Exercise 1 assuming σ_X and σ_Y are unknown (estimated to be $\sigma_X = \sigma_Y = 0.05$), $\sum (x_i - \bar{x})^2 = 4$, and $\sum (y_i - \bar{y})^2 = 5$. *Ans.* $|t_{\text{data}}| = 0.995$, accept H_0.

3. Repeat Exercise 1 assuming σ_X and σ_Y are unknown and not necessarily equal, $s_x^2 = 0.042$, and $s_y^2 = 0.049$. *Ans.* $|t_{\text{data}}| = 0.994$, accept H_0.

4. Assume in Exercise 1 that it is desired to detect with probability 0.90 a difference $\mu_X - \mu_Y = \sqrt{0.0041}$. Determine minimum sample size. *Ans.* $n = 15$.

7.11 Test That Two Means Are Equal, Paired Observations

The test of equality of the means is treated in a more sophisticated manner by "analysis of variance" in Chap. 11. However, a common technique usually presented independent of the analysis-of-variance topic is the case of paired comparisons. Each pair of observations (X_i, Y_i) is taken under the same experimental conditions, with the conditions varying from pair to pair.

Consider an experiment comparing the tread loss of different types of tires after a specified mileage. A criterion for judging tire type is mean tread loss difference $\mu_{(X-Y)}$, where μ_X is mean tread loss of type X tire and μ_Y is mean tread loss of type Y tire. If $\mu_{(X-Y)}$ is significantly different from zero, it may be concluded that tire types X and Y differ from the standpoint of mean tread loss. Obtaining a sample difference $\bar{x} - \bar{y}$ which is a good estimate of $\mu_X - \mu_Y$ is no simple statistical matter if the experiment is an uncontrolled test under actual road conditions. Observations made under driving conditions are affected by a number of factors contributing to tread loss—differences between cars, drivers, and driving conditions, types of wheel mounting, wheel positions on cars, etc. A sample mean difference in tire tread loss can easily reflect not a difference between tire types but a difference between cars.

Control and randomization are two ways of guarding against experimental bias. Two examples of "control" are as follows. In the tire-tread-loss experiment, a physical method of control is to rotate tire positions on cars at specified mileage periods during the experiment, thus controlling the wheel-position effect on the observations. A statistical method of control is to mount tire types X and Y equally to two cars being used in the experiment, two X type and two Y type tires on each car. Where a car goes, both tire types go, and to a certain extent this equalizes or "averages out" the effects of driving conditions on the observations.

In the engineering and technology environment, a typical experiment involves a number of factors affecting the observations. Some factors are physically controllable. Other factor effects are statistically controllable by proper design of the experiment (Chap. 11). The effects of other uncontrollable factors contribute to experimental error. In practice, this error is reduced somewhat by randomization methods. The term *randomization* refers to methods of obtaining observations in a random sequence. Randomization of the order of experimentation tends to average out the effects of uncontrolled variables on the observations.*

The paired-observations method involves one restriction on complete randomization, that is, one factor is experimentally or statistically controlled and presumably all other factors are partially controlled by randomization. This is the simplest example of a class of experimental designs that are later (Chap. 11) called "randomized complete block designs." The assumptions underlying the paired-comparisons method are identical to that of Sec. 7.10, i.e., independent random observations, normal populations of responses, and variances equal—in this case, $\sigma_X = \sigma_Y$.

In a test of the equality of means when observations are paired, the random variable is the difference $X_i - Y_i$ between the members of each pair. The null hypothesis being tested is that the mean difference $\mu_{(X-Y)}$ is zero. Thus, this hypothesis test is a special case of the single parameter test of Sec. 7.9, a test of the mean with standard deviation unknown, where the variable X_i is now replaced by the variable $X_i - Y_i$. For the test of Sec. 7.9, the random variable is X_i, the null hypothesis is H_0: $\mu_X = \mu_0$, and the statistic is

$$t_j = \frac{\bar{x}_j - \mu_0}{s_X\sqrt{n}}$$

Here, for the paired-comparisons test, the random variable is the difference $X_i - Y_i$, the null hypothesis is H_0: $\mu_{(X-Y)} = 0$, and the t-statistic becomes

$$t_j = \frac{\overline{(x - y)}_j - \mu_{(X-Y)}}{s_{(x-y)}/\sqrt{n}} = \frac{\overline{(x - y)}_j}{s_{(x-y)}/\sqrt{n}} \tag{7.12}$$

which is t-distributed with $n - 1$ $d.f.$ The mean difference is

$$\overline{(x - y)} = \frac{1}{n}\sum_{i=1}^{n}(x_i - y_i) \tag{7.13}$$

and the sample variance is

$$s^2_{(x-y)} = \frac{1}{n-1}\sum_{i=1}^{n}[(x_i - y_i) - \overline{(x - y)}]^2 \tag{7.14}$$

The OC curves for the test of Sec. 7.9 are also appropriate here; only the

*For example, in the tire-tread-loss experiment, if position effect is not physically controlled by identical tire rotation for the two tire brands being tested, the tires can be randomly assigned to wheel positions for the various replications of the experiment (i.e., several tires of each brand being tested). Hopefully, this procedure will tend to average out wheel-position effect on the observations.

variable is changed. For the test of Sec. 7.9,

$$d = \frac{\mu_1 - \mu_0}{\sigma_X}$$

where σ_X is estimated from historical or physical knowledge. For the paired-comparison test, changing the variable,

$$d = \frac{|\mu_{(X-Y)}| - 0}{\sigma_{(X-Y)}}$$

where the first numerator term is the alternative hypothesis value H_i: $\mu_{(X-Y)} \neq 0$ and the second numerator term follows from the null hypothesis H_0: $\mu_{(X-Y)} = 0$. Since the X_i and Y_i are assumed to be independent, from Eq. (5.10), the denominator term becomes

$$\sigma_{(X-Y)} = \sqrt{\sigma_X^2 + \sigma_Y^2}$$

and

$$d = \frac{|\mu_{(X-Y)}|}{\sqrt{\sigma_X^2 + \sigma_Y^2}} \tag{7.15}$$

with it being understood that the numerator term is the alternative hypothesis H_1-value.

EXAMPLE 7.11.1

The wear of two guide rails on a product item is an important factor determining the functional life of the item. A test of two material types X and Y for the guide rails is conducted as follows.

One type X and one type Y material specimens are used on each product unit and the specimens are randomly assigned to left-hand and right-hand rail positions. The test is performed at a significance level $\alpha = 0.01$. From previous experience, the standard deviation of guide-rail wear is estimated to be 0.000118 in. It is desired to detect a wear difference of 0.0003 in. with probability 0.90. The test and test conditions are:

$$H_0: \quad \mu_{(X-Y)} = 0 \qquad \alpha = 0.01$$
$$H_1: \quad |\mu_{(X-Y)}| = 0.0003, \qquad \beta = 0.10$$

Using the estimated standard deviation of guide-rail wear for the d-computation,

$$\sigma_X = \sigma_Y = 0.00018$$

and

$$d = \frac{0.0003}{\sqrt{2(0.00018)^2}} = 1.179$$

Entering Fig. 7.18 with $d = 1.8$ and $\beta = 0.10$, sample size is read to be 7. The $t_{\alpha,\nu}$-value, for $\alpha = 0.01$ and $d.f. = 7 - 1$, read from Appendix F is 3.143. Therefore, the decision rule is

$$t_{\text{data}} \geq t_{\alpha,n-1} = 3.143, \qquad \text{reject } H_0$$
$$\text{otherwise,} \qquad \text{accept } H_0$$

Table 7.2 gives the test results for seven paired observations (x_i, y_i) and computations of the mean difference and difference variance. The mean difference is

$$\overline{(x - y)}_{\text{data}} = \frac{1}{7} \sum_{i=1}^{7} (x_i - y_i) = 0.4$$

and the sample variance is

$$s^2_{(x-y)} = \frac{1}{6} \sum_{i=1}^{7} [(x_i - y_i) - \overline{(x - y)}]^2 = \frac{1}{6}(0.84) = 0.14$$

The data value for the test statistic is

$$t_{\text{data}} = \frac{0.4}{\sqrt{\frac{0.14}{7}}} = 2.83$$

and since

$$t_{\text{data}} = 2.83 < t_{\alpha,n-1} = 3.143$$

the hypothesis is accepted and the conclusion is that the wear difference between materials X and Y is not significant.

Table 7.2. EXAMPLE 7.11.1: GUIDE-RAIL WEAR IN UNITS OF 0.001 IN., PAIRED OBSERVATIONS ON MATERIALS X AND Y

x_i	y_i	$x_i - y_i$	$[(x_i - y_i) - \overline{(x - y)}]^2$
3.8	3.3	0.5	0.01
3.2	2.7	0.5	0.01
3.0	3.0	0.0	0.16
3.7	3.1	0.6	0.04
0.8	1.0	−0.2	0.36
2.9	2.0	0.9	0.25
2.9	2.4	0.5	0.01
		2.8	0.84

Summary—Testing That Two Means Are Equal, Paired Comparisons

$X_i - Y_i$ is the random variable, paired observations (x_i, y_i)
X and Y: independent random variables.
X- and Y-distributions normal, or $n_x = n_y \geq 30$.
σ_X and σ_Y unknown, but equal.

$$H_0:\ \mu_{(X-Y)} = 0 \qquad H_1:\ \mu_{(X-Y)} \neq 0$$

Use single parameter t-test, Sec. 7.9, replacing X_i by $X_i - Y_i$ and

$$t_j = \frac{\bar{x}_j - \mu_0}{s_x/\sqrt{n}} \quad \text{by} \quad t_j = \frac{|(\overline{x-y})_j|}{s_{(x-y)}/\sqrt{n}}$$

and

$$d = \frac{\mu_1 - \mu_0}{\sigma_X} \quad \text{by} \quad d = \frac{|\mu_{(X-Y)}|}{\sqrt{\sigma_X^2 + \sigma_Y^2}}$$

where $\sigma = \sigma_X = \sigma_Y$ is estimated and $|\mu_{(X-Y)}|$ is understood to be the absolute value of the alternative hypothesis H_1-value.
Decision Rule:

$$t_{\text{data}} \geq t_{\alpha, n-1}, \qquad \text{reject } H_0$$

where

$$t_{\text{data}} = \frac{|(\overline{x-y})_{\text{data}}|}{s_{(x-y)}/\sqrt{n}}$$

EXERCISES

1. The experiment of Example 7.11.1 is repeated yielding the following paired observations (X_i, Y_i):

X	2.9	3.1	3.5	1.3	2.3	2.9	2.6	1.8	2.2	2.6
Y	2.9	2.9	3.2	1.1	2.2	2.8	2.5	1.3	1.9	2.4

Test the hypothesis that the mean wear difference is zero.

Ans. $t_{\text{data}} = 4.472$, reject H_0.

2. Surface finish reading in microinches for two types of materials X and Y are:

X	40	30	30	40	40	30	20	20
Y	30	30	50	30	40	20	40	20

Test the hypothesis that the mean surface finish difference is zero.

Ans. $t_{\text{data}} = 0.2835$, accept H_0.

7.12 Test of the Standard Deviation of a Normal Population

To this point, the hypothesis tests have been concerned with inferences regarding the mean of an unknown population. It is natural to continue now with tests concerning the other important parameter: the variance (or standard deviation). In the case of testing the mean, two test types are important: (1) testing that the mean has a specified value, and (2) testing that two means are equal. Following the same pattern, a first consideration is that of testing that the standard deviation has a specified value, and then testing for the equality of two standard deviations.

In Sec. 6.6, the observation is made that comparisons between sample and population variances can be made in terms of differences $(s_x^2 - \sigma_X^2)$ or ratios (s_x^2/σ_X^2), using distributions of these measures. A common measure with a known distribution function is the ratio

$$\frac{(n-1)s_x^2}{\sigma_X^2}$$

which is chi-square distributed with $n - 1$ *d.f.*

If interest is in whether or not the standard deviation of an unknown normal population is equal to a specified value σ_0, the following fact can be utilized. If X is $N(\mu, \sigma_0^2)$, then the statistic $(n-1)s_x^2/\sigma_0^2$ is chi-square distributed with $n - 1$ *d.f.* Thus, if the hypothesis $H_0: \sigma_X = \sigma_0$ is true, the chi-square distribution model is appropriate for formulating decision rules based on risk values α of incurring a Type I decision error. For the three types of one-and two-sided tests, the decision rules are

$$\begin{aligned}
&H_1:\ \sigma_X = \sigma_1 > \sigma_0, && \chi^2_{\text{data}} \geq \chi^2_{\alpha,n-1}, && \text{reject } H_0\\
&H_1:\ \sigma_X = \sigma_1 < \sigma_0, && \chi^2_{\text{data}} \leq \chi^2_{1-\alpha,n-1}, && \text{reject } H_0\\
&H_1:\ \sigma_X = \sigma_1 \gtrless \sigma_0, && \left.\begin{aligned}\chi^2_{\text{data}} &\geq \chi^2_{\alpha/2,n-1}\\ &\leq \chi_{1-(\alpha/2),n-1}\end{aligned}\right\}, && \text{reject } H_0
\end{aligned}$$

In the tests of the mean, the abscissa values for *OC* curves are in terms of differences between hypotheses values (e.g., $\mu_1 - \mu_0$). For tests concerning the standard deviation, the abscissa values are ratios of the null and alternative hypothesis values. Corresponding to Eq. (7.3), the ratio

$$\lambda = \frac{\sigma_1}{\sigma_0} \tag{7.16}$$

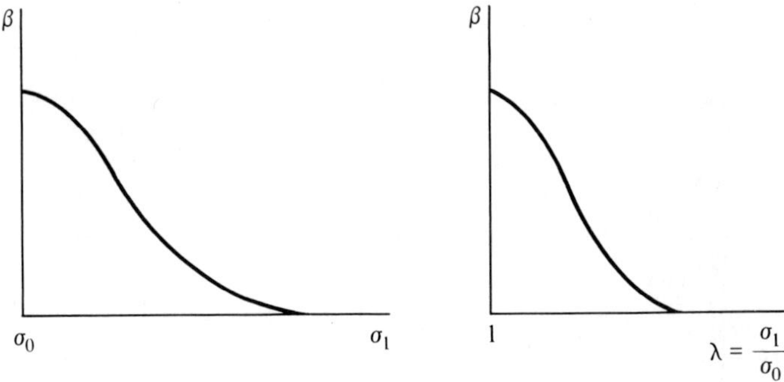

Figure 7.22. *OC* curves, definition and standard form, for a chi-square test.

converts an *OC* curve in definitive form to a standard units form (review Sec. 7.4) as shown in Fig. 7.22.

OC curves for one-and two-sided tests of a hypothesis $H_0: \ \sigma_X = \sigma_0$ are given in Figs. 7.23–7.26.

EXAMPLE 7.12.1

An existing automatic machine fills 5-lb boxes with product. A new machine is considered for this job and will be approved if the variability is compatible with the standard for the old machine (i.e., $\sigma_X = 0.05$ lb). A test of the demonstration model of the new machine is proposed. The test is to be conducted at a 1% significance level. If σ_X is as large as 0.10 lb, the probability of detecting this fact should be 0.95. Summarizing the test conditions,

$$H_0: \ \sigma_X = 0.05, \qquad \alpha = 0.01$$
$$H_1: \ \sigma_X = 0.10, \qquad \beta = 0.05$$

To determine sample size, the ratio of the null and alternative hypothesis values is computed:

$$\lambda = \frac{\sigma_1}{\sigma_0} = \frac{0.10}{0.05} = 2$$

Entering Fig. 7.23 with $\lambda = 2$ and $\beta = 0.05$, sample size is read to be approximately 18. The decision rule is

$$\chi^2_{\text{data}} \geq \chi^2_{0.01,17} = 33.41, \qquad \text{reject } H_0$$
$$\text{otherwise,} \qquad \text{accept } H_0$$

A random sample of 18 box fills is weighed, yielding a variance $s_x^2 = 0.008$

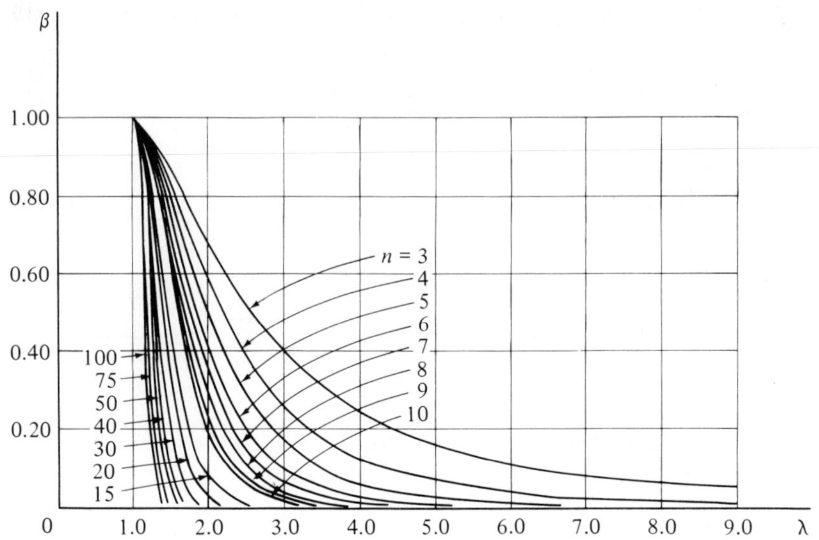

Level of significance $\alpha = 0.01$

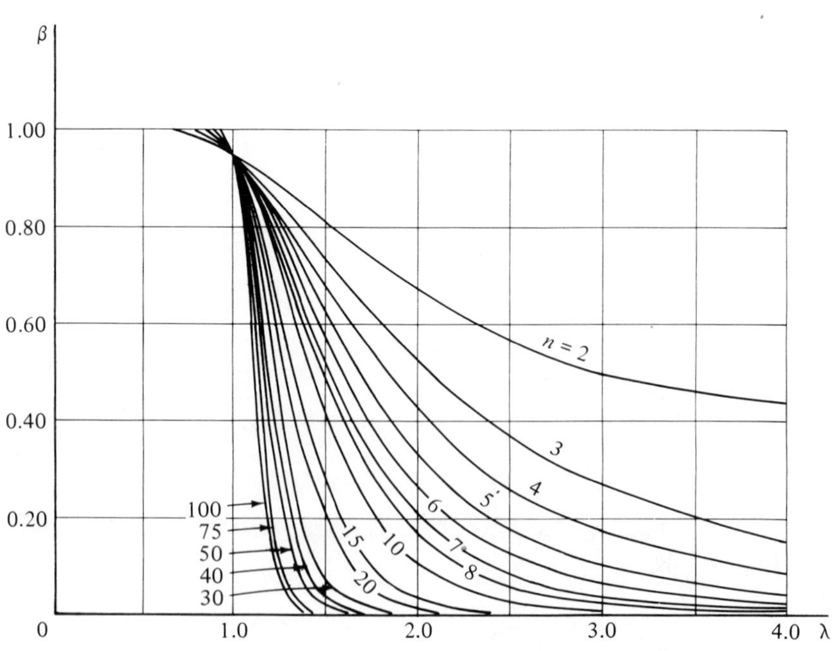

Level of significance $\alpha = 0.05$

Figure 7.23. *OC* curves for a one-sided (upper tail) chi-square test.

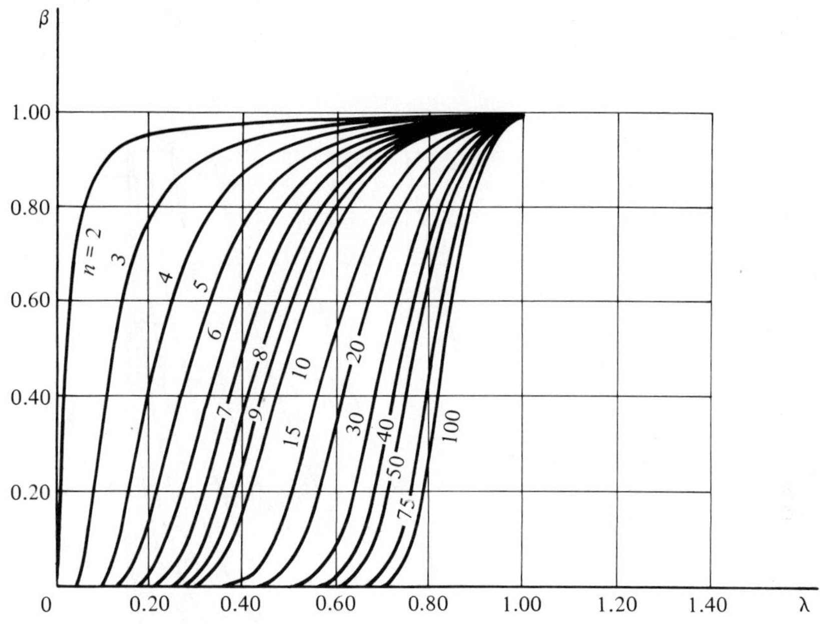

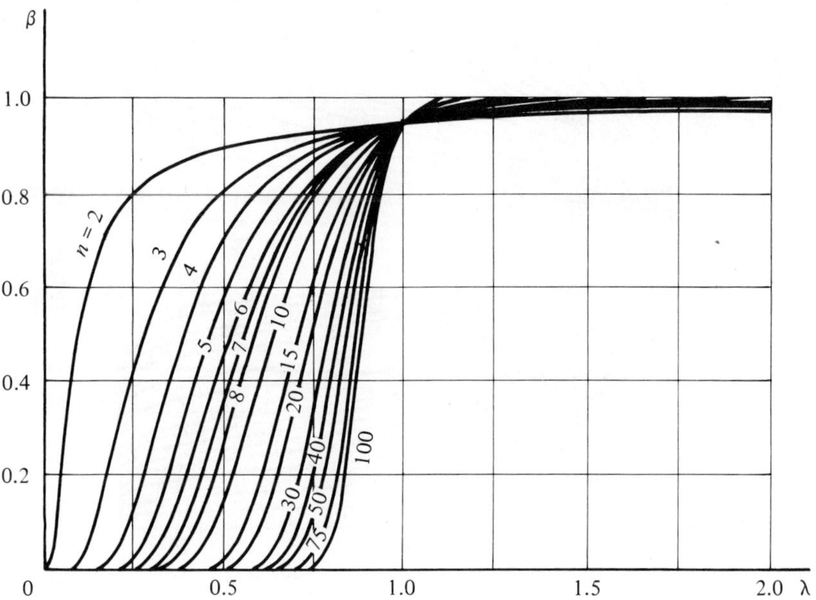

Figure 7.24. *OC* curves for a one-sided (lower tail) chi-square test.

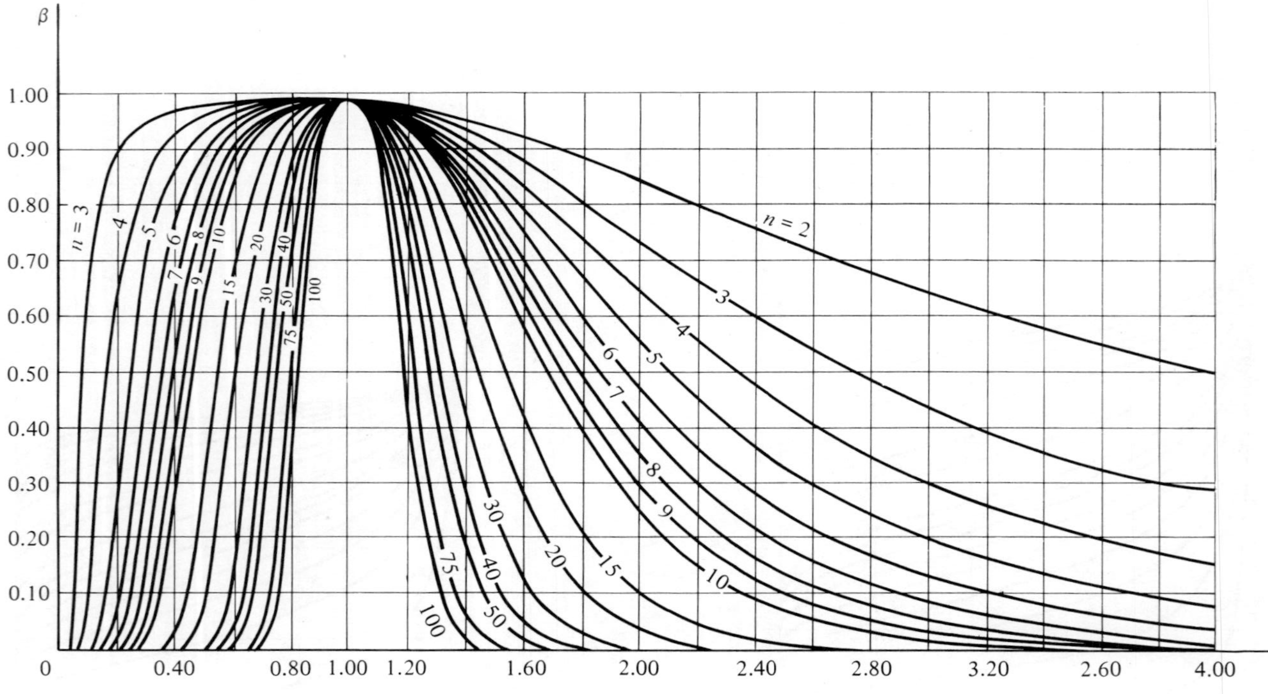

Figure 7.25. *OC* curves for a two-sided chi-square test, $\alpha = 0.01$.

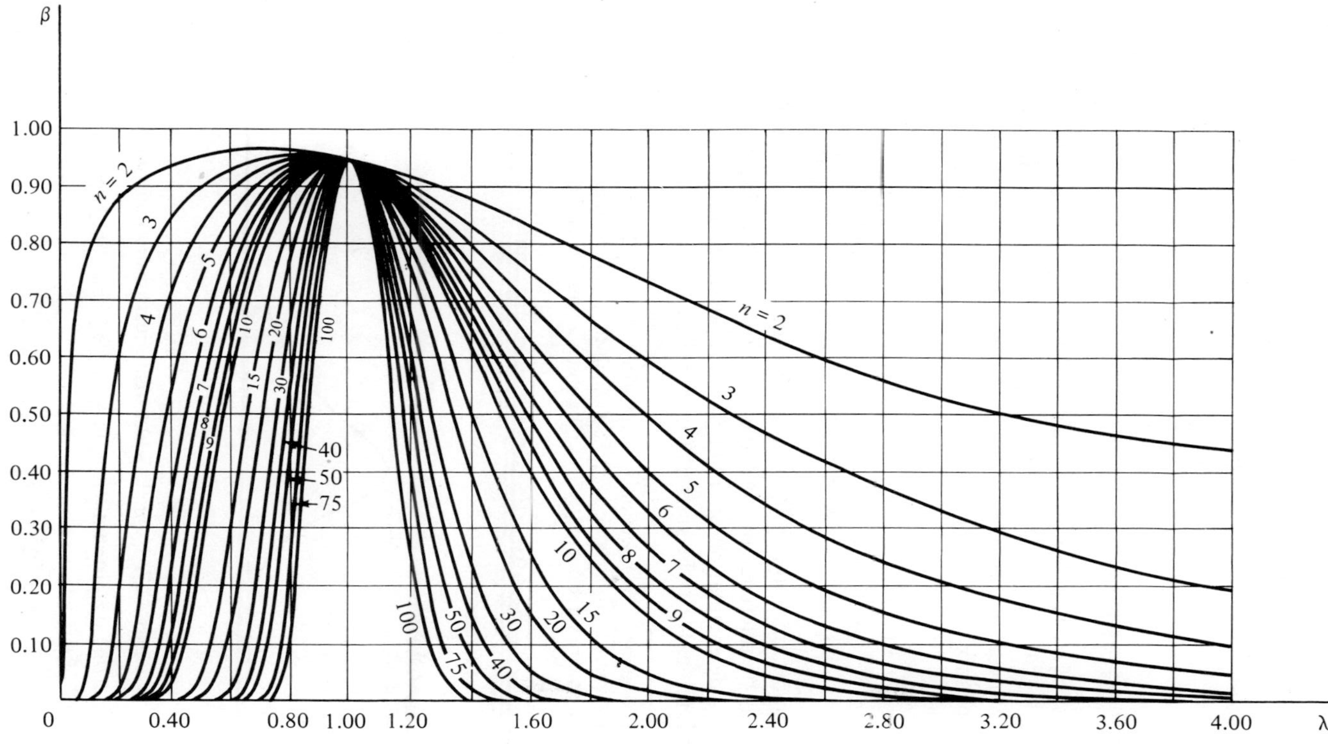

Figure 7.26. *OC* curves for a two-sided chi-square test, $\alpha = 0.05$.

for the 18 observations. The test statistic is

$$\chi^2_{\text{data}} = \frac{(n-1)s_x^2}{\sigma_0^2} = \frac{17(0.008)}{(0.05)^2} = 54.4$$

and since

$$\chi^2_{\text{data}} = 54.4 > \chi^2_{\alpha,n-1} = 33.41$$

the hypothesis is rejected.*

A test of σ_X is usually an upper one-sided test, since applications problems are usually concerned with the possibility that variation may be too great. However, if a two-sided test is desired, the decision rule is obtained from Appendix G as follows. For the conditions of Example 7.12.1, the decision rule is

$$\left.\begin{aligned}\chi^2_{\text{data}} &\geq \chi^2_{0.005,17} = 35.718\\ &\leq \chi^2_{0.995,17} = 5.697\end{aligned}\right\}, \qquad \text{reject } H_0$$

The most efficient applications procedure for handling the β and n combination is by reference to available *OC* curves. It is of interest, however, to note that a determination of either β or n is possible simply by exploring Appendix E table. It can be shown† that for the case of $H_1: \sigma_X = \sigma_1 > \sigma_0$

*The argument for rejection is summarized in Fig. 7.27. If H_0 is true, by chance alone a chi-square value as large as 33.41 is obtained, on the average, only 1% of the time. Yet, the chi-square value computed from the observations is 60.8. Therefore, H_0 is rejected.

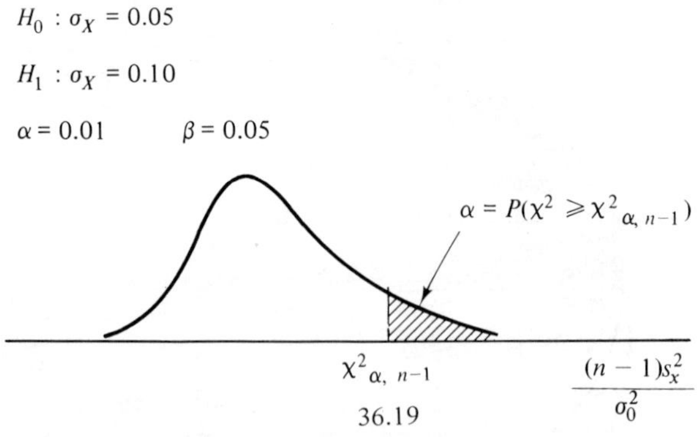

$H_0 : \sigma_X = 0.05$

$H_1 : \sigma_X = 0.10$

$\alpha = 0.01 \qquad \beta = 0.05$

Figure 7.27. Example 7.12.1—chi-square distribution model.

†Considering the distributions of $(n-1)s_x^2/\sigma_X^2$ under the hypotheses $H_0: \sigma_X = \sigma_0$ and $H_1: \sigma_X = \sigma_1 > \sigma_0$,

$$\frac{(n-1)s_x^2}{\sigma_0^2} = \chi^2_{\alpha,n-1} \quad \text{and} \quad \frac{(n-1)s_x^2}{\sigma_1^2} = \chi^2_{1-\beta,n-1}$$

Constructing the ratio of these two quantities,

$$\lambda^2 = \frac{\chi^2_{\alpha, n-1}}{\chi^2_{1-\beta, n-1}}$$

Thus, for the conditions of Example 7.12.1,

$$\lambda^2 = 2^2 = 4 = \frac{\chi^2_{0.01, n-1}}{\chi^2_{0.95, n-1}}$$

and, exploring the 0.01 and 0.95 columns of the chi-square table (Appendix E) for combinations of the above dividend and divisor that yield a quotient near 4, the combination yielding a quotient closest to 4 is found in the row for *d.f.* equal to 16. Thus, *n* is exactly 17.

Summary—Test of the Standard Deviation of a Normal Population

---

X-distribution normal.

$$H_0:\ \sigma_X = \sigma_0 \qquad H_1:\ \sigma_X = \sigma_1 > \sigma_0$$

$$\lambda = \frac{\sigma_1}{\sigma_0} \qquad n:\ \text{enter Fig. 7.23 with } \lambda,\ \beta.$$

Decision Rule:

$$\chi^2_{\text{data}} \geq \chi^2_{\alpha, n-1}, \qquad \text{reject } H_0$$

where

$$\chi^2_{\text{data}} = \frac{(n-1)s_x^2}{\sigma_0^2}$$

$$H_0:\ \sigma_X = \sigma_0 \qquad H_1:\ \sigma_X = \sigma_1 < \sigma_0$$

$$\lambda = \frac{\sigma_1}{\sigma_0} \qquad n:\ \text{enter Fig. 7.24 with } \lambda,\ \beta.$$

Decision Rule:

$$\chi^2_{\text{data}} \leq \chi^2_{1-\alpha, n-1}, \qquad \text{reject } H_0$$

$$\frac{\chi^2_{\alpha, n-1}}{\chi^2_{1-\beta, n-1}} = \frac{[(n-1)s_x^2]/\sigma_0^2}{[(n-1)s_x^2]/\sigma_1^2} = \frac{1}{\sigma_0^2} \cdot \frac{\sigma_1^2}{1} = \lambda^2$$

In a similar fashion, for the case of $H_1: \sigma_X = \sigma_1 < \sigma_0$,

$$\lambda^2 = \frac{\chi^2_{1-\alpha, n-1}}{\chi^2_{\beta, n-1}}$$

For a two-sided test, sample size is obtained approximately from

$$\lambda^2 = \frac{\chi^2_{\alpha/2, n-1}}{\chi^2_{1-\beta, n-1}}$$

when the probability $P(\chi^2 \leq \chi^2_{1-\alpha/2, n-1}/\lambda^2)$ is small compared to β, or from

$$\lambda^2 = \frac{\chi^2_{1-\alpha/2, n-1}}{\chi^2_{\beta, n-1}}$$

when the probability $P(\chi^2 \geq \chi^2_{\alpha/2, n-1}/\lambda^2)$ is small compared to β.

$$H_0: \ \sigma_X = \sigma_0 \qquad H_1: \ \sigma_X = \sigma_1 \gtreqless \sigma_0$$

$$\lambda = \frac{\sigma_1}{\sigma_0} \qquad n: \ \text{enter Figs. 7.25, 7.26 with } \lambda, \beta.$$

Decision Rule:

$$\left.\begin{aligned} \chi^2_{\text{data}} &\geq \chi^2_{(\alpha/2),n-1} \\ &\leq \chi^2_{1-(\alpha/2),n-1} \end{aligned}\right\} \text{ reject } H_0$$

EXERCISES

1. A machine fills boxes with product. Concern is with the variation of box weights. Assume that box weight X is normally distributed. Test the hypothesis H_0: $\sigma_X = 0.1$ lb against an alternative H_1: $\sigma_X = 0.2$ lb at a 1% significance level. The probability of detecting variation such that $\sigma_X = 0.2$ lb should be 0.80. A random sample of 10 boxes yields the following weights: 5.5, 5.0, 5.2, 5.2, 5.0, 4.5, 4.9, 4.9, 4.8, 5.6 lb. *Ans.* $\chi^2_{\text{data}} = 96.40$, reject H_0.

2. Perform the *OC* curve reference to confirm the sample size given in Exercise 1.

3. State the decision rule if the hypothesis test in Exercise 1 is a two-sided test. *Ans.* $\chi^2_{\text{data}} \geq 23.589$ or ≤ 1.735, reject H_0.

7.13 Test That Two Standard Deviations Are Equal (Normal Populations)

In the case of manufacturing processes, the mean can usually be set by proper machine adjustments, but the variability cannot be fixed. In applications of this sort, a frequent objective is to test that the variation of two processes is similar.

Comparisons of two population variances σ_X^2 and σ_Y^2 can be made in terms of a difference or a ratio. For example, if the ratio σ_X^2/σ_Y^2 is near unity, the variances differ only slightly. If the comparison is by means of a ratio σ_X^2/σ_Y^2, then since neither variance is usually known it is natural to compare the ratio s_x^2/s_y^2 of the estimated variances.

If X and Y are independent normally distributed random variables, from Eq. (6.14), the ratio

$$\frac{s_x^2/\sigma_X^2}{s_y^2/\sigma_Y^2}$$

is F-distributed with $n_x - 1$ and $n_y - 1$ *d.f.* Thus, if the hypothesis H_0: $\sigma_X = \sigma_Y$ is of interest, it follows that, if H_0 is true, the statistic s_x^2/s_y^2 is F-distributed with $n_x - 1$ and $n_y - 1$ *d.f.*, and this known distribution function can be utilized as a model to formulate the decision rule for this hypothesis test.

Since the *F*-distribution is nonsymmetrical, it is not possible to have equivalent lower and upper one-sided tests. The variable with the larger variability is denoted by X and the decision rules are

$$\begin{aligned} &H_1\colon\ \sigma_X > \sigma_Y, \qquad F_{\text{data}} \geq F_{\alpha, n_x-1, n_y-1}, \qquad \text{reject } H_0 \\ &H_1\colon\ \sigma_X \gtrless \sigma_Y, \qquad \left.\begin{aligned} F_{\text{data}} &\geq F_{\alpha/2, n_x-1, n_y-1} \\ &\leq F_{1-(\alpha/2), n_x-1\ n_y-1} \end{aligned}\right\}, \qquad \text{reject } H_0 \end{aligned}$$

where, from Eq. (6.10),

$$F_{1-(\alpha/2), n_x-1, n_y-1} = \frac{1}{F_{\alpha/2, n_y-1, n_x-1}}$$

Estimates of σ_X and σ_Y (based on historical or physical data) are required to compute $\lambda = \sigma_X/\sigma_Y$ and for identical sample sizes $n = n_x = n_y$* the *OC* curves of Figs. 7.28–7.31 may be used.

An examination of the *OC* curves, Figs. 7-28–7-31, indicates that the *F*-statistic for testing $H_0\colon \sigma_X = \sigma_Y$ is relatively insensitive for λ-values less than 2.0. Very large sample sizes are required to detect small differences between σ_X and σ_Y. Also, if the equality of more than two variances is in question, this test is not appropriate. To test the equality of several variances, a procedure given by Cochran's test† is available.

EXAMPLE 7.13.1

Production requirements indicate the need for utilizing several machines for the same processing operation. The variance of a product characteristic generated on two similar machines is of interest. Six product units each from machine X and machine Y yield variances $s_x^2 = 0.0044$ in. and $s_y^2 = 0.0011$ in. Is this difference significant at a 5% significance level?

A summary of the hypothesis test conditions is

$$H_0\colon\ \sigma_X = \sigma_Y, \qquad H_1\colon\ \sigma_X > \sigma_Y, \qquad \alpha = 0.05$$

---

*For $n_x \neq n_y$, *OC* curves can be found in Bowker, A. H. and Lieberman, G. J. (1972), *Engineering Statistics*, 2nd ed., Prentice-Hall, Inc., Englewood Cliffs, N. J., pp. 398–402.

†This test, due to W. G. Cochran, utilizes a test statistic

$$C = \frac{\text{largest } s_j^2}{s_1^2 + s_2^2 + \ldots + s_k^2}$$

where the s_j^2 are the respective estimates of σ_j^2. The interested student can refer to Dixon, W. J. and Massey, F. J. (1969), *Introduction to Statistical Analysis*, 3rd ed., McGraw-Hill Book company, N. Y., pp. 310, 536.

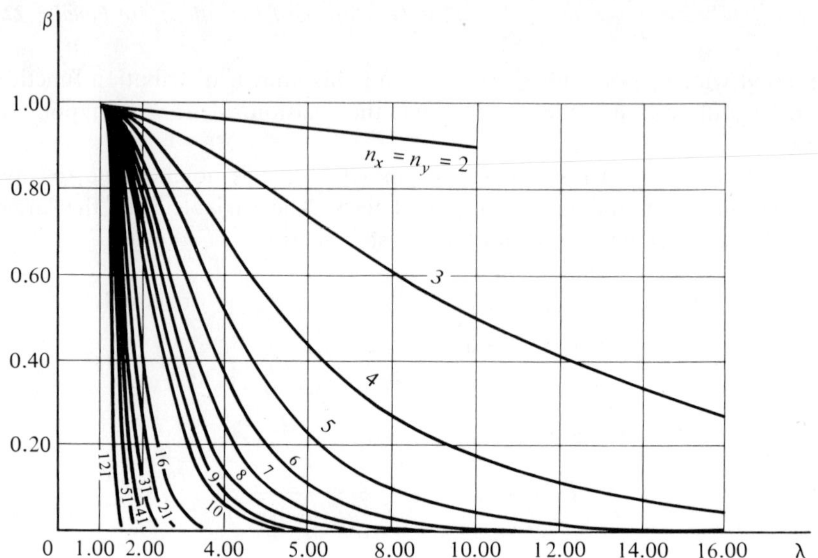

Figure 7.28. *OC* curves for a one-sided *F*-test, $\alpha = 0.01$.

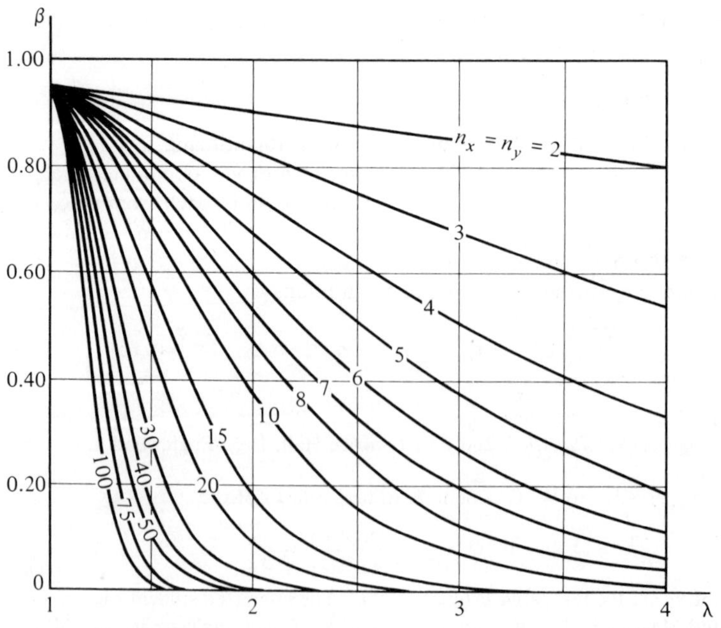

Figure 7.29. *OC* curves for a one-sided *F*-test, $\alpha = 0.05$.

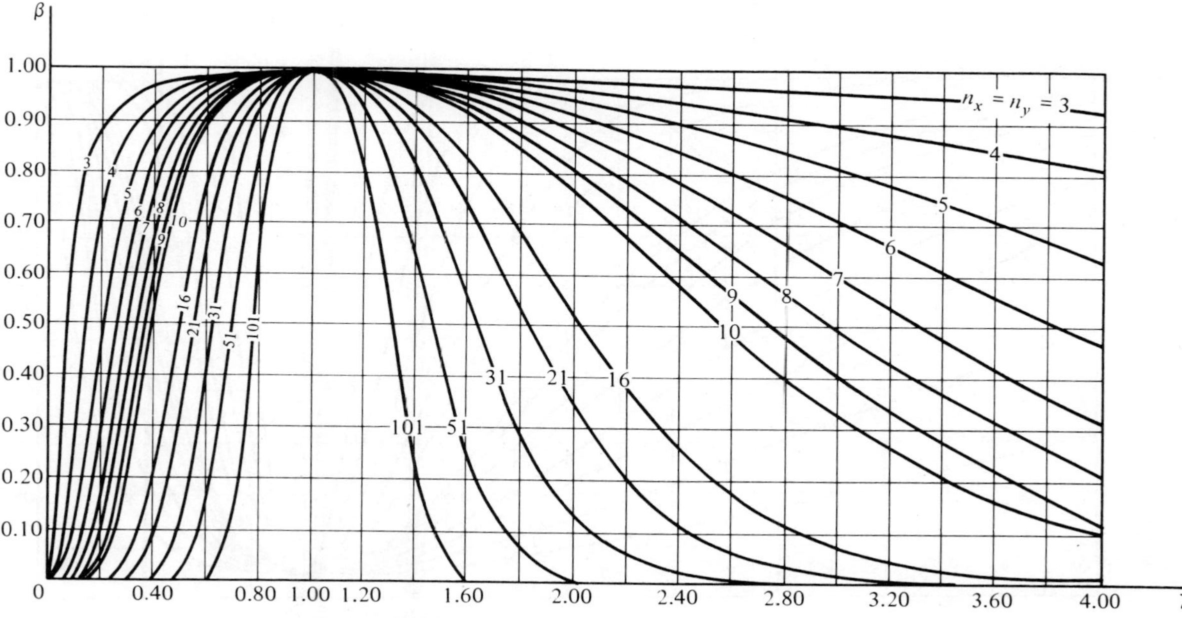

Figure 7.30. *OC* curves for a two-sided *F*-test, $\alpha = 0.01$.

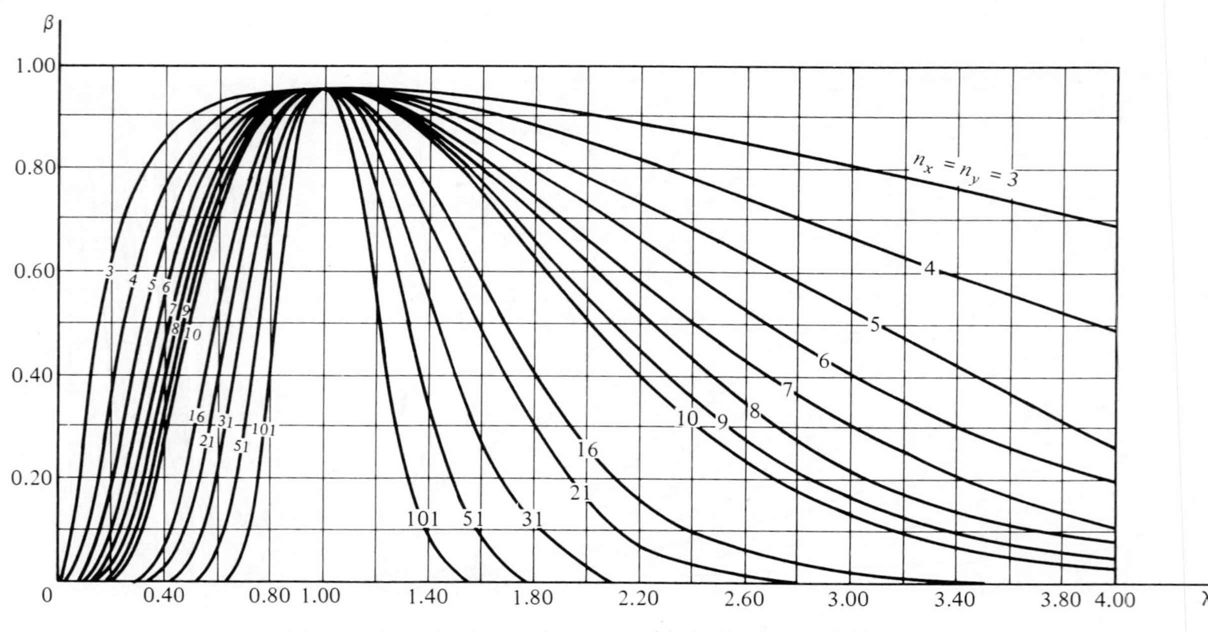

Figure 7.31. *OC* curves for a two-sided *F*-test, $\alpha = 0.05$.

and the decision rule is

$$F_{\text{data}} \geq F_{0.05,5,5} = 5.05, \qquad \text{reject } H_0$$

The F_{data}-value is

$$\frac{s_x^2}{s_y^2} = \frac{0.0044}{0.0011} = 4.0$$

and since

$$F_{\text{data}} = 4.0 < F_{\alpha, n_x - 1, n_y - 1} = 5.05$$

the hypothesis is accepted.

EXAMPLE 7.13.2

It is desired to explore further the situation in Example 7.13.1 by designing a formal experiment. Identical sample sizes are to be used. The test is to be conducted at a 1% significance level. If σ_X is as much as twice the value of σ_Y, this fact is to be detected with a probability 0.90. Determine the required sample size.

$$\lambda = \frac{\sigma_X}{\sigma_Y} = 2$$

Entering Fig. 7.28 with $\lambda = 2$ and $\beta = 0.10$, sample size is read to be 31.

Summary—Test That Two Standard Deviations Are Equal, Normal Populations

---

X- and Y-distributions normal.

$$H_0: \ \sigma_X = \sigma_Y \qquad H_1: \ \sigma_X > \sigma_Y$$

$$\lambda = \frac{\sigma_X}{\sigma_Y} \qquad n: \ \text{enter Figs. 7.28, 7.29 with } \lambda, \beta.$$

$$n = n_x = n_y$$

Decision Rule:

$$F_{\text{data}} \geq F_{\alpha, n_x - 1, n_y - 1}, \qquad \text{reject } H_0$$

where

$$F_{\text{data}} = \frac{s_x^2}{s_y^2}$$

$$H_0: \ \sigma_X = \sigma_Y \qquad H_1: \ \sigma_X \gtrless \sigma_Y$$

$$\lambda = \frac{\sigma_X}{\sigma_Y} \qquad n: \ \text{enter Figs. 7.30, 7.31 with } \lambda, \beta.$$

$$n = n_x = n_y$$

Decision Rule:

$$\left.\begin{aligned} F_{\text{data}} &\geq F_{(\alpha/2), n_x - 1, n_y - 1} \\ &\leq F_{1-(\alpha/2), n_x - 1, n_y - 1} \end{aligned}\right\}, \qquad \text{reject } H_0$$

7.14 Test of the Proportion

In technology and engineering, two very common measures are average and proportion (and percentage, which is essentially a proportion). The *proportion* of a population, designated by π, is the ratio of the number of units X that possess a certain attribute to the total number of units in the population, that is,

$$\pi = \frac{X}{N} \tag{7.17}$$

The corresponding sample parameter is the proportion

$$p = \frac{x}{n}^{*} \tag{7.18}$$

where x is the number of sample units possessing the specific attribute.

If possession of the specified attribute is designated a success, then X refers to the number of successes in n Bernoulli trials and X is a binomial random variable. Possible sample proportions are $0/n$, $1/n$, $2/n$, . . . , $(n - 1)/n$, n/n. From Eq. (4.20), the respective probabilities of $X = 0, 1, 2, \ldots, n - 1, n$ are given by

$$\binom{n}{X} \pi^X (1 - \pi)^{n-X}$$

and thus the respective probabilities of the possible proportion values are also given by the same binomial expression.

Typically, interest is in testing if the population proportion π is at some satisfactory minimum level π_0 (e.g., proportion defective of a manufacturing lot) against an alternative that it is at a nonacceptable level $\pi_1 > \pi_0$. Thus, the usual form of a hypothesis test is

$$H_0: \pi = \pi_0, \qquad H_1: \pi = \pi_1 > \pi_0$$

---

*Some confusion develops by using standard symbols. The symbol p as introduced in Chap. 4 refers to the probability of a success at a single Bernoulli trial. A common designation for a sample proportion is also p where $p = x/n$ and x refers to the number of sample units possessing the attribute of interest. However, if the possession of the attribute of interest is considered to be a success, then the population proportion π is also the probability of a success at a single Bernoulli and the binomial model can be written as

$$\binom{n}{X} \pi^X (1 - \pi)^{n-X}$$

and the symbol p can uniquely stand for the sample proportion.

For the hypothesis tests in Secs. 7.8–7.13, the decision rule in each case is based on a known continuous distribution model for which there are convenient tables of probabilities (Appendices D–G). A hypothesis test based on a discrete distribution function such as the binomial model usually involves the construction of a portion of the table of probabilities—unless an extensive table of binomial or Poisson probabilities is available. The following tables are presented primarily for purposes of explanation.

If the hypothesis $H_0\colon \ \pi = \pi_0$ is true, the respective probabilities of the possible proportion values are given by Table 7.3.

Table 7.3. BINOMIAL PROBABILITIES OF $p = \frac{x}{n}$ IF H_0 IS TRUE

$\frac{x}{n}$	$P\left(\frac{x}{n}\right) = \binom{n}{X}\pi_0^X(1-\pi_0)^{n-X}$
$\frac{0}{n}$	$\binom{n}{0}\pi_0^0(1-\pi_0)^{n-0}$
$\frac{1}{n}$	$\binom{n}{1}\pi_0^1(1-\pi_0)^{n-1}$
$\frac{2}{n}$	$\binom{n}{2}\pi_0^2(1-\pi_0)^{n-2}$
$\vdots$	$\vdots$
$\frac{k}{n}$	$\binom{n}{k}\pi_0^k(1-\pi_0)^{n-k}$
$\vdots$	$\vdots$
$\frac{n}{n}$	$\binom{n}{n}\pi_0^n(1-\pi_0)^{n-n}$

A choice of α determines a proportion k/n such that

$$\alpha = \sum_{X=k}^{n} \binom{n}{X}\pi_0^X(1-\pi_0)^{n-X} \tag{7.19}$$

and thus also a decision rule

$$p_{\text{data}} \geq \frac{k}{n}, \qquad \text{reject } H_0$$

Of course, since k is necessarily an integer, the exact value of α is restricted

Table 7.4. BINOMIAL PROBABILITIES OF $p = \frac{x}{n}$ IF H_1 IS TRUE

$\frac{x}{n}$	$P\left(\frac{x}{n}\right) = \binom{n}{X} \pi_1^X (1 - \pi_1)^{n-X}$
$\frac{0}{n}$	$\binom{n}{0} \pi_1^0 (1 - \pi_1)^{n-0}$
$\frac{1}{n}$	$\binom{n}{1} \pi_1^1 (1 - \pi_1)^{n-1}$
$\frac{2}{n}$	$\binom{n}{2} \pi_1^2 (1 - \pi_1)^{n-2}$
⋮	⋮
$\frac{k}{n}$	$\binom{n}{k} \pi_1^k (1 - \pi_1)^{n-k}$
⋮	⋮
$\frac{n}{n}$	$\binom{n}{n} \pi_1^n (1 - \pi_1)^{n-n}$

by Eq. (7.19). Table 7.4 gives the respective probabilities of the possible proportion values if the alternative hypothesis H_1: $\pi = \pi_1 > \pi_0$ is true. A choice of α and n determines k and then β is given by the expression

$$\beta = \sum_{X=0}^{k-1} \binom{n}{X} \pi_1^X (1 - \pi_1)^{n-X} \tag{7.20}$$

The sample proportion p is a random variable that can assume values $0/n, 1/n, 2/n, \ldots, n/n$. To obtain a figure summary analogous to those of Secs. 7.3–7.10, continuous distributions are used *for illustration only*. A summary of the preceding argument is given in Fig. 7.32.

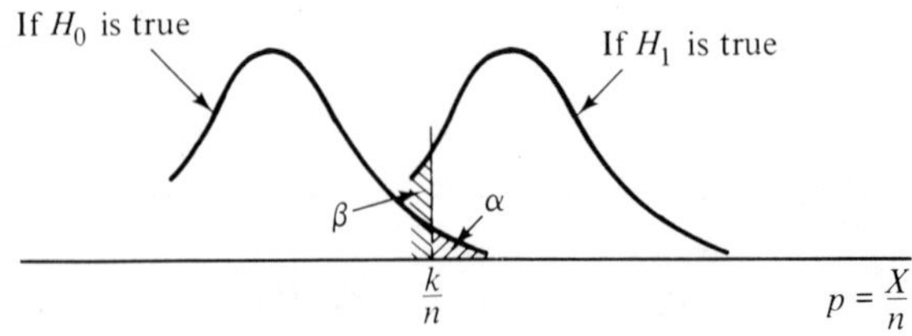

Figure 7.32. Testing H_0: $\pi = \pi_0$ against an alternative H_1: $= \pi = \pi_1 > \pi_0$.

EXAMPLE 7.14.1

If X is a Bernoulli variable, test the hypothesis $H_0: \pi = \frac{1}{4}$ against the alternative $H_1: \pi > \frac{1}{4}$ using a significance level $\alpha = 0.01$. A random sample of 10 observations yields $p_{\text{data}} = 0.4$.

Using the binomial model, computing the respective probabilities of $p = \frac{10}{10}, \frac{9}{10}, \frac{8}{10}, \ldots$ the following probabilities are obtained:

$\frac{x}{n}$	$P\left(\frac{x}{n}\right) = \binom{10}{X}\left(\frac{1}{4}\right)^X\left(\frac{3}{4}\right)^{10-X}$
⋮	⋮
$\frac{k}{10} = \frac{6}{10}$	0.01622
$\frac{7}{10}$	0.00309
$\frac{8}{10}$	0.00039
$\frac{9}{10}$	0.00003
$\frac{10}{10}$	0.00000
	0.01973

Summing probabilities, k is determined to be 6, that is, for $k = 6$ the following sum approximates $\alpha = 0.01$:

$$\alpha = 0.01973 = \sum_{X=6}^{10} \binom{10}{X}\left(\frac{1}{4}\right)^X\left(\frac{3}{4}\right)^{10-X}$$

Therefore, the decision rule is

$$P_{\text{data}} \geq \frac{k}{n} = \frac{6}{10}, \qquad \text{reject } H_0$$

and since

$$p_{\text{data}} = 0.4 < \frac{k}{n} = 0.6$$

the hypothesis is accepted.

If an extensive table of binomial probabilities is available, the decision rule is easily determined by exploring the table for the proper k-value. Referring to

Appendix A, when $k = 6$ the sum

$$\alpha = \sum_{X=6}^{10} \binom{10}{X} \left(\frac{1}{4}\right)^X \left(\frac{3}{4}\right)^{10-X}$$

is

$$1 - 0.9803 = 0.0197*$$

EXAMPLE 7.14.2

To consider the probability of a Type II error, an alternative hypothesis statement is necessary. Assume that in Example 7.14.1 the alternative hypothesis is $H_1{:}\ \ \pi = \frac{1}{2}$. The value of β is then given by

$$\beta = \sum_{X=0}^{5} \binom{10}{X} \left(\frac{1}{2}\right)^X \left(\frac{1}{2}\right)^{10-X}$$

which, read from Appendix A, is 0.6230.†

In Sec. 5.4, a normal distribution model is used to approximate binomial probabilities. With large n, this approximation is satisfactory. This procedure is utilized in the following example.

---

*If a binomial table is not available (or if the required proportion value is not given in the table), a Poisson approximation is possible. From Sec. 4.6, if binomial conditions prevail (i.e., constant probability of success from trial to trial), the probability of a success is small (less than 0.1), and n is sufficiently large (greater than 16), a Poisson approximation is appropriate.

Clearly, these requirements are not satisfied by the conditions of Example 7.14.1. However, to illustrate the method, a Poisson solution of Example 7.14.1 is given here.

$$n\pi_0 = 10(\tfrac{1}{4}) = 2.5$$

The Poisson table (Appendix B) yields the following sum which most closely approximates $\alpha = 0.01$:

$$\alpha = 0.015 = \sum_{X=7}^{10} \frac{e^{-10/4}(\frac{10}{4})^X}{X!}$$

which, read from Appendix B, is

$$1 - 0.985 = 0.015$$

and the decision rule is

$$p_{\text{data}} \geq \frac{k}{n} = \frac{7}{10}, \quad \text{accept } H_0$$

†The β-value is, of course, very large. Also, the decision rule is severe, requiring a relatively large p_{data}-value for rejection of H_0. This follows from the sample size being much too small. A small sample was selected to minimize arithmetic associated with the explanation.

Exploring Appendix A, for $H_0{:}\ \ \pi = \frac{1}{4}$, if $n = 100$ and $\alpha = 1 - 0.9906 = 0.01$, the decision rule is

$$p_{\text{data}} \geq 0.36, \qquad \text{reject } H_0$$

and, for $H_1{:}\ \ \pi = \frac{1}{2}$, β is 0.0018.

EXAMPLE 7.14.3

If X is a Bernoulli variable, test the hypothesis $H_0: \pi = \frac{1}{2}$ against the alternative $H_1: \pi > \frac{1}{2}$ using a significance level $\alpha = 0.05$. A random sample of 400 observations yields $p_{\text{data}} = 0.57$.

If H_0 is true, the mean and variance are given by

$$\mu = n\pi_0 = 400(\tfrac{1}{2}) = 200$$

and

$$\sigma_X^2 = n\pi_0(1 - \pi_0) = 400(\tfrac{1}{2})(\tfrac{1}{2}) = 100$$

whence

$$\sigma_X = 10$$

Figure 7.33 summarizes the hypothesis test using a normal model. The decision rule is

$$z_{\text{data}} \geq Z_\alpha = 1.645, \qquad \text{reject } H_0$$

and

$$z_{\text{data}} = \frac{np_{\text{data}} - \mu}{\sigma_X}$$

$$= \frac{228 - 200}{10} = 2.8$$

Therefore, since

$$z_{\text{data}} = 2.8 > Z_\alpha = 1.645$$

the hypothesis is rejected.

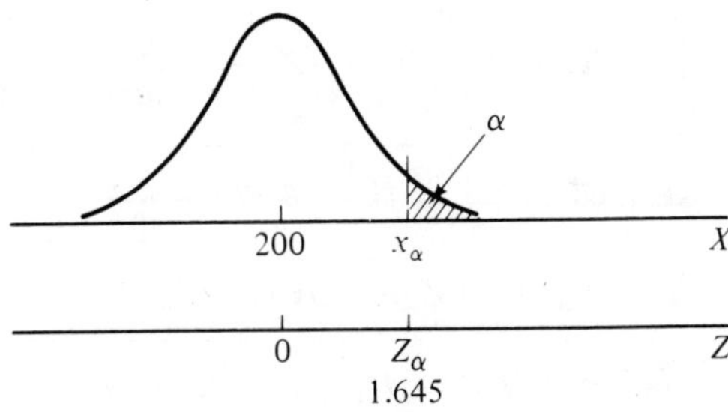

Figure 7.33. Example 7.14.3—normal model for testing $H_0: \pi = \pi_0$ against an alternative $H_1: \pi = \pi_1 > \pi_0$.

Summary—Test of the Proportion

$\pi = \dfrac{X}{N}$ is the population proportion.

The sample proportion $p = \dfrac{x}{n}$ is the random variable.

$$H_0\colon \ \pi = \pi_0 \qquad H_1\colon \ \pi = \pi_1 > \pi_0$$

Decision Rule:

$$P_{\text{data}} = \frac{x}{n} \geq \frac{k}{n}, \qquad \text{reject } H_0$$

where k is a value such that

$$\alpha = \sum_{X=k}^{n} \binom{n}{X} \pi_0^X (1 - \pi_0)^{n-X}$$

and

$$\beta = \sum_{X=0}^{k-1} \binom{n}{X} \pi_1^X (1 - \pi_1)^{n-X}$$

Normal approximation and test using a Z-statistic.

---

n large, and π near 0.5 (see Table 5.2)

Replace $p_{\text{data}} = \dfrac{x}{n}$ by $n\, p_{\text{data}}$.

$$H_0\colon \ \pi = \pi_0 \qquad H_1\colon \ \pi = \pi_1 > \pi_0$$

Decision Rule:

$$z_{\text{data}} \geq Z_\alpha, \qquad \text{reject } H_0$$

where

$$z_{\text{data}} = \frac{n p_{\text{data}} - n\pi_0}{\sqrt{n\pi_0(1 - \pi_0)}}$$

and

$$\beta = \int_{-\infty}^{Z_\beta} f(Z)\, dZ \quad \text{where} \quad Z_\beta = \frac{x_\alpha - n\pi_1}{\sqrt{n\pi_1(1 - \pi_1)}}$$

$$H_0\colon \ \pi = \pi_0 \qquad H_1\colon \ \pi = \pi_1 < \pi_0$$

Decision Rule:

$$z_{\text{data}} \leq -Z_\alpha, \qquad \text{reject } H_0$$

$$H_0\colon \ \pi = \pi_0 \qquad H_1\colon \ \pi = \pi_1 \gtrless \pi_0$$

Decision Rule:

$$|z_{\text{data}}| \geq \frac{Z_\alpha}{2}, \qquad \text{reject } H_0$$

EXERCISES

1. Repeat Example 7.14.1 at a significance level $\alpha = 0.05$.
Ans. $p_{\text{data}} < 0.5$, accept H_0.

2. A random variable X is a Bernoulli variable. Test the hypothesis $H_0: \pi = 0.4$ against the alternative $H_1: \pi > 0.4$ at a significance level $\alpha = 0.01$. A random sample of 20 observations yields a p-value of 0.7.
Ans. $p_{\text{data}} = 0.7 > 0.65$, reject H_0.

3. Determine β for Exercise 2 assuming $H_1: \pi_1 = 0.5$. *Ans.* $\beta = 0.8684$.

4. Assume 50 observations for Exercise 2 and repeat the solution using a Poisson approximation to the binomial probabilities. *Ans.* $p_{\text{data}} < 0.62$, reject H_0.

5. Repeat Exercise 4 using a normal model. *Ans.* $z_{\text{data}} = 4.330$, reject H_0.

7.15 Test of Goodness of Fit

Sections 7.3–7.14 are concerned with a hypothesis that a population parameter has a specified value. Another important type of hypothesis test is one that deals with an inference concerning the type of *D.F.* that describes the population. The first such test considered is the chi-square test of goodness of fit.

Let the X-axis be divided by points a_j into k intervals, the first and last being infinite intervals. The points are chosen such that the *C.D.F.* is continuous at the a_j points (for a discrete variable, the a_j are selected as points of zero probability).

For a specified $f(X)$, the probability of X being in the jth interval is

$$P_j = P(a_{j-1} < X < a_j) = F(a_j) - F(a_{j-1})$$

and the expected number e_j of observations occurring in this interval is

$$e_j = nP_j$$

Designating the actual number of sample observations in this interval by o_j, consider the quantity

$$\sum_{j=1}^{k} \frac{(o_j - e_j)^2}{e_j} \tag{7.21}$$

which is simply a relative difference between sample frequencies and expected frequencies. If this sum is small, sample frequencies agree with the expected frequencies computed under the hypothesis that $f(X)$ is the *D.F.* Conversely, if the sum is large, it indicates too large a difference between sample and expected frequencies and suggests a possible rejection of the hypothesis that $f(X)$ is the population *D.F.*

The question of how large a difference is required to indicate reasons other than sampling error for obtaining such a difference is handled in the usual manner. If the k intervals are chosen such that all $P_j > 0$, then the limiting distribution of the expression given by Eq. (7.21) as $n \longrightarrow \infty$ is a chi-square distribution with $k - 1$ *d.f.* Thus, the chi-square distribution function can be used as a model to construct the decision rule for a hypothesis test that X has a specified *D.F.* The decision rule is

$$\chi^2_{\text{data}} \geq \chi^2_{\alpha,k-1}, \qquad \text{reject } H_0$$
$$\text{otherwise,} \qquad \text{accept } H_0$$

where

$$\chi^2_{\text{data}} = \sum_{j=1}^{k} \frac{(o_j - e_j)^2}{e_j}$$

In applications, it is necessary to assume that n is large enough to make the approximation by the chi-square distribution sufficiently accurate. Two conditions are imposed: (1) $e_j > 5$ for all j, and (2) $k < 20$. The first condition can usually be satisfied by lumping together the small frequencies at the ends of a distribution.

Another practical consideration is that the chi-square test can be accurately applied only if allowance is made for the number of parameters determined from the sample in reconstructing the population. The *d.f.* is equal to the number k of intervals less one, minus the number of parameters estimated from the sample (see Sec. 6.1). For example, in testing a hypothesis that the distribution function is normal (but with unknown mean and variance), the *d.f.* is $k - 3$.

EXAMPLE 7.15.1

Using a significance level $\alpha = 0.05$, it is desired to test the hypothesis that a die is fair. Since the probability P_j of each of the six numbers of a fair die is $\frac{1}{6}$, the expected frequency is

$$e_j = nP_j = n(\tfrac{1}{6})$$

Thus, the hypothesis test is:

H_0: $f(X)$ is a uniform distribution.

H_1: $f(X)$ is not a uniform distribution.

A random sample of 60 tosses of the die yields the observed frequencies:

x_j	1	2	3	4	5	6	
o_j	8	7	10	11	9	15	60

The expected frequency e_j is $60(\frac{1}{6}) = 10$ and

$$\chi^2_{\text{data}} = \frac{(-2)^2}{10} + \frac{(-3)^2}{10} + \frac{0^2}{10} + \frac{1^2}{10} + \frac{(-1)^2}{10} + \frac{5^2}{10}$$
$$= 4$$

and since

$$\chi^2_{\text{data}} = 4 < \chi^2_{0.05,\,5} = 11.07$$

the hypothesis is accepted.

EXAMPLE 7.15.2

Test at a 5% significance level that the following sample is from a binomial population

x_j	0	1	2	3	4	5	
o_j	35	40	15	5	3	2	100

From Eq. (4.23), for a binomial variable $X = 0, 1, 2, \ldots, n = 5$, the mean is

$$np = 5p$$

Computing the sample mean,

$$\bar{x} = \tfrac{1}{100}[35(0) + 40(1) + 15(2) + 5(3) + 3(4) + 2(5)]$$
$$= 1.07$$

and thus an estimate of p from the sample data is obtained from

$$5p = 1.07$$
$$p = 0.21$$

Using $p = 0.2$ and reading the binomial probabilities from Appendix A, Table 7.5 is constructed. The o_j for $X = 4, 5$ are combined into one interval, and hence there are $k = 5$ intervals and $5 - 1 - 1$ *d.f.* The chi-square data value (from Table 7.5) is

$$\sum_{j=1}^{5} \frac{(o_j - e_j)^2}{e_j} = 33.92$$

and since

$$\chi^2_{\text{data}} = 33.92 > \chi^2_{0.05,3} = 7.815$$

the hypothesis is rejected.

Table 7.5. EXAMPLE 7.15.2: TESTING THE HYPOTHESIS THAT X IS BINOMIALLY DISTRIBUTED

X	P_j	$e_j = nP_j$	o_j	$\frac{(o_j - e_j)^2}{e_j}$
0	0.3277	32.8	35	0.15
1	0.4096	41.0	40	0.02
2	0.2048	20.5	15	1.48
3	0.0512	5.1	5	0.00
4	0.0064	0.6 } 0.6	3 } 5	32.27
5	0.0003	0.0	2	
				33.92

EXAMPLE 7.15.3

Test at a 5% significance level that the sample data given in Table 7.6 is from a normal population.

Table 7.6. EXAMPLE 7.15.3: TESTING THE HYPOTHESIS THAT X IS NORMALLY DISTRIBUTED

x_j	*Mid*-x_j	o_j
0–4	2	1
5–9	7	6
10–14	12	22
15–19	17	32
20–24	22	26
25–29	27	8
30–34	32	5
		100

Estimates of the mean and variance computed from the sample data in Table 7.6 are $\bar{x} = 18$ and $s_x = 6.24$. Thus, the null hypothesis is H_0: X is $N(18, 6.24^2)$.

To obtain the expected frequencies e_j for a continuous distribution, it is necessary to obtain the Z-values for the class limits. For any class (or interval) in

Table 7.7. EXAMPLE 7.15.3: TESTING THE HYPOTHESIS THAT X IS NORMALLY DISTRIBUTED

Class Limits X_j	*Class Limits* Z_j	*Cumulative Probability*	P_j	$e_j = nP_j$	o_j	$\frac{(o_j - e_j)^2}{e_j}$
$-\infty$	$-\infty$	0				
			0.0154	1.5 } 8.7	1 } 7	
4.5	−2.16	0.0154				0.33
			0.0715	7.2	6	
9.5	−1.36	0.0869				
			0.2008	20.1	22	0.18
14.5	−0.56	0.2877				
			0.3071	30.7	32	0.06
19.5	+0.24	0.5948				
			0.2560	25.6	26	0.01
24.5	+1.04	0.8508				
			0.1163	11.6	8	1.12
29.5	+1.84	0.9671				
			0.0329	3.3	5	0.88
$+\infty$	$+\infty$	1.0000				
			1.0000	100.0	100	2.58

Table 7.7,

$$Z_{\text{class limit}} = \frac{X_{\text{class limit}} - 18}{6.24}$$

Also, the first and last classes are treated as being infinite intervals.

The o_j for the first two classes are combined into one interval, and hence there are $k = 6$ intervals and $6 - 1 - 2$ *d.f.* The chi-square data value (from Table 7.7) is

$$\sum_{j=1}^{6} \frac{(o_j - e_j)^2}{e_j} = 2.58$$

and since

$$\chi^2_{\text{data}} = 2.58 < \chi^2_{0.05,3} = 7.815$$

the hypothesis is accepted.

Summary—Chi-Square Test of Goodness of Fit

X is either a discrete or continuous random variable.

H_0: X has a specified density function $f(X)$.
Decision Rule:

$$\chi^2_{\text{data}} \geq \chi^2_{\alpha,k-1-m}, \qquad \text{reject } H_0$$

where the X-range is divided into $k < 20$ intervals and m is the number of population parameters estimated from the sample and

$$\chi^2_{\text{data}} = \sum_{j=1}^{k} \frac{(o_j - e_j)^2}{e_j}$$

where o_j is the observed frequency and $e_j > 5$ the expected frequency in the jth interval.

Another goodness-of-fit test, the Kolmogorov-Smirnov test, is based on the statistic

$$D_n = \text{maximum} \, | F_n(X) - F(X) | \tag{7.22}$$

where $F(X)$ is the hypothesized *C.D.F.*, and $F_n(X)$ is the corresponding sample distribution function (see Sec. 3.5) evaluated at the observed values in the sample.

The population $F(X)$ is assumed to be continuous. The distribution of the sample statistic is independent of the hypothesized distribution of X and its only parameter is sample size n. The hypothesis (H_0: X has a specified *D.F.*) is tested against the alternative that the distribution of X is other than that specified, and the decision rule is

$$\begin{aligned} D_{\text{data}} \geq \text{maximum} \, | F_n(X) - F(X) |, \qquad & \text{reject } H_0 \\ \text{otherwise}, \qquad & \text{accept } H_0 \end{aligned}$$

Values of the test statistic are given in Table 7.8. The principal advantage of the Kolmogorov-Smirnov test is that it compares all of the data in unaltered form, whereas the chi-square test lumps data and compares discrete data

Table 7.8. STATISTIC VALUES FOR THE KOLMOGOROV-SMIRNOV GOODNESS-OF-FIT TEST

Sample Size	α		
n	0.01	0.05	0.10
5	0.67	0.56	0.51
6	0.62	0.52	0.47
7	0.58	0.48	0.44
8	0.54	0.45	0.41
9	0.51	0.43	0.39
10	0.49	0.41	0.37
11	0.47	0.39	0.35
12	0.45	0.37	0.34
13	0.43	0.36	0.32
14	0.42	0.35	0.31
15	0.40	0.34	0.30
20	0.35	0.29	0.26
25	0.32	0.26	0.24
30	0.29	0.24	0.22
40	0.25	0.21	0.19
Large n	$\frac{1.63}{\sqrt{n}}$	$\frac{1.36}{\sqrt{n}}$	$\frac{1.22}{\sqrt{n}}$

categories. A disadvantage is that the Kolmogorov-Smirnov test is strictly valid only for continuous distribution models and only where the model is hypothesized wholly independently of the data.

EXAMPLE 7.15.4

Test at a 5% significance level that the sample observations 6, 9, 4, 10, 2, 3, 11, 15 are from a normal population with mean 8 and variance 4.

From Appendix D, the hypothesized values $F(X)$ in Table 7.9 are constructed. The class intervals are $x < 2$, $2 \leq x < 3$, $3 \leq x < 4$, ... Since each sample x-value has a frequency one, the empirical probability of X being in each class interval is $\frac{1}{8}$ and thus the $F_n(X)$ values are $\frac{1}{8}, \frac{2}{8}, \frac{3}{8}, \ldots$.

Both $F(X)$ and $F_n(X)$ are increasing functions, the latter being an increasing step function. Therefore, the differences to be examined are the differences that occur at the jumps in $F_n(X)$. Column no. 5 (Table 7.9) gives differences $F_n(X) - F(X)$ at the right end point of each interval of the step function, and column no. 6 gives differences at the left point of each interval.*

From Table 7.8, the decision rule for this test is

$$D_{\text{data}} \geq D_{\alpha,n} = 0.41, \qquad \text{reject } H_0$$

and since

$$D_{\text{data}} = 0.352 < D_{0.05,8} = 0.45$$

the hypothesis is accepted.

Table 7.9. EXAMPLE 7.15.4: TESTING THE HYPOTHESIS THAT X IS NORMALLY DISTRIBUTED

				$F_n(X) - F(X)$	
X-range	*Z-range*	$F(X)$	$F_n(X)$	No. 5	No. 6
$X < 2$	$Z < -3.0$	0.000	0.000	0.000	0.125
$X < 3$	$Z < -2.5$	0.006	0.125	0.119	0.244
$X < 4$	$Z < -2.0$	0.023	0.250	0.227	0.352 ← D
$X < 6$	$Z < -1.0$	0.159	0.375	0.216	0.341
$X < 9$	$Z < +0.5$	0.691	0.500	−0.191	−0.066
$X < 10$	$Z < +1.0$	0.841	0.625	−0.216	−0.091
$X < 11$	$Z < +1.5$	0.933	0.750	−0.183	−0.058
$X < 15$	$Z < +3.5$	1.000	0.875	−0.125	0.000
$+\infty$	$+\infty$	1.000	1.000	0.000	0.000

*The differences in column no. 6 are the differences between the values joined by the diagonal line segments in Table 7.9. It is suggested that the student plot $F(X)$ and $F_n(X)$ and observe that there are two differences to be considered at each jump point of the step function.

Summary—Kolmogorov-Smirnov Test of Goodness of Fit

X population distribution function is continuous.

H_0: X has a specified density function $f(X)$.
Decision Rule:

$$D_{\text{data}} \geq D_{\alpha, n}, \qquad \text{reject } H_0$$

where

$$D_{\text{data}} = \text{maximum } |F_n(X) - F(X)|$$

with $F(X)$ being the hypothesized $C.D.F.$ and $F_n(X)$ the corresponding sample distribution function.
For k intervals on X, there are $2k$ number of differences to be examined in arriving at D_{data}.

EXERCISES

1. A random sample of observations yields the following numbers of defective product units manufactured on three shifts:

Shift	1	2	3	
o_j	25	42	33	100

A question is whether the variation shown is due only to chance or to a significant difference between shifts. Using the chi-square statistic, test at a 1% significance level that the same frequency of defectives exists on all shifts [i.e., that $f(X)$ is uniform]. *Ans.* $\chi^2_{\text{data}} = 4.345$, accept H_0.

2. Test at a 5% significance level that the following sample is from a Poisson population. Use a chi-square test, grouping the frequencies in intervals $x = 3$, 4, 5 into one interval.

x	0	1	2	3	4	5	
o_j	51	30	14	4	1	0	100

Ans. $\chi^2_{\text{data}} = 1.403$, accept H_0.

Both goodness-of-fit tests given in this section are called *nonparametric tests*—tests that do not depend on an assumption of normality. Research in statistics is increasingly in the direction of development of nonparametric tests. Many such tests, tailored for specific applications, are now available in advanced statistics texts.*

7.16 The Estimation Problem

Consider a random variable X whose population frequency function is $f(X, \theta_1, \theta_2, \ldots, \theta_k)$ where θ_j denotes the jth parameter. In applications, a random sample $x_1, x_2, \ldots, x_n$ from the population is the basis for computing an estimate $\hat{\theta}$ of a specified population parameter θ. The estimator $\hat{\theta}$ is a function of the x_i of the sample and thus is itself a random variable. From sample to sample, $\hat{\theta}$ assumes different values. For all possible samples of size n from a given population, the corresponding values generated for $\hat{\theta}$ constitute the *sampling distribution of the statistic* $\hat{\theta}$.

EXAMPLE 7.16.1

In Sec. 5.7, a sampling distribution of the mean is introduced. A number of hypothesis tests in this section use the sampling distribution of the mean as a model to formulate the decision rule.

The estimator $\hat{\theta}$ is the sample mean $\bar{x}$, and $\bar{x}$ is an estimate of the population parameter $\theta = \mu$. The sample mean $\bar{x}$ is a function of the x_i of the sample, that is,

$$\bar{x} = \frac{1}{n} \sum_{i=1}^{n} x_i$$

and, since X is a random variable, so also is $\bar{x}$.

Frequently, there exist different statistics for estimating the same population parameter. For example, Chap. 1 gives three averages, mean, median, and mode, any one of which can be used to estimate the population mean. When a choice of estimator exists, the natural objective is to seek the best estimator for the task. Figure 7.34 gives the sampling distributions of three

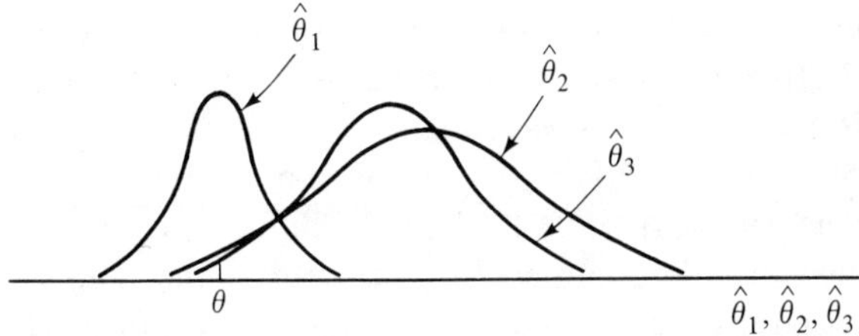

Figure 7.34. Distributions of estimators of θ.

*A brief introduction to nonparametric tests is given in Dixon, W. J. and Massey, F. J. (1969), *Introduction to Statistical Analysis*, 3rd ed., McGraw-Hill Book Company, N. Y., Chap. 17.

statistics, $\hat{\theta}_1, \hat{\theta}_2$, and $\hat{\theta}_3$, used to estimate the same population parameter θ. The statistic $\hat{\theta}_1$ is superior from the standpoint that the values assumed by $\hat{\theta}_1$ are distributed symmetrically about the parameter θ being estimated. Although $\hat{\theta}_2$ and $\hat{\theta}_3$ are inferior estimators, a choice between the two is possible using another criterion. The values assumed by $\hat{\theta}_3$ exhibit less variation than do the values taken by $\hat{\theta}_2$ (note the distribution ranges).

A number of criteria are used in theoretical statistics to evaluate estimators, the principal ones being unbiasedness, efficiency, consistency, and sufficiency. The first two are relevant to statistical applications.

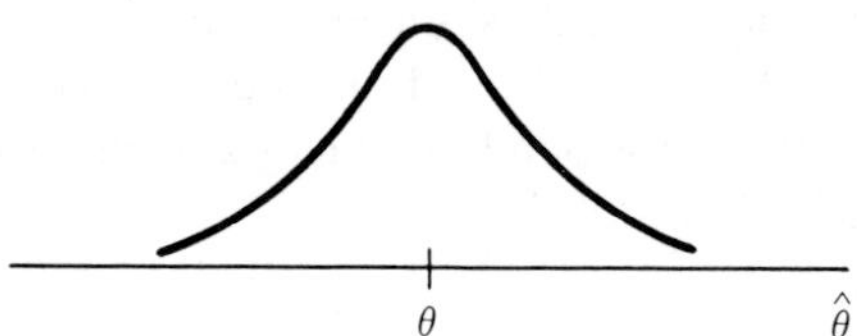

Figure 7.35. Unbiased estimator $\hat{\theta}$.

An estimator $\hat{\theta}$ is an *unbiased* estimator of θ if $E(\hat{\theta}) = \theta$ as shown in Fig. 7.35. Unbiasedness is a desirable property for an estimator. It is reassuring to know that the $\hat{\theta}$-values (from sample to sample) distribute about θ as a mean. Some examples of estimators being biased or unbiased are:

1. The sample mean $\bar{x}$ is an unbiased estimate of the population mean μ.
2. The quantity $(1/n) \sum (x - \mu)^2$ is an unbiased estimate of the population variance σ_X^2.
3. The quantity $(1/n) \sum (x - \bar{x})^2$ is a biased estimate of the population variance σ_X^2 (usually it is smaller then σ_X^2).
4. The quantity $[1/(n - 1)] \sum (x - \bar{x})^2$ is an unbiased estimate of the population variance σ_X^2.

An *efficient* estimator is an estimator with small variation as measured by $E[(\hat{\theta} - \theta)^2]$. In Fig. 7.34, $\hat{\theta}_3$ is a more efficient estimator than is $\hat{\theta}_2$. A larger sample size is required for a less efficient estimator in order to obtain the same degree of accuracy of estimate given by a more efficient estimator. Thus, the word "efficiency" also refers to the maximum use of limited information available from a smaller sample. In recent texts, the language "minimum variance unbiased estimator" is common. This description refers to an estimator that is both unbiased and efficient, the latter term meaning that its variance $E[(\hat{\theta} - \theta)^2]$ is at least as small as that for any other estimator.

A *point estimate* of a parameter θ is a single value used to estimate θ. For example, $\bar{x} = 20$ is a point estimate of μ. An *interval estimate* is an interval (a, b) used to estimate θ. Thus, $18 \leq \mu \leq 22$ is an interval (18, 22) estimate of μ. A *confidence interval estimate* of θ gives both an interval (a, b) and the probability of θ occurring in an interval so computed. The probability $P(a \leq \theta \leq b) = 0.95$ is a 95% confidence interval estimate of θ. In applications, a and b are a function of the sample x_i and hence are random variables. Thus, the interval (a, b) assumes different values from sample to sample. The meaning of a statement $P(18 \leq \mu \leq 22) = 0.95$ is that if the sample x_i is from a population whose mean is μ, then 95% of the samples yield intervals (a, b) that contain μ and (18, 22) is one of these intervals.*

7.17 Rationale of Estimation

Estimation is analogous to hypothesis testing. In testing hypotheses, a decision rule is generated from the probability α, whereas in estimation the confidence interval is based on the probability $1 - \alpha$. To illustrate the analogy between hypothesis testing and estimation, the hypothesis test of Sec. 7.8 is reexamined here. The argument and the graphic summary (Fig. 7.36) is for the two-sided hypothesis test $H_0: \mu = \mu_0$, $H_1: \mu = \mu_1 \gtrless \mu_0$.

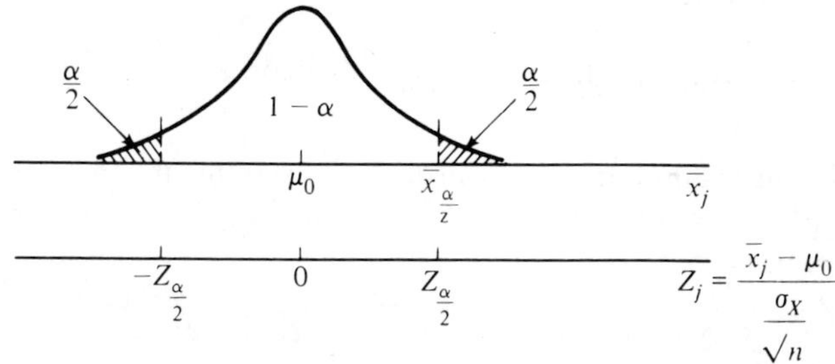

Figure 7.36. Distribution model for a test of the mean.

The probability α of incurring a Type I decision error is given by

$$\alpha = P(\text{reject } H_0, H_0 \text{ true})$$

From Fig. 7.36,

$$\alpha = P(Z_{\alpha/2} \leq Z_j \leq -Z_{\alpha/2})$$

$$= P\left(Z_{\alpha/2} \leq \frac{\bar{x}_j - \mu_0}{\sigma_X/\sqrt{n}} \leq -Z_{\alpha/2}\right)$$

*It is important that the student reflect on the fact that this confidence interval statement does not mean that 95% of the time μ is between 18 and 22. The "95% confidence" means that for 95% of the possible samples the interval so computed contains μ.

and the decision rule can be stated either as

$$|z_{\text{data}}| \geq Z_{\alpha/2}, \quad \text{reject } H_0$$

or

$$|\bar{x}_{\text{data}}| \geq \bar{x}_{\alpha/2}, \quad \text{reject } H_0$$

However, an equivalent decision rule can be developed in terms of acceptance (instead of rejection) and the probability $1 - \alpha$, that is,

$$\begin{aligned} 1 - \alpha &= P(\text{accept } H_0, H_0 \text{ true}) \\ &= P(-Z_{\alpha/2} \leq Z_j \leq Z_{\alpha/2}) \\ &= P\left(-Z_{\alpha/2} \leq \frac{\bar{x}_j - \mu_0}{\sigma_X/\sqrt{n}} \leq Z_{\alpha/2}\right) \\ &= P\left\{-Z_{\alpha/2}\left(\frac{\sigma_X}{\sqrt{n}}\right) \leq \bar{x}_j - \mu_0 \leq Z_{\alpha/2}\left(\frac{\sigma_X}{\sqrt{n}}\right)\right\} \\ &= P\left\{\mu_0 - Z_{\alpha/2}\left(\frac{\sigma_X}{\sqrt{n}}\right) \leq \bar{x}_j \leq \mu_0 + Z_{\alpha/2}\left(\frac{\sigma_X}{\sqrt{n}}\right)\right\} \end{aligned}$$

and the decision rule is

$$\text{Accept } H_0 \text{ if } \mu_0 \pm Z_{\alpha/2}\left(\frac{\sigma_X}{\sqrt{n}}\right) \text{ contains } z_{\text{data}}.$$

In a similar fashion, the $1 - \alpha$ probability statement can be algebraicaly manipulated* to generate an interval, dependent on $\bar{x}_{\text{data}}$, that contains μ_0 if H_0 is true. That is,

$$1 - \alpha = P\left(-Z_{\alpha/2} \leq \frac{\bar{x}_j - \mu_0}{\sigma_X/\sqrt{n}} \leq Z_{\alpha/2}\right)$$

or

$$1 - \alpha = P\left\{\bar{x}_j - Z_{\alpha/2}\left(\frac{\sigma_X}{\sqrt{n}}\right) \leq \mu_0 \leq \bar{x}_j + Z_{\alpha/2}\left(\frac{\sigma_X}{\sqrt{n}}\right)\right\}$$

---

*Each inequality, left-hand and right-hand, can be operated on as follows: (1) multiplication by $\sigma_X/\sqrt{n}$, (2) subtraction of $\bar{x}_j$, and (3) multiplication by -1. For example, considering the left-hand inequality,

$$\begin{aligned} -Z_{\alpha/2} &\leq \frac{\bar{x}_j - \mu_0}{\sigma_X\sqrt{n}} \\ -Z_{\alpha/2}\left(\frac{\sigma_X}{\sqrt{n}}\right) &\leq \bar{x}_j - \mu_0 \\ -Z_{\alpha/2}\left(\frac{\sigma_X}{\sqrt{n}}\right) - \bar{x}_j &\leq -\mu_0 \\ Z_{\alpha/2}\left(\frac{\sigma_X}{\sqrt{n}}\right) + \bar{x}_j &\geq \mu_0 \end{aligned}$$

and, for $\bar{x}_j = \bar{x}_{\text{data}}$,

$$1 - \alpha = P\left\{\bar{x}_{\text{data}} - Z_{\alpha/2}\left(\frac{\sigma_X}{\sqrt{n}}\right) \leq \mu_0 \leq \bar{x}_{\text{data}} + Z_{\alpha/2}\left(\frac{\sigma_X}{\sqrt{n}}\right)\right\}$$

Thus, if the sample x_i are from a population $N(\mu_0, \sigma_X^2)$, a $100(1 - \alpha)\%$ confidence interval estimate of μ_0 is

$$\bar{x}_{\text{data}} \pm Z_{\alpha/2}\left(\frac{\sigma_X}{\sqrt{n}}\right)$$

This is a two-sided confidence interval estimate of μ. A lower one-sided confidence interval is

$$\bar{x}_{\text{data}} - Z_{\alpha}\left(\frac{\sigma_X}{\sqrt{n}}\right) \quad +\infty$$

and an upper one-sided confidence interval is

$$-\infty, \bar{x}_{\text{data}} + Z_{\alpha}\left(\frac{\sigma_X}{\sqrt{n}}\right)$$

Analogous to hypothesis tests, there are both one-sided and two-sided confidence intervals. The one-sided intervals are developed in the same manner. Summarizing, if X is $N(\mu, \sigma_X^2)$ and sampling is from an infinite population (or with replacement from a finite population), confidence interval estimates of the population mean are:

$$\text{2-sided:} \quad \bar{x}_{\text{data}} \pm Z_{\alpha/2}\left(\frac{\sigma_X}{\sqrt{n}}\right)$$

$$\text{1-sided:} \quad -\infty, \quad \bar{x}_{\text{data}} + Z_{\alpha}\left(\frac{\sigma_X}{\sqrt{n}}\right)$$

$$\bar{x}_{\text{data}} - Z_{\alpha}\left(\frac{\sigma_X}{\sqrt{n}}\right) \quad +\infty$$

If sampling is without replacement from a finite population of size N, then $\sigma_X/\sqrt{n}$ is replaced by

$$\frac{\sigma_X}{\sqrt{N}}\sqrt{\frac{N-n}{N-1}}$$

in each of the above confidence interval expressions.

There are two general types of applications problems: (1) The required interval (a, b) is specified; determine the probability $P(a \leq \theta \leq b)$, and (2) the probability α is specified; determine a and b such that the probability $P(a \leq \theta \leq b) = \alpha$.

EXAMPLE 7.17.1

In Example 7.4.1, the test is to detect a possible increase in breaking strength of the product. Using the same numerical values of this example, an experiment involving 20 random observations yields a mean breaking strength of 565 lb.

A 95% lower one-sided confidence interval estimate of mean breaking strength μ is

$$= \bar{x}_{\text{data}} - Z_{\alpha}\left(\frac{\sigma_X}{\sqrt{n}}\right) \quad +\infty$$

$$= 565 - 1.645\left(\frac{5}{\sqrt{20}}\right) \quad +\infty$$

$$= 563.16, \quad +\infty$$

EXAMPLE 7.17.2

To compensate for measurement error, four repeat measurements of a product attribute are taken yielding a mean measurement of 0.215 in. Previous measurement error analyses indicate that repeat measurements of this magnitude are normally distributed with standard deviation 0.0005 in.

A 95% two-sided confidence interval estimate of the mean measurement in question is

$$\bar{x}_{\text{data}} \pm Z_{\alpha/2}\left(\frac{\sigma_X}{\sqrt{n}}\right) = 0.215 \pm 1.96\left(\frac{0.0005}{\sqrt{4}}\right)$$

$$= 0.2145, \quad 0.2155 \text{ in.}$$

EXAMPLE 7.17.3

Assume a confidence interval half as large as that in Example 7.17.2 is required. How many measurements are required?

$$Z_{\alpha/2}\left(\frac{\sigma_X}{\sqrt{n}}\right) = 1.96\left(\frac{0.0005}{\sqrt{n}}\right) = \frac{0.0005}{2}$$

whence

$$n = \left[\frac{2(1.96)(0.0005)}{0.0005}\right]^2 = 15.36$$

Confidence interval estimates are all routinely computed in the same manner. The same statistic that is used in each hypothesis test (Sec. 7.8–7.14) appears in each corresponding confidence interval expression. A summary of these confidence interval expressions is given here.

Summary—100(1 − α)% Confidence Interval Estimate of:

Mean μ:
X distribution normal, or $n \geq 30$.
σ_X known.

$$\text{2-sided:}\quad \bar{x}_{\text{data}} \pm Z_{\alpha/2}\left(\frac{\sigma_X}{\sqrt{n}}\right)$$

$$\text{1-sided:}\quad -\infty,\quad \bar{x}_{\text{data}} + Z_{\alpha}\left(\frac{\sigma_X}{\sqrt{n}}\right) \qquad \text{(upper)}$$

$$\bar{x}_{\text{data}} - Z_{\alpha}\left(\frac{\sigma_X}{\sqrt{n}}\right),\quad +\infty \qquad \text{(lower)}$$

Mean μ:
X distribution normal, or $n \geq 30$.
σ_X unknown.

$$\text{2-sided:}\quad \bar{x}_{\text{data}} \pm t_{\alpha/2,n-1}\left(\frac{s_x}{\sqrt{n}}\right)$$

$$\text{1-sided:}\quad -\infty,\quad \bar{x}_{\text{data}} + t_{\alpha,n-1}\left(\frac{s_x}{\sqrt{n}}\right) \qquad \text{(upper)}$$

$$\bar{x}_{\text{data}} - t_{\alpha,n-1}\left(\frac{s_x}{\sqrt{n}}\right),\quad +\infty \qquad \text{(lower)}$$

Mean difference $\mu_X - \mu_Y$.
X and Y: independent random variables.
X and Y distributions normal, or $n_x \geq 30$ and $n_y \geq 30$.
σ_X and σ_Y known.

$$\text{2-sided:}\quad (\bar{x} - \bar{y})_{\text{data}} \pm Z_{\alpha/2}\sqrt{\frac{\sigma_X^2}{n_x} + \frac{\sigma_Y^2}{n_y}}$$

$$\text{1-sided:}\quad -\infty,\quad (\bar{x} - \bar{y})_{\text{data}} + Z_{\alpha}\sqrt{\frac{\sigma_X^2}{n_x} + \frac{\sigma_Y^2}{n_y}} \qquad \text{(upper)}$$

$$(\bar{x} - \bar{y})_{\text{data}} - Z_{\alpha}\sqrt{\frac{\sigma_X^2}{n_x} + \frac{\sigma_Y^2}{n_y}},\quad +\infty \qquad \text{(lower)}$$

Mean difference $\mu_X - \mu_Y$.
X and Y: independent random variables.
X- and Y-distributions normal, or $n_x \geq 30$ and $n_y \geq 30$.
σ_X and σ_Y unknown, but equal.

$$\text{2-sided:}\quad (\bar{x} - \bar{y})_{\text{data}} \pm t_{\alpha/2,n_x+n_y-2}I$$

$$\text{1-sided:}\quad -\infty,\quad (\bar{x} - \bar{y})_{\text{data}} + t_{\alpha,n_x+n_y-2}I \qquad \text{(upper)}$$

$$(\bar{x} - \bar{y})_{\text{data}} - t_{\alpha,n_x+n_y-2}I,\quad +\infty \qquad \text{(lower)}$$

where

$$I = \sqrt{\frac{1}{n_x} + \frac{1}{n_y}}\sqrt{\frac{\sum (x_i - \bar{x})^2 + \sum (y_i - \bar{y})^2}{n_x + n_y - 2}}$$

Standard deviation σ_X.
X-distribution normal.

2-sided: $s_x\sqrt{\dfrac{n-1}{\chi^2_{\alpha/2,n-1}}}, \quad s_x\sqrt{\dfrac{n-1}{\chi^2_{1-(\alpha/2),n-1}}}$

1-sided: $0, \quad s_x\sqrt{\dfrac{n-1}{\chi^2_{1-\alpha,n-1}}}$ (upper)

$s_x\sqrt{\dfrac{n-1}{\chi^2_{\alpha,n-1}}}, \quad +\infty$ (lower)

Ratio of two standard deviations $\dfrac{\sigma_X}{\sigma_Y}$.
X- and Y-distributions normal.

2-sided: $\dfrac{s_x}{s_y}\sqrt{\dfrac{1}{F_{\alpha/2,n_x-1,n_y-1}}}, \quad \dfrac{s_x}{s_y}\sqrt{F_{\alpha/2,n_y-1,n_x-1}}$

1-sided: $0, \quad \dfrac{s_x}{s_y}\sqrt{F_{\alpha,n_y-1,n_x-1}}$ (upper)

$\dfrac{s_x}{s_y}\sqrt{\dfrac{1}{F_{\alpha,n_x-1,n_y-1}}}, \quad +\infty$ (lower)

Proportion p.
n large (see Sec. 5.4).

2-sided: $p_{\text{data}} \pm Z_{\alpha/2}\sigma_p$

where σ_p can be estimated from the sample by

$$\sqrt{\frac{p_{\text{data}}(1-p_{\text{data}})}{n}}$$

1-sided: $0, \quad p_{\text{data}} + Z_\alpha\sigma_p$ (upper)

$p_{\text{data}} - Z_\alpha\sigma_p, \quad 1$ (lower)

EXERCISES

1. Repeat Example 7.17.1 assuming that σ_X is unknown and its estimate is $s_x = 5$. *Ans.* 563.07, $+\infty$.

2. Using the information of Example 7.10.1, assuming an experiment where 20 random observations yield sample means $\bar{x} = 565$ lb. and $\bar{y} = 563$ lb, determine a 95% two-sided confidence interval estimate of the population mean difference $\mu_X - \mu_Y$. *Ans.* -1.10, 5.10.

3. Using the data of Example 7.12.1, with 18 random observations yielding a sample variance $s_x^2 = 0.008$, determine a 99% two-sided confidence interval estimate of the population standard deviation σ_X. *Ans.* 0.0055, 0.0138.

GLOSSARY OF TERMS

Null hypothesis
Alternative hypothesis
Decision rule
Test
Significance level
OC curve
Power curve
Goodness of fit
Expected frequency
Empirical frequency
Nonparametric test
Unbiased estimator
Efficient estimator
Point estimate
Interval estimate
Confidence interval estimate

GLOSSARY OF FORMULAS

Page

(7.1) Probabilities of decision errors. 182

$$\alpha = P(\text{reject } H_0, H_0 \text{ true})$$
$$\beta = P(\text{accept } H_0, H_0 \text{ false, and } H_1 \text{ true})$$

(7.2) Probabilities of correct decisions. 182

$$1 - \alpha = P(\text{accept } H_0, H_0 \text{ true})$$
$$1 - \beta = P(\text{reject } H_0, H_0 \text{ false, and } H_1 \text{ true})$$

(7.3) Standard *OC* curve units—test of the mean. 186

$$d = \frac{\mu_1 - \mu_0}{\sigma_X}$$

(7.4) *OC*-power curve relationship. 198

OC curve ordinate = 1 − power curve ordinate

(7.5) *Z*-statistic—test of equality of means. 207

$$Z_j = \frac{(\bar{x}_j - \bar{y}_j) - (\mu_X - \mu_Y)}{\sqrt{(\sigma_X^2/n_x) + (\sigma_Y^2/n_y)}}$$

(7.6) *Z*-statistic—test of equality of means. 207

$$Z_j = \frac{\bar{x}_j - \bar{y}_j}{\sqrt{(\sigma_X^2/n_x) + (\sigma_Y^2/n_y)}}$$

(7.7) Standard *OC* curve units—test of equality of means. 207

$$d = \frac{\mu_X - \mu_Y}{\sqrt{\sigma_X^2 + \sigma_Y^2}} \qquad n = n_x = n_y$$

(7.8) *n*-formula when sample sizes are unequal. 209

$$n = \frac{\sigma_X^2 + \sigma_Y^2}{(\sigma_X^2/n_x) + (\sigma_Y^2/n_y)}$$

(7.9) *t*-statistic—test of equality of means. 209

$$t_j = \frac{\bar{x}_j - \bar{y}_j}{\sqrt{(1/n_x) + (1/n_y)}\sqrt{(1/n_x + n_y - 2)[\sum (x_i - \bar{x})^2 + \sum (y_i - \bar{y})^2]}}$$

(7.10) t-statistic—test of equality of means. 209

$$t_j = \frac{\bar{x}_j - \bar{y}_j}{\sqrt{(s_x^2/n_x) + (s_y^2/n_y)}}$$

(7.11) $d.f.$ for t-statistic—Eq. (7.10). 209

$$\nu = \frac{[(s_x^2/n_x) + (s_y^2/n_y)]^2}{\dfrac{(s_x^2/n_x)^2}{n_x + 1} + \dfrac{(s_y^2/n_y)^2}{n_y + 1}} - 2$$

(7.12) t-statistic—paired comparisons test. 214

$$t_j = \frac{\overline{(x - y)}_j - \mu_{(X-Y)}}{s_{(x-y)}\sqrt{n}} = \frac{\overline{(x - y)}_j}{s_{(x-y)}\sqrt{n}}$$

(7.13) Mean difference—paired comparisons. 214

$$\overline{(x - y)} = \frac{1}{n}\sum_{i=1}^{n} (x_i - y_i)$$

(7.14) Sample variance—paired comparisons. 214

$$s_{(x-y)}^2 = \frac{1}{n-1}\sum_{i=1}^{n} [(x_i - y_i) - \overline{(x - y)}]^2$$

(7.15) Standard OC curve units—paired comparisons test. 215

$$d = \frac{|\mu_{(X-Y)}|}{\sqrt{\sigma_X^2 + \sigma_Y^2}}$$

(7.16) Standard OC curve units—test of the standard deviation. 218

$$\lambda = \frac{\sigma_1}{\sigma_0}$$

(7.17) Population proportion. 232

$$\pi = \frac{X}{N}$$

(7.18) Sample proportion. 232

$$p = \frac{x}{n}$$

(7.19) Type I error probability—test of the proportion. 233

$$\alpha = \sum_{X=k}^{n} \binom{n}{X} \pi_0^X (1 - \pi_0)^{n-X}$$

(7.20) Type II error probability—test of the proportion. 234

$$\beta = \sum_{X=0}^{k-1} \binom{n}{X} \pi_1^X (1 - \pi_1)^{n-X}$$

(7.21) Chi-square statistic—goodness-of-fit test. 239

$$\sum_{j=1}^{k} \frac{(o_j - e_j)^2}{e_j}$$

(7.22) Kolmogorov-Smirnov statistic—goodness-of-fit test. 244

$$D_n = \text{maximum}\,|F_n(X) - F(X)|$$

REFERENCES

BOWKER, A. H. AND LIEBERMAN, G. J. (1972), *Engineering Statistics*, 2nd ed., Prentice-Hall, Inc., Englewood Cliffs, N. J., Chaps. 6–8.

DIXON, W. J. AND MASSEY, F. J. (1969), *Introduction to Statistical Analysis*, 3rd ed., McGraw-Hill Book Company, New York, Chaps. 6–8.

FERRIS, C. L., GRUBBS, F. E. AND WEAVER, C. L. (June 1946), "Operating Characteristics for the Common Statistical Tests of Significance," *Annals of Mathematical Statistics*, Vol. 17, pp. 178–226.

HINES, W. W. AND MONTGOMERY, D. C. (1972), *Probability and Statistics in Engineering and Management Science*, The Ronald Press Company, New York, Chap. 10.

HUNTSBERGER, D. V. (1967), *Elements of Statistical Inference*, 2nd ed., Allyn and Bacon, Inc., Boston, Mass., Chaps. 6–10.

MILLER, I. AND FREUND, J. E. (1965), *Probability and Statistics for Engineers*, Prentice-Hall, Inc., Englewood Cliffs, N. J., Chaps. 8–11.

SPIEGEL, M. R. (1961), *Schaum's Outline of Statistics*, Schaum Publishing Co., New York, Chaps. 9–12.

PROBLEMS

1. A participant in a crap game adopts a decision rule: if his opponent rolls four or more consecutive sevens, he will reject the dice and terminate the game. What is the probability of a Type I decision error?

2. Define the following quantities: (a) α and β, and (b) $1 - \alpha$ and $1 - \beta$. Which of these values is called the *significance level*?

3. One random observation x is the basis for testing a hypothesis that the observation is from a population described by $f(X) = \frac{1}{2}$, $1 \leq X \leq 3$. The decision rule is: reject the hypothesis if $x \geq 2.75$. Determine the probability of a Type I decision error.

4. Determine a decision rule for Problem 3 such that $\alpha = 0.10$.

5. Suppose for the conditions of Problem 3 that the hypotheses being tested are:

$$H_0\colon \quad x \text{ belongs to a population } f_0(X) = \tfrac{1}{2},\ 1 \leq X \leq 3$$
$$H_1\colon \quad x \text{ belongs to a population } f_1(X) = \tfrac{1}{3},\ 2 \leq X \leq 5$$

and the decision rule is: $x \geq 2.5$, reject H_0. Determine the values of α and β.

6. A manufacturer of rope claims that average breaking strength of his product has increased from 276 to 282 lb. Breaking strength X is normally distributed, and σ_X is known to be 4 lb. Test this claim at a significance level $\alpha = 0.05$, using the following information. A random sample of 4 specimen rope pieces is tested yielding a mean breaking strength of 280.2 lb. (Summarize the test information on a distribution sketch.)

7. Repeat Problem 6 at a 1% significance level.

8. What is the probability of a Type II decision error in: (a) Problem 7, and (b) Problem 6?

9. Assume that a 6 lb. increase in mean breaking strength (Problem 6) is actually true, and a hypothesis test (Problem 7) results in an acceptance of H_0. In effect, this is a rejection of a valid claim. Explain this contradiction in terms of β and n.

10. Suppose in Problem 6 that the claim is an increase from 276 to 280 lb. (a) What is the probability of a Type II decision error? (b) What conclusion can you draw regarding the detection of small differences between H_0 and H_1 values? (c) How large a sample is required here to decrease the Type II decision error to 0.10 or less?

11. Minimum mean breaking strength of manufactured rope is 276 lb. The quality of a shipment of rope is in question. The standard deviation of breaking strength X is known to be 4 lb. Test at a 1% significance level the possibility that mean breaking strength is less than 276 lb., using the following information. A random sample of 36 rope specimens from the shipment is tested yielding a mean breaking strength of 274 lb.

12. A normal $\bar{x}$-distribution is a required condition in a hypothesis test of the mean. If breaking strength X is not normally distributed in Problem 11, what can you say regarding the normality of the $\bar{x}$-distribution model?

13. If the null hypothesis of Problem 11 is being tested against an alternative hypothesis H_1: $\mu = 274$, what is the probability of a Type II decision error? Suppose that a 95% protection against a Type II error is required (i.e., if H_1 is in fact true, the probability of detecting this fact is 0.95). What sample size is required to assure this protection, on the average?

14. The quality of a shipment of brass-wire springs is in question. It is desired to test the hypothesis that mean shear stress of the springs is 52,000 psi. against the alternative that mean shear stress differs from this value (in either direction) by as much as 1,250 psi. The standard deviation of shear stress X is known to be 2,500 psi. Agreed risks of Type I and II decision errors are 0.05 and 0.10 respectively. (a) Determine sample size, and (b) state the decision rule.

15. Perform the hypothesis test for Problem 14, using the following information. For economic reasons, sample size is 30, thus changing the agreed upon β-value. The significance level $\alpha = 0.05$ is unchanged. A random sample of 30 springs is tested yielding a mean shear stress of 51,000 psi. (Summarize the test information on a distribution sketch.)

16. A hypothesis test concerns the setting of an automatic machine that fills 2 lb. boxes. Test that the mean box fill is 2 lb. against the alternative that mean fill differs by as much as 0.1 lb. from 2 lb. If the mean fill is actually 2 lb., it is desired to reach this conclusion with probability 0.95. Also, the probability of detecting a difference of 0.1 lb. must not be less than 0.90. The standard deviation of individual box fills is 0.13. The sample observations yield a mean weight of 2.06 lb. (Summarize the test information on a distribution sketch.)

17. A hypothesis that a mean process setting is $\mu = 0.625$ in. is tested against an alternative that the mean setting differs from 0.625 in. by 0.001 in. The test is made at a 1% significance level with σ_X being known and equal to 0.0015 in. The probability of detection of a difference in mean setting as large as 0.001 in. is 0.90. Perform this hypothesis test assuming that sample observations of production output from this machine yield a mean of 0.6254 in.

18. For a two-sided hypothesis test of the mean of a normally distributed characteristic of interest, the *d*-ratio is 1.0. The nature of the test indicates that Type I and II decision errors are equally critical and their probabilities should both be small. However, economic reasons restrict sample size to a maximum of 10 items. A tentative proposal is to test at a 1% significance level. Is this satisfactory? What is your suggestion?

19. Using expected cost as a criterion, how would you handle the test design question of Problem 18, using the following information. The significance level is most critical, and α is fixed at 0.01. The dollar cost of a Type II decision error is $2,000. Increasing sample size from 10 to 15 units increases the test cost by $20 a unit. (See Problem 4.43 to review the concept of expected cost.)

20. A manufacturer of baseball bats receives a shipment of Oregon ash wood. A compression test (parallel to the grain) usually yields a mean strength of 5,840 psi. A random sample of 36 wood pieces from this shipment yields a mean strength of 5,795 psi. Test the hypothesis that mean strength of the shipment is 5,840 psi. Use a 1% significance level, $\sigma_X = 88$ psi., and disregard the Type II decision error.

21. Mean breaking strength of a certain manufactured cable is 1,600 lb. and σ_X is 100 lb. A new manufacturing treatment is expected to increase breaking strength. A random sample of 50 cable specimens from the new treatment is tested and the observed mean breaking strength is 1,660 lb. Do you conclude that there is, or is not, significant empirical evidence of a strength increase at a 1% significance level?

22. The mean life of a sample of 80 light bulbs is observed to be 1,170 hours, and σ_X is known to be 100 hours. Test at a 1% significance level that the mean life of all such bulbs is 1,200 hours against the alternative that mean life is less than 1,200 hours.

23. What is your conclusion for Problem 22 if $\sigma_X = 140$ hours?

24. Problem 1.14 gives 25 observations of nitrogen dioxide levels (ppm) in urban air at a high density traffic location. Perform a two-sided test of a hypothesis that the actual mean nitrogen dioxide level is 1.17 ppm, based on the sample information repeated here. Use a 1% significance level for the test, and assume that σ_X is known to equal 0.20 ppm.

x_j	0.4	0.6	0.8	1.0	1.2	1.4	1.6	1.8	2.0	25
f_j	1	3	5	6	4	2	2	1	1	25

25. Problem 1.19 gives 10 observations of the transverse strength X of bricks. Transverse strength is approximately normally distributed and σ_X is known to be 2 psi. Test at a 5% significance level that this sample is from a population of bricks whose mean transverse strength is 571.6 psi.

x_j	568	570	572	574	
f_j	4	3	2	1	10

26. Problem 1.31 gives 20 observations of breaking strength X of 0.104 in. hard-drawn copper wire. The standard deviation of individual breaking strengths is 3 lb. Test at a 5% significance level that this sample is from a copper-wire population whose mean breaking strength is 573.2 lb.

x_j	562	564	566	572	576	580	582	
f_j	1	3	4	5	3	2	2	20

27. Problem 1.36 gives 100 observations of nitrogen 14 level X (ppm) of lake water. Assuming σ_X is known and equal to 4 ppm, test at a 5% significance level that this sample is from a population whose mean is 22.25 ppm.

x_j	9–12	13–16	17–20	21–24	25–28	29–32	33–36	
f_j	1	6	22	32	26	8	5	100

28. A sample of 20 measurements of oxygen level at the bottom of a lake gives the following values. Assuming σ_X is known and equal to 0.06 ppm, test at a 5% significance level that this sample is from a population whose mean oxygen level is 2 ppm.

x_j	1.8	1.9	2.0	2.1	2.2	
f_j	2	4	6	5	3	20

29. A null hypothesis that the mean breaking strength of a certain rope is 300 lb. is tested against the alternative that mean breaking strength is 310 lb. The significance level is 1%, sample size is 64, and σ_X is 24 lb. Using the method of Sect. 7.7, analytically determine the probability of a Type II decision error and compare your result with the probability as read from Fig. 7.4.

30. Repeat Problem 29 for various alternative hypothesis values, $\mu_1 = 300$, 305, 310, 315, and 320, and plot the *OC* curve for this hypothesis test.

31. Suppose that, for the hypothesis test of Problem 29, $\alpha = 0.01$, $\beta = 0.20$, and $\sigma_X = 20$ lb. Using the method of Sect. 7.7, analytically determine sample size n and compare your result with the n-value read from Fig. 7.4.

32. A hypothesis H_0: $\mu_0 = 19.5$ is tested against H_1: $\mu_1 = 19.0$. Decision error risks are $\alpha = 0.05$ and $\beta = 0.10$, and $\sigma_X = 2.0$. Analytically determine sample size n and compare your result with the n-value read from Fig. 7.4.

33. In 100 trials of a fatigue test, 42 failures occur. Test at a 5% significance level that this sample is from a population where failure and success are equally likely. Use the normal approximation to the binomial probability, with mean np and variance npq.

34. Suppose for Problem 33, as stated, that a decision rule is adopted as follows: $65 \leq$ no. of failures ≤ 35, reject H_0. What is the probability of a Type I decision error?

35. Suppose for Problem 34, as stated, that the hypothesis test is against an alternative that the population mean is either 0.30 or 0.70. What is the probability of a Type II decision error?

36. A water processing plant makes daily measurements regarding the number of algae cells per milliliter of lake water. A sample of 36 measurements yields a mean algae cell count of 2,475 and a variance $s_x^2 = 50^2$. Test at a 1% significance level that the sample measurements are from a population with mean algae cell count of 2,500.

37. Repeat the hypothesis test of Problem 24 assuming that σ_X is unknown and estimated from the sample variance.

38. Repeat the hypothesis test of Problem 26 assuming that σ_X is unknown and estimated from the sample variance.

39. Repeat the hypothesis test of Problem 27 assuming that σ_X is unknown and estimated from the sample variance.

40. A sample of 80 observations of lip opening pressure (L.O.P.) on an automotive seal are given in Problem 1.1 (Chap. 1). Test at a 5% significance level that this sample is from a shipment of seals with mean L.O.P. equal to 11.30 psi.

41. For identical values of the *d*-ratio (1.2), and α (0.01), and β (0.20), and approximately equal sample sizes ($n = 8$–10), a $z_{\text{data}} \geq 2.326$ or a $t_{\text{data}} \geq 2.821$ is required for a rejection of H_0: $\mu = \mu_0$. Explain this difference in magnitude between $Z_\alpha = 2.326$ and $t_{\alpha, n-1} = 2.821$.

42. Replica machines automatically fill 10 lb. boxes. Based on the following sample data, test at a 1% significance level that there is no difference in mean box fill for the two machines X_1 and X_2 being studied. Individual fill amounts are each normally distributed with respective variances $\sigma_1^2 = 0.015^2$ and $\sigma_2^2 = 0.018^2$.

x_{1j}	9.96	9.98	9.99	10.01	10.02	10.03	10.04	10.05	
f_j	1	1	1	1	2	1	1	2	10

x_{2j}	9.96	9.97	9.99	10.00	10.01	10.02	10.03	10.04	
f_j	1	1	1	1	2	2	1	1	10

43. Assuming equal sample sizes, what sample size is required in Problem 42 to detect with probability 0.95 a population mean difference as large as 0.025 lb.?

44. Repeat the hypothesis test of Problem 42 assuming that σ_1 and σ_2 are unknown but equal.

45. Repeat the hypothesis test of Problem 42 assuming that σ_1 and σ_2 are unknown and not necessarily equal.

46. In Problem 42, allowing the machines to continue running when their true mean fills differ significantly is more costly than the error of unnecessary shutdown and adjustment of a machine. Which probability α or β should be fixed at a small value?

47. Suppose in Problem 42 that β is fixed at 0.10, and the *d*-ratio is 1.0. A Type I decision error cost (unnecessary machine shutdown and adjustment) is \$50. If unit weight inspection cost is \$0.20, and there are two possibilities, $\alpha = 0.01$ and $\alpha = 0.05$, what is the economic α-value?

48. Two shipments of Oregon ash are received (see Problem 20). A sample of 36 specimen pieces is randomly selected from each shipment and subjected to a compression test yielding means $\bar{x}_1 = 5{,}800$ psi. and $\bar{x}_2 = 5{,}750$ psi. Within shipment variation is usually approximately the same. Thus, σ_X and σ_Y are known: $\sigma_X = \sigma_Y = 100$ psi. Test at a 1% significance level that the shipments do not differ significantly relative to their mean strengths.

49. Assuming equal sample sizes, what sample size is required in Problem 48 to detect with probability 0.95 a mean strength difference of 100 psi.?

50. Test the hypothesis $\mu_X = \mu_Y$ against the alternative $\mu_X > \mu_Y$, where $\sigma_X = 3$, $\sigma_Y = 4$, $\alpha = 0.14$, and $n_x = n_y = 16$. Sample data yields $\bar{x} = 3$ and $\bar{y} = 1$.

51. Thirty observations x of nitrogen dioxide air content at a high density traffic location yield the following data (in ppm):

x_j	0.6	0.8	1.0	1.2	1.4	1.6	1.8	
f_j	2	4	4	8	6	5	1	30

Test at a 5% significance level the hypothesis H_0: $\sigma_X = 0.25$ ppm against the alternative H_1: $\sigma_X > 0.25$ ppm.

52. Suppose in Problem 51 that a null hypothesis H_0: $\sigma_X = 0.2$ ppm is tested against H_1: $\sigma_X = 0.3$ ppm. What sample size is required to assure a 95% protection against a Type II decision error?

53. An existing gage is unsatisfactory. In use, its standard deviation is 50 microinches. The claim for a proposed replacement gage is that it will reduce application error to $\sigma_X = 25$ microinches. Test at a 5% significance level that σ_X for the new gage is 50 microinches against the alternative that σ_X is 25 microinches. If the claim for the new gage is valid, this conclusion should be reached with probability 0.95. Assume that the appropriate number of test observations yields $s_x = 32$ microinches.

54. Give an analysis with numerical comparisons for Problem 53 assuming that the hypotheses are framed in an opposite manner, that is, H_0: $\sigma_X = 25$ microinches and H_1: $\sigma_X = 50$ microinches.

55. Test at a 1% significance level that the variation in box fill from an automatic filling operation is such that $\sigma_X = 3$ ounces. If the variation is more, or less, to an extent that σ_X differs by as much as 2.4 ounces from 3 ounces, it is desired to detect this fact with probability 0.95. Assume that the appropriate number of test observations yields $s_x = 5.0$ ounces.

56. In the past, the standard deviation of certain 40 ounce packages was 0.3 ounces. A test is made to determine if the variation in weight of these packages is still essentially unchanged. A random sample of 17 packages yields $s_x = 0.5$ ounces. Test at a 5% significance level that σ_X is 0.3 ounces against the alternative that $\sigma_X > 0.3$ ounces.

57. What sample size is required in Problem 56 to detect an increase in σ_X of 0.2 ounces with probability 0.90 of doing so?

58. Observations of breaking strength of copper-wire specimens from two production treatments X and Y are given as follows. Test at a 5% significance level the hypothesis H_0: $\sigma_X = \sigma_Y$ against the alternative that $\sigma_X \gtrless \sigma_Y$.

x_j	570	572	574	576	578	580	
f_j	1	2	3	2	1	1	10

y_j	568	570	572	576	578	584	
f_j	1	3	3	1	1	1	10

59. What sample size is required in Problem 58 to detect with probability 0.90 a difference such that one standard deviation is twice as large as the other?

60. Five trials of a laboratory experiment involving an X-treatment yield a variance $s_x^2 = 0.00045$, and six trials involving a Y-treatment give a variance $s_y^2 = 0.00039$. Test at a 1% significance level the hypothesis $H_0: \sigma_X = \sigma_Y$ against the alternative that $\sigma_X > \sigma_Y$.

61. Repeat Problem 60 considering that σ_X may be either greater than or less than σ_Y.

62. A physical test has two outcomes—either the test specimen fails or it does not fail. Test at a 1% significance level the hypothesis that the probability π of a failure is 0.3 against the alternative that $\pi > 0.3$. A random sample of 10 specimens is tested and 5 failures are observed. (Use the binomial probabilities in Appendix A.)

63. Suppose in Problem 62 that a random sample of 100 specimens is tested. What is the decision rule?

64. If a random sample of 100 specimens is tested in Problem 62, and the null hypothesis $H_0: \pi = 0.3$ is tested against a specific alternative $H_1: \pi = 0.5$, what is the probability of a Type II decision error?

65. Using a normal model to approximate the binomial probabilities in Problem 64, what is the decision rule and the probability of a Type II decision error?

66. A study is made regarding the malfunction of an assembly unit. For each assembly unit examined, malfunction is due either to a fit that is too tight or a fit that is too loose. Regarding the proportion of tight and loose fits, it is felt that they are equally likely. Let π denote the probability of a tight fit. Test at a 5% significance level the hypothesis that $\pi = 0.5$ against the alternative that $\pi > 0.5$, using a normal model to approximate the binomial probabilities. A random sample of 400 assembly units examined yields $p = 0.57$.

67. The manufacturer of a special lubricant claims that his lubricant is 80% effective in preventing a certain product malfunction. The lubricant is tested on 200 product items and is successful 143 times. Can the lubricant manufacturer's claim be supported at a 1% significance level? (Use a normal model.)

68. An old process machine is used for a particular operation. The distribution of product sizes from this operation is uniformly distributed over an interval (0.500, 0.504) in. A new type of tooling setup is tested in an attempt to obtain more concentration of product sizes near the nominal specification value of 0.502 in. A random sample of output product from the new setup yields the following frequencies. Perform an appropriate hypothesis test, at a 1% significance level, to determine if the new tooling setup is a significant improvement.

x_j	0.500	0.501	0.502	0.503	0.504	
f_j	7	11	12	11	9	50

69. Four machines are operated three shifts each day. The number of breakdowns by machine and shift are recorded as follows. An effort is being made to identify breakdowns with machine operators to possibly determine cause. Test the hypothesis that breakdowns are independent of the shift. Use a 1% significance level.

Shift	Machines			
	A	*B*	*C*	*D*
1	4	5	3	4
2	6	3	2	3
3	3	4	4	4

70. The number of defectives X observed in 10 random samples from a process output are recorded as follows. Test at a 10% significance level the hypothesis that the number of defectives in samples of size 5 follows a binomial distribution.

x_j	0	1	2	3	4	5	
f_j	1	3	4	2	0	0	10

71. Machine breakdown frequency per day is being studied. Test at a 5% significance level that the number of breakdowns X per day is Poisson distributed. Base the test on the following sample observations.

x_j	0	1	2	3	4	
f_j	46	37	13	3	1	100

72. A life test of 100 randomly selected tubes yields the following data (life X is measured in hours for the intervals shown). Test at a 1% significance level that life follows an exponential *D.F.*: $f(X) = e^{-0.0005X}$.

x_j	0, 100	101, 200	201, 300	300, > 300	
f_j	40	22	16	22	100

73. Test at a 1% significance level that the following sample is from a normally distributed population. The data represents 100 random observations of time to failure X in units of 100 hours.

x_j	0–2	3–5	6–8	9–11	12–14	15–17	18–20	21–23	
f_j	1	7	15	27	26	16	6	2	100

74. Fifty random observations of resistance X in ohms of wire specimens are recorded as follows. Test at a 5% significance level that this sample is from a normally distributed population.

x_j	8.4	8.8	9.2	9.6	10.0	10.4	10.8	11.2	
f_j	1	3	7	14	13	8	3	1	50

75. Test at a 5% significance level that the following sample observations are from a population whose *D.F.* is $f(X) = 1 - \frac{X}{2}$.

x_j	0	0.4	0.8	1.2	1.6	
f_j	11	7	6	5	1	30

76. The *D.F.* in Problem 75 is of the form $f(X) = a_0 + a_1 X$. Suppose the a_0 and a_1 values have been established from sample data. What is the decision rule for the chi-square test now?

77. Define: (a) *unbiased* estimator, and (b) *efficient* estimator.

78. Define: (a) *point* estimate, (b) *interval* estimate, and (c) *confidence interval* estimate.

79. When testing $H_0: \mu = \mu_0$ against the alternative $H_1: \mu = \mu_1 \gtrless \mu_0$, the decision rule is $|z_{\text{data}}| \geq Z_{\alpha/2}$, reject H_0. State an equivalent decision rule in terms of acceptance and $1 - \alpha$.

80. For an upper one-sided test of the mean μ, an equivalent decision rule in terms of acceptance is: accept H_0 if the interval $-\infty$, $\mu_0 + Z_\alpha(\sigma_X/\sqrt{n})$ contains z_{data}. Demonstrate this rule numerically for Example 7.4.1.

81. Determine a 95% confidence interval estimate of the mean μ for: (a) Problem 7.6, (b) Problem 7.11, and (c) Problem 7.16.

82. Determine a 95% confidence interval estimate of the mean μ, assuming that σ_X is unknown and estimated by s_x for: (a) Problem 7.24, (b) Problem 7.26, and (c) Problem 7.27.

83. Determine a two-sided 99% confidence interval estimate of the mean difference $\mu_X - \mu_Y$ for: (a) Problem 7.42, and (b) Problem 7.42, assuming that σ_X and σ_Y are unknown but equal.

84. Prepare confidence interval estimates of σ_X for Problem 7.51 as follows: (a) two-sided interval, (b) lower one-sided interval, and (c) upper one-sided interval.

85. Prepare confidence interval estimates of the ratio σ_X/σ_Y for Problem 7.58 as follows: (a) two-sided interval, (b) lower one-sided interval, and (c) upper one-sided interval.

86. Prepare confidence interval estimates of the proportion π for Problem 7.66 as follows: (a) two-sided interval, (b) lower one-sided interval, and (c) upper one-sided interval.

QUALITY CONTROL

The hypothesis testing techniques of Chap. 7 are applicable to three principal quality control activities: (1) process capability evaluation, (2) process control, and (3) sampling inspection. Knowledge of the population mean and variance is basic to the development of capability measures. Statistical process control and sampling inspection procedures are hypothesis tests. Further, the first two activities are necessary preliminaries to rational sampling inspection procedures.

8.1 Process Capability

There are three sources of variation in any process: the operator, the material, and the process machine. The variation from the first two sources can be reduced to a minimum by using a skilled operator and homogeneous material. A measure of the remaining variation is viewed as the *capability* or *natural tolerance* of the process.

Process capability values are useful to: (1) product designers in arriving at realistic specifications that are compatible with manufacturing capabilities, (2) production supervisors in assigning production jobs to specific machines,

8

and (3) quality control staff in designing control procedures and sampling inspection plans. The six σ_X natural tolerance values for overlapping specification tolerances (see Sec. 5.9) coincide with process capability values. In both cases, reference is to the best effort of the process in the sense that assignable causes of variation (e.g., process operator, product material, process setup, process adjustment, etc.) have been either eliminated or at least minimized.

A common measure of process capability is six σ_X. The product quality characteristic is denoted by X and the output X from a given process operation is described by a *D.F.* Thus, the range of X represents the X-values occurring a high percentage of the time, i.e., the range represents expected variation and hence the term "capability." In Fig. 8.1, X is $N(\mu, (0.001/6)^2)$ and X occurs in the designated six σ_X interval 99.73% of the time.

From Eq. (1.20) and Sec. 7.16, an unbiased estimate of the population variance is given by

$$s_x^2 = \frac{1}{n-1}\sum_{i=1}^{n}(x_i - \bar{x})^2$$

and the estimated process capability is $6s_x$. Determining capability in this

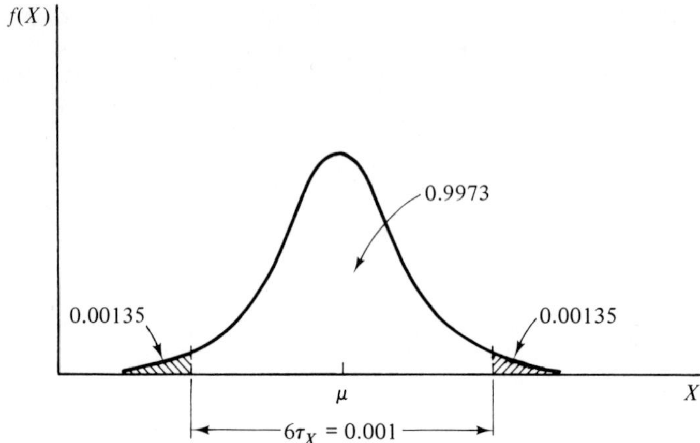

Figure 8.1. Process capability.

manner assumes that the study yielding s_x is conducted under controlled conditions obtainable in the production environment. Specifically, this means that assignable variation causes are minimized for the period of observation of X. If historical data or physical knowledge is such that σ_X is known, the capability value is taken to be $6\sigma_X$.

Another method of determining capability, often preferred because of less tedious computation involved, is based on an estimate of σ_X obtained from sample range values (see Sec. 1.13). Let r_j denote the range of the jth sample. Then, for k samples,

$$\bar{r} = \frac{1}{k}\sum_{j=1}^{k} r_j \quad j = 1, 2, \ldots, k \tag{8.1}$$

A quantity d_2, dependent on sample size n, can be determined such that

$$E\left(\frac{r}{\sigma_X}\right) = d_2$$

or

$$\frac{E(r)}{d_2} = \sigma_X$$

and, replacing $E(r)$ by its estimate $\bar{r}$, an estimate of σ_X is

$$\frac{\bar{r}}{d_2} \tag{8.2}$$

Values of d_2 corresponding to various n-values are given in Appendix H.

EXAMPLE 8.1.1

Fifty random observations of X under controlled production conditions are summarized by the coded data of Table 8.1. The coding is described by

$$u_j = \frac{x_j - 0.5000}{0.0001}$$

Computing the mean range,

$$\begin{aligned}\bar{r} &= \tfrac{1}{10} \sum_{j=1}^{10} r_j \\ &= \tfrac{1}{10}(52) = 5.2\end{aligned}$$

Estimating σ_U,

$$\begin{aligned}s_u &= \frac{\bar{r}}{d_2} \\ &= \frac{5.2}{2.326} = 2.2356\end{aligned}$$

Converting to x-units,

$$\begin{aligned}s_x &= 0.0001 s_u \\ &= 0.000224\end{aligned}$$

and an estimate of process capability is

$$\begin{aligned}6s_x &= 6(0.000224) \\ &= 0.0013\end{aligned}$$

or approximately 0.001 in.

Table 8.1. CODED PROCESS CAPABILITY DATA, WHERE

$$u_j = \frac{x_j - 0.5000}{0.0001} \text{ IN.}$$

Sample Number	1	2	3	4	5	6	7	8	9	10	
u_1	0	−2	−2	−3	0	−2	+1	+1	−3	+4	
u_2	−3	+2	0	−1	+1	0	−1	−2	0	−4	
u_3	−2	−1	+1	+2	−2	+5	+3	+2	+1	−1	
u_4	−1	−2	−2	+2	−4	+4	0	−2	−2	−1	
u_5	0	−3	+3	−1	+2	+1	−2	−1	0	+1	
Maximum u	0	+2	+3	+2	+2	+5	+3	+2	+1	+4	
Minimum u	−3	−3	−2	−3	−4	−2	−2	−2	−3	−4	
r_j	3	5	5	5	6	7	5	4	4	8	52

A prerequisite condition for a process-capability study is that a controlled state of the process exists. Clearly, this means that assignable process variation causes are minimized, leaving for the most part only random causes of variation. A physical interpretation of random causes of variation is: (1) A process machine is a mechanical device composed of many mating component parts. The clearances, end-plays, and other error effects due to the fits of mating parts determine the variability of the machine: (2) Each error effect is small in magnitude. (3) These error effects combine in a random manner to determine the overall machine variation. In this sense, a controlled process is composed mainly of random causes of variation.*

EXERCISES

1. Determine an estimate of the process capability for the L.O.P. data of Problem 1.1 (Chap. 1). Base the estimate on the sample range, and consider the data given to represent 16 samples of 5 observations each.

Ans. $s_x = 0.834$ and $6s_x = 5.00$ psi.

2. Repeat Exercise 1 basing the estimate of process capability on the sample variance, the sample being the 80 observations given in Problem 1.1.

Ans. $6s_x = 5.24$ psi.

8.2 Process Control

An $\bar{x}$-*control chart* is a statistical tool for detecting shifts in the process average under the assumption that process variation remains approximately constant. This control chart method is a hypothesis test, where

$$H_0: \text{ process is in control at a mean level } \mu_0$$
$$\alpha: \ P(\text{reject } H_0, H_0 \text{ true}) = 0.0027$$

---

*The student should not view the procedure of establishing a capability value as a routine matter of simply obtaining a good estimate of σ_X. Other practical criteria may dominate the specific production situation. It may be difficult or impossible to physically maintain a continuous state of control under production conditions. It may even be uneconomic to maintain control.

To illustrate, in Example 8.1.1, the statistical capability value is approximately 0.001 in. This is the *best* expected production effort of the process in the absence of any fundamental process design change. The final capability value for production planning purposes may be a value slightly larger than 0.001 in., depending on practical production considerations.

Also, it should be clear that as a process machine ages, its variability naturally increases, thus changing the capability value.

and the decision rule is accept H_0 if the sample mean $\bar{x}$ occurs in the interval $\mu_0 \pm 3\,\sigma_{\bar{x}}$ and reject H_0 if $\bar{x}$ is outside of this interval.

The process operation is monitored by repeating the hypothesis test at appropriate time points $1, 2, \ldots, j, \ldots, k$. A rejection of H_0 at any time j is equivalent to a refusal to accept only random causes as an explanation for the unusual event of a sample mean occurring outside of the interval $\mu_0 \pm 3\sigma_{\bar{x}}$. Accepting H_0 is equivalent to considering the sample mean observation as being due only to random causes of variation.*

Confidence estimates of the control limits $\mu_0 \pm 3\sigma_{\bar{x}}$ are used since both μ_0 and $\sigma_{\bar{x}}$ are not usually known. An unbiased estimate of μ_0 (usually based initially on $k = 25$ samples) is

$$\bar{\bar{x}} = \frac{1}{k}\sum_{j=1}^{k} \bar{x}_j \quad \text{[see Eq. (1.12)]}$$

Each sample mean is a random event and the standard deviation of the mean is

$$\sigma_{\bar{x}} = \frac{\sigma_x}{\sqrt{n}}$$

If H_0 is true, a 99.73% confidence interval estimate of the mean μ_0 (called upper and lower control limits, UCL and LCL) is obtained by considering

$$\text{UCL, LCL} = \mu_0 \pm 3\sigma_{\bar{x}}$$

or

$$\text{UCL, LCL} = \mu_0 \pm 3\left(\frac{\sigma_x}{\sqrt{n}}\right)$$

and, replacing μ_0 by its unbiased estimate $\bar{\bar{x}}$,

$$\text{UCL, LCL} = \bar{\bar{x}} \pm 3\left(\frac{\sigma_x}{\sqrt{n}}\right)$$

From Eq. (8.2), replacing σ_X by its estimate $\bar{r}/d_2$, this expression becomes

---

*This is the same argument given in Secs. 7.1–7.8. However, in this application context a rejection of H_0 is an inference that the process is "out of control," whereas an acceptance of H_0 indicates control. The rejection decision is a signal for practical corrective action relative to the process, and an acceptance decision is interpreted as no call for action.

Thus, in terms of corrective action taken or not taken, a Type I decision error is followed by unjustified corrective action and a Type II decision error by failure to take required action. The first error usually leads to overadjustments of the process, an action that tends to increase rather than decrease variation. The second error results in the manufacture of poor quality product that could have been prevented.

$$\text{UCL, LCL} = \bar{\bar{x}} \pm \frac{3\bar{r}}{d_2\sqrt{n}}$$

Appendix H summarizes coefficients occurring in quality control computations. The quantity $3/d_2\sqrt{n}$ is combined under the designation A_2 and for various n-values is read from this table. Thus, a computation of control limits is simply

$$\text{UCL, LCL} = \bar{\bar{x}} \pm A_2\bar{r} \tag{8.3}$$

where $\bar{\bar{x}}$ and $\bar{r}$ are computed from the sample data and A_2 is read from Appendix H.

EXAMPLE 8.2.1

Ten samples of five observations each yield individual measurements, means, and ranges as shown in Table 8.2. An estimate of μ_0 is given by

$$\bar{\bar{x}} = \tfrac{1}{10}\sum_{j=1}^{10}\bar{x}_j$$
$$= \tfrac{1}{10}(5.030) = 0.503$$

The mean range is

$$\bar{r} = \tfrac{1}{10}\sum_{j=1}^{10} r_j$$
$$= \tfrac{1}{10}(0.040) = 0.004$$

Entering Appendix H with $n = 5$, the coefficient A_2 is read to be 0.577 and thus,

Table 8.2. EXAMPLE 8.2.1: $\bar{x}$- AND r-CONTROL CHART DATA.

Sample No.	x_1	x_2	x_3	x_4	x_5	$\bar{x}_j$	r_j
1	0.505	0.503	0.501	0.502	0.500	0.5022	0.005
2	0.501	0.499	0.504	0.503	0.505	0.5024	0.006
3	0.501	0.503	0.505	0.502	0.502	0.5026	0.004
4	0.499	0.503	0.503	0.502	0.503	0.5020	0.004
5	0.504	0.505	0.505	0.504	0.503	0.5042	0.002
6	0.503	0.503	0.504	0.503	0.503	0.5032	0.001
7	0.505	0.504	0.504	0.505	0.505	0.5046	0.001
8	0.504	0.505	0.501	0.504	0.507	0.5042	0.006
9	0.503	0.505	0.501	0.500	0.502	0.5022	0.005
10	0.499	0.505	0.503	0.504	0.501	0.5024	0.006
						5.0300	0.040

from Eq. (8.3),

$$\begin{aligned} \text{UCL, LCL} &= \bar{\bar{x}} \pm A_2\bar{r} \\ &= 0.503 \pm 0.577(0.004) \\ &= 0.5053, 0.5007 \end{aligned}$$

and the resulting control chart is given in Fig. 8.2.

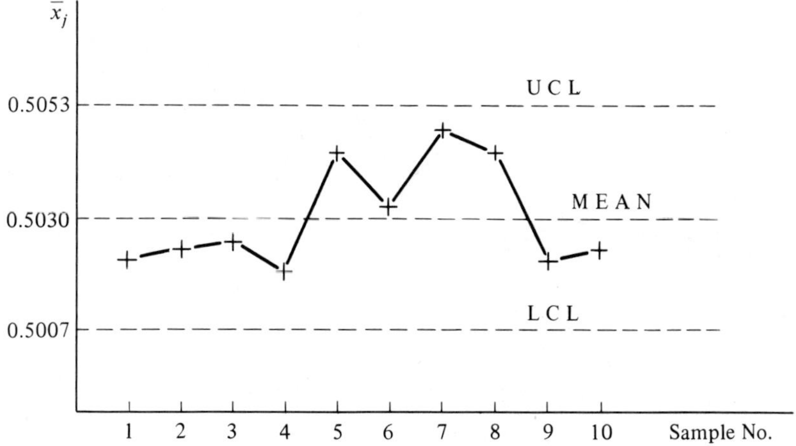

Figure 8.2. Example 8.2.1—x-control chart.

Changes in process variation are not effectively detected by the $\bar{x}$-chart (see Fig. 8.3). A *range chart* (or *r-chart*) is used to supplement the $\bar{x}$-chart.

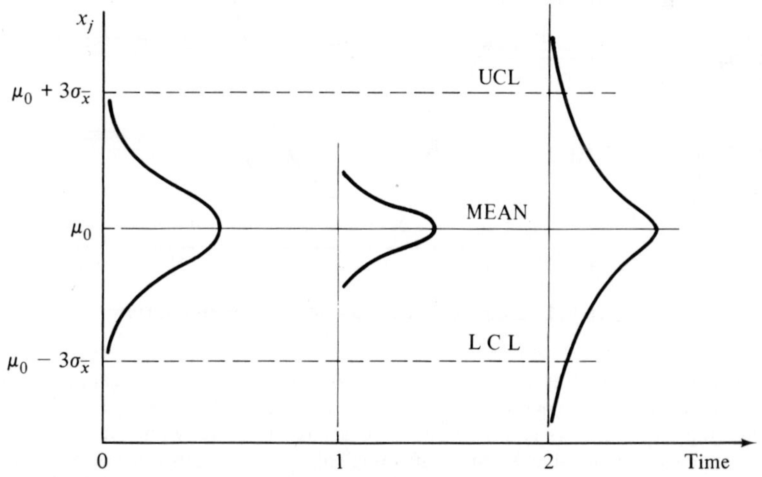

Figure 8.3. Variation change from time 0 to times 1 and 2, with new populations (at 1 and 2) having the same mean as that at time 1.

Whereas the $\bar{x}$-chart is a hypothesis test of the process mean, the r-chart is a test of a hypothesis concerning the variation as measured by the range.*

Sample range values are approximately chi-square distributed. Control limits based on 99+ % confidence interval estimates of $E(r)$ are $E(r) \pm 3\sigma_r$. It can be shown that $E(r) = d_2\sigma_X$ and $\sigma_r = d_3\sigma_X$, where the coefficients d_2

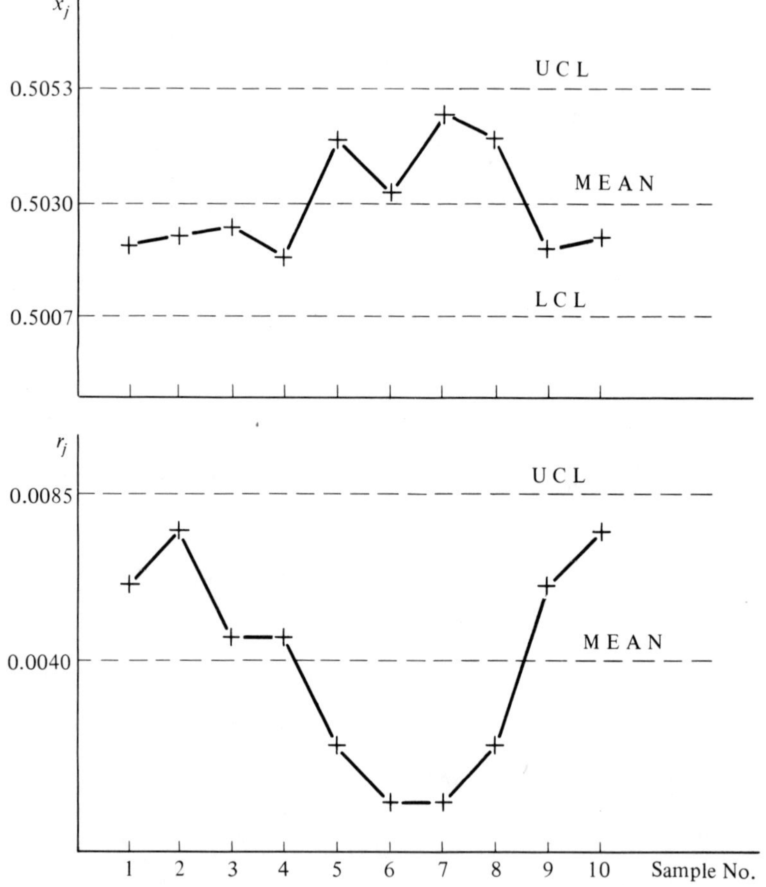

Figure 8.4. Example 8.2.2—$\bar{x}$- and r-control charts.

*Stated differently, the $\bar{x}$-control chart operates to disclose between-sample variation (consistent sources of assignable variation tending to affect all sample observations and thus also the sample mean), and the r-control chart acts to detect within-sample variation (erratic sources of assignable variation affecting some observations but not others in the same sample).

Examples of consistent sources of variation are tool wear and change of stock. On the other hand, a loose tool block, chuck, bearing, etc., are erratic variation causes.

and d_3 depend on sample size n and are tabulated for various n-values (Appendix H). As in the $\bar{x}$-control chart case, coefficients are combined to simplify computation. The resulting computational form of the control limits is

$$\text{UCL, LCL} = D_4\bar{r},\ D_3\bar{r}^* \tag{8.4}$$

EXAMPLE 8.2.2

For the data of Example 8.2.1 (see Table 8.2) control limits for the range r are obtained from

$$\begin{aligned} D_4\bar{r} &= 2.115(0.004) \\ &= 0.0085 \\ D_3\bar{r} &= 0 \end{aligned}$$

where the D_4- and D_3-values, for $n = 5$, are read from Appendix H. The central line on the range chart corresponds to $E(r)$ and is estimated by $\bar{r}$. The $\bar{x}$- and r-control charts are usually used together, as shown in Fig. 8.4.

In the exceptional case, where σ_X and thus also $\sigma_{\bar{x}}$ are known, either from physical information or historical data, control limits for the $\bar{x}$-control chart are computed from

$$\text{UCL, LCL} = \bar{\bar{x}} \pm A\sigma_X \tag{8.5}$$

where

$$A = \frac{3}{\sqrt{n}}^{\dagger}$$

---

*A 99+% confidence interval estimate of $E(r)$ is obtained by considering

$$E(r) \pm 3\sigma_r$$

or

$$E(r) \pm 3d_3\sigma_X$$

and, replacing $E(r)$ by its sample estimate $\bar{r}$ and σ_X by its estimate $\bar{r}/d_2$, this expression becomes

$$\bar{r} \pm \frac{3d_3\bar{r}}{d_2}$$

or

$$\bar{r}\left(1 \pm \frac{3d_3}{d_2}\right)$$

where the coefficients $1 + (3d_3/d_2)$ and $1 - (3d_3/d_2)$ are denoted by D_4 and D_3, respectively, in Appendix H.

†A 99.73% confidence interval estimate of μ_0 is obtained by considering

$$\mu_0 \pm 3\sigma_{\bar{x}}$$

and, replacing μ_0 by its estimate $\bar{\bar{x}}$ and $\sigma_{\bar{x}}$ by its equivalent,

$$\bar{\bar{x}} \pm 3\left(\frac{\sigma_X}{\sqrt{n}}\right)$$

or

$$\bar{\bar{x}} \pm A\sigma_X$$

where A designates $3/\sqrt{n}$.

Sample size for the $\bar{x}$- and r-control charts combination is usually 4 or 5, except in the case where a lower control limit not equal to zero is desired for the range chart. To detect between-sample variation, small samples (over as short a time period as possible) increase the possibility of an abrupt process change of short duration occurring between samples rather than within a sample. Further, if the process is subject to slow and continuous change, a small sample increases the possibility that the variation within a sample is small relative to that between samples. Thus, small samples are desired. On the other hand, sample size should be at least 4 in order that samples from a nonnormal X-population will generate an approximately normal $\bar{x}$-population (see Example 5.7.1).

Another important consideration in sampling is to preserve the process identity of the samples (i.e., sample identification over time with process conditions). Identification of samples with process conditions (e.g., process adjustments, regrinding of cutting tools, etc.) makes possible the recognition of assignable causes of variation. For this reason, samples are taken in the order of production at regular time intervals.

The $\bar{x}$- and r-control charts are appropriate for *variables measurement* of product items, that is, any inspection operation where the gage indicates deviations from the product specification over a continuous scale of possible values. However, because it is usually less expensive, the inspection operation is frequently an *attributes measurement* where the gage merely classifies the product into a few discrete categories (e.g., effective and defective product). Two common control charts used for attributes measurement are: (1) p-chart, and (2) c-chart.

A *p-chart* is a hypothesis test of the population proportion defective π where the hypothesis is $H_0\colon \pi = \pi_0$ and $\alpha = 0.0027$. The control limits are set at ± 3 standard deviations from π_0, that is,

$$\text{UCL, LCL} = \pi_0 \pm 3\sigma_{X/n}$$

Since $\sigma_{X/n} = (1/n)\sigma_X$ and X is a Bernoulli random variable (i.e., there are two outcomes of X, defective or effective), the control limits can be expressed as

$$\begin{aligned}\text{UCL, LCL} &= \pi_0 \pm \frac{3}{n}\sqrt{n\pi_0(1-\pi_0)} \\ &= \pi_0 \pm 3\sqrt{\frac{\pi_0(1-\pi_0)}{n}}\end{aligned}$$

If π_0 is estimated by

$$\bar{p} = \frac{1}{k}\sum_{j=1}^{k} p_j \quad \text{(identical sample size, } k \text{ samples)} \tag{8.6}$$

or

$$\bar{p} = \frac{1}{N} \sum_{j=1}^{k} x_j, \quad N = \sum_{j=1}^{k} n_j \quad \text{(unequal sample size } n_j) \tag{8.7}$$

with x_j being the observed number of defectives in the jth sample, then

$$\text{UCL, LCL} = \bar{p} \pm 3\sqrt{\frac{\bar{p}(1-\bar{p})}{n}} \tag{8.8}$$

If sample sizes are unequal, the n-value in Eq. (8.8) is the average sample size.

EXAMPLE 8.2.3

Ten samples of 50 observations each yield the following numbers of defectives: 5, 1, 1, 2, 3, 3, 1, 4, 2, 3. Establish initial control limits for a p-chart.

$$\bar{p} = \tfrac{1}{10}(0.10 + 0.02 + 0.02 + \cdots + 0.06) = 0.05$$

$$\text{UCL, LCL} = 0.05 \pm 3\sqrt{\frac{0.05(0.95)}{50}}$$

$$= 0.14, 0$$

The resulting p-chart is given in Fig. 8.5.

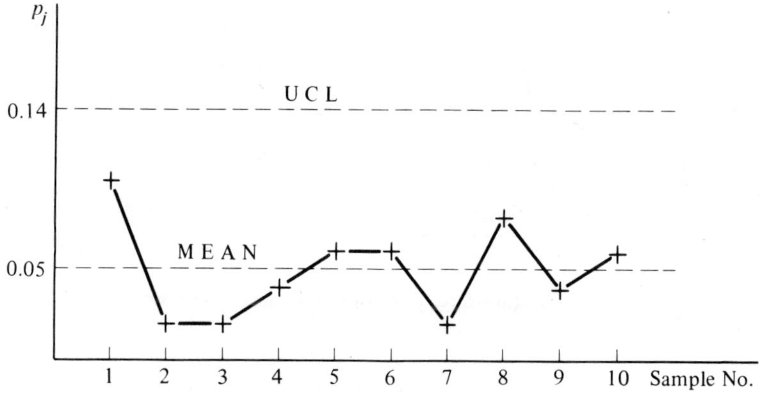

Figure 8.5. Example 8.2.3—a p-control chart.

A *c-chart* is a control chart for the number of defects c' per inspection unit. The distribution of c follows approximately a Poisson *D.F.* with mean c'. The control limits for a c-chart are

$$\text{UCL, LCL} = \bar{c} \pm 3\sqrt{\bar{c}} \tag{8.9}$$

where c' is being estimated by $\bar{c}$, the mean number of observed defects per inspection unit, that is,

$$\bar{c} = \frac{1}{k}\sum_{j=1}^{k} c_j, \quad j = 1, 2, \ldots, k \text{ inspection units} \tag{8.10}$$

In applications, an "inspection unit" is usually an assembly or subassembly, where for each unit there are a very large number of quality characteristics that can be defective. Thus, the Poisson conditions are satisfied, that is, the Poisson distribution model describes a population wherein the opportunity for defects is large yet the actual occurrence of defects tends to be small (see Sec. 4.6). An example of a *c-control* chart is shown in Fig. 8.6.

There are, of course, a number of sophisticated criteria regarding control chart design that are considered to be beyond the scope of this introductory text. These considerations are usually covered in any standard quality control textbook.*

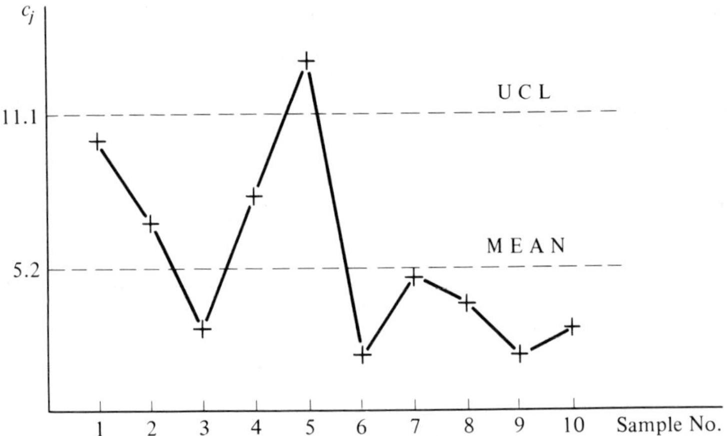

Figure 8.6. A *c*-control chart.

EXERCISES

1. Ten samples of five observations each are recorded in the following table. Compute control limits and prepare an $\bar{x}$- and r-control chart combination using Eq. (8.2) to estimate σ_X.

Ans. $\bar{x}$ limits; 0.2783, 0.2729 in.; r limits: 0.0097, 0 in.

---

*Selected references are: Duncan, A. J. (1965), *Quality Control and Industrial Statistics*, 3rd ed., Richard D. Irwin, Inc., Homewood, Ill., Chaps. 18–23, and Grant, E. L. and Leavenworth, R. S. (1972), *Statistical Quality Control*, McGraw-Hill Book Company, N. Y., Chaps. 3–11.

1	2	3	4	5	6	7	8	9	10
0.278	0.273	0.272	0.275	0.275	0.279	0.275	0.278	0.274	0.271
0.272	0.274	0.276	0.277	0.276	0.276	0.276	0.277	0.275	0.270
0.278	0.273	0.274	0.277	0.274	0.276	0.277	0.276	0.275	0.273
0.277	0.276	0.276	0.274	0.273	0.278	0.275	0.279	0.276	0.271
0.280	0.279	0.277	0.278	0.280	0.281	0.277	0.275	0.275	0.273

2. Compute the process capability value for Exercise 1. *Ans.* 0.012 in.

3. Fifteen samples of 50 observations each yield the following proportions defective p_j: 0.08, 0.04, 0.02, 0.08, 0.09, 0.02, 0.02, 0.04, 0.06, 0.06, 0.02, 0.08, 0.04, 0.04, 0.06. Compute control limits and prepare a p-control chart. *Ans.* 0.14, 0.

4. Twenty-five samples of 50 product items each are inspected yielding 34 defective product items. Compute control limits for a p-control chart. *Ans.* 0.096, 0.

5. Twenty assemblies are inspected yielding the following number c_j of defects per assembly: 10, 17, 16, 20, 10, 14, 7, 14, 19, 16, 21, 10, 13, 11, 25, 15, 11, 12, 8, 23. Compute control limits and prepare a c-control chart. *Ans.* 26.1, 3.1.

8.3 Sampling Inspection

The alternatives to 100% inspection of manufactured product are: (1) process control of the product at a satisfactory quality level, in which case formal inspection may be avoided, and (2) sampling inspection. When a product lot undergoes a *sampling inspection*, the decision to accept or reject the entire lot is based on an examination of a random sample of n product items taken from the lot. Product lots rejected by sampling inspection are usually 100% inspected.

A sampling inspection *plan*, in its most simple form, is a systematic rule specifying lot size N, sample size n, and acceptance number c. If the observed number of defectives d from the sample is less than or equal to c, the lot is accepted. If d is greater than c, the lot is rejected.

Development of criteria for designing and evaluating sampling plans depends on probability of acceptance computations. In Example 4.7.2 the probability of acceptance P_a of a manufacturing lot whose proportion defective is p under a sampling plan (N, n, c) is computed using the hypergeometric distribution model. Thus, the exact probability of acceptance P_a is given by

$$P_a = \sum_{d=0}^{c} \frac{\binom{Np}{d}\binom{Nq}{n-d}}{\binom{N}{n}} \tag{8.11}$$

where p is the proportion defective, q is the proportion effective, and $p + q = 1$.

To avoid the tedious computation associated with Eq. (8.11), binomial and Poisson *D.F.*'s are used to approximate P_a. If N is large relative to n, the binomial approximation is adequate. In this case, P_a depends only on n and c. The probability of acceptance P_a of a lot whose proportion defective is p under a sampling plan (n, c) is given by

$$P_a = \sum_{d=0}^{c} \binom{n}{d} p^d q^{n-d} \tag{8.12}$$

In addition to the size requirement for N, if p is small the Poisson model can be used to obtain an approximation of the binomial P_a. The resulting Poisson probability is

$$P_a = \sum_{d=0}^{c} \frac{e^{-np}(np)^d}{d!} \tag{8.13}$$

EXAMPLE 8.3.1

If (n, c) is (30, 1), the probability of acceptance of a 4% defective lot is given by the binomial expression

$$\begin{aligned} P_a &= \sum_{d=0}^{1} \binom{30}{d} (0.04)^d (0.96)^{30-d} \\ &= (0.96)^{30} + \frac{30!}{1!\,29!}(0.04)(0.96)^{29} \\ &= 0.661 \end{aligned}$$

and, using a Poisson approximation,

$$\begin{aligned} P_a &= \sum_{d=0}^{1} \frac{e^{-1.2}(1.2)^d}{d!} \\ &= 0.663 \end{aligned}$$

The 0.663 value is obtained from Appendix B by entering the table with

$$\mu = np = 30(0.04) = 1.20$$

and $c = 1$.

8.4 Operating Characteristic Curve

A decision-making aid in selecting an appropriate sampling plan is an *operating characteristic curve* (*OC curve*). An *OC* curve gives the probability of acceptance P_a for various proportion defective quality levels p under the

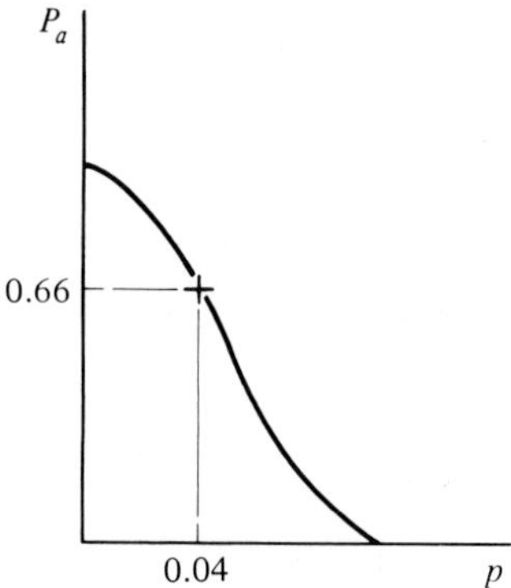

Figure 8.7. Example 8.3.1—sampling plan $(n, c) = (30,1)$.

given sampling plan. The computation of Example 8.3.1 generates one point (0.04, 0.66) of the *OC* curve shown in Fig. 8.7.

The economics of inspection depends in part on various combinations of inspection and failure costs as indicated by the simplified conditions of Example 8.4.1 that follows. In this example, if a defective product unit is not detected by an inspection operation, a subsequent production loss of \$10 is incurred (i.e., failure cost). Unit cost of inspection is \$0.15. The expected proportion defective is $p = 0.02$.

EXAMPLE 8.4.1

To establish whether or not a 100% inspection is economic, unit inspection cost is compared with the expected failure cost, which is

$$0.02(\$10) = \$0.20$$

Inspection is economic since there is a unit gain from inspection of

$$\$0.20 - \$0.15 = \$0.05$$

The proportion-defective point of no gain or loss, called the *break-even* quality level, is

$$p = \frac{\$0.15}{\$10.00} = 0.015$$

A sampling plan should discriminate product lots around the break-even quality level. An ideal sampling plan, ideal in the sense that it discriminates perfectly, is described by Fig. 8.8, which shows the *OC* curve for a perfect plan keyed to the conditions of Example 8.4.1. Product lots of quality better than $p = 0.015$ have a probability of acceptance equal to 1.0. Product lots of quality worse than $p = 0.015$ have zero probability of acceptance.

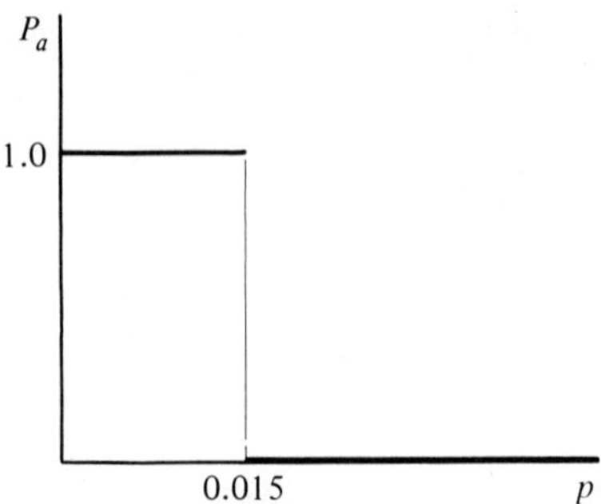

Figure 8.8. Example 8.4.1—*OC* curve for an ideal sampling plan.

Of course, no ideal sampling plan exists since any plan involves decision errors that follow from a failure to obtain perfectly random or representative samples. However, examination of the *OC* curve (Fig. 8.8) suggests that an *OC* curve, for a sampling plan with good discrimination in the vicinity of the break-even quality level, should be steep in the region of the break-even point as shown by Fig. 8.9. A sampling plan's discrimination can be increased by

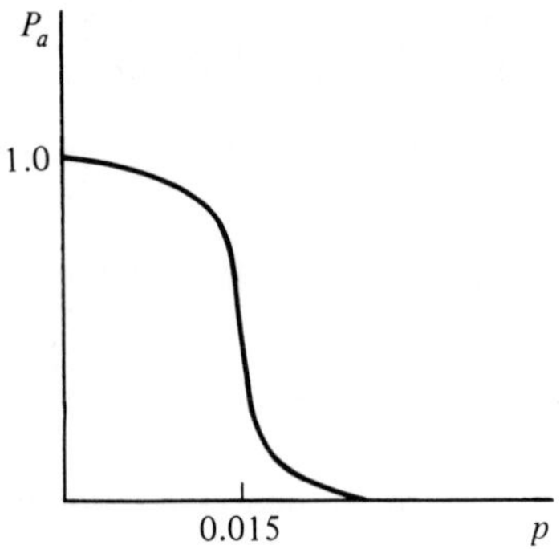

Figure 8.9. Example 8.4.1—*OC* curve for an optimum sampling plan.

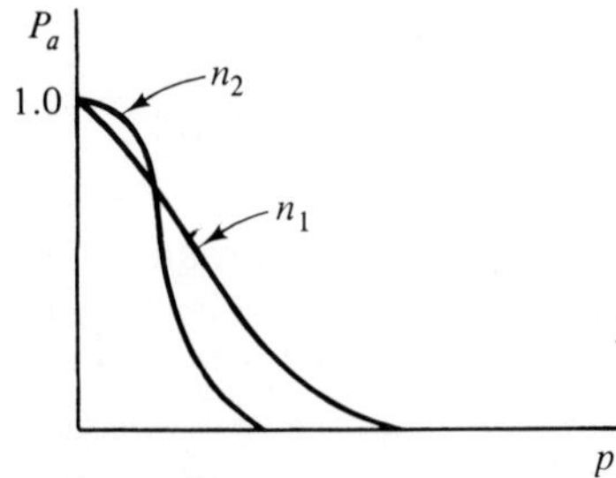

Figure 8.10. Increasing sample size from n_1 to n_2.

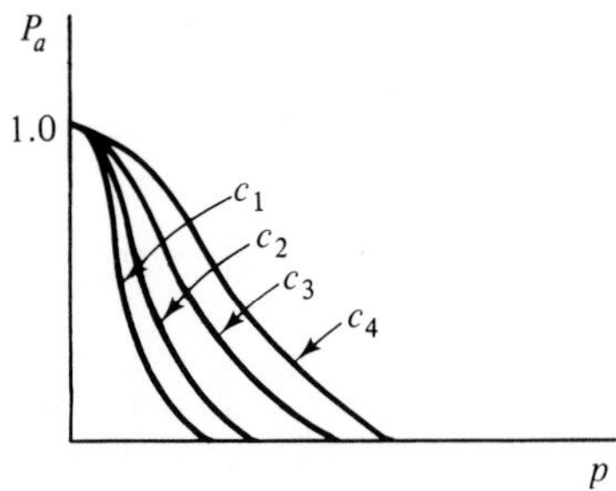

Figure 8.11. Increasing the acceptance number from c_1 to c_4.

increasing n and c proportionately. Increasing n increases the steepness of an OC curve (Fig. 8.10), whereas altering c moves the curve away from or towards the origin (Fig. 8.11).

EXERCISES

1. Write the expression for the hypergeometric probability of acceptance of a 5% defective product lot for a sampling plan $N = 500$, $n = 75$, $c = 1$.
2. Assuming that N is large relative to n, write the expression for the binomial probability of acceptance of a 5% defective product lot under a sampling plan $n = 80$, $c = 2$.
3. Compute the Poisson probability of acceptance for the conditions of Exercise 2 (use Appendix B). *Ans.* 0.238.
4. Unit inspection cost is \$0.20 and failure cost is \$15. What is the break-even proportion-defective quality level? *Ans.* 0.0133.

8.5 Sampling Plan Evaluation

Standard published sampling plans are indexed according to certain criteria that are summarized in this section. OC curves constitute a graphic means of applying these criteria when selecting a sampling plan for a particular application.

A sampling plan is a test of a hypothesis that the product lot is acceptable. The hypothesis may be

$$H_0: \ p \leq \frac{C_i}{C_d}$$

where C_i designates unit inspection cost, C_d the unit failure cost due to an undetected defective, and the ratio C_i/C_d the break-even quality level. The decision rule is

$$\frac{d}{n} \leq \frac{C_i}{C_d}, \qquad \text{accept } H_0$$

$$\frac{d}{n} > \frac{C_i}{C_d}, \qquad \text{reject } H_0$$

where d is the observed number of defectives in a random sample of size n.

A more comprehensive hypothesis, however, is one involving an *acceptable quality level* (AQL) and a *lot tolerance fraction-defective level* ($LTFD$). The null and alternative hypotheses are

$$H_0\colon\ p \leq AQL \text{ fraction defective}$$
$$H_1\colon\ p \geq LTFD \text{ fraction defective}$$

An AQL is a "good" quality level which, from the producer's viewpoint, should be accepted most of the time by the sampling plan. The $LTFD$ quality is a "bad" quality level which, from the consumer's viewpoint, should be rejected most of the time by the sampling plan. The usual probability of acceptance values are P_a equal to 0.95 for AQL quality lots and equal to 0.10 for $LTFD$ quality lots. In sampling inspection applications, the probabilities of incurring Type I and Type II decision errors, called *producer's risk* and *consumer's risk*, respectively, are

$$\alpha = P \text{ (sampling plan reject an } AQL \text{ lot)}$$
$$\beta = P \text{ (sampling plan accept an } LTFD \text{ lot)}$$

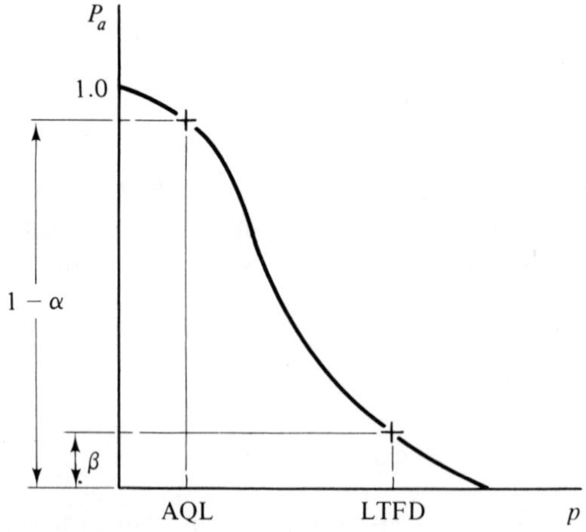

Figure 8.12. Producer's and consumer's risks.

Choice of a sampling inspection plan, using the *AQL* and *LTFD* criteria, involves selection of an *OC* curve passing through the points (AQL, $1 - \alpha$) and ($LTFD$, β). In practice, a sampling plan choice is usually made by referring to standard plans* and selecting an *OC* curve that best satisfies this two-point condition (see Fig. 8.12).

8.6 Sampling Plan Design

Designing a sampling plan that satisfies the probability conditions for the hypothesized quality levels *AQL* and *LTFD* involves a simultaneous solution of two equations for n and c, given the desired *AQL* and *LTFD* quality levels and the probabilities α and β.

The method is the same regardless of the distribution model that is used. To avoid tedious computation, however, Poisson conditions are assumed. Then, from Eq. (8.13),

$$1 - \alpha = \sum_{d=0}^{c} \frac{e^{-np_1}(np_1)^d}{d!}$$

$$\beta = \sum_{d=0}^{c} \frac{e^{-np_2}(np_2)^d}{d!}$$

where p_1 designates *AQL* quality and p_2 denotes *LTFD* quality. These equations are solved for n and c using the Poisson table, Appendix B.

EXAMPLE 8.6.1

A sampling plan is required such that $p_1 = 0.02$, $\alpha = 0.05$ and $p_2 = 0.08$, $\beta = 0.10$.

For various trial values of c, Appendix B is explored for a p_1n-value approximately equal to $1 - \alpha = 0.95$ and a p_2n-value approximately equal to $\beta = 0.10$, For example, when $c = 1$, $p_1n = 0.35$ and $1 - \alpha$ is 0.951, $p_2n = 3.90$ and $\beta = 0.10$ (using straight-line interpolation).

This procedure is repeated for $c = 0, 1, 2, \ldots$ until a ratio p_2n/p_1n is obtained that approximates as closely as possible the specified $p_2/p_1 = 0.08/0.02 = 4$. This occurs at $c = 4$, whence

$$\frac{p_2n}{p_1n} = \frac{8.00}{1.96} = 4.1$$

---

*For example, see Military Standard MIL-STD-105D (April 1963), *Sampling Procedures and Tables for Inspection by Attributes*, Supt. of Documents, U. S. Govt. Printing Office, Wash., D. C.

In addition to an indexing of plans by *AQL* and *LTFD* criteria, plans are also classified in terms of *average outgoing quality* (*AOQ*) and *average outgoing quality limit* (*AOQL*). The *AOQ* measure is an expected average quality of product obtained by averaging the proportion defective of lots accepted by the sampling plan and rejected lots that are 100% inspected. The maximum *AOQ* possible under a given sampling plan is called *AOQL*.

Taking c to be 4,

$$p_1 n = 1.96$$
$$(0.02)n = 1.96$$
$$n = 98$$

and

$$p_2 n = 0.08(98)$$
$$= 7.80$$

For $p_2 n = 7.80$ and $c = 4$, Appendix B gives $\beta = 0.112$, which is sufficiently close to the specified $\beta = 0.10$. Thus, the required sampling plan is $(n, c) = (98, 4)$.*

EXERCISES

1. Repeat Example 8.6.1 determining n from the $p_2 n$ equation. What is the sampling plan (n, c) and the resulting probabilities α and β?
Ans. $(n, c) = (100, 4)$; $\alpha = 0.053$, $\beta = 0.10$.

2. Determine a sampling plan (n, c) and the Poisson probabilities α and β (α is more critical) for the following conditions: $p_1 = 0.01$, $\alpha = 0.05$ and $p_2 = 0.06$, $\beta = 0.10$.
Ans. $(n, c) = (82, 2)$, $\alpha = 0.05$, $\beta = 0.132$.

The sampling plans discussed in Secs. 8.3–8.6 are single sampling plans for inspection by attributes. A decision to accept or reject the lot is made on the basis of one random sample from the manufacturing lot. Sampling plans involving successive samples to reach a decision are called *sequential sampling* plans. The advantage of plans involving more than one sample is that, for a given quality protection, the average total inspections are reduced. Sequential plans are not widely used, except for the double plan which appears in most standard published plans. A double plan is defined by (n_1, n_2, c_1, c_2). If d_1 denotes the observed number of defectives in a first sample and d_2 the observed number of defectives in a second sample, the decision rule is:

First sample of n_1 product items:

$d_1 \le c_1$,	accept the lot
$d_1 > c_2$,	reject the lot
$c_1 < d_1 \le c_2$,	take a second sample

*Since n and c are necessarily integers, this method yields an approximate solution to the $1 - \alpha$ and β pair of equations. If β is considered to be more critical than α for a given application, sample size n is determined from the equation $p_2 n = 8.00$, yielding a β-value closer to that specified.

Second sample of n_2 items:

$$d_1 + d_2 \le c_2, \qquad \text{accept the lot}$$
$$d_1 + d_2 > c_2, \qquad \text{reject the lot}$$

8.7 Variables Sampling Plans

Sampling plans for inspection by variables offer equivalent quality protection with smaller sample size and provide additional diagnostic information for process control purposes. The following simplified example* is presented here to illustrate the variables sampling plan concept.

Consider a product characteristic defined by the specification 0.500 $\pm$ 0.004 in., where σ_X is 0.0015 for the process operation involved. A sampling plan is required such that *AQL* quality is $p_1 = 0.01$ with probability of acceptance $P_a = 0.95$ and *LTFD* quality is $p_2 = 0.03$ with $P_a = 0.10$. Figure 8.13 shows optimum, *AQL*, and *LTFD* process settings, the latter two settings being determined by trial-and-error downward process shifts to points where the proportion defective is 0.01 and 0.03, respectively.

EXAMPLE 8.7.1

Summarizing the conditions of Fig. 8.13 (and rounding off to four places), *AQL* and *LTFD* mean process settings are:

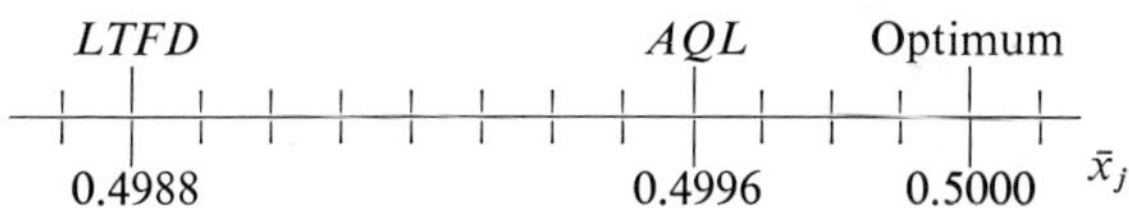

Considering only downward process shifts first, a value $\bar{x}_c$ is required to formulate a decision rule

$$\bar{x}_{\text{data}} \le \bar{x}_c, \qquad \text{reject } H_0$$
$$\text{otherwise}, \qquad \text{accept } H_0$$

where the null and alternative hypothesis values are

$$H_0\colon\ p \le p_1, \qquad \frac{\alpha}{2} = 0.025$$

$$H_1\colon\ p \ge p_2, \qquad \frac{\beta}{2} = 0.05$$

with the probabilities α and β being halved to cover both possibilities, upward and downward process shifts.

---

*This example has been abstracted from Kirkpatrick, E. G. (1970), *Quality Control for Managers and Engineers*, John Wiley & Sons, Inc., New York, pp. 244–247.

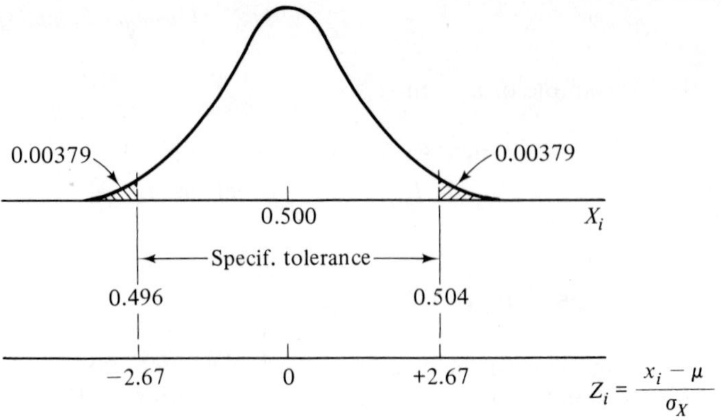

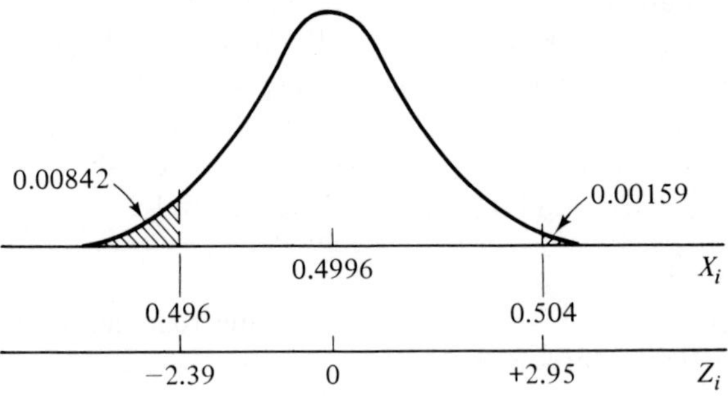

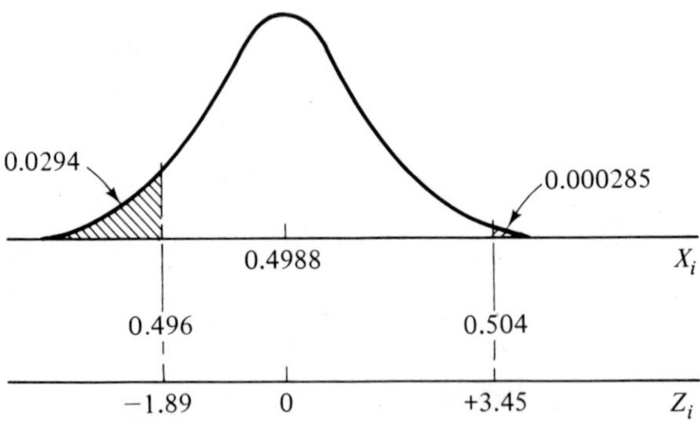

Figure 8.13. Process settings at optimum, *AQL*, and *LTFD* quality levels. Optimum level, $\mu = 0.500$: $p = 2(0.00379) = 0.00758$. *AQL* level, $\mu = 0.4996$: $p_1 = 0.00842 + 0.00159 = 0.01001$. *LTFD* level, $\mu = 0.4988$: $p_2 = 0.0294 + 0.000285 = 0.029685$.

A graphical summary of a solution for $\bar{x}_c$ and n is given by Fig. 8.14. Two equations with unknowns $\bar{x}_c$ and n result from Z-transformations about 0.4996 and 0.4988 as origins.

$$Z_{\alpha/2} = -1.96 = \frac{\bar{x}_c - 0.4996}{0.0015/\sqrt{n}}$$

$$Z_{\beta/2} = +1.645 = \frac{\bar{x}_c - 0.4988}{0.0015/\sqrt{n}}$$

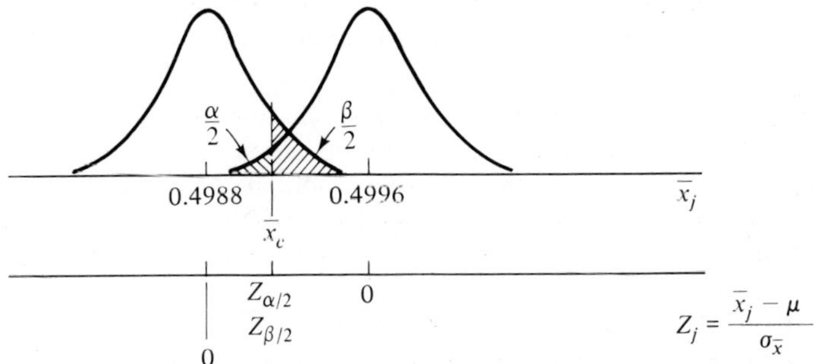

Figure 8.14. Example 8.7.1—solution for $\bar{x}_c$ when mean process settings are μ_{AQL} and μ_{LTFD}.

Subtracting the first equation from the second, $\bar{x}_c$ vanishes and

$$\sqrt{n} = \frac{3.605(0.0015)}{0.0008} = 6.76$$

whence

$$n = 45.69$$

Taking n to be 46, a solution of either equation yields

$$\bar{x}_c = 0.4992$$

Symmetry considerations yield an $\bar{x}_c$-value for upward process shifts equal to

$$0.5000 + 0.0008 = 0.5008$$

Thus, the required sampling plan is take one sample of $n = 46$ product items and if

$$\left.\begin{array}{l} \bar{x}_{\text{data}} \leq \bar{x}_c = 0.4992 \\ \bar{x}_{\text{data}} \geq \bar{x}_c = 0.5008 \end{array}\right\}, \quad \text{reject the lot}$$

$$\text{otherwise,} \quad \text{accept the lot}$$

To obtain the same degree of quality protection using an attributes sampling plan requires a considerably larger sample. Determining an attributes plan for the conditions of Example 8.7.1 is left as an exercise for the student.

EXERCISES

1. Using the values for p_1, p_2, α, and β given in Example 8.7.1, determine an attributes sampling plan (n, c) by the method of Sec. 8.6.

   Ans. $(n, c) = (398, 7)$.

2. A product characteristic is defined by a specification 0.250 ± 0.0025 in. The standard deviation for the manufacturing operation is $\sigma_X = 0.001$ in. Design a variables sampling plan for the following conditions: *AQL* quality is $p_1 = 0.02$ with a probability of acceptance $P_a = 0.90$. *LTFD* quality is $p_2 = 0.05$ with $P_a = 0.10$. *Ans.* $n = 56$, $\bar{x}_{\text{data}} \leq 0.24937$ or ≥ 0.25063, reject H_0.

Control activities incur extra expenditures. These costs are considered to be avoidable. Thus, there are always the questions: "is control feasible?" and "how much control is economic?"

Production personnel are routinely engaged in the technology of process control. This is a basic production function and is viewed as an unavoidable cost of manufacture. Formal control assistance by quality control staff and inspectors is, however, an extra effort involving avoidable costs that must be weighed against expected benefits from this enlarged control activity. Extra control costs are mainly due to quality control staff time, patrol-inspection time, and any other than normal setup and operator time required for making extra measurements, maintaining a control chart, and similar activities. Also, if the formal control procedure restricts process yield, this cost should be considered (i.e., output per unit time being reduced because of more than normal machine downtime due to the operator being involved in the extra control effort).

In Sec. 5.8, an optimum tolerance is determined by the minimum point of a total cost curve (Fig. 5.21), where total cost is the sum of: (1) a decreasing process cost function, and (2) an increasing discovery-and-correction-of-defective-assembly cost function. Many specific quality control applications follow generally this situation of compromising between two essentially opposing sets of costs. For example, the question of whether or not to formally control a process is a matter of determining which is more economic, process cost plus control cost (yielding acceptable output quality) or dispensing with formal control and incurring a process cost plus rectifying inspection cost (also yielding acceptable output quality).

A comprehensive treatment of quality costs is beyond the scope of an introductory statistics text. The student is referred to a standard quality control textbook for detailed coverage of this topic.* However, some specific cost factors related to Secs. 8.2–8.7 are identified here.

A quality control chart application is justified only on operations where routine physical control is inadequate and the opportunity for cost savings is significant. Even though the application is feasible, careful consideration should be given to the extra cost of the control effort—staff time, extra inspection cost, and possible restriction of process yield. Further, there are economic questions involved in the design of the control plan. Type I and Type II decision errors lead to production expenses associated with looking for trouble that does not exist and failing to look for trouble that does exist. Any economic balancing of these two fundamental costs depends on three factors: the control limits, sampling size, and sampling interval. An interesting analysis of control chart design from this standpoint is reported by Fetter.†

An analogous situation prevails in the economic selection and design of a sampling inspection plan. Standard published sampling plans develop minimum sample sizes for given degrees of quality protection. However, minimizing sample size may reduce only a relatively small cost, small in relation to the costs generated by decision errors. The total real cost of sampling inspection includes: (1) cost of obtaining and testing the sample, and (2) losses occasioned by the operation of producer's and consumer's risks. Clearly, the larger the sample, the more costly is inspection and test. Yet, risks and their attendant costs are reduced by large samples. Usually there is a relationship between size and cost of the sample that makes possible a minimization of the total cost of sampling inspection.‡

---

*See Feigenbaum, A. V. (1961), *Total Quality Control*, McGraw-Hill Book Company, New York, Chap. 5 and Kirkpatrick, E. G. (1970), *Quality Control for Managers and Engineers*, John Wiley & Sons, Inc., New York, Chaps. 1, 3, 9, 10.

†See Fetter, R. B. (1967), *The Quality Control System*, Richard D. Irwin, Inc., Homewood, Ill., pp. 67–70 and Appendix II.

‡See Ellner, H. and Savage, I. R. (June 1957), *Sampling for Destructive or Expensive Testing by Attributes*, Army Science Conf., West Point, N. Y.; Mandelson, J. (March 1967), "Sampling Plans for Destructive or Expensive Testing," *Industrial Quality Control*, Vol. 23, No. 9, pp. 440–450; and Smith, B. E. (July 1961), *Some Economic Aspects of Quality Control*, Tech. Report No. 53, Applied Mathematics and Statistics Laboratories Stanford University, Stanford, Calif.

GLOSSARY OF TERMS

Process capability
Natural tolerance
Process control
Attributes inspection
Variables inspection
Control chart
Control limits
$\bar{x}$-control chart
r-control chart
p-control chart
c-control chart
Sampling inspection plan
Operating characteristic (OC) curve
Break-even quality level
Acceptable quality level (AQL)
Lot tolerance fraction defective ($LTFD$)
Producer's risk
Consumer's risk
Average outgoing quality (AOQ)
Average outgoing quality limit ($AOQL$)
Single sampling plan
Double sampling plan

GLOSSARY OF FORMULAS

Page

(8.1) Mean range. 268

$$\bar{r} = \frac{1}{k} \sum_{j=1}^{k} r_j \quad j = 1, 2, \ldots, k$$

(8.2) Estimate of σ_X. 268

$$\frac{\bar{r}}{d_2}$$

(8.3) Control limits for $\bar{x}$-chart (σ_X estimated). 272

$$\text{UCL, LCL} = \bar{\bar{x}} \pm A_2\bar{r}$$

(8.4) Control limits for r-chart. 275

$$\text{UCL, LCL} = D_4\bar{r},\ D_3\bar{r}$$

(8.5) Control limits for $\bar{x}$-chart (σ_X known). 275

$$\text{UCL, LCL} = \bar{x} \pm A\sigma_X$$

(8.6) Estimate of π_0 (identical sample sizes). 276

$$\bar{p} = \frac{1}{k} \sum_{j=1}^{k} p_j$$

(8.7) Estimate of π_0 (unequal sample sizes). 277

$$\bar{p} = \frac{1}{N} \sum_{j=1}^{k} x_j, \qquad N = \sum_{j=1}^{k} n_j$$

(8.8) Control limits for p-chart. 277

$$\text{UCL, LCL} = \bar{p} \pm 3\sqrt{\frac{\bar{p}(1 - \bar{p})}{n}}$$

(8.9) Control limits for c-chart. 277

$$\text{UCL, LCL} = \bar{c} \pm 3\sqrt{\bar{c}}$$

(8.10) Estimate of c'. 278

$$\bar{c} = \frac{1}{k}\sum_{j=1}^{k} c_j$$

(8.11) Hypergeometric probability of acceptance. 279

$$P_a = \sum_{d=0}^{c} \frac{\binom{Np}{d}\binom{Nq}{n-d}}{\binom{N}{n}}$$

(8.12) Binomial probability of acceptance. 280

$$P_a = \sum_{d=0}^{c} \binom{n}{d} p^d q^{n-d}$$

(8.13) Poisson probability of acceptance. 280

$$P_a = \sum_{d=0}^{c} \frac{e^{-np}(np)^d}{d!}$$

REFERENCES

DUNCAN, A. J. (1965), *Quality Control and Industrial Statistics*, 3rd ed., Richard D. Irwin, Inc., Homewood, Ill., Parts II and IV.

ELLNER, H. AND SAVAGE, I. R. (June 1957), *Sampling for Destructive or Expensive Testing by Attributes*, Army Science Conf., West Point, N. Y.

FEIGENBAUM, A. V. (1961), *Total Quality Control*, McGraw-Hill Book Company, New York, Chap. 5.

FETTER, R. B. (1967), *The Quality Control System*, Richard D. Irwin, Inc., Homewood, Ill., pp. 67–70 and Appendix II.

GRANT, E. L. AND LEAVENWORTH, R. S. (1972), *Statistical Quality Control*, McGraw-Hill Book Company, New York, Chaps. 3–11.

KIRKPATRICK, E. G. (1970), *Quality Control for Managers and Engineers*, John Wiley & Sons, Inc., New York, pp. 244–247.

MANDELSON, J. (March 1967), "Sampling Plans for Destructive or Expensive Testing," *Industrial Quality Control*, Vol. 23, No. 9, pp. 440–450.

MILITARY STANDARD MIL*STD*105D (April 1963), *Sampling Procedures and Tables for Inspection by Attributes*, Supt. of Documents, U. S. Govt. Printing Office, Wash., D. C.

SMITH, B. E. (July 1961), *Some Economic Aspects of Quality Control*, Tech. Report No. 53, Applied Mathematics and Statistics Laboratories, Stanford University, Stanford, Calif.

PROBLEMS

1. State the three principal sources of variation in a manufacturing process. In terms of these variation sources, explain what is meant by the *capability* or *natural tolerance* of a process operation. What is the statistical measure of capability?

2. State two unbiased estimates of σ_X that are used in determining process capability values.

3. Determine an estimate of the process capability from the following sample observations (10 samples of 5 observations each) by two methods: (a) based on the sample variance, and (b) based on the sample range.

1	2	3	4	5	6	7	8	9	10
0.571	0.581	0.576	0.578	0.571	0.577	0.578	0.573	0.575	0.575
0.573	0.572	0.577	0.576	0.579	0.576	0.572	0.575	0.577	0.577
0.577	0.578	0.575	0.573	0.576	0.570	0.576	0.574	0.573	0.576
0.578	0.576	0.576	0.579	0.574	0.576	0.580	0.575	0.577	0.578
0.579	0.576	0.575	0.574	0.574	0.577	0.574	0.575	0.573	0.575

4. An $\bar{x}$-control chart application is a hypothesis test. State the hypothesis and give the usual probaility of a Type I decision error value that is used.

5. Give expressions for the population central line and control limit lines on an $\bar{x}$-chart and their respective sample estimates.

6. Compute control limits and prepare an $\bar{x}$-control chart for the lip opening pressure data of Problem 1.1 (Chap. 1). Consider this data to represent 16 samples of 5 observations each.

7. Compute control limits and prepare an r-control chart for Problem 6.

8. Summarize the argument for sample size usually being approximately 4 or 5 on the $\bar{x}$- and r-control chart combination.

9. Compute control limits and prepare a p-chart for the following sample data obtained from an inspection of manufactured television tubes. Sample size is 100.

Sample	1	2	3	4	5	6	7	8	9	10
No. of defectives	11	12	8	4	7	5	5	9	12	8

10. Inspection and test data for a particular subassembly unit disclose the following numbers of defects per unit. Compute control limits and prepare a c-control chart.

Sample	1	2	3	4	5	6	7	8	9	10
No. of defects	8	10	15	6	4	8	11	15	7	6

11. An objective of control-chart applications is to minimize two kinds of errors: (1) looking for trouble that does not exist, and (2) failing to look for trouble that does exist. Relate and identify these errors with Type I and II decision errors. What are the usual physical consequences of these decision errors relative to a manufacturing process?

12. A c-control chart is based on a Poisson distribution model. State the two principal conditions underlying the use of this model. Based on these conditions, explain why the typical c-chart application is an assembly inspection operation (and not component part inspection).

13. A process is in control at a mean setting $\mu_0 = 0.5000$ in. The standard deviation σ_X is 0.0004 in. Sample size is 4, and $\pm 3\sigma$ control limits are being used. Assume that the only process change is the mean setting. What is the probability that an $\bar{x}$-chart will detect a process shift upward to $\mu = 0.5009$ in. on: (a) the first sample after the process shift occurs, and (b) the first two samples after the shift occurs?

14. Repeat Problem 13 for the following conditions: $\mu_0 = 0.5006$ in., and the process shifts downward to a mean setting of 0.5002 in.

15. A process has been in control at an estimated proportion-defective level $p = 0.10$. The process shifts to an estimated level $p = 0.20$. Using the information available (i.e., control chart limits for samples of $n = 100$), determine an estimated probability of detection of the process shift on the first sample after the shift occurs?

16. An assembly operation has been in control at $\bar{c} = 16$. The assembly operation deteriorates to a $\bar{c} = 25$. Based on the existing control chart limits (for $\bar{c} = 16$), determine the probability of detection of the assembly operation shift to $\bar{c} = 25$ on the first sample after the shift occurs.

17. A sampling inspection plan is described by $N = 1{,}000$, $n = 240$, and $c = 2$. Set up an expression (do not compute) for the exact hypergeometric probability of acceptance P_a of a 2.5% defective lot under this sampling plan.

18. Set up an expression for the binomial probability of acceptance P_a in Problem 17. Then, using a Poisson distribution model as an approximation, determine the probability of acceptance P_a from Appendix B table.

19. Plot an OC curve for the sampling plan of Problem 17 using the following p-values for abscissa values: 0.002, 0.005, 0.008, 0.01, 0.02, 0.03, and 0.04.

20. Define: (a) *AQL* and *LTFD* quality levels, and (b) producer's risk and consumer's risk.
21. Determine from the *OC* curve plotted in Problem 19: (a) consumer's risk for an *LTFD* quality level $p_2 = 0.03$, and (b) producer's risk for an *AQL* quality level $p_1 = 0.005$. Verify these numerical values by reference to Appendix B table.
22. Determine, from the *OC* curve plotted in Problem 19, the *AQL* and *LTFD* quality levels corresponding to producer and consumer risks of 0.20 and 0.10 respectively. Verify your results by reference to Appendix B table.
23. Unit inspection cost is \$0.30 and unit failure cost is \$40 (i.e., failure due to an undetected defective). Determine the break-even proportion-defective quality level and sketch an *OC* curve that describes an ideal sampling plan for this break-even level.
24. Determine a sampling plan (n, c) and the Poisson probabilities α and β for the following conditions: *AQL* quality is $p_1 = 0.03$, $\alpha = 0.10$ and *LTFD* quality is $p_2 = 0.10$, $\beta = 0.20$. It is important to obtain a plan that agrees closely with the specified α-value.
25. The solution to Problem 24 yields a sampling plan with satisfactory α and a β slightly less than the specified $\beta = 0.20$. Repeat the solution, determining sample size from a fixed β to see if either α or n (or both) can be reduced.
26. Using only the available (n, c) combinations and binomial probabilities obtainable from Appendix A table, determine a sampling plan (n, c) for the following conditions: *AQL* quality is $p_1 = 0.10$, $\alpha = 0.07$ and *LTFD* quality is $p_2 = 0.20$, $\beta = 0.08$.
27. Using binomial probabilities (Appendix A table), determine α and β for a sampling plan $(n, c) = (25, 4)$. ($p_1 = 0.10$ and $p_2 = 0.20$)
28. Suppose that β is critical and is fixed at approximately 0.20 in Problem 27. Determine the acceptance number and producer's risk, assuming that sample size is 25.
29. Specifications for a product characteristic are 0.500 ± 0.025 in. Process variation is known and measured by $\sigma_X = 0.010$ in. Design a variables sampling plan for the following conditions: *AQL* quality is $p_1 = 0.015$, $\alpha = 0.05$ and *LTFD* quality is $p_2 = 0.04$, $\beta = 0.05$.
30. Determine an attributes sampling plan to satisfy the conditions of Problem 29 and compare sample sizes required for the two plans.
31. Standard sampling plans are classified or indexed in terms of average outgoing quality (*AOQ*). Determine the *AOQ* under a sampling plan $(n, c) = (200, 3)$ if the manufacturing lot size is 10,000 and expected proportion defective is $p = 0.01$. For $p = 0.01$, the *OC* curve indicates a probability of acceptance approximately equal to 0.86.
32. Other *AOQ* values for various expected proportion defectives p can be obtained for Problem 31. Plotting p vs. *AOQ* generates an *AOQ* curve. Sketch such a

curve and identify average outgoing quality limit (*AOQL*). Explain the meaning of this *AOQL* value relative to Problem 31.

33. Assuming a large manufacturing lot size and using a Poisson distribution model, compute the probability of acceptance of a 1% defective product lot under a double sampling plan, $n_1 = 35$, $c_1 = 0$, and $n_2 = 60$, $c_2 = 3$.

CURVE FITTING

Engineering problems often require a presentation of data showing the observed relationship between variables. Data values (x_i, y_i) *are plotted and it is desired to obtain an analytic expression* $y = f(x)$ *for the functional relationship suggested by the empirical data. The usual method of obtaining* $y = f(x)$ *is to fit a line to the plotted points in such a manner that an "average" functional relationship is derived. This general problem is called* curve fitting.

Although this topic is related to other statistical procedures and, in fact, is preparatory for Chap. 10, this chapter is presented independently of statistical methods (with the exception of Sec. 9.6). No distinction is made between population and sample, no population parameters are being tested or estimated, and no statistical assumptions are made. The methods of the following sections are appropriate for fitting a curve to any set of observations (x_i, y_i). *Equations obtained in this manner are called empirical to distinguish them from the rational expressions of pure mathematics derived by induction.*

9

9.1 Method of Moments

Given the functional form of a relationship $f(x, a_0, a_1, \ldots, a_k)$, the parameters a_j can be determined by obtaining expressions for as many moments of the function $y = f(x)$ as there are parameters in the function and equating these to numerical moments of corresponding order of the observed y-values.*

EXAMPLE 9.1.1

A straight line $y = a_0 + a_1x$ is fitted to the following data:

x_i	0	1	2	3	4	5
y_i	1.7	7.7	17.3	27.1	28.0	35.6

*Analogous to Eqs. (1.31) and (1.32), for n values of (x_i, y_i) the rth moment of y is defined by the expression

$$\frac{1}{n}\sum_{i=1}^{n} y_i x_i^r$$

The zeroth moment is

$$\frac{1}{n}\sum_{i=1}^{n} y_i x_i^0 = \frac{1}{n}\sum_{i=1}^{n} (a_0 + a_1 x_i)$$

and the first moment is

$$\frac{1}{n}\sum_{i=1}^{n} y_i x_i = \frac{1}{n}\sum_{i=1}^{n} (a_0 + a_1 x_i)x_i$$

Equating these functional moments to the corresponding numerical moments of the observed y-values,

a_0	$+$	a_1x_i	$=$	y_i
a_0	$+$	$0a_1$	$=$	1.7
a_0	$+$	$1a_1$	$=$	7.7
a_0	$+$	$2a_1$	$=$	17.3
a_0	$+$	$3a_1$	$=$	27.1
a_0	$+$	$4a_1$	$=$	28.0
a_0	$+$	$5a_1$	$=$	35.6
$6a_0$	$+$	$15a_1$	$=$	117.4
$\frac{1}{6}(6a_0$	$+$	$15a_1)$	$=$	$\frac{117.4}{6}$

$(a_0$	$+$	$a_1x_i)x_i$	$=$	y_ix_i
a_0	$+$	a_1	$=$	7.7
$2a_0$	$+$	$4a_1$	$=$	34.6
$3a_0$	$+$	$9a_1$	$=$	81.3
$4a_0$	$+$	$16a_1$	$=$	112.0
$5a_0$	$+$	$25a_1$	$=$	178.0
$15a_0$	$+$	$55a_1$	$=$	413.6
$\frac{1}{6}(15a_0$	$+$	$55a_1)$	$=$	$\frac{413.6}{6}$

A solution of the two expressions for the zeroth and first moments, respectively,

$$\frac{1}{6}(6a_0 + 15a_1) = \frac{117.4}{6}$$

$$\frac{1}{6}(15a_0 + 55a_1) = \frac{413.6}{6}$$

yields $a_0 = 2.41$ and $a_1 = 6.86$ and thus the fitted line

$$y = 2.41 + 6.86x$$

is an "average" line since the a_0-values have been averaged to yield $a_0 = 2.41$ and an averaging of the a_1-values has produced $a_1 = 6.86$.

The zeroth and first moments in Example 9.1.1 can be solved directly to obtain formulas for a_0 and a_1, respectively. The zeroth moment is

$$\frac{1}{n}\sum y_i = \frac{1}{n}\sum (a_0 + a_1 x_i)$$

$$= a_0 + \frac{a_1}{n}\sum x_i$$

or

$$\sum y_i = na_0 + a_1 \sum x_i$$

and the first moment is

$$\frac{1}{n} \sum y_i x_i = \frac{1}{n} \sum (a_0 + a_1 x_i) x_i$$

$$= \frac{a_0}{n} \sum x_i + \frac{a_1}{n} \sum x_i^2$$

or

$$\sum y_i x_i = a_0 \sum x_i + a_1 \sum x_i^2$$

A simultaneous solution for a_0 and a_1 of these two expressions gives

$$\left.\begin{aligned} a_0 &= \frac{\sum x_i \sum y_i x_i - \sum x_i^2 \sum y_i}{(\sum x_i)^2 - n \sum x_i^2} \\ a_1 &= \frac{\sum x_i \sum y_i - n \sum y_i x_i}{(\sum x_i)^2 - n \sum x_i^2} \end{aligned}\right\} \tag{9.1}$$

EXAMPLE 9.1.2

Using Eqs. (9.1) and repeating Example 9.1.1, squares x_i^2 and product $y_i x_i$ are computed:

x_i	y_i	x_i^2	$y_i x_i$
0	1.7	0	0
1	7.7	1	7.7
2	17.3	4	34.6
3	27.1	9	81.3
4	28.0	16	112.0
5	35.6	25	178.0
15	117.4	55	413.6

and a_0 and a_1 are computed by formula:

$$a_0 = \frac{15(413.6) - 55(117.4)}{(15)^2 - 6(55)}$$

$$= \frac{-253}{-105} = 2.41$$

$$a_1 = \frac{15(117.4) - 6(413.6)}{(15)^2 - 6(55)}$$

$$= \frac{-720.6}{-105} = 6.86$$

9.2 Method of Least Squares

The same equations for a_0 and a_1 in the preceding example can be developed by the method of *least squares*, a procedure that generates some convenient memory devices for writing formulas when fitting polynomials of any degree and also exponential functions.

Let e_i represent the difference between the ordinate value of an observed y_i and the corresponding functional value of y as shown in Fig. 9.1. Thus,

$$e_i = y_i - y = y_i - (a_0 + a_1 x_i)$$

the observed x_i and the functional x being identical for any i.

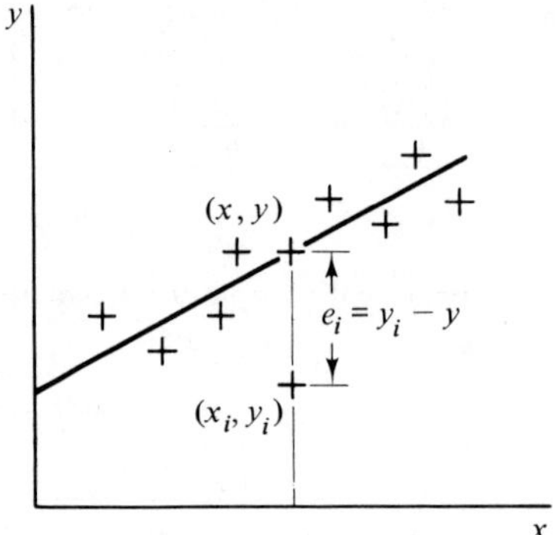

Figure 9.1. Differences e_i when fitting $y = a_0 + a_1 x$.

It is desired to determine a_0 and a_1 such that the sum of the e_i^2 is a minimum. Differentiating $\sum e_i^2$ partially with respect to a_0 and a_1, respectively, and equating the results to zero, the following expressions are obtained:

$$\begin{aligned}
\frac{\partial}{\partial a_0} \sum e_i^2 &= \sum \frac{\partial}{\partial a_0}(e_i^2) \\
&= \sum \frac{\partial}{\partial a_0}[y_i - (a_0 + a_1 x_i)]^2 \\
&= \sum 2(y_i - a_0 - a_1 x_i)(-1) = 0 \\
\frac{\partial}{\partial a_1} \sum e_i^2 &= \sum \frac{\partial}{\partial a_1}(e_i^2) \\
&= \sum \frac{\partial}{\partial a_1}[y_i - (a_0 + a_1 x_i)]^2 \\
&= \sum 2(y_i - a_0 - a_1 x_i)(-x_i) = 0
\end{aligned}$$

Separating sums and rearranging terms, these two expressions reduce to

$$\left.\begin{aligned} \sum y_i &= \sum a_0 + a_1 \sum x_i \\ \sum y_i x_i &= a_0 \sum x_i + a_1 \sum x_i^2 \end{aligned}\right\} \tag{9.2}$$

which are the same equations obtained by the method of moments. Equations (9.2) are called *least squares normal equations* (or *LSNE*), the word "normal" having no reference to the distribution model of Chap. 5.

A solution for a_0 and a_1 again involves Eqs. (9.1). However, the computation can be greatly simplified by a linear transformation on x. If the x_i are equally spaced* such that ordered x_i differ by an amount $c = x_{i+1} - x_i$, the transformation

$$u_i = \frac{x_i - \bar{x}}{c}, \qquad c = x_{i+1} - x_i \tag{9.3}$$

causes any term of Eqs. (9.1) having a coefficient $\sum x_i$ to vanish, since the sum of deviations from the mean is zero.

The result of the transformation, Eq. (9.3), is that Eqs. (9.1) reduce to

$$\left.\begin{aligned} a_0' &= \frac{1}{n} \sum y_i = \bar{y} \\ a_1' &= \frac{\sum y_i u_i}{\sum u_i^2} \end{aligned}\right\} \tag{9.4}$$

and the fitted line $y = a_0 + a_1 x$ with slope a_1 and intercept a_0 has been replaced by $y = a_0' + a_1' u$ with slope a_1' and intercept a_0' (see Fig. 9.2). The fitted line is unchanged; only the abscissa units are changed; and a_0' and a_1' are easily converted to a_0 and a_1, respectively, in order to return to x-units.

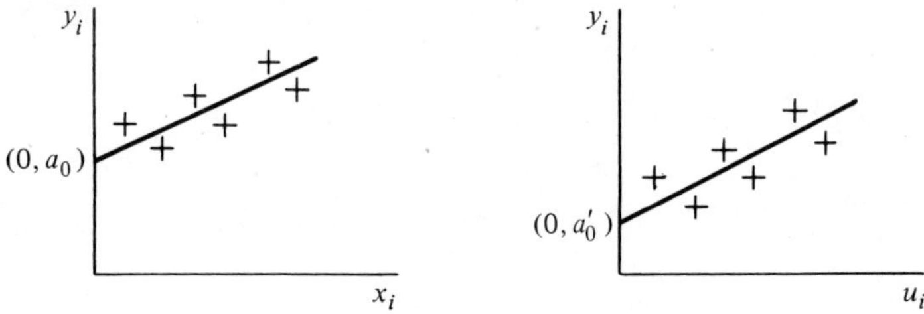

Figure 9.2. Fitting a line $y = a_0' + a_1' u$.

*This is not an overly generous assumption. The independent variable is x and presumably x is controlled and assigned values by the experimenter.

EXAMPLE 9.2.1

Equations (9.4) are used for a solution of Example 9.1.2. A transformation

$$u_i = \frac{x_i - \bar{x}}{(c/2)} = \frac{x_i - 2.5^*}{0.5}$$

is applied before computing squares u_i^2 and products $y_i u_i$:

x_i	y_i	u_i	u_i^2	$y_i u_i$
0	1.7	−5	25	−8.5
1	7.7	−3	9	−23.1
2	17.3	−1	1	−17.3
3	27.1	+1	1	+27.1
4	28.0	+3	9	+84.0
5	35.6	+5	25	+178.0
	117.4		70	240.2

From Eqs. (9.4),

$$a'_0 = \frac{1}{n} \sum y_i$$

$$= \tfrac{1}{6}(117.4) = 19.56$$

$$a'_1 = \frac{\sum y_i u_i}{\sum u_i^2}$$

$$= \frac{240.2}{70} = 3.43$$

Therefore,

$$y = 19.56 + 3.43u$$

or

$$y = 19.56 + 3.43\left(\frac{x - 2.5}{0.5}\right)$$

$$= 19.56 + 6.86x - 17.15$$

$$= 2.41 + 6.86x$$

*With an even number $n = 6$ of x_i-values, and thus the mean $\bar{x} = 2.5$ occurring between $x_i = 2, 3$, a transformation $(x_i - \bar{x})/c$ produces u-values $\pm 0.5, 1.5, 2.5$. In this case, changing the denominator of the transformation to $c/2$ produces integer u-values.

EXERCISES

1. Fit a line $y = a_0 + a_1x$ by the method of moments to the observations (x_i, y_i): (1, 5), (2, 8), (3, 9), and (4, 10). *Ans.* $y = 4 + 1.6x$.

2. Using Eqs. (9.4) and the method of least squares, repeat Exercise 1.

3. Using Eqs. (9.4) and the method of least squares, fit $y = a_0 + a_1x$ to: (6, 5), (7, 5), (7, 4), (8, 5), (8, 4), (8, 3), (9, 4), (9, 3), and (10, 3). *Ans.* $y = 8 - 0.5x$.

9.3 Fitting a Polynomial

The method of Sec. 9.2 can be applied to a polynomial of any degree. Minimizing the sum of the e_i^2 proceeds in the same manner, taking as many partial derivatives as there are parameters to be determined. For example, consider a parabolic trend described by

$$y = a_0 + a_1x + a_2x^2$$

The sum of the squared differences is

$$\begin{aligned} \sum e_i^2 &= \sum (y_i - y)^2 \\ &= \sum [y_i - (a_0 + a_1x_i + a_2x_i^2)]^2 \end{aligned}$$

Three partial derivatives of $\sum e_i^2$ with respect to a_0, a_1, and a_2, respectively, give

$$\begin{gathered} \sum 2(y_i - a_0 - a_1x_i - a_2x_i^2)(-1) \\ \sum 2(y_i - a_0 - a_1x_i - a_2x_i^2)(-x_i) \\ \sum 2(y_i - a_0 - a_1x_i - a_2x_i^2)(-x_i^2) \end{gathered}$$

Equating each expression to zero, separating sums, and rearranging terms, yields

$$\left.\begin{aligned} \sum y_i &= \sum a_0 + a_1 \sum x_i + a_2 \sum x_i^2 \\ \sum y_ix_i &= a_0 \sum x_i + a_1 \sum x_i^2 + a_2 \sum x_i^3 \\ \sum y_ix_i^2 &= a_0 \sum x_i^2 + a_1 \sum x_i^3 + a_2 \sum x_i^4 \end{aligned}\right\} \tag{9.5}$$

Assuming equally spaced x_i, and applying the transformation

$$u_i = \frac{x_i - \bar{x}}{c}, \qquad c = x_{i+1} - x_i$$

Eqs. (9.5) are expressed in (u_i, y_i) units and, since the sum of u_i to any odd power is zero (i.e., $\sum u_i = 0$, $\sum u_i^3 = 0$, etc.), Eqs. (9.5) reduce to

$$\left.\begin{aligned} \sum y_i &= \sum a'_0 + a'_2 \sum u_i^2 \\ \sum y_i u_i &= a'_1 \sum u_i^2 \\ \sum y_i u_i^2 &= a'_0 \sum u_i^2 + a'_2 \sum u_i^4 \end{aligned}\right\} \tag{9.6}$$

EXAMPLE 9.3.1

A parabola $y = a_0 + a_1x + a_2x^2$ is fitted to the following data

x_i	0	1	2	3	4	5	6
y_i	5	11	28	55	93	141	198

A transformation

$$u_i = \frac{x_i - \bar{x}}{c} = \frac{x_i - 3}{1}$$

is applied before computing u_i^2 and u_i^4 and products yu_i and yu_i^2:

x_i	y_i	u_i	u_i^2	u_i^4	y_iu_i	$y_iu_i^2$
0	5	−3	9	81	−15	45
1	11	−2	4	16	−22	44
2	28	−1	1	1	−28	28
3	55	0	0	0	0	0
4	93	+1	1	1	+93	93
5	141	+2	4	16	+282	564
6	198	+3	9	81	+594	1,782
	531		28	196	904	2,556

From Eqs. (9.6),

$$\begin{aligned} \sum y_i u_i &= a'_1 \sum u_i^2 \\ 904 &= a'_1(28) \\ a'_1 &= 32.286 \end{aligned}$$

$$\begin{aligned} \sum y_i &= \sum a'_0 + a'_2 \sum u_i^2 \\ 531 &= 7a'_0 + a'_2(28) \\ a'_0 &= \frac{531 - 28a'_2}{7} \end{aligned}$$

$$\sum y_i u_i^2 = a_0' \sum u_i^2 + a_2' \sum u_i^4$$

$$2{,}556 = \frac{531 - 28a_2'}{7}(28) + a_2'(196)$$

$$a_2' = 5.143$$

$$a_0' = \frac{531 - 28a_2'}{7}$$

$$= \frac{531 - 28(5.143)}{7}$$

$$a_0' = 55.285$$

The fitted line is

$$y = 55.285 + 32.286u + 5.143u^2$$

or

$$\begin{aligned} y &= 55.285 + 32.286(x - 3) + 5.143(x - 3)^2 \\ &= 4.71 + 1.43x + 5.14x^2 \end{aligned}$$

An examination of Eqs. (9.5) suggests a pattern for writing the *LSNE* for a polynomial of any degree without going through the steps of minimizing the sum of the e_1^2. Thus, for a polynomial of degree k, there are $k + 1$ equations:

$$\left.\begin{aligned} \sum y_i &= \sum a_0 + a_1 \sum x_i + a_2 \sum x_i^2 + \ldots + a_k \sum x_i^k \\ \sum y_i x_i &= a_0 \sum x_i + a_1 \sum x_i^2 + a_2 \sum x_i^3 + \ldots + a_k \sum x_i^{(k+1)} \\ \sum y_i x_i^2 &= a_0 \sum x_i^2 + a_1 \sum x_i^3 + a_2 \sum x_i^4 + \ldots + a_k \sum x_i^{(k+2)} \\ &\;\;\vdots \\ \sum y_i x_i^k &= a_0 \sum x_i^k + a_1 \sum x_i^{(k+1)} + a_2 \sum x_i^{(k+2)} + \ldots + a_k \sum x_i^{2k} \end{aligned}\right\} \qquad (9.7)$$

EXERCISES

1. Fit a parabola $y = a_0 + a_1x + a_2x^2$ to the following observations: (2, 27), (4, 93), (6, 198), (8, 348), and (10, 532). *Ans.* $y = 0.60 + 3.04x + 5.02x^2$.
2. Write the *LSNE* for $y = a_0 + a_1x + a_2x^2 + a_3x^3$.

9.4 Fitting an Exponential

Two common exponential functions can be fitted to empirical data using the pattern of Eqs. (9.2). Consider an exponential trend described by

$$y = ab^x$$

Taking logs of both sides of this equation,

$$\log y = \log a + x \log b$$

which is of the linear form described in Secs. 9.1–9.2, and, from Eqs. (9.1), replacing y by $\log y$,*

$$\left.\begin{aligned} \log a &= \frac{\sum x_i \sum (x_i \cdot \log y_i) - \sum x_i^2 \sum \log y_i}{(\sum x_i)^2 - n \sum x_i^2} \\ \log b &= \frac{\sum x_i \sum \log y_i - n \sum (x_i \cdot \log y_i)}{(\sum x_i)^2 - n \sum x_i^2} \end{aligned}\right\} \tag{9.8}$$

Again assuming equally spaced x_i and applying the customary transformation, Eqs. (9.8) can be expressed in u-units and reduce to

$$\left.\begin{aligned} \log a' &= \frac{1}{n} \sum \log y_i = \overline{\log y} \\ \log b' &= \frac{\sum (u_i \cdot \log y_i)}{\sum u_i^2} \end{aligned}\right\} \tag{9.9}$$

EXAMPLE 9.4.1

An exponential $y = ab^x$ is fitted to the following data:

x_i	0	4	8	12	16	20	24	28
y_i	107.0	88.0	74.2	66.8	51.0	41.6	33.6	26.9

A transformation

$$u_i = \frac{x_i - \bar{x}}{(c/2)} = \frac{x_i - 14}{2}$$

*It should be noted that the solutions for a and b are not then strictly least squares solutions. To obtain the *LSNE*, it is necessary to minimize

$$\sum e_i^2 = \sum (\log y_i - \log y)^2$$

which is an awkward procedure. However, the fit obtained is not seriously affected using Eqs. (9.8).

is applied before computing squares u_i^2 and products $u_i \cdot \log y_i$:

x_i	y_i	u_i	u_i^2	$\log y_i$	$u_i \cdot \log y_i$
0	107.0	−7	49	2.0294	−14.2058
4	88.0	−5	25	1.9445	−9.7225
8	74.2	−3	9	1.8704	−5.6112
12	66.8	−1	1	1.8248	−1.8248
16	51.0	+1	1	1.7076	+1.7076
20	41.6	+3	9	1.6191	+4.8573
24	33.6	+5	25	1.5263	+7.6315
28	26.9	+7	49	1.4298	+10.0086
			168	13.9519	−7.1593

From Eqs. (9.9),

$$\log a' = \frac{1}{n} \sum \log y_i$$

$$= \tfrac{1}{8}(13.9519) = 1.74399$$

$$\log b' = \frac{\sum (u_i \cdot \log y_i)}{\sum u_i^2}$$

$$= \frac{-7.1593}{168} = -0.04261$$

The fitted line is

$$\log y = 1.74399 + u(-0.04261)$$

or

$$\log y = 1.74399 + \left(\frac{x - 14}{2}\right)(-0.04261)$$

$$= 2.04226 + x(-0.02130)$$

The anti-log of 2.04226 is 110.22 and the anti-log of −0.02130 (or 0.97870 − 1) is 0.9521. Thus, the fitted exponential is

$$y = 110.22(0.9521)^x$$

Another common curve often fitted to empirical data is the power function

$$y = ax^b$$

Taking logs of both sides of this equation,

$$\log y = \log a + b \log x$$

which again is of the linear form described in Secs. 9.1–9.2. Replacing y by $\log y$ and x by $\log x$ in Eqs. (9.1),

$$\left.\begin{aligned} \log a &= \frac{\sum \log x_i \sum (\log y_i \cdot \log x_i) - \sum (\log x_i)^2 \sum \log y_i}{(\sum \log x_i)^2 - n \sum (\log x_i)^2} \\ b &= \frac{\sum \log x_i \sum \log y_i - n \sum (\log y_i \cdot \log x_i)}{(\sum \log x_i)^2 - n \sum (\log x_i)^2} \end{aligned}\right\} \tag{9.10}$$

No transformation on x is possible here and Eqs. (9.10) must be used although the computation is certainly tedious.

EXERCISES

1. Fit an exponential $y = ab^x$ to the following observations: (1, 1.5), (2, 4.6), (3, 12.0), (4, 39.1), and (5, 124.2). *Ans.* $y = 0.4912(2.996)^x$.

2. Fit a power function $y = ax^b$ to the following observations: (2, 0.01), (4, 0.1), (8, 1), (16, 10), and (32, 100). *Ans.* $y = 0.001x^{3.3219}$.

9.5 Choice of Model

Selecting the functional form of the relationship suggested by the observations can be done by plotting the (x_i, y_i) points. A suggested procedure is:

1. Plot the (x_i, y_i) on regular graph paper and, if a linear trend is obtained, fit $y = a_0 + a_1x$.
2. If the trend is nonlinear, perhaps a parabola or a higher degree polynomial is appropriate.
3. If these first two possibilities are not promising, plot (x_i, y_i) on semilog paper and, if a linear trend develops, fit $y = ab^x$.
4. If the trend in step 3 is nonlinear, plot (x_i, y_i) on log-log paper and, if the plot exhibits a linear trend, fit $y = ax^b$.

If plotting the (x_i, y_i) is not convenient, the following steps are suggested.

1. If the y_i differences are approximately constant, fit $y = a_0 + a_1x$.
2. If the second differences* in the y_i are approximately constant, fit a parabola $y = a_0 + a_1x + a_2x^2$.
3. If the $\log y_i$ differences are approximately constant, fit $y = ab^x$.

*Second differences are differences of differences. In Example 9.3.1, first differences are 6, 17, 27, 38, 48, and 57. The differences of these differences are 11, 10, 11, 10, and 9.

There are other possible models, of course, for specific situations. Some examples follow:

1. $y = ae^{bx}$ or $\log y = \log a + (b \log e)x$, which is a linear model, and the Eqs. (9.1) pattern is appropriate.
2. $y = a/b^x$ or $\log y = \log a - x \log b$, thus fitting a function of the form $y = ab^{-x}$.
3. $1/y = a_0 + a_1x$. Let $y' = 1/y$ and fit a line $y' = a_0 + a_1x$.

9.6 Fitting a Distribution Function

Occasionally, an application involves the fitting of a distribution function to observations x_i. This is a slightly different problem of plotting x against the functional $f(x)$ values, where the differences between respective $f(x)$ values and corresponding observed frequencies are analogous to the previous differences e_i in Secs. 9.1–9.5. Functional parameters are estimated from sample statistics instead of being determined by least squares procedure.

EXAMPLE 9.6.1

In the hypothesis test of Example 7.15.3, Table 7.7 is constructed to generate expected frequencies $e_j = nP_j$. This procedure is an example of fitting a distribution function to a set of observations x_i.

The hypothesis H_0: X is $N(18, 6.24^2)$ is tested and accepted, thus establishing the functional form to be normal. Using sample estimates of the population mean and standard deviation in Z-transformations to obtain tabled normal probabilities P_j of x occurring in the jth class interval (see Tables 7.6–7.7), expected frequencies [or $f(x)$ values] are obtained from $e_j = nP_j$.

A curve is fitted to the expected frequency points, which are assumed to coincide with mid-x values of the respective class intervals as shown in Fig. 9.3.

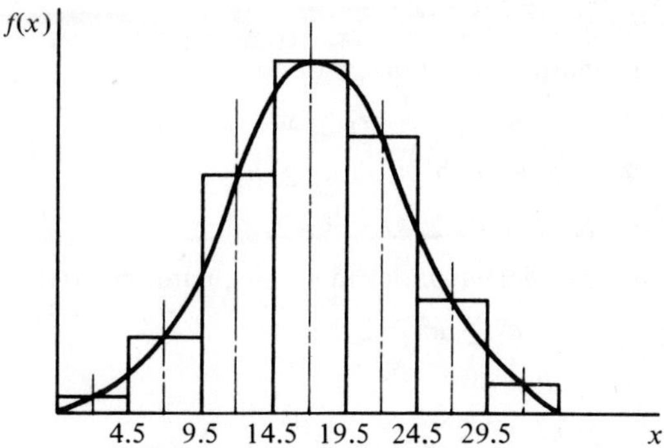

Figure 9.3. Example 9.6.1—fitting a normal curve.

GLOSSARY OF TERMS

Curve fitting
Method of moments
Average line
Method of least squares
Least squares normal equations (*LSNE*)
First and second differences

GLOSSARY OF FORMULAS

		Page
(9.1)	Equations for parameters, linear model.	301

$$a_0 = \frac{\sum x_i \sum y_i x_i - \sum x_i^2 \sum y_i}{(\sum x_i)^2 - n \sum x_i^2}$$

$$a_1 = \frac{\sum x_i \sum y_i - n \sum y_i x_i}{(\sum x_i)^2 - n \sum x_i^2}$$

(9.2) Least squares normal equations, linear model. 303

$$\sum y_i = \sum a_0 + a_1 \sum x_i$$

$$\sum y_i x_i = a_0 \sum x_i + a_1 \sum x_i^2$$

(9.3) Transformation *x*- to *u*-units. 303

$$u_i = \frac{x_i - \bar{x}}{c}, \qquad c = x_{i+1} - x_i$$

(9.4) Equations for parameters, linear model in *u*-units. 303

$$a_0' = \frac{1}{n} \sum y_i = \bar{y}$$

$$a_1' = \frac{\sum y_i u_i}{\sum u_i^2}$$

(9.5) Least squares normal equations, parabolic trend. 305

$$\sum y_i = \sum a_0 + a_1 \sum x_i + a_2 \sum x_i^2$$

$$\sum y_i x_i = a_0 \sum x_i + a_1 \sum x_i^2 + a_2 \sum x_i^3$$

$$\sum y_i x_i^2 = a_0 \sum x_i^2 + a_1 \sum x_i^3 + a_2 \sum x_i^4$$

(9.6) Least squares normal equations in *u*-units, parabolic trend. 306

$$\sum y_i = \sum a_0' + a_2' \sum u_i^2$$

$$\sum y_i u_i = a_1' \sum u_i^2$$

$$\sum y_i u_i^2 = a_0' \sum u_i^2 + a_2' \sum u_i^4$$

(9.7) Least squares normal equations for a polynomial of any degree. 307

$$\sum y_i = \sum a_0 + a_1 \sum x_i + a_2 \sum x_i^2 + \ldots + a_k \sum x_i^k$$
$$\sum y_i x_i = a_0 \sum x_i + a_1 \sum x_i^2 + a_2 \sum x_i^3 + \ldots + a_k \sum x_i^{(k+1)}$$
$$\sum y_i x_i^2 = a_0 \sum x_i^2 + a_1 \sum x_i^3 + a_2 \sum x_i^4 + \ldots + a_k \sum x_i^{(k+2)}$$
$$\vdots$$
$$\sum y_i x_i^k = a_0 \sum x_i^k + a_1 \sum x_i^{(k+1)} + a_2 \sum x_i^{(k+2)} + \ldots + a_k \sum x_i^{2k}$$

(9.8) Least squares normal equations, exponential model $y = ab^x$. 308

$$\log a = \frac{\sum x_i \sum (x_i \cdot \log y_i) - \sum x_i^2 \sum \log y_i}{(\sum x_i)^2 - n \sum x_i^2}$$

$$\log b = \frac{\sum x_i \sum \log y_i - n \sum (x_i \cdot \log y_i)}{(\sum x_i)^2 - n \sum x_i^2}$$

(9.9) Equations for parameters, exponential model $y = ab^x$ in u-units. 308

$$\log a' = \frac{1}{n} \sum \log y_i = \overline{\log y}$$

$$\log b' = \frac{\sum (u_i \cdot \log y_i)}{\sum u_i^2}$$

(9.10) Equations for parameters, exponential model $y = ax^b$. 310

$$\log a = \frac{\sum \log x_i \sum (\log y_i \cdot \log x_i) - \sum (\log x_i)^2 \sum \log y_i}{(\sum \log x_i)^2 - n \sum (\log x_i)^2}$$

$$b = \frac{\sum \log x_i \sum \log y_i - n \sum (\log y_i \cdot \log x_i)}{(\sum \log x_i)^2 - n \sum (\log x_i)^2}$$

REFERENCES

Guttman, I. and Wilks, S. S. (1971), *Introductory Engineering Statistics*, John Wiley & Sons, Inc., New York, pp. 341–346.

Miller, I. and Freund, J. E. (1965), *Probability and Statistics for Engineers*, Prentice-Hall, Inc., Englewood Cliffs, N. J., pp. 226–231.

Volk, W. (1969), *Applied Statistics for Engineers*, 2nd ed., McGraw-Hill Book Company, New York, pp. 261–266.

PROBLEMS

1. Fit a line $y = a_0 + a_1 x$ by the method of moments to the following observations:

x_i	0	3	6	9	12	15
y_i	1.8	3.6	3.4	5.5	5.7	7.4

2. Using the least squares method, repeat Problem 1 working in x-units and using Eqs. (9.2).

3. Repeat Problem 2 working in u-units and using Eqs. (9.4).

4. Fit a line $y = a_0 + a_1 x$ by the method of least squares to the following observations:

x_i	0	3	6	9	12	15	18
y_i	7.2	5.9	4.9	4.3	2.8	1.9	1.1

5. Fit $y = a_0 + a_1 x + a_2 x^2$ by the method of least squares to the following observations:

x_i	0	2	4	6	8	10
y_i	4	24	90	186	335	510

6. Fit $y = ae^{bx}$ by the method of least squares to the following observations:

x_i	0	1	2	3	4	5
y_i	0.5	1.6	4.5	13.8	40.2	125.0

7. Fit $y = ab^x$ by the method of least squares to the following observations:

x_i	-3	-2	-1	0	1	2	3
y_i	1.1	1.8	4.2	7.9	16.3	31.7	64.4

8. Fit $y = e^{-ax}$ by the method of least squares to the following observations:

x_i	100	200	300	400	500
y_i	0.61	0.37	0.22	0.14	0.08

9. Use Eqs. (9.10) to fit $y = ax^b$ to the following ideal x, y values:

x_i	1	10	100	1,000	10,000
y_i	0.0001	0.001	0.01	0.1	1.0

10. Graph and then fit a best fitting line to the following shear strain x (in 0.0001 in.) and shear stress y (in 100 psi) observations:

x_i	2	4	6	8	10	12
y_i	41	76	130	162	220	261

11. Graph and then fit a best fitting curve to the following observations:

x_i	-4	0	4	8	12	16
y_i	2	8	9	11	8	5

12. Graph and then fit a best fitting curve to the following tensile test observations, where x is load (in tons) and y is elongation (in 0.0001 in.)

x_i	1	2	3	4	5
y_i	14	27	40	55	68

13. Graph and then fit a best fitting curve to the following observations:

x_i	0	5	10	15	20
y_i	1,000	100	10	1	0.1

14. In an analytical chemistry experiment, samples containing known weights x (in milligrams) of a chemical are analyzed by a standard procedure yielding measured weights y. Graph and fit a best fitting curve to the following observations. What can you say regarding the chemical analysis procedure?

x_i	4.0	8.4	12.5	16.0	20.0	25.0	31.0
y_i	3.7	7.8	12.1	15.6	19.8	24.5	30.7

15. Determine the relationship between fahrenheit and centrigrade temperature scales for the following observations. Consider centrigrade readings to be the independent variable.

c_i	0	5	10	15	20	25
f_i	32	41	50	59	68	77

16. Determine the least squares normal equations (LSNE) for a model $y = a_0x + a_1x^2$.

17. Determine the LSNE for $y = a_0 + a_1x_1 + a_2x_2$.

18. Determine the LSNE for $y = a_0 + a_1x_1 + a_2x_1x_2 + a_3x_2$.

19. Cost-output relationships experienced in a multiple product firm are being studied. One relationship of interest is that of variations in total variable cost y for each product with output x for that product. Graph and fit a best fitting line for each relationship—that for Product A and also that for Product B. The data is coded with each output x-unit representing 100 product units and each cost y-unit representing $1,000.

A	x_i	2	4	6	8	10	12	14	16	18	20
	y_i	3	7	9	8	10	14	15	19	20	23

B	x_i	10	14	18	22	26	30	34	38	42	46
	y_i	7	10	12	9	11	16	12	17	15	19

20. Determine the LSNE for $y = a/x$.

21. The shearing stress y (in psi) of a rivet and cross-sectional area x (in in.^{2}) of the rivet are related by $y = a/x$ where a is the load (in lb). Determine a by the method of least squares for the following shear-stress observations.

x_i	0.110	0.196	0.307	0.442	0.601	0.785
y_i	45,300	25,500	16,300	11,300	8,300	6,400

22. Graph and fit a best fitting line to the following observations of warp breaks y in fabric weaving and relative humidities x.

x_i	68	70	72	74	76	78
y_i	1	6	12	16	21	26

23. Thickness (in 0.00001 in.) of nonmagnetic coatings of galvanized zinc on steel pieces is measured either by a standard destructive method y or a proposed nondestructive magnetic method x. If there is a good relationship between x and y, the less costly inspection procedure x may be used with y being estimated from x. Graph the following observations and determine the relationship. Each (x_i, y_i) represents two measurements on the same specimen piece—one by method x, and then one by method y.

x_i	85	105	115	120	121	127	155	250	310	443	630
y_i	104	116	114	132	139	129	174	312	338	465	720

24. Two methods x and y of measuring surface finish (in microinches) are being considered. Method y is standard; method x is cheaper and more rapidly performed. If there is a good relationship between x and y, the less costly method x can be utilized with standard values y being estimated from x. Graph the following observations and determine the relationship.

x_i	60	62	64	66	68	70	72
y_i	54.0	55.2	55.5	58.3	58.0	58.8	61.3

25. The following observations represent a sample distribution of tool life (in centiminutes) for a 5° negative rake accelerated life test. Fit a normal distribution curve to these observations.

x_j	11–15	16–20	21–25	26–30	31–35	36–40	41–45
f_j	1	2	4	5	12	19	26

46–50	51–55	56–60	61–65	66–70	71–75	76–80
19	11	9	8	10	6	3

26. The following observations represent a sample distribution of tool life (in centiminutes) for a 10° positive rake accelerated life test. Fit a normal distribution curve to these observations.

x_i	26–50	51–75	76–100	101–125	126–150	151–175
y_i	2	4	9	13	10	6

176–200	201–225	226–250
3	2	1

REGRESSION AND CORRELATION

All of the statistical topics of Chaps. 1–8 deal with a sample $x_1, x_2, \ldots, x_i, \ldots, x_n$ *from a population of* X*-values. In this case, each* x_i *represents both the element of the sample and a single observation of the element. Attention is now given to a situation where two measurements (observations) are made on each sample element. The sample consists of* n *elements on which pairs of observations* $(x_1, y_1), (x_2, y_2), \ldots, (x_n, y_n)$ *are made.* A basic objective is to make inferences regarding the nature of the population relationship between* X *and* Y*, based on sample information* (x_i, y_i)*. One method of doing this is by means of regression.*

10.1 The Regression Problem

In Chap. 9, relationships between samples (x_i, y_i) are summarized by fitting a line or curve to the plotted points. A functional form of the relation-

*Consider a population of all of the male students in a school. A random sample of 200 students is selected. The height X of each student is measured, x_i are the sample observations of height, and inferences can be made regarding population mean height, variance of

10

ship between the variables is selected: (1) by means of analytical or physical considerations, or (2) by studying a plot of the observations. A line or curve is fitted by the method of least squares.*

It is natural to desire to extend curve fitting to a basis whereby population relationships between X and Y can be inferred. For example, it is useful in applications to be able to test hypotheses and make confidence interval estimates of Y given a specified X-value, estimate variances, measure the degree of relationship between variables, and so forth. To develop methods for making such inferences requires a number of specific assumptions.

The term *regression* refers to the nature of the functional relationship, the term *regression model* identifies a specific functional relationship, and the *regression line* is the resulting line fitted to the plotted observations. The fol-

---

height, and so forth (i.e., Chap. 7 considerations). Suppose, however, that interest is with the relationship between height X and weight Y. Now, the sample elements are the 200 students, (x_i, y_i) are the sample observations, and inferences are relative to population height-weight relationship.

*In some respects, this procedure is analogous to that of descriptive statistics (Chap. 1), wherein samples x_i are summarized by frequency distributions with no assumptions being made and, of course, no rational inferences regarding the population being possible.

lowing introductory regression analysis sections deal with simple linear regression where the model (or population regression line) is

$$\mu_{Y|X} = A_0 + A_1 X \tag{10.1}$$

and the sample regression line is

$$y = a_0 + a_1 x \tag{10.2}$$

and solutions for a_0 and a_1 are given by Eqs. (9.2), which are obtainable from the *LSNE*:

$$\sum y_i = \sum a_0 + a_1 \sum x_i$$
$$\sum y_i x_i = a_0 \sum x_i + a_1 \sum x_i^2$$

Equation (10.1) reads "the population mean Y, given a specified X-value, is equal to $A_0 + A_1 X$." The regression line is viewed as a line joining the means of the respective Y-distributions corresponding to all possible values of X, as shown in Fig. 10.1. The first assumption is that the X-values are known constants controlled by the experimenter and that Y is a random variable.*

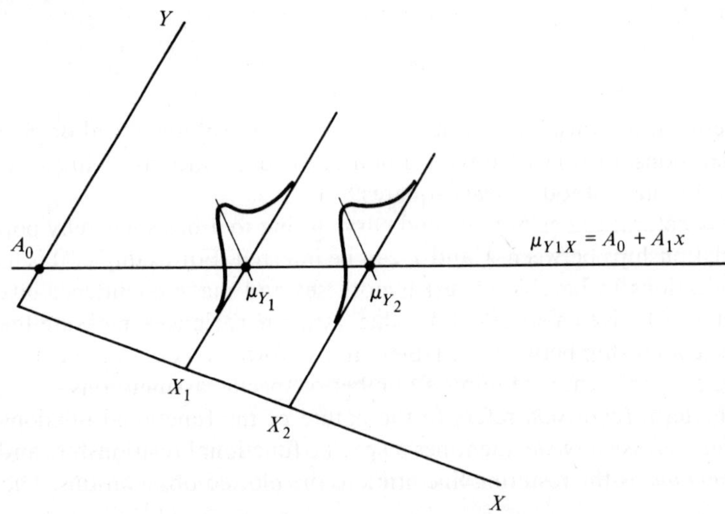

Figure 10.1. Regression line and Y- distributions.

*In applications, this is not quite true. For one thing, even though X is varied and controlled by the experimenter, the observations x_i are subject to measurement error. However, it is practically assumed that X-variation is small relative to the Y-variation.

Figure 10.1 suggests other assumptions that are made. All of the assumptions are summarized as follows:

1. The X-values are known constants.
2. For each value of X, the random variable Y is normally and independently distributed with mean

$$\mu_{Y|X} = A_0 + A_1 X$$

   and variance $\sigma^2_{Y|X}$ where A_0, A_1, and $\sigma^2_{Y|X}$ are unknown parameters.
3. The variances of Y are homogeneous for all X, that is,

$$\sigma^2_{Y_1} = \sigma^2_{Y_2} = \cdots = \sigma^2_{Y_k}$$

   where the subscripts are shorthand symbols for Y given a specified X. For example, $\sigma^2_{Y_1}$ is the variance of Y given $X = X_1$.

Whereas regression procedure is relatively simple, interpretation of results may be complex.* A fundamental precaution is emphasized here. A statistical indication of a certain functional relationship in no way implies a cause-and-effect relationship. Cause and effect can only be established by analysis based on physical laws and principles. A statistical analysis of relationship between X and Y is a study of *association* or *co-variation*—how X and Y vary together or jointly.

10.2 Significance of Regression

One objective of regression is to "explain" or "account for" as much of the variation in Y as possible. To illustrate, consider an ideal theoretical situation—explaining the variation of the volume Y of a cube, assuming no measurement error. The regression model is $Y = X^3$ and all of the variation in Y is explainable by the variation in edge-length X of the cube. The regression is *significant* in that the major part, in this case all, of the variation in Y

---

*Physical phenomena are not usually the result of a simple one-to-one cause-and-effect relationship between two variables X and Y. A Y-value is the result of the influences of a number of independent variables $X_1, X_2, X_3, \ldots$. For example, consider a two-variable statistical analysis of the relationship between X_1 and Y. For a specified value of X_1, the dependent variable Y may be the result of any or all of the following: (1) influence of X_1 on Y directly, (2) influence of other variables $X_2, X_3, \ldots$ that have not been monitored, (3) interaction influence of any or all of the $X_1, X_2, X_3, \ldots$ on Y, and (4) experimental error.

is identified as being "due to" the variation in X. Thus, if cubes are being manufactured, it is certainly desirable to control edge-length X.

To illustrate the actual (i.e., not ideal) situation faced by an experimenter, consider the shear strength Y of rivets whose diameter is X, a sample regression line $y = f(x)$, and an (x_i, y_i) plot of observations as shown in Fig. 10.2. The total variation of a single y_i-value is $y_i - \bar{y}$, which can be expressed as

$$y_i - \bar{y} = (y - \bar{y}) + (y_i - y)$$

where (x_i, y_i) is an observation, (x, y) is the corresponding point on the regression line, $x = x_i$, and $\bar{y}$ is the mean of the y_i. The first term of the right-hand member of the equation measures that part of the total variation explainable by regression, and the second right-hand term measures that part of the total variation that is unexplained.

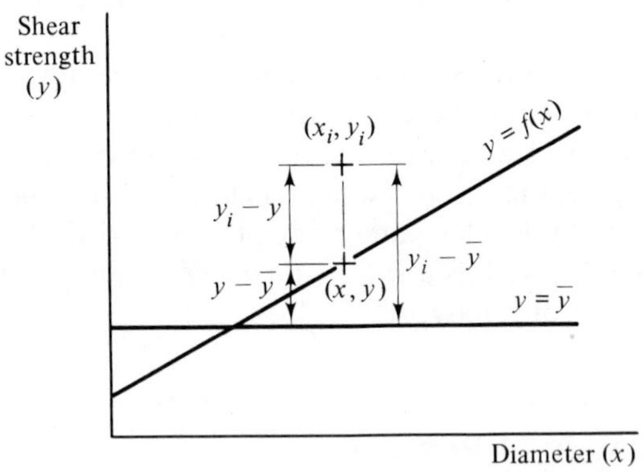

Figure 10.2. Explainable and unexplainable variation. [Only one observation (x_i, y_i) is shown. Assume n such observation points.]

Squaring both sides of the above expression and summing gives

$$\sum (y_i - \bar{y})^2 = \sum (y - \bar{y})^2 + \sum (y_i - y)^2 \tag{10.3}$$

where the total sum of squares for Y is partitioned into two parts, one part $\sum (y - \bar{y})^2$ being that part of the total sum of squares explainable by regression (e.g., that part of shear strength variation attributable to diameter change), the other part $\sum (y_i - y)^2$ measuring that part of the total sum of squares which is unexplained. Clearly, if the first part is large relative to the second part, regression is significant.

The degrees of freedom for each term of Eq. (10.3) is

$$n - 1 = 1 + (n - 2)^*$$

Computation of each member of Eq. (10.3) proceeds as follows. From Eq. (1.24), the left-hand member is

$$\sum (y_i - \bar{y})^2 = \sum y_i^2 - n\bar{y}^2$$

The second term of the right-hand member is obtained as follows. An algebraic expansion of $\sum (y_i - y)^2$ yields

$$\sum (y_i - y)^2 = \sum y_i^2 - a_0 \sum y_i - a_1 \sum y_i x_i \dagger \qquad (10.4)$$

and the first term of the right-hand member, the sum $\sum (y - \bar{y})^2$, is obtained from Eq. (10.3) by subtraction.

One example is traced through Chap. 10 to illustrate a number of regression operations. The data of Example 9.2.1 is used for this purpose. To place this data in an engineering context, X and Y designate the variables described in the following production cost problem. A new method of computing overhead cost Y is under study. Since most of the manufacturing operations are draw-die operations on bar-stock raw material, lineal feet of direct material X is being considered as a base for allocating overhead charges. If X and Y are approximately proportional (i.e., have a linear relationship),

---

*The sum $\sum (y_i - \bar{y})^2$ has $n - 1$ *d.f.* Since y depends on a_0 and a_1 (computed from sample observations), the sum $\sum (y_i - y)^2$ has $n - 2$ *d.f.* The *d.f.* for $\sum (y - \bar{y})^2$ is obtained by subtraction.

†Equation (10.4) is derived as follows. Substituting for y in $\sum (y_i - y)^2$ its equivalent from the linear model $y = a_0 + a_1 x$ and squaring,

$$\begin{aligned} \sum (y_i - y)^2 &= \sum (y_i - a_0 - a_1 x_i)^2 \\ &= \sum y_i^2 - 2a_1 \sum y_i x_i - 2a_0 \sum y_i + a_1^2 \sum x_i^2 + 2a_0 a_1 \sum x_i + na_0^2 \end{aligned}$$

An equivalent expression for the last three terms of this expression can be derived by multiplying the *LSNE* in Eqs. (9.2) by a_0 and a_1, respectively, and then adding. Thus,

$$\begin{aligned} \sum y_i &= \sum a_0 + a_1 \sum x_i && \text{multiplied by } a_0 \\ \sum y_i x_i &= a_0 \sum x_i + a_1 \sum x_i^2 && \text{multiplied by } a_1 \end{aligned}$$

and added gives

$$a_0 \sum y_i + a_1 \sum y_i x_i = a_1^2 \sum x_i^2 + 2a_0 a_1 \sum x_i + na_0^2$$

Substituting this equivalent expression for the last three terms of the original expression,

$$\begin{aligned} \sum (y_i - y)^2 &= \sum y_i^2 - 2a_1 \sum y_i x_i - 2a_0 \sum y_i + a_0 \sum y_i + a_1 \sum y_i x_i \\ &= \sum y_i^2 - a_0 \sum y_i - a_1 \sum y_i x_i \end{aligned}$$

X is considered to be an accurate base for charging overhead to jobs and the regression line can be used for estimating overhead charges prior to actual manufacture. Each X-unit corresponds to 500 lineal feet and each Y-unit to \$1,000 in the data of Example 9.2.1.

EXAMPLE 10.2.1

The data of Example 9.2.1 is repeated here along with squares y_i^2, cross products $y_i x_i$, and sums:

x_i	y_i	y_i^2	$y_i x_i$
0	1.7	2.89	0
1	7.7	59.29	7.7
2	17.3	299.29	34.6
3	27.1	734.41	81.3
4	28.0	784.00	112.0
5	35.6	1267.36	178.0
	117.4	3147.24	413.6

Computing the total sum of squares for Y,

$$\bar{y} = \frac{1}{n} \sum y_i$$
$$= \tfrac{1}{6}(117.4) = 19.57$$

and

$$\sum (y_i - \bar{y})^2 = \sum y_i^2 - n\bar{y}^2$$
$$= 3{,}147.24 - 6(19.57)^2$$
$$= 850.11$$

From Eq. (10.4),

$$\sum (y_i - y)^2 = \sum y_i^2 - a_0 \sum y_i - a_1 \sum y_i x_i$$
$$= 3{,}147.24 - 2.41(117.4) - 6.86(413.6)$$
$$= 27.01$$

where the values for a_0 and a_1 are from the solution of Example 9.2.1. By subtraction,

$$\sum (y - \bar{y})^2 = \sum (y_i - \bar{y})^2 - \sum (y_i - y)^2$$
$$= 850.11 - 27.01$$
$$= 823.10$$

These results are conveniently summarized as follows:

Source of Variation	*d.f.*	*SS*	*MS*
Due to regression	1	823.10	823.10
Deviations from regression	4	27.01	6.75
Total	5	850.11	

where *SS* designates sum of squares and *MS* denotes mean square (obtained by dividing *SS* by its *d.f.*).

From Sec. 6.4, the ratio of two mean square random variables is *F*-distributed. Thus, at a 1% significance level,

$$F_{\text{data}} = \frac{823.10}{6.75} = 121.9 > F_{0.01,1,4} = 21.2$$

indicates a significant regression. This suggests good proportionality (linearity) between lineal feet of direct material X and overhead expense Y. Thus, lineal feet of direct material becomes the new base for charging overhead, and the regression equation $y = 2.41 + 6.86x$ is used to estimate overhead charges.

EXERCISES

1. Fit $y = a_0 + a_1 x$ to the following observations (x_i, y_i) and check at an $\alpha = 0.01$ level for significance of regression: (0, 1), (1, 3), (2, 5), (3, 8), (4, 9), (5, 11), and (6, 12).

Ans. $\frac{100.33}{0.33} > F_{0.01,1,5}$, regression is significant.

2. What is the best fitting least squares line for the observations: (2, 2), (2, 4), (4, 2), (4, 4), (6, 2), and (6, 4)? Using Eq. (10.3), analyze this relationship. [What is the value of $\sum (y - \bar{y})^2$?]

Ans. $y = \bar{y} = 3$, $\sum (y - \bar{y})^2 = 0$.

3. Referring again to Eq. (10.3), what is the value of $\sum (y_i - y)^2$ for a line fitted to (2, 2), (6, 4)?

Ans. 0.

10.3 Hypothesis Tests—Slope and Intercept

From Eq. (10.3), an unbiased estimate of the total variance of Y is

$$s_y^2 = \frac{1}{n-1} \sum (y_i - \bar{y})^2 = \frac{1}{n-1} [\sum y_i^2 - n\bar{y}^2] \tag{10.5}$$

and, from Eq. (10.4), the variance of Y given X, assumed to be the same for all X, is estimated unbiasedly by the mean squared deviation about the regression line, that is, by

$$s_{y|x}^2 = \frac{1}{n-2} \sum (y_i - y)^2$$
$$= \frac{1}{n-2}[\sum y_i^2 - a_0 \sum y_i - a_1 \sum y_i x_i]^* \tag{10.6}$$

If the observations (x_i, y_i) are transformed such that the origin is shifted from $(0, 0)$ to $(\bar{x}, \bar{y})$, the second term of Eq. (10.6) vanishes and this expression simplifies to

$$s_{y|x}^2 = \frac{1}{n-2}[\sum y_i^2 - a_1 \sum y_i x_i] \tag{10.7}$$

The estimated variance of the sample regression coefficient a_1 is given by

$$s_{a_1}^2 = \frac{s_{y|x}^2}{\sum (x_i - \bar{x})^2} \tag{10.8}$$

and the variance of the estimator a_0 for the intercept is obtained from

$$s_{a_0}^2 = s_{y|x}^2\left[\frac{1}{n} + \frac{\bar{x}^2}{\sum (x_i - \bar{x})^2}\right] \tag{10.9}$$

Equations (10.8) and (10.9) are bases for hypothesis tests concerning the parameters A_0 and A_1 of the population regression model $\mu_{Y|X} = A_0 + A_1 X$. Under the normality assumption of Sec. 10.1, the statistics

$$\frac{a_1 - A_1}{s_{a_1}}$$

and

$$\frac{a_0 - A_0}{s_{a_0}}$$

are each t-distributed with $n - 2$ *d.f.* Using the t-distribution as a model for the hypothesis test, a test can be made regarding either parameter being equal to a specified value. The usual test of either parameter is that the parameter is zero, since interest is primarily on whether or not there is a relationship between the variables X and Y.

EXAMPLE 10.3.1

For Example 10.2.1 the hypotheses $H_0\colon A_0 = 0$ and $H_0\colon A_1 = 0$ are tested against alternatives that each is not equal to zero.

---

*This quantity is frequently called the *error variance*, and $s_{y|x}$ is called the *standard error of estimate*.

If H_0 is true, the statistic $(a_0 - A_0)/s_{a_0}$, where $A_0 = 0$, is t-distributed with $n - 2$ $d.f.$ and the decision rule at a 5% significance level is

$$|t_{\text{data}}| \geq t_{0.025,4} = 2.776, \qquad \text{reject } H_0$$

From Eq. (10.6),

$$s_{y/x}^2 = \frac{1}{n-2}\left[\sum y_i^2 - a_0 \sum y_i - a_1 \sum y_i x_i\right]$$
$$= \tfrac{1}{4}(27.01) = 6.75$$

and

$$s_{y/x} = 2.60$$

and, from Eq. (10.9),

$$s_{a_0}^2 = s_{y|x}^2\left[\frac{1}{n} + \frac{\bar{x}^2}{\sum (x_i - \bar{x})^2}\right]$$
$$= 6.75\left[\frac{1}{6} + \frac{(\frac{5}{2})^2}{\frac{70}{4}}\right]$$
$$= 3.536$$

and s_{a_0} is equal to 1.88.

Computing the value of the sample statistic,

$$|t_{\text{data}}| = \frac{a_0}{s_{a^0}} = \frac{2.41}{1.88} = 1.28$$

and thus H_0 is accepted. In a similar manner, H_0: $A_1 = 0$ is tested:

$$s_{a_1}^2 = \frac{6.75}{\frac{70}{4}} = 0.386$$
$$s_{a_1} = \sqrt{0.386} = 0.62$$

and

$$|t_{\text{data}}| = \frac{a_1}{s_{a_1}} = \frac{6.86}{0.62} = 11.1$$

and H_0 is rejected.

EXERCISES

1. Compute the mean square deviation about the regression line for Exercise 1, Sec. 10.2. *Ans.* $s_{y|x} = 0.5779$.

2. Test at an $\alpha = 0.05$ level the possibility of $A_0 = 0$, $A_1 = 0$ for Exercise 1, Sec. 10.2. *Ans.* $\left|\frac{1.32}{0.394}\right| > t_{0.025,5}$, $\left|\frac{1.89}{0.109}\right| > t_{0.025,5}$, reject both possibilities.

10.4 Confidence Interval Estimates—Slope and Intercept

Under the normality assumption of Sec. 10.1, the y_i's are normally distributed. Further, it can be shown that a_0 and a_1 are expressible as linear functions of the y_i's. Thus, from Eq. (5.11), the estimators a_0 and a_1 are normally distributed and, as in Sec. 7.17, since each is a normally distributed random variable with unknown population variance, the confidence interval is based on the t-distribution model. Thus, $100(1 - \alpha)$ confidence interval estimates of A_0 and A_1 are given by

$$\left.\begin{array}{l} a_0 \pm t_{(\alpha/2),n-2}(s_{a_0}) \\ a_1 \pm t_{(\alpha/2),n-2}(s_{a_1}) \end{array}\right\} \qquad (10.10)$$

EXAMPLE 10.4.1

Again referring to Example 10.2.1, 95% confidence interval estimates of A_0 and A_1, respectively, are

$$a_0 \pm t_{0.025,4}(s_{a_0})$$

or

$$2.41 \pm 2.776(1.88) = -2.8, +7.6^*$$

and

$$a_1 \pm t_{0.025,4}(s_{a_1})$$

or

$$6.86 \pm 2.776(0.62) = 5.1, 8.6$$

EXERCISES

1. Determine a 95% confidence interval estimate of A_0 for the regression example of Exercise 1, Sec. 10.2. *Ans.* (0.31, 2.33).
2. Determine a 95% confidence interval estimate of A_1 for the regression example of Exercise 1, Sec. 10.2. *Ans.* (1.61, 2.17).

---

*Some analysts use this kind of a result as a quick and rough criterion as to whether A_0 can easily be zero. This is the case here, with the interval (−2.8, +7.6) including zero.

10.5 Confidence Interval Estimate of $\mu_{Y|X}$

Given a specified value for X, the regression equation $y = a_0 + a_1 x$ can be used to estimate the corresponding mean $\mu_{Y|X}$. Let $X = X_0$ be the specified X-value. Then, a point estimate of $\mu_{Y|X}$ is given by

$$y_0 = a_0 + a_1 x_0$$

An estimate of the variance of Y, given $X = X_0$, is

$$s_{y|x}^2\left[\frac{1}{n} + \frac{(x_0 - \bar{x})^2}{\sum (x_i - \bar{x})^2}\right] \tag{10.11}$$

Again, the situation is that of estimating the expected value of a normally distributed random variable with unknown variance. Thus a $100(1 - \alpha)$ confidence interval estimate of $\mu_{Y|X}$ is based on the t-distribution model and is obtained from

$$a_0 + a_1 x_0 \pm t_{(\alpha/2),n-2}\left[s_{y|x}\sqrt{\frac{1}{n} + \frac{(x_0 - \bar{x})^2}{\sum (x_i - \bar{x})^2}}\right] \tag{10.12}$$

Since the estimated variance of Y, given $X = X_0$, depends on X_0, the confidence interval becomes wider with increasing deviation $(x_0 - \bar{x})$ as shown in Fig. 10.3. Therefore, better estimates of $\mu_{Y|X}$ are possible for specified X-values closer to the mean.

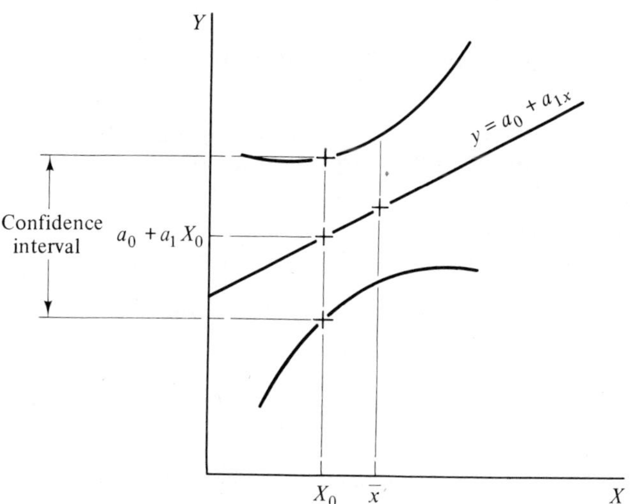

Figure 10.3. Confidence interval for the mean $\mu_{Y|X}$, given $X = X_0$.

EXAMPLE 10.5.1

Continuing with Example 10.2.1, a point estimate and a 95% confidence interval estimate of the mean of Y, given $X = 3.5$, is required. A point estimate is

$$\begin{aligned} y &= a_0 + a_1 x \\ &= 2.41 + 6.86(3.5) = 26.42 \end{aligned}$$

From Eq. (10.12), a 95% confidence interval estimate is given by

$$\begin{aligned} \mu_{Y|3.5} &= a_0 + a_1 x_0 \pm t_{0.025,4}\left[s_{y|x}\sqrt{\frac{1}{n} + \frac{(x_0 - \bar{x})^2}{\sum (x_i - \bar{x})^2}}\right] \\ &= 2.41 + 6.86(3.5) \pm 2.776\left[2.60\sqrt{\frac{1}{6} + \frac{(3.5 - 2.5)^2}{\frac{70}{4}}}\right] \\ &= 26.42 \pm 3.41 \\ &= 23.0, 29.8 \end{aligned}$$

EXERCISES

1. Compute a point estimate of the mean Y, given $X = 4$, for Exercise 1, Sec. 10.2. *Ans.* 8.89.
2. Compute a 95% confidence interval estimate of the mean Y, given $X = 4$, for Exercise 1, Sec. 10.2. *Ans.* (8.26, 9.52).
3. What is the increase in length of the confidence interval in Exercise 2 if, instead of $X = 4$, the specified X is 6? *Ans.* 0.77.

10.6 Confidence Interval Estimate of Y_0

A confidence interval estimate of an *individual* Y_0, given $X = X_0$, is also possible. However, this interval is larger than an interval used to estimate the mean of Y, given X_0, since the variance is that for individual Y's (which is larger than the variance of means).*

An estimate of the variance of Y_0, given $X = X_0$, is

$$s_{y_0}^2 = s_{y|x}^2\left[1 + \frac{1}{n} + \frac{(x_0 - \bar{x})^2}{\sum (x_i - \bar{x})^2}\right] \tag{10.13}$$

and a $100(1 - \alpha)$ confidence interval estimate of Y_0, given $X = X_0$, is

$$a_0 + a_1 x_0 \pm t_{(\alpha/2),n-2}\left[s_{y|x}\sqrt{1 + \frac{1}{n} + \frac{(x_0 - \bar{x})^2}{\sum (x_i - \bar{x})^2}}\right] \tag{10.14}$$

---

*This situation is analogous to Example 5.7.2, where the variance of means is $\frac{100}{144}$ and the variance of individual X's is $\frac{100}{12}$.

Again, because of the dependence on $x_0 - \bar{x}$, the interval widens, the more x_0 differs from $\bar{x}$.

EXAMPLE 10.6.1

A 95% confidence interval estimate of Y_0, given $X = 3.5$ is required (for Example 10.2.1). This estimate is given by

$$26.42 \pm 2.776\left[2.60\sqrt{1 + \frac{1}{6} + \frac{(3.5 - 2.5)^2}{\frac{70}{4}}}\right] = 18.4, 34.4$$

EXERCISES

1. Compute a 95% confidence interval estimate of an individual Y_0, given $X = 4$, for Exercise 1, Sec. 10.2. *Ans.* (7.28, 10.51).
2. Repeat Exercise 1 given $X = 6$. *Ans.* (10.88, 14.48).

Summary of Examples 10.2.1–10.6.1

Observations (x_i, y_i):

x_i	0	1	2	3	4	5
y_i	1.7	7.7	17.3	27.1	28.0	35.6

LSNE—Eqs. (9.2):

$$\sum y_i = \sum a_0 + a_1 \sum x_i \qquad \text{Solution: } a_0 = 2.41$$
$$\sum y_i x_i = a_0 \sum x_i + a_1 \sum x_i^2 \qquad a_1 = 6.86$$

Significance of regression:

$$SS_{\text{total}} = \sum y_i^2 - n\bar{y}^2 = 850.11$$
$$SS_{\text{error}} = \sum y_i^2 - a_0 \sum y_i - a_1 \sum y_i x_i = 27.01$$
$$SS_{\text{regression}} = SS_{\text{total}} - SS_{\text{error}} = 823.10$$

Source	*d.f.*	*SS*	*MS*
Regression	1	823.10	823.10
Deviations from regression	4	27.01	6.75
Total	5	850.11	

$$F_{\text{data}} = \frac{823.10}{6.75} = 121.9 > F_{0.01,1,4} = 21.2$$

Regression is significant.

Hypothesis tests:

$$\text{Eq. (10.6):}\quad s_{y|x}^2 = \frac{1}{n-2}\left[\sum y_i^2 - a_0 \sum y_i - a_1 \sum y_i x_i\right]$$

$$= 6.75$$

$$s_{y|x} = 2.60$$

$$\text{Eq. (10.8):}\quad s_{a_1}^2 = \frac{s_{y|x}^2}{\sum (x_i - \bar{x})^2}$$

$$= \frac{6.75}{\frac{70}{4}} = 0.386$$

$$s_{a_1} = 0.62$$

$$\text{Eq. (10.9):}\quad s_{a_0}^2 = s_{y|x}^2\left[\frac{1}{n} + \frac{\bar{x}^2}{\sum (x_i - \bar{x})^2}\right]$$

$$= 6.75\left[\frac{1}{6} + \frac{(\frac{5}{2})^2}{\frac{70}{4}}\right]$$

$$= 3.536$$

$$s_{a_0} = 1.88$$

H_0: $A_0 = 0$.

$$|t_{\text{data}}| = \frac{a_0}{s_{a_0}} = \frac{2.41}{1.88} = 1.28 < t_{0.025,4} = 2.776$$

Accept H_0.

H_0: $A_1 = 0$

$$|t_{\text{data}}| = \frac{a_1}{s_{a_1}} = \frac{6.86}{0.62} = 11.1 > t_{0.025,4} = 2.776$$

Reject H_0.

Confidence interval estimates of A_0 and A_1:

$$\text{Eqs. (10.10):}\quad a_0 \pm t_{0.025,4}(s_{a_0})$$

$$2.41 \pm 2.776(1.88) = -2.8, +7.6$$

$$a_1 \pm t_{0.025,4}(s_{a_1})$$

$$6.86 \pm 2.776(0.62) = 5.1, 8.6$$

Confidence interval estimates of $\mu_{Y|X}$:

$$\text{Eq. (10.12):}\quad a_0 + a_1 x_0 \pm t_{(\alpha/2),n-2}\left[s_{y|x}\sqrt{\frac{1}{n} + \frac{(x_0 - \bar{x})^2}{\sum (x_i - \bar{x})^2}}\right]$$

$$2.41 + 6.86(3.5) \pm 2.776\left[2.60\sqrt{\frac{1}{6} + \frac{(3.5 - 2.5)^2}{\frac{70}{4}}}\right]$$

$$= 23.0, 29.8$$

Confidence interval estimate of Y_0:

$$\text{Eq. (10.14): } a_0 + a_1 x_0 \pm t_{(\alpha/2), n-2}\left[s_{y|x}\sqrt{1 + \frac{1}{n} + \frac{(x_0 - \bar{x})^2}{\sum (x_i - \bar{x})^2}}\right]$$

$$2.41 + 6.86(3.5) \pm 2.776\left[2.60\sqrt{1 + \frac{1}{6} + \frac{(3.5 - 2.5)^2}{\frac{70}{4}}}\right]$$

$$= 18.4, 34.4$$

10.7 Correlation

The F-test for significance of regression (Sec. 10.2) measures the degree of relationship reflected by the model that is employed. Another more compact measure preferred by some experimenters is the correlation coefficient. Delaying a formal approach to this method, an intuitive geometric explanation is first presented to indicate the manner in which a correlation coefficient measures degree of relationship.

This measure is constructed from a sum of products of the deviations of x about $\bar{x}$ and y about $\bar{y}$:

$$\sum_{i=1}^{n} (x_i - \bar{x})(y_i - \bar{y})$$

where the X, Y plane is viewed as being divided into four quadrants by the perpendicular lines $x = \bar{x}$ and $y = \bar{y}$ as shown in Fig. 10.4.

Products $(x_i - \bar{x})(y_i - \bar{y})$ are positive for (x_i, y_i) points in quadrants I and III, and negative for (x_i, y_i) in quadrants II and IV. If there exists a good

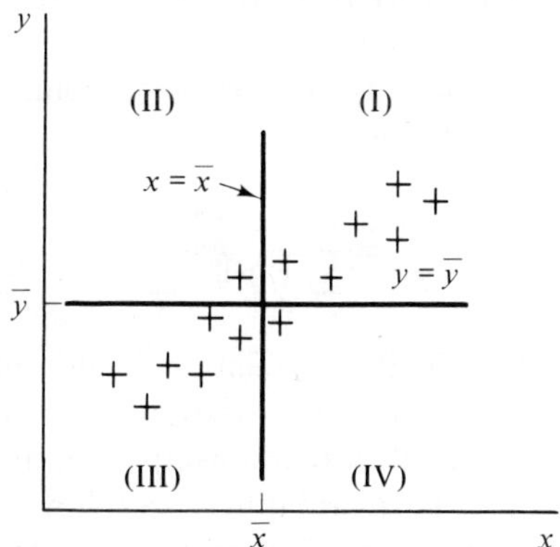

Figure 10.4. XY plane partitioned by $x = \bar{x}$ and $y = \bar{y}$ into four quadrants.

relationship between x and y (i.e., a trend of points, for the most part, either through quadrants I and III, or II and IV, the algebraic sum of the products $(x_i - \bar{x})(y_i - \bar{y})$ tends to be large in magnitude. There is only a comparatively small cancellation effect on the summation. If the (x_i, y_i) are scattered over all quadrants, the cancellation effect is larger when the products $(x_i - \bar{x})(y_i - \bar{y})$ are added, and thus the sum of the products is comparatively small. Even though the (x_i, y_i) are, for the most part, either in quadrants I and III or II and IV, if they are greatly dispersed about the regression line (a weak relationship) and thus occurring closer to $x = \bar{x}$ and $y = \bar{y}$, the magnitude of the individual deviations is smaller and, consequently, the sum of the products of the deviations is small.

The sum of products of deviations $x - \bar{x}$ and $y - \bar{y}$ is large for a strong linear relationship and small for a weak relationship. Thus, this sum measures degree of relationship. Division of the sum by n yields an average product, and division of this average product by s_x and s_y frees the measure from the original units. The resulting measure

$$r = \frac{\frac{1}{n}\sum_{i=1}^{n}(x_i - \bar{x})(y_i - \bar{y})}{s_x s_y} \tag{10.15}$$

is called the sample *correlation coefficient* and is designated by the symbol r. The corresponding population coefficient is denoted by ρ. The range of either ρ or r is $-1 \leq r \leq +1$. A positive correlation $(+r)$ indicates that as X increases, Y also increases. A negative correlation describes the relationship of Y decreasing with increasing X. If r is near one, the variables are said to be highly correlated. An r-value near zero indicates low correlation.

The correlation coefficient is unaffected by linear transformations. Thus, when it is advantageous to work in u-units (as in Example 9.2.1), the correlation coefficient may be computed in (u, y) units. Also, to avoid tedious computation involving deviations, the following computational form of the correlation coefficient is available:

$$r = \frac{n\sum x_i y_i - \sum x_i \sum y_i}{\sqrt{[n\sum x_i^2 - (\sum x_i)^2][n\sum y_i^2 - (\sum y_i)^2]}} \tag{10.16}$$

As in regression analysis, the correlation coefficient measures degree of relationship (or "association" or "co-variation") and in no way guarantees the existence of a cause-and-effect relationship. Clearly, regression and correlation are closely related. There are differences, however, in interpretation. Correlation analysis, independent of regression, assumes that both X and Y are random variables as indicated by Fig. 10.5 where the regression line is viewed as joining the mean values of regression *surfaces*. On the other hand,

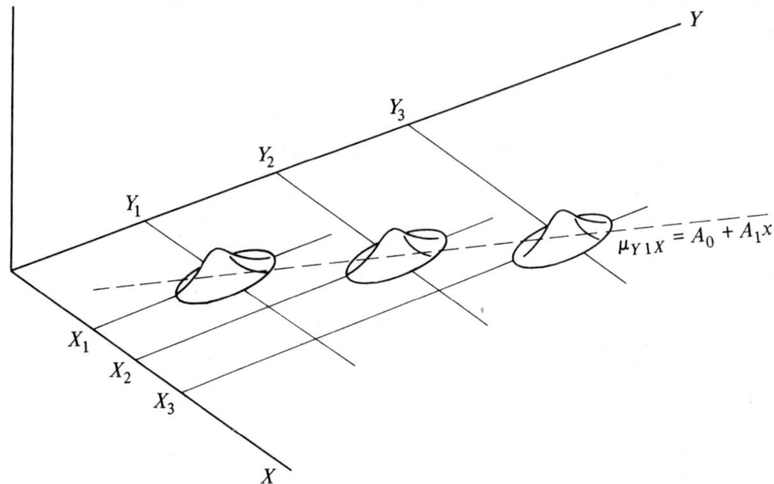

Figure 10.5. Regression surfaces.

when correlation is used jointly with regression analysis, the correlation coefficient is used as a measure of degree (or "significance") of relationship.

A useful relationship for expressing goodness of fit of the regression line, or degree of linear relationship between X and Y, is

$$r^2_{y|x} = 1 - \frac{s^2_{y|x}}{s^2_y} \tag{10.17}$$

where the estimates s^2_y and $s^2_{y|x}$ are defined by Eqs. (10.5) and (10.6), respectively, and the ratio

$$\frac{s^2_{y|x}}{s^2_y}$$

is an estimate of that fraction or proportion of the total variance not explainable by regression. To illustrate, consider a correlation coefficient $r = 0.6$. Then, from Eq. (10.17),

$$1 - (0.6)^2 = 0.64 = \frac{s^2_{y|x}}{s^2_y}$$

and

$$s_{y|x} = 0.8s_y$$

The standard deviation $s_{y|x}$ which measures unexplained variation (i.e., of the y_i from the corresponding y's on the regression line) has been reduced by an amount equal to only $0.2s_y$. The regression line is only slightly better than the line $y = \bar{y}$ for estimating Y, given a specified $X = X_0$.*

EXAMPLE 10.7.1

Referring to Example 10.2.1, $s^2_{\bar{y}|x} = 6.75$ and the total SS for Y is 850.11. Thus, from Eq. (10.17), the correlation coefficient is

$$r = \sqrt{1 - \frac{s^2_{\bar{y}|x}}{s^2_y}} = \sqrt{1 - \frac{6.75}{850.11/5}} = 0.98$$

EXERCISES

1. Compute a correlation coefficient for the data of Exercise 1, Sec. 10.2. *Ans.* 0.99.

2. What is the proportion of the total variance in Exercise 1 that is not explained by regression? *Ans.* 0.02.

10.8 Inferences Regarding

Common hypothesis tests concerning the correlation coefficient are H_0: $\rho = \rho_0 \neq 0$ (testing that ρ is as small as some specified value ρ_0) and H_0: $\rho = 0$. The latter hypothesis is equivalent to testing that the sample correlation coefficient $r \neq 0$ is due to sampling error and there is, in fact, no correlation between X and Y.

Assuming that X and Y are both random variables, jointly normally distributed, the quantity

$$z = 0.5 \log_e \frac{1 + r}{1 - r} \tag{10.18}$$

is approximately normally distributed with mean

$$\mu_Z = 0.5 \log_e \frac{1 + \rho}{1 - \rho} \tag{10.19}$$

and variance

$$\sigma^2_Z = \frac{1}{n - 3} \tag{10.20}$$

Thus, the statistic for testing the hypothesis H_0: $\rho = \rho_0 \neq 0$ is

*If there is no relationship between X and Y, the best estimate of Y for any specified X is $y = \bar{y}$ with variance s^2_y, admittedly a poor estimate.

The implication from this example ($r = 0.6$) is that a comparatively large value of r is required to give assurance of good estimates using the regression line and knowledge of X.

$$\frac{z_j - \mu_Z}{\sigma_Z} = \frac{z_j - 0.5 \log_e \dfrac{1 + \rho_0}{1 - \rho_0}}{\dfrac{1}{\sqrt{n - 3}}} \tag{10.21}$$

and the decision rule is

$$\left| \frac{z_{\text{data}} - \mu_Z}{\sigma_Z} \right| \geq Z_{\alpha/2} \qquad \text{reject } H_0{}^*$$

EXAMPLE 10.8.1

Given a sample correlation coefficient $r = 0.75$ based on 10 sample observations, test at a 5% significance level the hypothesis H_0: $\rho = 0.85$.

From Eqs. (10.18)–(10.21),

$$z_{\text{data}} = 0.5 \log_e \frac{1 + 0.75}{1 - 0.75} = 0.97296$$

$$\mu_Z = 0.5 \log_e \frac{1 + 0.85}{1 - 0.85} = 1.25602\dagger$$

and the sample statistic is

$$\left| \frac{0.97296 - 1.25602}{1/\sqrt{7}} \right| = 0.75$$

Since

$$0.75 < Z_{\alpha/2} = 1.96,$$

the hypothesis is accepted.

The test of a hypothesis H_0: $\rho = 0$ is a simpler test to execute. If $\rho = 0$, the statistic

$$t_j = r\sqrt{\frac{n - 2}{1 - r^2}} \tag{10.22}$$

is t-distributed with $n - 2$ *d.f.* and a standard t-test is appropriate. The decision rule is

$$|t_{\text{data}}| = \left| \frac{r}{\sqrt{1 - r^2}} \right| \sqrt{n - 2} \geq t_{(\alpha/2), n-2}, \qquad \text{reject } H_0.$$

EXAMPLE 10.8.2

Given $r = 0.6$ based on 10 observations, test at a 5% significance level the hypothesis H_0: $\rho = 0$.

---

*Since z [Eq. (10.18)] is only approximately normally distributed, this test is approximate. It is accurate for sample sizes $n \geq 50$. Closer approximations, some of which are reasonably accurate for n as small as 11, are found in Kendall, M. G. and Stuart, A. (1963), *The Advanced Theory of Statistics*, 2nd ed., Charles Griffin & Co. Ltd., London.

†If tables of natural logarithms are not available, use common logarithms and the following relationship:

$$\log_{10} N \cong 2.3026 \log_e N$$

From Eq. (10.22),

$$|t_{\text{data}}| = \left|\frac{0.6}{\sqrt{1-(0.6)^2}}\right|\sqrt{10-2} = 2.121$$

Since

$$t_{\text{data}} = 2.121 < t_{0.025,8} = 2.306$$

the hypothesis is accepted.*

The approximately normally distributed random variable given by Eq. (10.18) can be utilized to generate a $100(1-\alpha)$ confidence interval estimate of ρ. Thus,

$$0.5 \log_e \frac{1+r}{1-r} \pm t_{(\alpha/2),\infty}\left(\frac{1}{\sqrt{n-3}}\right) \tag{10.23}$$

affords a $100(1-\alpha)$ confidence interval estimate of μ_Z where

$$\mu_Z = 0.5 \log_e \frac{1+\rho}{1-\rho}$$

Equating this expression for μ_Z to the interval boundaries generated by Eq. (10.23) yields the corresponding confidence interval estimate of ρ.

EXAMPLE 10.8.3

Given $r = 0.80$ based on 19 sample observations, determine a 95% confidence interval estimate of ρ. The confidence interval estimate of μ_Z is

$$0.5 \log_e \frac{1+0.8}{1-0.8} \pm 1.96\left(\frac{1}{\sqrt{19-3}}\right) = 1.0986 \pm 0.49 = (0.6086, 1.5886)$$

and, if

$$0.5 \log_e \frac{1+\rho}{1-\rho} = 0.6086, 1.5886$$

then, a solution for ρ gives 95% confidence limits of 0.54 and 0.92.

EXERCISES

1. Given $r = 0.84$ based on 12 sample observations, test at a 1% significance level the hypothesis $H_0\colon \rho = 0.9$. *Ans.* $z_{\text{data}} = 0.75 < Z_{0.005}$, accept H_0.
2. Given $r = 0.36$ based on 18 observations, test at a 5% significance level the hypothesis $H_0\colon \rho = 0$. *Ans.* $t_{\text{data}} = 1.54 < t_{0.025,16}$, accept H_0.
3. Determine a 95% confidence interval estimate of ρ for Exercise 1. *Ans.* (0.51, 0.95).

---

*This conclusion indicates only that there is no simple linear correlation. There may be a multiple linear or even nonlinear correlation (see Secs. 10.9 and 10.11).

10.9 Multiple Linear Regression

Frequently, engineering applications involving regression require more than one independent variable. If Y depends linearly on several X-factors, it may be best to include all of them in a regression model of the form

$$\mu_{Y|X_1|\ldots|X_k} = A_0 + A_1X_1 + A_2X_2 + \ldots + A_kX_k \tag{10.24}$$

where the sample regression line is

$$y = a_0 + a_1x_1 + a_2x_2 + \ldots + a_kx_k \tag{10.25}$$

The estimates $a_0, a_1, \ldots, a_k$ of the population parameters are obtained by the least squares method and, if the regression equation is used for making inferences, then the same assumptions of Sec. 10.1 hold.

An expression analogous to Eq. (10.4) can be derived for the sum of squares of deviations about the regression line, Eq. (10.25). A derivation similar to that for Eq. (10.4) yields

$$\begin{aligned}\sum (y_i - y)^2 = \sum y_i^2 - a_0 \sum y_i - a_1 \sum y_i x_{1i} \\ - a_2 \sum y_i x_{2i} - \ldots - a_k \sum y_i x_{ki}\end{aligned} \tag{10.26}$$

and an unbiased estimate of that part of the total variance that is not explained by regression is

$$\begin{aligned}s^2_{y|x_1|\ldots|x_k} = \frac{1}{n-k-1}[\sum y_i^2 - a_0 \sum y_i - a_1 \sum y_i x_{1i} \\ - a_2 \sum y_i x_{2i} - \ldots - a_k \sum y_i x_{ki}]\end{aligned} \tag{10.27}$$

Also, the sample multiple linear correlation coefficient squared is given by

$$r^2_{y,x_1,\ldots,x_k} = 1 - \frac{s^2_{y|x_1|\ldots|x_k}}{s^2_y} \tag{10.28}$$

Longhand computations are very tedious for multiple regression. An example with few observations (and small-number values for the observations) is presented here to illustrate method. A computer solution of the same example appears in Chap. 12.

The overhead cost problem of Example 10.2.1 is developed further. In another department involving the same draw-die operations, overhead expenses are linearly related to both lineal feet of direct material X_1 and direct labor expense X_2. Again, each X_1-unit corresponds to 500 lineal feet of material, and each X_2-unit and Y-unit to \$1,000 of expense. It is desired to include both X_1 and X_2 in a regression equation for the purpose of estimating overhead charges for specific jobs.

EXAMPLE 10.9.1

The desired linear regression model is given by the equation

$$\mu_{Y|X_1|X_2} = A_0 + A_1 X_1 + A_2 X_2$$

and the sample regression line is

$$y = a_0 + a_1 x_1 + a_2 x_2$$

Following the least squares pattern given by Eq. (9.7), the *LSNE* are

$$\begin{aligned}
\sum y_i &= \sum a_0 + a_1 \sum x_{1i} + a_2 \sum x_{2i} \\
\sum y_i x_{1i} &= a_0 \sum x_{1i} + a_1 \sum x_{1i} x_{1i} + a_2 \sum x_{2i} x_{1i} \\
\sum y_i x_{2i} &= a_0 \sum x_{2i} + a_1 \sum x_{1i} x_{2i} + a_2 \sum x_{2i} x_{2i}
\end{aligned}$$

Referring x_{1i} and x_{2i} to their respective means, the following terms vanish:

$$\sum x_{1i} = 0, \qquad \sum x_{2i} = 0$$

and thus a longhand solution of the *LSNE* is not prohibitively laborious. The observations (x_{1i}, x_{2i}, y_i) are:

x_{1i}	x_{2i}	y_i
0	0	2.0
1	2	5.0
2	2	5.6
3	3	7.4
4	3	8.1
5	4	9.0

and sums of products required for a solution of the *LSNE* are:

u_1	u_2	y	u_1u_1	u_1u_2	u_2u_2	yu_1	yu_2
−5	−7	2.0	25	35	49	−10.0	−14.0
−3	−1	5.0	9	3	1	−15.0	−5.0
−1	−1	5.6	1	1	1	−5.6	−5.6
1	2	7.4	1	2	4	7.4	14.8
3	2	8.1	9	6	4	24.3	16.2
5	5	9.0	25	25	25	45.0	45.0
0	0	37.1	70	72	84	46.1	51.4

where the linear transformations are

$$u_{1i} = \frac{x_{1i} - 2.5}{0.5}, \qquad u_{2i} = \frac{x_{2i} - \frac{7}{3}}{\frac{1}{3}}$$

The *LSNE* and solutions for a_0', a_1', and a_2' are

$$a_0' = \frac{1}{n} \sum y_i = \frac{1}{6}(37.1) = 6.183$$

$$46.1 = 70a_1' + 72a_2'$$

$$51.4 = 72a_1' + 84a_2'$$

whence

$$a_0' = 6.183, \qquad a_1' = 0.246, \qquad a_2' = 0.401$$

and the regression equation in u-units is

$$y = 6.183 + 0.246u_1 + 0.401u_2$$

or

$$y = 6.183 + 0.246\left[\frac{x_1 - 2.5}{0.5}\right] + 0.401\left[\frac{x_2 - \frac{7}{3}}{\frac{1}{3}}\right]$$

$$= 2.146 + 0.493x_1 + 1.202x_2$$

The total sum of squares, from Eq. (1.21), is

$$\sum (y_i - \bar{y})^2 = \sum y_i^2 - n\bar{y}^2$$

$$= 261.73 - 6(6.183)^2$$

$$= 32.35$$

and, from Eq. (10.26),

$$\sum (y_i - y)^2 = \sum y_i^2 - a_0 \sum y_i - a_1 \sum yx_{1i} - a_2 \sum yx_{2i}$$

$$= 261.73 - 2.146(37.1) - 0.493(115.8) - 1.202(103.7)$$

$$= 0.38$$

A summary of the sums of squares and mean squares is

Source	*d.f.*	*SS*	*MS*
Regression	2	31.97	15.98
Deviations from regression	3	0.38	0.13
Total	5	32.35	

and since

$$F_{\text{data}} = \frac{15.98}{0.13} = 122.9 > F_{0.01,2,3} = 30.8$$

the regression is significant at an $\alpha = 0.01$ level.

The correlation coefficient is computed as follows. From Eq. (10.27),

$$s^2_{y|x_1|x_2} = \frac{1}{n-3} \sum (y_i - y)^2$$
$$= \tfrac{1}{3}(0.38) = 0.13$$

and, from Eq. (10.28),

$$r^2_{y,x_1,x_2} = 1 - \frac{s^2_{y|x_1|x_2}}{s^2_y}$$
$$= 1 - \frac{0.13}{32.35/5} = 0.98^*$$

10.10 Significance of Addition of a Variable to the Regression Equation

It is also possible to test the significance of the addition of another independent variable to the regression model. In Example 10.9.1, fitting a regression line $y = a_0 + a_1x_1 + a_2x_2$ produces the following sums of squares:

1. Total *SS*:

$$\sum (y_i - \bar{y})^2 = \sum y_i^2 - n\bar{y}^2$$

2. *SS* due to deviations from regression $y|x_1|x_2$:

$$\sum (y_i - y)^2 = \sum y_i^2 - a_0 \sum y_i - a_1 \sum y_i x_{1i} - a_2 \sum y_i x_{2i}$$

3. *SS* due to regression $y|x_1|x_2$:

$$\sum (y - \bar{y})^2 = \text{Total } SS - [\sum y_i^2 - a_0 \sum y_i - a_1 \sum y_i x_{1i} - a_2 \sum y_i x_{2i}]$$

Fitting a regression line $y = a_0 + a_1x_1$ to the same data yields the following sum of squares:

---

*Results such as this (very high correlation) are not to be expected in applications. Examples in this chapter have been designed with very small intrinsic variation, hence small deviations, and thus reasonable arithmetic in order to illustrate method. Generally, regression problems are not handled by longhand computation. Computer and calculator routines are followed since the computation is formidable.

4. *SS* due to regression $y|x_1$:

$$\sum (y - \bar{y})^2 = \text{Total } SS - [\sum y_i^2 - a_0 \sum y_i - a_1 \sum y_i x_{1i}]$$

The sum of squares due to the addition of X_2 to the regression model is (3) — (4) and if the ratio of mean squares

$$\frac{MS[(3) - (4)]}{MS(2)} > F_{\alpha, 1, n-3}$$

the addition of X_2 to the regression equation is significant. This argument is conveniently summarized in a sum of squares table in the following example.

EXAMPLE 10.10.1

Assume that a regression line $y = a_0 + a_1x_1$ is fitted to the observations of Example 10.9.1. The resulting sums of squres are

Source	*d.f.*	*SS*	*MS*
Regression $y\|x_1$	1	30.37	30.37
Deviations from regression	4	1.98	0.49
Total	5	32.35	

The sum of squares due to the addition of X_2 to the regression model is

$$31.97 - 30.37 = 1.60$$

and this is also the amount of reduction in the *SS* due to deviations from regression. A summary of the results of the addition of X_2 to the regression model are

Source	*d.f.*	*SS*	*MS*
Regression $y\|x_1$	1	30.37	
Addition of x_2	1	1.60	1.60
Regression $y\|x_1\|x_2$	2	31.97	
Deviations from regression	3	0.38	0.13
Total	5	32.35	

and

$$\frac{1.60}{0.13} > F_{0.05,1,3} = 10.1$$

and thus the addition of X_2 to the regression model is significant at an $\alpha = 0.05$ level.

As in simple linear regression and correlation, hypotheses are tested and confidence interval estimates are possible for the parameters a_j, the means $\mu_{Y|X_j}$, and the correlation coefficient ρ. However, the computations are tedious and are usually handled by calculator and computer routines. A calculator readout strip of the complete solution of the small numbers problem of Example 10.9.1 is given in Chap. 12.

EXERCISES

1. Perform the regression $y|x_1$ analysis and verify the tabled results for Example 10.10.1. *Ans.* $y = 2.8905 + 1.3171x$.

2. Fit $y = a_0 + a_1x_1$ and $y = a_0 + a_1x_1 + a_2x_2$ to the following data. Test at an $\alpha = 0.05$ level for significance of regression in each case. Test for significance of addition of X_2 to the regression model.

x_1	0	1	2	3	4	5
x_2	0	2	2	3	3	4
y	2.0	5.0	4.6	8.4	6.2	10.0

Ans. $\frac{4.72}{1.31} = F_{0.05,1,3} = 10.1$, addition of X_2 is not significant.

10.11 Additional Remarks—Regression and Correlation

Polynomials of degree k (see Sec. 9.3) occur frequently in engineering applications. A common regression model is a polynomial in one variable,

$$\mu_{Y|X} = A_0 + A_1X + A_2X^2 + \ldots + A_kX^k$$

or, the model may be of degree $k > 1$ and multiple in the variables. For example, a second-degree polynomial model in two variables is

$$\mu_{Y|X_1|X_2} = A_0 + A_1X_1 + A_2X_2 + A_3X_1^2 + A_4X_2^2 + A_5X_1X_2$$

Although models of this type are nonlinear in the X_j, they are linear in the unknown parameters A_j and least squares estimates of the A_j (and inferences) are obtainable by the methods of Sec. 10.9. Many such relationships can be analyzed by transforming the original variables into new ones that are related linearly.*

---

*For example, a cubic regression model

$$\mu_{Y|X} = A_0 + A_1X + A_2X^2 + A_3X^3$$

is transformed by letting $X_1' = X$, $X_2' = X^2$, and $X_3' = X^3$. The X_1', X_2', and X_3' are not statistically independent, but independence is not necessary in multiple regression analysis.

In many applications, the experimenter either does not know the exact form of the regression model beforehand or he is uncertain as to how many of the independent variables to include in the model. There are a number of computer techniques known as *stepwise regression* procedures which select independent variables one at a time for inclusion in the regression model, their selection being dependent on their correlation with the response variable. For example, inclusion of the $(k + 1)$th independent variable into the regression equation is desirable if its inclusion increases the value of the multiple correlation coefficient over that for a model with only k independent variables included. Generally, a variable X_{k+1} is a potential improvement for the regression equation if it is not highly correlated with any of the k independent variables already in the equation, and is correlated with Y. However, there is always the chance that the addition of X_{k+1} to the equation does not reduce the sum of squares of deviations from regression sufficiently to offset the loss of a degree of freedom (due to the inclusion of X_{k+1}).

A graphing of the observations is an indispensable tool in regression analysis, particularly if sample size n is small. Erratic and isolated observations can exert undue influence on the least squares fitting procedure. For example, the relationship exhibited by Fig. 10.6 may, in fact, follow approximately a uniform model. Yet the isolated observation influences the fitted line away from the line $y = \bar{y}$ for a uniform distribution (also, to exaggerate, recall from a previous exercise, that a perfect fit with $r = 1$ is obtained with $n = 2$ observations).*

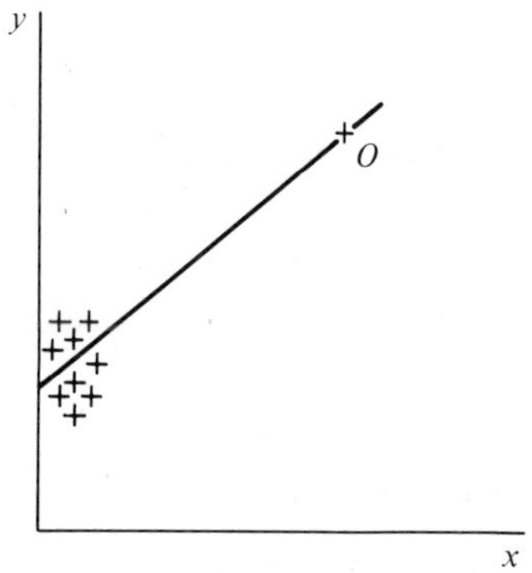

Figure 10.6. An isolated observation point O and the regression line.

*The sample sizes for the illustrative examples of this chapter have been impractically small. However, small sample sizes permitted illustration of method without the encumbrance of tedious arithmetic.

GLOSSARY OF TERMS

Regression
Regression model
Regression line
Simple linear regression
Association
Co-variation
Significance of regression
Sum of squares (*SS*)
Mean square (*MS*)
Total sum of squares
SS due to regression
SS due to deviations from regression
Error variance
Standard error of estimate
Correlation
Correlation coefficient
Multiple linear regression
Stepwise regression

GLOSSARY OF FORMULAS

Page

(10.1) Simple linear regression model. 320

$$\mu_{Y|X} = A_0 + A_1 X$$

(10.2) Sample regression line. 320

$$y = a_0 + a_1 x$$

(10.3) Partitioning the total sum of squares. 322

$$\sum (y_i - \bar{y})^2 = \sum (y - \bar{y})^2 + \sum (y_i - y)^2$$

(10.4) Sum of squares due to deviations from regression. 323

$$\sum (y_i - y)^2 = \sum y_i^2 - a_0 \sum y_i - a_1 \sum y_i x_i$$

(10.5) Unbiased estimate of the total variance of Y. 325

$$s_y^2 = \frac{1}{n-1} \sum (y_i - \bar{y})^2 = \frac{1}{n-1} [\sum y_i^2 - n\bar{y}^2]$$

(10.6) Unbiased estimate of the variance of Y, given X. 326

$$s_{y|x}^2 = \frac{1}{n-2} \sum (y_i - y)^2 = \frac{1}{n-2} [\sum y_i^2 - a_0 \sum y_i - a_1 \sum y_i x_i]$$

(10.7) Eq. (10.6) with the origin shifted from (0, 0) to $(\bar{x}, \bar{y})$. 326

$$s_{y|x}^2 = \frac{1}{n-2} [\sum y_i^2 - a_1 \sum y_i x_i]$$

(10.8) Estimated variance of sample regression coefficient a_1. 326

$$s_{a_1}^2 = \frac{s_{y|x}^2}{\sum (x_i - \bar{x})^2}$$

(10.9) Estimated variance of sample regression coefficient a_0. 326

$$s_{a_0}^2 = s_{y|x}^2 \left[\frac{1}{n} + \frac{\bar{x}^2}{\sum (x_i - \bar{x})^2}\right]$$

(10.10) A $100(1-\alpha)$ confidence interval estimates of A_0 and A_1. 328

$$a_0 \pm t_{(\alpha/2),n-2}(s_{a_0})$$
$$a_1 \pm t_{(\alpha/2),n-2}(s_{a_1})$$

(10.11) Estimate of the variance of Y, given $X = X_0$. 329

$$s^2_{y|x}\left[\frac{1}{n} + \frac{(x_0 - \bar{x})^2}{\sum (x_i - \bar{x})^2}\right]$$

(10.12) A $100(1-\alpha)$ confidence interval estimate of $\mu_{Y|X}$. 329

$$a_0 + a_1 x_0 \pm t_{(\alpha/2),n-2}\left[s_{y/x}\sqrt{\frac{1}{n} + \frac{(x_0 - \bar{x})^2}{\sum (x_i - \bar{x})^2}}\right]$$

(10.13) Estimate of the variance of Y_0, given $X = X_0$. 330

$$s^2_{y_0} = s^2_{y|x}\left[1 + \frac{1}{n} + \frac{(x_0 - \bar{x})^2}{\sum (x_i - \bar{x})^2}\right]$$

(10.14) A $100(1-\alpha)$ confidence interval estimate of Y_0, given $X = X_0$. 330

$$a_0 + a_1 x_0 \pm t_{(\alpha/2),n-2}\left[s_{y|x}\sqrt{1 + \frac{1}{n} + \frac{(x_0 - \bar{x})^2}{\sum (x_i - \bar{x})^2}}\right]$$

(10.15) Sample correlation coefficient. 334

$$r = \frac{\frac{1}{n}\sum (x_i - \bar{x})(y_i - \bar{y})}{s_x s_y}$$

(10.16) Computational form of Eq. (10.15). 334

$$r = \frac{n\sum x_i y_i - \sum x_i \sum y_i}{\sqrt{[n\sum x_i^2 - (\sum x_i)^2][n\sum y_i^2 - (\sum y_i)^2]}}$$

(10.17) Degree of linear relationship between x and y. 335

$$r^2_{y,x} = 1 - \frac{s^2_{y|x}}{s^2_y}$$

(10.18) A normally distributed random variable. 336

$$z = 0.5 \log_e \frac{1+r}{1-r}$$

(10.19) Mean of the random variable, Eq. (10.18). 336

$$\mu_Z = 0.5 \log_e \frac{1+\rho}{1-\rho}$$

(10.20) Variance of the random variable, Eq. (10.18). 336

$$\sigma^2_Z = \frac{1}{n-3}$$

(10.21) Statistic for testing H_0: $\rho = \rho_0 \neq 0$ 337

$$\frac{z_j - \mu_Z}{\sigma_Z} = \frac{z_j - 0.5 \log_e \frac{1+\rho_0}{1-\rho_0}}{\frac{1}{\sqrt{n-3}}}$$

(10.22) Statistic for testing H_0: $\rho = 0$. 337

$$t_j = r\sqrt{\frac{n-2}{1-r^2}}$$

(10.23) A $100(1-\alpha)$ confidence interval estimate of μ_Z. 338

$$0.5 \log_e \frac{1+r}{1-r} \pm t_{(\alpha/2),\infty}\left(\frac{1}{\sqrt{n-3}}\right)$$

(10.24) Multiple linear regression model. 339

$$\mu_{Y|X_1|\dots|X_k} = A_0 + A_1X_1 + A_2X_2 + \dots + A_kX_k$$

(10.25) Sample regression line. 339

$$y = a_0 + a_1x_1 + a_2x_2 + \dots + a_kx_k$$

(10.26) Sum of squares due to deviations from regression. 339

$$\sum (y_i - y)^2 = \sum y_i^2 - a_0 \sum y_i - a_1 \sum y_i x_{1i} - a_2 \sum y_i x_{2i} - \dots - a_k \sum y_i x_{ki}$$

(10.27) Unbiased estimate of the variance of Y, given X. 339

$$s^2_{y|x_1|\dots|x_k} = \frac{1}{n-k-1}\left[\sum y_i^2 - a_0 \sum y_i - a_1 \sum y_i x_{1i} - a_2 \sum y_i x_{2i} - \dots - a_k \sum y_i x_{ki}\right]$$

(10.28) Sample multiple linear correlation coefficient. 339

$$r_{y,x_1,\dots,x_k} = \sqrt{1 - \frac{s^2_{y|x_1|\dots|x_k}}{s_y^2}}$$

REFERENCES

DUNCAN, A. J. (1965), *Quality Control and Industrial Statistics*, 3rd ed., Richard D. Irwin, Inc., Homewood, Ill., Chaps. 32–34.

GUTTMAN, I. AND WILKS, S. S. (1971), *Introductory Engineering Statistics*, John Wiley & Sons, Inc., New York, Chap. 15.

HUNTSBERGER, D. V. (1967), *Elements of Statistical Inference*, 2nd ed., Allyn and Bacon, Inc., Boston, Mass., Chap. 11.

LI, JEROME, C. R. (1964), *Statistical Inference*, Edwards Brothers, Inc., Ann Arbor, Mich., Chaps. 16–17.

VOLK, W. (1969), *Applied Statistics for Engineers*, 2nd ed., McGraw-Hill Book Company, New York, Chap. 9.

PROBLEMS

1. Explain what is being measured by each term of the following expression:

$$\sum (y_i - \bar{y})^2 = \sum (y - \bar{y})^2 + \sum (y_i - y)^2$$

2. Compute the respective *SS*'s for Problem 9.15. What part of the total *SS* is explainable by regression? What is the physical meaning of this result?

3. Prepare a summary table for the respective *SS*'s in Problem 9.4 and test for significance of regression at a 1% significance level.

4. Prepare a summary table for the respective *SS*'s in Problem 9.5 and test at a 5% level for significance of regression.

5. Compute the estimated variances $s_{a_0}^2$ and $s_{a_1}^2$ for Problem 9.4.

6. Test at a 10% significance level, for Problem 9.4, the following hypotheses: $H_0: A_1 = 0$ and $H_1: A_1 = 0$.

7. Determine 95% confidence interval estimates of A_0 and A_1 for Problem 9.4.

8. Determine a 95% confidence interval estimate of $\mu_{Y|X}$ for $x = 8$ in Problem 9.4.

9. Determine a 95% confidence interval estimate of $\mu_{Y|X}$ for $x = 2$ in Problem 9.4. Explain the difference in interval magnitude between that of the preceding problem and the one here.

10. Determine a 95% confidence interval estimate of an individual Y: (a) given that $x = 2$, and (b) given that $x = 8$ for Problem 9.4.

11. Using Eq. (10.15), compute the correlation coefficient value for Problem 9.15. Explain the meaning of your result.

12. Compute an r-value for Problem 9.4 using: (a) Eq. (10.16), (b) Eq. (10.17).

13. What is the proportion of the total variance in Problem 9.4 that is not explainable by regression?

14. Test at a 5% significance level, for Problem 9.4, the following hypothesis: $H_0: \rho = 0$.

15. Prepare summary tables for the respective *SS*'s and test at a 5% level for significance of regression for the two cost-output relationships of Problem 9.19.

16. Compute r and determine the proportion of the total variance in Problem 9.19 that is explainable by regression. Test at a 5% significance level the hypothesis $H_0: \rho = 0$. Compare your analysis of Problem 15 with that here.

17. Determine 95% confidence intervals for $\mu_{Y/X}$ in Problem 9.19. Estimate the respective mean costs for products A and B if both are at the same output level $x = 18$. Which cost estimate do you consider to be the better one? State your criterion and explain.

18. Test at a 1% level for significance of regression in Problem 9.23. Do you feel that the less costly inspection method x can be safely substituted for the standard procedure?

19. Test at a 1% level for significance of regression in Problem 9.24. What is your conclusion regarding choice of measurement procedure?

20. An experimental comparison of tool life, with and without compensation of thermal E.M.F., yields the following two sets of observations. Graph each relationship between cutting time x_i (in min.) and wear land y_i (10^{-2} mm). Compute a correlation coefficient for each relationship.

A	x_i	1	2	3	4	8	10	12	15
	y_i	8	13	18	21	34	42	49	56

B	x_i	2	4	7	15	20
	y_i	9	12	18	25	28

21. Write the *LSNE* for a sample regression line

$$y = a_0 + a_1x_1 + a_2x_2 + a_3x_3.$$

22. Write the computational expression for $\sum (y_i - y)^2$, for Problem 21, corresponding to Eq. (10.4) for simple linear regression.

23. Write the expression for the variance of y given x, for Problem 21, corresponding to Eq. (10.7) for simple linear regression. Then, state the computational expression for r^2 corresponding to Eq. (10.17) for simple linear regression.

24. Two production factors X_1 and X_2 are being considered as a basis for estimating overhead charges Y (see Example 10.9.1). Accuracy of overhead estimates depends on the degree of relationship between X_1, X_2 and Y. The following data is coded in dollar units.

x_{1i}	0	0	1	1	2	2	3	3	4	4
x_{2i}	1	2	1	2	2	3	2	3	3	4
y_i	0.6	0.7	1.1	1.3	2.0	2.2	2.7	3.1	3.4	3.5

(a) Fit a relationship $y = a_0 + a_1x_1$, test for significance of regression at an $\alpha = 0.05$ level, and compute r^2.

(b) Fit a relationship $y = a_0 + a_1x_1 + a_2x_2$, test for significance of regression, and compute r^2.

25. Prepare a summary table for Problem 24, developing the appropriate *SS*'s, *MS*'s, and *F* test for significance of the addition of x_2 to the regression equation.

26. Cost relationships between expenses X_1 and X_2 and total variable cost Y are being studied. Fit a relationship $y = a_0 + a_1x_1$ to the following coded cost data. Test for significance of regression at an $\alpha = 0.05$ level. Then, add x_2 to the regression equation and again test for significance of regression.

x_{1i}	0.05	0.05	0.06	0.06	0.07	0.09	0.10	0.10	0.11	0.13
x_{2i}	1.0	1.0	1.1	1.2	1.2	1.4	1.5	1.2	1.1	1.2
y_i	0.30	0.34	0.39	0.36	0.41	0.48	0.46	0.50	0.56	0.61

27. Prepare a summary table for Problem 26, developing the appropriate *SS*'s, *MS*'s, and *F* test for significance of the addition of x_2 to the regression equation.

28. Production variables X_1 and X_2 seem to be related to a certain quality improvement Y. Assess this multiple linear relationship by fitting $y = a_0 + a_1x_1 + a_2x_2$ to the following coded data and testing at an $\alpha = 0.05$ level for significance of regression.

x_{1i}	0.5	0.5	0.5	1.0	1.5	1.5	2.0	2.5	2.5	2.5
x_{2i}	6.5	7.5	9.0	10.0	5.5	12.5	7.0	8.0	8.0	11.0
y_i	0	3	39	75	1	57	12	21	46	27

29. Prepare a summary table for an *F* test of the significance of addition of x_2 to the regression equation in Problem 28.

30. One or both of two nondestructive tests X_1 and X_2 are being considered as a replacement for a costly standard destructive test Y. One test (either X_1 or X_2) is cheaper than using both. However, using both as a replacement for Y is still considerably cheaper than Y. If Y is replaced, good estimates of Y are required, however, for certification purposes. What is your decision based on the following coded data?

x_{1i}	0.2	0.2	0.2	0.2	1.0	1.0	1.0	1.0	1.8	1.8	1.8	1.8
x_{2i}	10	11	12	13	10	11	12	13	10	11	12	13
y_i	4.51	4.43	6.40	6.71	4.42	4.64	5.85	6.98	4.75	4.90	6.08	7.45

ANALYSIS OF VARIANCE

The analysis of variance *is an arithmetic device for partitioning the total variation exhibited by sample observations according to the various sources of variation that are present, thus permitting hypothesis tests regarding the respective variation sources. An analysis of variance results in a summary table similar to that for Example 10.2.1. In fact, this example is an analysis of variance for simple linear regression.*

Testing the significance of regression is based on partitioning the total sum of squares. For example, in regression analysis the partitioning is

$$\sum (y_i - \bar{y})^2 = \sum (y - \bar{y})^2 + \sum (y_i - y)^2 \qquad (10.3)$$

where (x_i, y_i) *designates a pair of observations on a sample element and* (x, y) *denotes the corresponding point on the regression line. The sum of squares* $\sum (y - \bar{y})^2$ *measures that part of the total* SS *explainable by regression, and* $\sum (y_i - y)^2$ *represents that part of the total* SS *that is not explained.*

11

11.1 Single Variable, Completely Randomized Design

The most simple analysis-of-variance model is a completely randomized, one-variable design. With this model, attention is on a comparison of the *effects of treatments* T_j.* Letting x_{ij} designate the ith observed effect of the jth treatment, $\bar{x}_{.j}$ the sample mean of the jth treatment effect, and $\bar{x}_{..}$ the mean of all observations x_{ij}, the total variation as measured by the total SS can be partitioned as follows:

$$\sum_j \sum_i (x_{ij} - \bar{x}_{..})^2 = \sum_j \sum_i (\bar{x}_{.j} - \bar{x}_{..})^2 + \sum_j \sum_i (x_{ij} - \bar{x}_{.j})^2 \qquad (11.1)$$

Equations (10.3) and (11.1) differ with respect to experiment objectives.

---

*In analysis-of-variance language, *treatments* are any procedures, methods, types, etc. (independent variables) whose *effects* on a response variable (dependent variable) are to be measured and compared.

Treatments in one experiment are different process machines, in another they are machine operators, in another they can be different raw material types, and so forth.

Whereas Eq. (10.3) is concerned with improvement of regression-line estimates (Secs. 10.2 and 10.10 considerations), Eq. (11.1) is a basis for testing the significance between treatments T_j. However, the two equations reflect remarkably similar analyses. The sum of squares

$$\sum_j \sum_i (\bar{x}_{\cdot j} - \bar{x}_{\cdot\cdot})^2$$

is called sum of squares *between treatments* and measures "explained error" (i.e., that part of the total variation due to differences between treatment effects) and is analogous to the SS representing "explained variation" in the regression model. Similarly, the sum of squares

$$\sum_j \sum_i (x_{ij} - \bar{x}_{\cdot j})^2$$

is called sum of squares *within treatments* or *error SS* and is analogous to the SS due to unexplained variation in the regression model.

An objective common to regression and analysis of variance is a reduction of the SS representing unexplained error or variation. In Sec. 10.1, the statement is made that physical phenomena are characterized by a response variable Y depending on a number of independent variables $X_1, X_2, \ldots, X_k$ (some of which are either unknown to the experimentor or are not monitored in the experiment). In Example 10.10.1, an independent variable X_2 is added to the regression equation, hopefully to improve the regression and thereby increase the explained variation and reduce the unexplained variation. Likewise, in analysis of variance an objective is through proper design of the experiment to reduce the error SS and thus identify as many significant sources of variation as is possible.

Table 11.1 describes an experiment for comparing the effects of k treatments, each of which is assigned to n experimental units. Generally, the same number of experimental units receive each treatment. Thus, for each of kn units a number from 1 to kn is selected at random to decide which treatment is applied to that experimental unit.* An experiment in which no restrictions are imposed on the randomization (except for the minor one of equal number of units per treatment) is called a *completely randomized design.*

An example of the design in Table 11.1 is an experiment involving k types of heat treatment of a given steel on kn specimen steel pieces, with the response variable x_{ij} being surface finish obtainable on a given processing

*In the introduction to Sec. 7.11, a distinction is made between physical (or experimental) control and randomization for guarding against experimental bias. Some independent variables are controlled through proper design of the experiment, others are randomized. A randomization of the order of experimentation tends to average out the effects of uncontrolled variables on the observations.

Table 11.1. DESIGN AND NOMENCLATURE: SINGLE VARIABLE, COMPLETELY RANDOMIZED EXPERIMENT.

Treatment	1	2	. . .	j	. . .	k	
	x_{11}	x_{12}		x_{1j}		x_{1k}	
	x_{21}	x_{22}		x_{2j}		x_{2k}	
	.	.		.		.	
	.	.		.		.	
	.	.		.		.	
	x_{i1}	x_{i2}		x_{ij}		x_{ik}	
	.	.		.		.	
	.	.		.		.	
	.	.		.		.	
	x_{n_1}	x_{n_2}		x_{n_j}		x_{n_k}	Totals
Totals	$T_{\cdot 1}$	$T_{\cdot 2}$		$T_{\cdot j}$		$T_{\cdot k}$	$T_{\cdot\cdot}$
Number	n	n		n		n	N
Means	$\bar{x}_{\cdot 1}$	$\bar{x}_{\cdot 2}$		$\bar{x}_{\cdot j}$		$\bar{x}_{\cdot k}$	$\bar{x}_{\cdot\cdot}$

operation. The heat procedure is the "treatment," specimen steel pieces are "experimental units," and surface finishes are "effects." The convenient dot notation is borrowed from Hicks* and other recent texts that employ essentially the same symbolism. A summary of the notation is:

x_{ij}: ith observed effect of the jth treatment

n: number of observations on jth treatment effect

N: total number of observations for the experiment

$$N = \sum_{j=1}^{k} n = kn$$

$T_{\cdot j}$: sum of observations, jth treatment effect

$$T_{\cdot j} = \sum_{i=1}^{n} x_{i\cdot} \quad \text{for specified } j$$

$T_{\cdot\cdot}$: total of observations for the experiment.

$$T_{\cdot\cdot} = \sum_{j=1}^{k} \sum_{i=1}^{n} x_{ij} = \sum_{j=1}^{k} T_{\cdot j}$$

*An excellent introduction to analysis of variance, oriented to engineering applications, is Hicks, C. R. (1965), *Fundamental Concepts in the Design of Experiments*, Holt, Rinehart & Winston, Inc., New York.

$\bar{x}_{.j}$: observed mean of the jth treatment effect

$$\bar{x}_{.j} = \frac{1}{n} \sum_{i=1}^{n} x_{i.} \quad \text{for specified } j$$

$\bar{x}_{..}$: observed mean of all observations x_{ij}

$$\bar{x}_{..} = \frac{1}{N} \sum_{j=1}^{k} \sum_{i=1}^{n} x_{ij} = \frac{1}{N} \sum_{j=1}^{k} n\bar{x}_{.j}$$

(Generally, replacement of the letter by a dot identifies the index of summation, e.g., $i.$ denotes sum over $i = 1, 2, \ldots, n$.)

The model for a single variable, completely randomized design is

$$X_{ij} = \mu + T_j + e_{ij} \quad i = 1, 2, \ldots, n \quad j = 1, 2, \ldots, k \tag{11.2}$$

which states that any observed x_{ij} is equal to the overall mean μ for the composite of k populations, plus the deviation T_j of the jth population mean from the overall mean μ, plus a random deviation e_{ij} from the mean of the jth population.*

The symbol T_j is a shorthand designation for $\mu_{.j} - \mu$, where μ is the population correspondent of the sample mean $\bar{x}_{..}$ and $\mu_{.j}$ the population value analogous to the sample mean $\bar{x}_{.j}$. Thus, Eq. (11.2) can be expressed as

---

*For example, consider the observed effects x_{ij} of $k = 4$ treatments on $kn = 4 \cdot 5 = 20$ experimental units as shown, with means $\bar{x}_{.j} = 1.6, 2.0, 2.4$, and 2.0, and overall mean $\bar{x}_{..} = 2.0$.

T_1	T_2	T_3	T_4	
1	1	2	1	
2	2	3	3	$x_{4,3}$
1	1	2	2	
2	3	(3)	3	
2	3	2	1	
8	10	12	10	40
1.6	2.0	2.4	2.0	2.0

The sample expression for Eq. (11.2) is

$$x_{ij} = \bar{x}_{..} + (\bar{x}_{.j} - \bar{x}_{..}) + (x_{ij} - \bar{x}_{.j})$$

and, for example, if $x_{ij} = x_{4,3}$

$$3 = 2.0 + (2.4 - 2.0) + (3 - 2.4)$$

$$X_{ij} = \mu + (\mu._j - \mu) + (X_{ij} - \mu._j)$$

or

$$X_{ij} - \mu = (\mu._j - \mu) + (X_{ij} - \mu._j)$$

Squaring both members of this expression and summing over i and j gives*

$$\sum_j \sum_i (X_{ij} - \mu)^2 = \sum_j \sum_i (\mu._j - \mu)^2 + \sum_j \sum_i (X_{ij} - \mu._j)^2 \qquad (11.3)$$

which is the population correspondent of Eq. (11.1) and is analogous to the partitioning of the total *SS* in regression analysis.

A summary analysis-of-variance table, similar to that of Sec. 10.2 for regression analysis, can be constructed either from Eq. (11.1) or Eq. (11.3). Using sample *SS* values, a summary is given by:

Source of Variation	*d.f.**	*SS*
Between treatments	$k - 1$	$\sum_j \sum_i (\bar{x}._j - \bar{x}..)^2$
Within treatments	$N - k$	$\sum_j \sum_i (x_{ij} - \bar{x}._j)^2$
Total	$N - 1$	$\sum_j \sum_i (x_{ij} - \bar{x}..)^2$

*The within-treatments *SS* is the pooled *SS* of deviations of the x_{ij} within each group of treatment effects. Each group is composed of n values, and thus there are $n - 1$ *d.f.* among these values. Since there are k such groups, the pooled *SS* has $k(n - 1)$ *d.f.* or $kn - k = N - k$ *d.f.* The between-treatments *SS* is a sum of squares of deviations for k quantities and therefore has $k - 1$ *d.f.*

As in Sec. 10.2, the sums of squares are not usually computed as stated in Eq. (11.1). Computation formulas, which do not involve deviations, can be obtained by squaring and simplifying the separate *SS*-expressions given in

---

*The cross product term from the squaring operation is

$$2 \sum_j \sum_i (\mu._j - \mu)(X_{ij} - \mu._j)$$

and, for the first summation over i, the coefficient $(\mu._j - \mu)$ is a constant which when factored out gives

$$2 \sum_j (\mu._j - \mu) \sum_i (X_{ij} - \mu._j)$$

and since

$$\sum_i (X_{ij} - \mu._j) = 0$$

the entire cross product expression vanishes.

Eq. (11.1).* These formulas are summarized as follows:

Source of Variation	*d.f.*	*SS*	*MS*
Between treatments	$k-1$	$\sum_j \frac{T_{\cdot j}^2}{n} - \frac{T_{\cdot\cdot}^2}{N}$	$\frac{SS}{k-1}$
Within treatments	$N-k$	$\sum_j \sum_i x_{ij}^2 - \sum_j \frac{T_{\cdot j}^2}{n}$	$\frac{SS}{N-k}$
Total	$N-1$	$\sum_j \sum_i x_{ij}^2 - \frac{T_{\cdot\cdot}^2}{N}$	

(11.4)

Analysis of variance is also operationally analogous to regression. Testing the significance of regression, if there is zero regression (i.e., the *SS* due to regression is zero), the regression line is $y = \bar{y}$ as shown in Fig. 11.1 and estimates $y = \bar{y}$ of $\mu_{Y|X}$ are identical for all X. On the other hand, if the *SS* due to regression is not zero and the sample regression line is $y = a_0 + a_1 x$, then estimates of $\mu_{Y|X}$ for $X = X_1, X_2, \ldots,$ are unequal. In analysis of variance, the equality of means is tested directly. The hypothesis is

$$H_0\colon\quad \mu_{\cdot 1} = \mu_{\cdot 2} = \ldots = \mu_{\cdot j} = \ldots = \mu_{\cdot k}$$

where $\mu_{\cdot j}$ is the population mean† of the jth treatment effect or, simply,

$$H_0\colon\quad T_j = 0$$

that is, if H_0 is true, then there is no treatment effect and each $x_{ij} = \mu + e_{ij}$.

*For example, performing the indicated squaring operation for the total *SS*,

$$\begin{aligned}
\sum_j \sum_i (x_{ij} - \bar{x}_{\cdot\cdot})^2 &= \sum_j \sum_i (x_{ij}^2 - 2x_{ij}\bar{x}_{\cdot\cdot} + \bar{x}_{\cdot\cdot}^2) \\
&= \sum_j \sum_i x_{ij}^2 - 2\sum_j \sum_i x_{ij}\bar{x}_{\cdot\cdot} + \sum_j \sum_i \bar{x}_{\cdot\cdot}^2 \\
&= \sum_j \sum_i x_{ij}^2 - 2\bar{x}_{\cdot\cdot} \sum_j \sum_i x_{ij} + N\bar{x}_{\cdot\cdot}^2 \\
&= \sum_j \sum_i x_{ij}^2 - 2\bar{x}_{\cdot\cdot}(N\bar{x}_{\cdot\cdot}) + N\bar{x}_{\cdot\cdot}^2 \\
&= \sum_j \sum_i x_{ij}^2 - N\bar{x}_{\cdot\cdot}^2 \\
&= \sum_j \sum_i x_{ij}^2 - \frac{T_{\cdot\cdot}^2}{N}
\end{aligned}$$

†It is important to appreciate the following correspondence. In regression, X is the independent variable and Y, the dependent variable, is a random variable possessing distribution properties. Its expected value, given X, is $\mu_{Y|X}$. In analysis of variance, the treatment T_j is the independent variable, and the response variable X_{ij} is a random variable with expected value $\mu_{\cdot j}$. Roughly speaking, the $\mu_{\cdot j}$ of analysis of variance correspond to the $\mu_{Y|X}$ of regression.

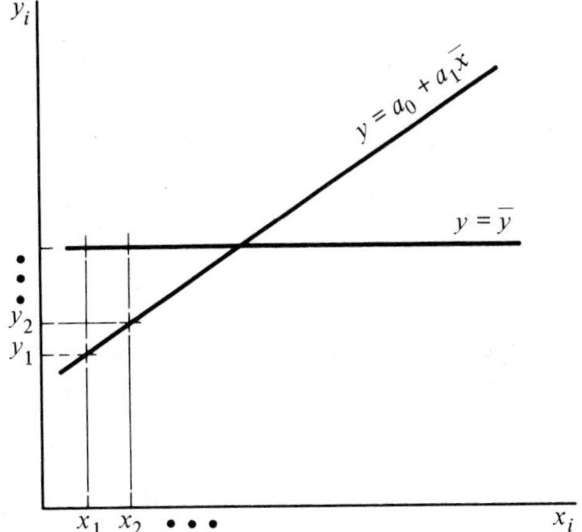

Figure 11.1. Regression line estimates.

For the single variable, completely randomized analysis-of-variance model, it is assumed* that:

1. μ is an unknown parameter.
2. The T_j are unknown constants.
3. The e_{ij} are normally and independently distributed with mean zero and variance σ_e^2.

The second assumption is appropriate for a *fixed model*, that is, where the populations being sampled constitute the whole set of populations of interest. In cases where the populations being sampled are themselves a sample of the populations that might be employed, the T_j are assumed to be random variables which are normally and independently distributed with mean zero and variance σ_T^2. This latter case is called a *random model*.† In this introductory text, only simple analysis of variances are considered, and only fixed models are employed.

---

*Note the similarity of assumptions for analysis of variance with that for regression (see Sec. 10.1). In regression, the X's are assumed to be unknown constants and the variances of Y for all X are assumed to be homogeneous. In analysis of variance, the T_j are assumed to be unknown constants (for this model) and the variances of the e_{ij} for all k treatment levels are assumed to be homogeneous.

†A simple example of fixed and random models is an application where the treatment factor is temperature and the treatment levels are 100°, 120°, and 140°. If these levels constitute all of the temperatures under consideration in the application, the model is "fixed." On the other hand, if these three levels are the result of a random selection from all possible temperature levels, the model is "random." If a model is random, different hypothesis test procedures are employed and different interpretations are made of the results.

EXAMPLE 11.1.1

Heat-treatment-of-material effect T_j on surface finish X_{ij} (obtainable on a given processing operation) is being studied. Treatment T_1 is applied to a random selection of 4 of 16 specimen steel pieces, T_2 to a random selection of 4 of the 16 pieces, and so forth, as follows:

3(0), 1(4), 4(2), 4(0), 1(8), 1(5), 2(5), 3(3),

4(2), 3(3), 2(4), 2(5), 1(4), 4(4), 2(0), 3(1).

This example is repeated for other experiment designs in subsequent sections. The notation used here is continued throughout all sections, that is, the first of each pair of numbers refers to treatment type T_j and the second number in parentheses is the response value x_{ij}. The responses can be viewed as coded data; actually they are hypothetical small numbers selected to illustrate method and argument.

The data is collected in the following table for computation purposes.

Treatment	1	2	3	4	
	4	5	0	2	
	8	4	3	0	
	5	5	3	2	
	4	0	1	4	$T_{..}$
$T_{.j}$	21	14	7	8	50

The null hypothesis to be tested is

$$H_0: \quad \mu_{.1} = \mu_{.2} = \mu_{.3} = \mu_{.4}$$

and the test is at an $\alpha = 0.05$ significance level. Using Eqs. (11.4), the following SS's are computed. The total SS is given by

$$\sum_j \sum_i x_{ij}^2 - \frac{T_{..}^2}{N} = 230 - \frac{50^2}{16}$$
$$= 230 - 156.25 = 73.75$$

The between-treatments SS is obtained from

$$\sum_j \frac{T_{.j}^2}{n} - \frac{T_{..}^2}{N} = \frac{21^2 + 14^2 + 7^2 + 8^2}{4} - \frac{50^2}{16}$$
$$= 187.50 - 156.25 = 31.25$$

The within-treatments SS is obtained by subtraction and is equal to

$$73.75 - 31.25 = 42.50$$

A summary analysis-of-variance is

Source of Variation	*d.f.*	*SS*	*MS*
Between treatments (T_j)	3	31.25	10.42
Within treatments (e_{ij})	12	42.50	3.54
Total	15	73.75	

The mean squares represent two independent, chi-square distributed, unbiased estimates of the population variance σ_X^2 when H_0 is true.* Thus, from Sec. 6.4, their ratio is F-distributed with 3 and 12 *d.f.* If H_0 is true, by chance alone this F-ratio is expected to exceed $F_{0.05,3,12} = 3.49$ only 5% of the time. Since

$$\frac{10.42}{3.54} = 2.94 < F_{0.05,3,12} = 3.49$$

the hypothesis is accepted. The conclusion is that no treatment effect exists and any surface finish observation x_{ij} is equal to $\mu + e_{ij}$.

EXERCISES

1. Test the following observations for treatment effect. The significance level is $\alpha = 0.05$.

Treatment	1	2	3	4
	5	6	8	13
	4	11	3	18
	0	4	4	20

Ans. $\frac{116}{10} > F_{0.05,3,8}$, reject H_0: $T_j = 0$.

*It can be intuitively argued that, if H_0 is true, the observations x_{ij} are from identical populations (identical means and variances assumed to be identical) and therefore a common population with mean μ and variance σ_X^2.

If such is the case, either mean square is an unbiased estimate of the population variance σ_X^2 and thus the ratio of the mean squares should differ from one by an amount due only to sampling error, i.e., by an amount equal to 3.49 only 5% of the time.

2. Test the following observations for treatment effect. The significance level is $\alpha = 0.01$.

Treatment	1	2	3
	50	67	50
	48	72	44
	53	71	43
	48	74	45
	51	66	43

Ans. $\frac{875}{8.2} > F_{0.01,2,12}$, reject $H_0: T_j = 0$.

11.2 Randomized Complete Block Design

The within-treatments or error SS is relatively large in Example 11.1.1. Assume that there exists another independent variable B_i possibly affecting the response variable X_{ij} in this example. Although complete randomization of the order of experimentation in Example 11.1.1 tends to average out the effects of other independent variables such as B_i, there is no guarantee that the "averaging-out" effect operates well. The unexplained error e_{ij} may include, in addition to experimental error, variation traceable to a B_i effect.

One objective of analysis of variance is to reduce the unexplained variation. An experiment design for removing a B_i effect (if it exists) from the error SS is called a *randomized complete block design.* In this design, each T_j treatment level appears once with each B_i level (or *block*), and randomization of the order of experimentation is only within each block. The model is described by

$$X_{ij} = \mu + B_i + T_j + e_{ij} \tag{11.5}$$

where μ is the overall mean, B_i is the ith block effect on the response variable X_{ij}, T_j is the jth treatment effect on X_{ij}, and e_{ij} is the deviation of each X_{ij} from its expected value.

An equation analogous to Eq. (11.3) can be obtained to describe this model. Again, $\mu_{.j} - \mu$ represents the treatment effect T_j. In addition, $\mu_{i.} - \mu$ measures the block effect B_i. Thus, e_{ij} is measured by

$$(X_{ij} - \mu) - (\mu_{i.} - \mu) - (\mu_{.j} - \mu) = X_{ij} - \mu_{i.} - \mu_{.j} + \mu$$

and, from Eq. (11.5),

$$X_{ij} = \mu + (\mu_{i\cdot} - \mu) + (\mu_{\cdot j} - \mu) + (X_{ij} - \mu_{i\cdot} - \mu_{\cdot j} + \mu)$$

or,

$$X_{ij} - \mu = (\mu_{i\cdot} - \mu) + (\mu_{\cdot j} - \mu) + (X_{ij} - \mu_{i\cdot} - \mu_{\cdot j} + \mu)$$

The corresponding sample estimate is

$$x_{ij} - \bar{x}_{\cdot\cdot} = (\bar{x}_{i\cdot} - \bar{x}_{\cdot\cdot}) + (\bar{x}_{\cdot j} - \bar{x}_{\cdot\cdot}) + (x_{ij} - \bar{x}_{i\cdot} - \bar{x}_{\cdot j} + \bar{x}_{\cdot\cdot})$$

Squaring both members of this expression and summing over i and j gives

$$\sum_j \sum_i (x_{ij} - \bar{x}_{\cdot\cdot})^2 = \sum_j \sum_i (\bar{x}_{i\cdot} - \bar{x}_{\cdot\cdot})^2 + \sum_j \sum_i (\bar{x}_{\cdot j} - \bar{x}_{\cdot\cdot})^2 + \sum_j \sum_i (x_{ij} - \bar{x}_{i\cdot} - \bar{x}_{\cdot j} + \bar{x}_{\cdot\cdot})^{2*} \qquad (11.6)$$

Equation (11.6) demonstrates a partitioning of the total *SS* into three parts: (1) *SS* due to differences between blocks, (2) *SS* due to differences between treatments, and (3) *SS* due to deviations of the x_{ij} from their expected values. This partition is summarized by:

Source of Variation	*d.f.*	*SS*
Between blocks (B_i)	$n - 1$	$\sum_j \sum_i (\bar{x}_{i\cdot} - \bar{x}_{\cdot\cdot})^2$
Between treatments (T_j)	$k - 1$	$\sum_j \sum_i (\bar{x}_{\cdot j} - \bar{x}_{\cdot\cdot})^2$
Error (e_{ij})	$(k - 1)(n - 1)$	$\sum_j \sum_i (x_{ij} - \bar{x}_{i\cdot} - \bar{x}_{\cdot j} + \bar{x}_{\cdot\cdot})^2$
Total	$kn - 1$	$\sum_j \sum_i (x_{ij} - \bar{x}_{\cdot\cdot})^2$

where the degrees of freedom for the error component e_{ij} has been reduced, that is, the *d.f.* are

$$(kn - 1) - (n - 1) - (k - 1) = kn - n - k + 1$$
$$= (k - 1)(n - 1)$$

*As in the development of Eq. (11.3), the cross products from the squaring operation vanish under algebraic manipulation.

A similar development to that of Eqs. (11.4) leads to computation formulas, Eqs. (11.7), summarized as follows:

Source of Variation	*d.f.*	*SS*	*MS*
Between blocks (B_i)	$n - 1$	$\sum_i \frac{T_{i.}^2}{k} - \frac{T_{..}^2}{N}$	$\frac{SS}{n-1}$
Between treatments (T_j)	$k - 1$	$\sum_j \frac{T_{.j}^2}{n} - \frac{T_{..}^2}{N}$	$\frac{SS}{k-1}$
Error (e_{ij})	$(k-1)(n-1)$	$\sum_j \sum_i x_{ij}^2 - \sum_i \frac{T_{i.}^2}{k} - \sum_j \frac{T_{.j}^2}{n} + \frac{T_{..}^2}{n}$	$\frac{SS}{(k-1)(n-1)}$
Total	$kn - 1$	$\sum_j \sum_i x_{ij}^2 - \frac{T_{..}^2}{N}$	

(11.7)

The same assumptions as that for the model of Sec. 11.1 hold, and the hypothesis of primary interest is still H_0: $T_j = 0$, although a hypothesis regarding block effects is also possible.

EXAMPLE 11.2.1

Example 11.1.1 is repeated here, assuming the same responses, but using a randomized block design.

Assume that prior to the experimentation leading to the response values of Example 11.1.1 the experiment is designed to also provide for detection of material-type effect B_i on surface finish response X_{ij}. The 16 specimens consist of four material types (four pieces to each type). The two independent variables T_j and B_i are linked by the experiment design such that each treatment T_j appears once with each material B_i, yielding the previous responses as shown below. The T_j are randomly applied to the four specimens at each B_i level.

Blocks (B_i)				
1	2(5)	3(3)	1(8)	4(4)
2	4(2)	3(3)	2(5)	1(5)
3	1(4)	2(4)	4(2)	3(1)
4	3(0)	4(0)	2(0)	1(4)

The first number of each pair of numbers identifies the treatment level and the second number is the response. The responses are summarized as follows for computation purposes.

B_i \ T_j	1	2	3	4	$T_{i\cdot}$
1	8	5	3	4	20
2	5	5	3	2	15
3	4	4	1	2	11
4	4	0	0	0	4
$T_{\cdot j}$	21	14	7	8	$T_{\cdot\cdot} = 50$

Using Eqs. (11.7), the following SS's are computed. The total SS is again

$$\sum_j \sum_i x_{ij}^2 - \frac{T_{\cdot\cdot}^2}{N} = 230 - \frac{50^2}{16}$$
$$= 230 - 156.25 = 73.75$$

The between-treatments SS is obtained from

$$\sum_j \frac{T_{\cdot j}^2}{n} - \frac{T_{\cdot\cdot}^2}{N} = \frac{21^2 + 14^2 + 7^2 + 8^2}{4} - \frac{50^2}{16}$$
$$= 187.50 - 156.25 = 31.25$$

The between-blocks SS is computed from

$$\sum_i \frac{T_{i\cdot}^2}{k} - \frac{T_{\cdot\cdot}^2}{N} = \frac{20^2 + 15^2 + 11^2 + 4^2}{4} - \frac{50^2}{16}$$
$$= 190.50 - 156.25 = 34.25$$

The error SS is obtained by subtraction and is equal to

$$73.75 - 31.25 - 34.25 = 8.25$$

A summary analysis-of-variance is

Source of Variation	*d.f.*	*SS*	*MS*
T_j	3	31.25	10.42
B_j	3	34.25	11.42
e_{ij}	9	8.25	0.92
Total	15	73.75	

Each mean square is an independent, chi-square distributed, unbiased estimate of the population variance σ_X^2 when the hypothesis of identical means is true. Each ratio, MS treatments/MS error and MS blocks/MS error, is F-distributed and the decision rule at an $\alpha = 0.05$ significance level is

$$F_{\text{data}} \geq F_{0.05,3,9} = 3.86, \qquad \text{reject } H_0$$

Testing the hypothesis of interest, $H_0\colon\ T_j = 0$

$$F_{\text{data}} = \frac{10.42}{0.92} = 11.33$$

and thus H_0 is rejected and it is concluded there is a significant difference between treatment means.

The hypothesis $H_0\colon\ B_i = 0$ can also be tested. Since

$$F_{\text{data}} = \frac{11.42}{0.92} = 12.41$$

it is also concluded that the block means differ significantly and hence that material type also affects surface finish.

EXERCISES

1. Test the following observations for both treatment and block effects. The significance level is $\alpha = 0.05$.

Blocks B_i			
1	2(0)	1(1)	3(1)
2	3(1)	2(0)	1(2)
3	1(1)	3(2)	2(1)
4	1(2)	2(1)	3(2)

Ans. $\dfrac{1.33}{0.22} > F_{0.05,2,6}$, reject $H_0\colon T_j = 0$. $\dfrac{0.56}{0.22} < F_{0.05,3,6}$, accept $H_0\colon B_i = 0$.

2. Test the following observations for both treatment and block effects. The significance level is $\alpha = 0.05$.

Blocks B_i				
1	3(2.0)	2(2.7)	1(3.0)	4(2.6)
2	4(2.8)	1(3.0)	3(3.2)	2(2.5)
3	3(2.0)	4(2.0)	2(3.5)	1(3.0)

Ans. $\dfrac{0.27}{0.28} < F_{0.05,3,6}$, accept $H_0\colon T_j = 0$. $\dfrac{0.10}{0.28} < F_{0.05,2,6}$, accept $H_0\colon B_i = 0$.

11.3 Latin Square Design

The randomized block design of Example 11.2.1 reduced the error *SS* from 42.50 (Example 11.1.1) to 8.25, thus accomplishing one objective of analysis of variance—reducing the unexplained variation. If the number of levels for each of the independent variables under study is identical, it is possible to vary the randomized block design slightly and include a third independent variable in the analysis of variance model. The result is a *Latin square* design, an experiment wherein each treatment level T_j appears once and only once with each level of two independent variables B_i and C_h. The model is

$$X_{ijh} = \mu + T_j + B_i + C_h + e_{ijh} \tag{11.8}$$

and C_h, where $h = 1, 2, \ldots, m$, represents the effect (if it exists) of the additional independent variable.

By the methods of Secs. 11.1–11.2, it can be shown that the *SS* due to C_h is

$$SS_C = \sum_h \frac{T^2_{..h}}{m} - \frac{T^2_{...}}{N} \tag{11.9}$$

EXAMPLE 11.3.1

Example 11.1.1 is repeated here, assuming the same responses, but using a Latin square design to include a processing variable C_h.

The 16 specimen steel pieces are, as in Example 11.2.1 four pieces of material type B_1, four pieces of B_2, and so forth, with treatment types T_j still being of primary interest.

Randomization is now restricted to choosing at random from several possible Latin squares of a 4 × 4 size. Tables of such squares can either be constructed or found in Fisher and Yates.*

The experiment design, together with the responses, is given by

B_i \ C_h	1	2	3	4
1	3(3)	2(5)	1(8)	4(4)
2	4(2)	3(3)	2(5)	1(5)
3	1(4)	4(2)	3(1)	2(4)
4	2(0)	1(4)	4(0)	3(0)

where each T_j level appears once and only once with each B_i level and with each

*See Fisher, R. A. and Yates, F. (1953), *Statistical Tables for Biological, Agricultural, and Medical Research*, 4th ed., Oliver & Boyd, Ltd., London.

C_h level. The responses summary for computation purposes is

B_i \ C_h	1	2	3	4	$T_{i..}$
1	3	5	8	4	20
2	2	3	5	5	15
3	4	2	1	4	11
4	0	4	0	0	4
$T_{..h}$	9	14	14	13	$T_{...} = 50$
T_j	1	2	3	4	
$T_{.j.}$	21	14	7	8	

The *SS* due to the C_h effect is

$$\sum_h \frac{T_{..h}^2}{m} - \frac{T_{...}^2}{N} = \frac{9^2 + 14^2 + 14^2 + 13^2}{4} - \frac{50^2}{16}$$
$$= 160.50 - 156.25 = 4.25$$

From the results for Example 11.2.1,

$$SS_{\text{total}} = 73.75, \qquad SS_T = 31.25, \qquad SS_B = 34.25$$

and thus

$$SS_e = 73.75 - 31.25 - 34.25 - 4.25$$
$$= 4.0$$

The analysis-of-variance summary is:

Source of Variation	*d.f.*	*SS*	*MS*
T_j	3	31.25	10.42
B_i	3	34.25	11.42
C_h	3	4.25	1.42
e_{ijh}	6	4.00	0.67
Total	15	73.75	

Although the error *SS* has been reduced (from 8.25 to 4.00), there is no apparent gain since the C_h effect on X_{ijh} is small and certainly not significant. This, of course, could have been anticipated by examining the totals $T_{..h}$, which differ only slightly.*

EXERCISES

1. Prepare a Latin square analysis for the following observations. The significance level is $\alpha = 0.05$.

$B_i \backslash C_h$	1	2	3	4
1	3(8)	2(4)	4(9)	1(3)
2	2(7)	3(6)	1(7)	4(6)
3	1(4)	4(1)	2(6)	3(5)
4	4(1)	1(1)	3(0)	2(1)

Ans. $\frac{0.73}{3.98} < F_{0.05,3,6}$, accept $H_0\colon T_j = 0$. $\frac{27.23}{3.98} > F_{0.05,3,6}$, reject $H_0\colon B_i = 0$.
$\frac{5.23}{3.98} < F_{0.05,3,6}$, accept $H_0\colon C_h = 0$.

2. Construct a Latin square design to investigate 5 treatments. How many degrees of freedom are assigned to the variables T_j, B_i, and C_h, and to error e_{ijh}?

11.4 Introduction to Factorial Design

Assume an experiment involving two independent variables A_i and B_j with levels $i = 1, 2, 3$ and $j = 1, 2$. Each level of one variable is combined with each level of the other variable in the experiment. There are $3 \cdot 2 = 6$ such combinations. There are three observations on each combination.

Viewing the six combinations as six treatments—A_1B_1, A_1B_2, A_2B_1, A_2B_2, A_3B_1, and A_3B_2—a single variable, completely randomized design (Sec. 11.1) can be arranged.

EXAMPLE 11.4.1

Eighteen combinations A_iB_j are randomized in the manner of Example 11.1.1 and summarized together with the response values as follows:

---

*As in regression analysis (see Sec. 10.11), there is always the chance that the addition of another independent variable to the model does not reduce the error *SS* sufficiently to offset the loss of *d.f.* for the error *SS*. Depending on the relative magnitude of the various sums of squares, a loss of *d.f.* for the error *SS* may result in a large *MS* error and hence an *F*-ratio that is not a reasonable yardstick for evaluating other independent variable effects.

i \ T_j	A_1B_1	A_1B_2	A_2B_1	A_2B_2	A_3B_1	A_3B_2	
1	5	1	8	5	4	0	
2	6	2	7	3	5	1	
3	5	3	6	2	6	1	$T_{..}$
$T_{.j}$	16	6	21	10	15	2	70

The respective sums of squares are:

$$SS_{\text{total}} = \sum_j \sum_i x_{ij}^2 - \frac{T_{..}^2}{N} = 366 - \frac{70^2}{18}$$

$$= 366 - 272.22 = 93.78$$

$$SS_{\text{treatments}} = \sum_j \frac{T_{.j}^2}{n} - \frac{T_{..}^2}{N}$$

$$= \frac{16^2 + 6^2 + 21^2 + 10^2 + 15^2 + 2^2}{3} - \frac{70^2}{18}$$

$$= 354 - 272.22 = 81.78$$

$$SS_{\text{error}} = 93.78 - 81.78 = 12.00$$

Example 11.4.1 can be viewed more compactly and efficiently as a *factorial design*, an experiment which comprises every combination of the different levels of the independent variables. This example is a two-variable factorial and, since there are three levels of A_i and two levels of B_j, it is called a 3×2 *factorial design*. The model is

$$X_{ijk} = \mu + A_i + B_j + AB_{ij} + e_{k(ij)} \tag{11.10}$$

where μ is the overall mean, A_i is the ith effect of this variable on the response variable X_{ijk}, B_j is the jth effect of this variable on X_{ijk}, AB_{ij} is the "interaction" effect of the two variables on X_{ijk}, and $e_{k(ij)}$ is the deviation of each X_{ijk} from its expected value. The latter symbol is a shorthand designation referring to k observations for each i, j combination of the variables.

An *interaction effect* is an effect on the response variable produced jointly by two or more independent variables, an effect that goes beyond the total of their individual effects. Two or more independent variables "click" together in an optimum fashion and affect the response variable. Examples of interaction are common in chemistry where, for example, the properties of a mixture may differ widely from the properties of the mixture's components.

EXAMPLE 11.4.2

Example 11.4.1 is repeated here as a factorial design. The order of executing the experimental combinations is randomized yielding the following order together with the responses:

$$
\begin{array}{lll}
A_3B_1(4), & A_3B_2(0), & A_2B_2(5) \\
A_1B_2(1), & A_2B_2(3), & A_3B_1(5) \\
A_3B_1(6), & A_1B_2(2), & A_1B_1(5) \\
A_3B_2(1), & A_3B_2(1), & A_1B_1(6) \\
A_1B_1(5), & A_2B_1(8), & A_2B_2(2) \\
A_2B_1(7), & A_2B_1(6), & A_1B_2(3)
\end{array}
$$

The responses are summarized for computation as follows:

B_j \ A_i	1	2	3	$T_{\cdot j \cdot}$
1	5 6 (16) 5	8 7 (21) 6	4 5 (15) 6	52
2	1 2 (6) 3	5 3 (10) 2	0 1 (2) ← $T_{ij\cdot}$ 1	18
$T_{i\cdot\cdot}$	22	31	17	$T_{\cdots} = 70$

The sums of squares between the 18 treatments are partitioned into respective SS's due to A_i, B_j, and AB_{ij} effects.

$$SS_A = \sum_i \frac{T_{i\cdot\cdot}^2}{n} - \frac{T_{\cdots}^2}{N} = \frac{22^2 + 31^2 + 17^2}{6} - \frac{70^2}{18}$$

$$= 289 - 272.22 = 16.78$$

$$SS_B = \sum_j \frac{T_{\cdot j \cdot}^2}{n} - \frac{T_{\cdots}^2}{N} = \frac{52^2 + 18^2}{9} - \frac{70^2}{18}$$

$$= 336.44 - 272.22 = 64.22^*$$

**Note:* The divisor n in each expression is the number of observations that are totaled. For $T_{i\cdot\cdot}$, $n = 6$, and for $T_{\cdot j\cdot}$, $n = 9$.

$$SS_{AB} = SS_{\text{treatments}} - SS_A - SS_B$$
$$= 81.78 - 16.78 - 64.22$$
$$= 0.78$$

The analysis-of-variance summary is:

Source of Variation	*d.f.*	*SS*		*MS*
A_i	2	16.78		8.39
B_j	1	64.22	81.78	64.22
AB_{ij}	2*	0.78		0.39
$e_{k(ij)}$	12	12.00		1.00
Total	17	93.78		

Three hypotheses can be tested:

H_1: $A_i = 0$

$$F_{\text{data}} = \frac{8.39}{1.00} = 8.39 > F_{0.01,2,12} = 6.93$$

Reject H_1.

H_2: $B_j = 0$

$$F_{\text{data}} = \frac{64.22}{1.00} = 64.22 > F_{0.01,1,12} = 9.33$$

Reject H_2.

H_3: $AB_{ij} = 0$

$$F_{\text{data}} = \frac{0.39}{1.00} = 0.39 < F_{0.01,2,12} = 6.93$$

Accept H_3.

There are many special and sophisticated factorial designs. A thorough treatment of this efficient method of experimentation is beyond the scope of an introductory text. The interested student is referred to advanced statistics and design of experiments textbooks for further study in this area.

---

*To obtain the *d.f.* for the AB_{ij} source of variation, observe that there are 6 treatments A_iB_j with $6 - 1 = 5$ *d.f.* Thus, the *d.f.* for AB_{ij} is 5 minus 2 *d.f.* for A_i minus 1 *d.f.* for B_j equals 2 *d.f.*

EXERCISES

1. Complete an analysis of variance for the following observations using a single variable, completely randomized design model. The significance level is = 0.05.

$A_2B_2(1.0)$,	$A_1B_1(4.8)$,	$A_1B_2(5.4)$,	$A_3B_2(0.6)$
$A_2B_1(3.3)$,	$A_1B_2(6.2)$,	$A_2B_2(1.4)$,	$A_3B_1(0.7)$
$A_3B_1(1.5)$,	$A_2B_1(2.8)$,	$A_1B_1(5.8)$,	$A_3B_2(0.9)$

2. Repeat Exercise 1 as a 3 × 2 factorial design.

Ans. A_i and AB_{ij} effects are significant.

GLOSSARY OF TERMS

Analysis of variance
Treatment
Between-treatments *SS*
Within-treatments *SS*
Completely randomized design
Fixed model
Random model
Randomized complete block design
Latin square design
Factorial design
Interaction effect

GLOSSARY OF FORMULAS

Page

(11.1) Partitioning the total *SS*. 353

$$\sum_j \sum_i (x_{ij} - \bar{x}_{..})^2 = \sum_j \sum_i (\bar{x}_{.j} - \bar{x}_{..})^2 + \sum_j \sum_i (x_{ij} - \bar{x}_{.j})^2$$

(11.2) Single variable completely randomized model. 356

$$X_{ij} = \mu + T_j + e_{ij}$$

(11.3) Partitioning the total *SS*. 357

$$\sum_j \sum_i (X_{ij} - \mu)^2 = \sum_j \sum_i (\mu_{.j} - \mu)^2 + \sum_j \sum_i (X_{ij} - \mu_{.j})^2$$

(11.4) *SS*'s for single variable, completely randomized design. 358

Due to T_j: $\sum_j \frac{T_{.j}^2}{n} - \frac{T_{..}^2}{N}$

Due to e_{ij}: $\sum_j \sum_i x_{ij}^2 - \sum_j \frac{T_{.j}^2}{n}$

Total *SS*: $\sum_j \sum_i x_{ij}^2 - \frac{T_{..}^2}{N}$

(11.5) Randomized complete block model. 362

$$X_{ij} = \mu + B_i + T_j + e_{ij}$$

(11.6) Partitioning the total *SS*. 363

$$\sum_j \sum_i (x_{ij} - \bar{x}_{..})^2 = \sum_j \sum_i (\bar{x}_{i.} - \bar{x}_{..})^2 + \sum_j \sum_i (\bar{x}_{.j} - \bar{x}_{..})^2 + \sum_j \sum_i (x_{ij} - \bar{x}_{i.} - \bar{x}_{.j} + \bar{x}_{..})^2$$

(11.7) *SS*'s for randomized complete block model. 364

Due to B_i: $\sum_i \frac{T_{i.}^2}{k} - \frac{T_{..}^2}{N}$

Due to T_j: $\sum_j \frac{T_{.j}^2}{n} - \frac{T_{..}^2}{N}$

Due to e_{ij}: $\sum_j \sum_i x_{ij}^2 - \sum_i \frac{T_{i.}^2}{k} - \sum_j \frac{T_{.j}^2}{n} + \frac{T_{..}^2}{N}$

Total *SS*: $\sum_j \sum_i x_{ij}^2 - \frac{T_{..}^2}{N}$

(11.8) Latin square model. 367

$$X_{ijh} = \mu + T_j + B_i + C_h + e_{ijh}$$

(11.9) *SS* for Latin square model. 367

$$SS_C = \sum_h \frac{T_{..h}^2}{m} - \frac{T_{...}^2}{N}$$

(11.10) 3×2 factorial design. 370

$$X_{ijk} = \mu + A_i + B_j + AB_{ij} + e_{k(ij)}$$

REFERENCES

CHOU, YA-LUN (1969), *Statistical Analysis*, Holt, Rinehart & Winston, Inc., New York, Chap. 13.

DIXON, W. J. AND MASSEY, F. J. (1969), *Introduction to Statistical Analysis*, 3rd ed., McGraw-Hill Book Company, New York, Chap. 10.

HICKS, C. R. (1965), *Fundamental Concepts in the Design of Experiments*, Holt, Rinehart & Winston, Inc., New York, Chaps. 3–6.

LIPSON, C. AND SHETH, N. J. (1973), *Statistical Design and Analysis of Engineering Experiments*, McGraw-Hill Book Company, New York, Chap. 6.

RAY, W. S. (1960), *An Introduction to Experimental Design*, The Macmillan Company, New York, Chap. 11.

PROBLEMS

1. Explain what each term measures in the following expression of the model for a single variable, completely randomized design:

$$X_{ij} = \mu + T_j + e_{ij}$$

2. Describe the similarities between analysis of variance and regression analysis. Compare the *SS* measuring unexplained variation in regression analysis with the *SS* within treatments in analysis of variance.

3. Problem 7.58 gives observations of breaking strength of copper-wire specimens from two production treatments X and Y. Assume that 4 levels of production treatment X yield the following coded responses (breaking strengths). Test at a 5% significance level the hypothesis H_0: $T_j = 0$ that production treatment X has no effect on breaking strength.

Treatment Level	1	2	3	4
	2	2	0	0
	4	3	2	1
	3	3	3	2
	2	1	1	1

4. Prepare a summary analysis-of-variance table for Problem 3.

5. Suppose that in addition to the treatment variable T_j in Problem 3 another production variable B_i is suspected to have an effect on breaking strength of the manufactured wire. Prior to experimentation leading to the same responses given in Problem 3, the experiment is designed such that each T_j level appears once and only once with each B_i level. Test at a 5% significance level the hypotheses H_0: $T_j = 0$ and H_0: $B_i = 0$ for this randomized block design and prepare a summary analysis-of-variance table. The coded responses are as follows:

		T_j			
		1	2	3	4
	1	4	3	3	2
B_i	2	3	3	2	1
	3	2	2	1	1
	4	2	1	0	0

6. Describe and compare the randomization procedures for Problems 3 and 5.

7. As in regression analysis, one objective of analysis of variance is to reduce unexplained variation. Explain this procedure relative to adding the B_i variable to the model for Problem 5. Suppose there is a B_i variable (present but not monitored) affecting the responses in Problem 3. Which term of the model (or which SS) reflects the B_i effect?

8. A randomized block experiment is designed for the overhead cost situation of Problem 10.24 yielding the following responses. Denote the X_1 production factor by T_j with $j = 3$ levels of this factor, and the X_2 production factor by B_i with $i = 3$ levels of this factor. Test at a 1% significance level for production factor effect on overhead cost.

		T_j	
	1	2	3
B_i 1	0.6	1.1	2.0
2	0.7	1.3	2.2
3	0.7	1.6	2.3

9. A randomized block experiment is used for the quality gain situation in Problem 10.28. Test at 5% significance levels hypotheses regarding quality gain effects from production variables X_1 and X_2.

		x_1	
	1	2	3
x_2 1	0	75	21
2	3	1	46
3	39	57	27

10. Test at a 1% significance level for differences between machines and between shifts in Problem 7.69.

11. The effects of 4 coolant types on surface finish (microinches) is being studied. The order of testing is completely randomized. Analyze at a 5% significance level the following responses as a single variable, completely randomized design.

	Type		
1	2	3	4
20	24	25	23
24	25	26	23
20	24	28	25

12. Repeat Problem 11, treating the problem from a randomized block design viewpoint, with an additional variable, tool type, being a part of the experiment.

		Coolant Type			
		1	2	3	4
Tool Type	1	18	22	24	20
	2	31	27	34	29
	3	15	20	21	22

13. Three methods of measurement of nitrogen-14 levels of a lake water yield the following responses (in ppm). Do the measurement methods agree or disagree significantly? Test at a 1% α-level for significant variation in mean measurements.

Measurement Method		
1	2	3
12	10	9
16	15	16
26	22	20
18	16	19
9	8	7

14. Four levels of a production factor are being tested for effect on breaking strength of the product. Analyze at a 5% significance level the following coded responses as a completely randomized design.

Factor			
1	2	3	4
8	9	11	16
7	14	6	21
3	7	7	23

15. Nitrogen dioxide emissions by motor vehicles are being studied. Speculation is that nitrogen dioxide levels of urban air may vary considerably with the days of the week (due to varying traffic loads) and also with the time of the day. A randomized block experiment yields the following responses (in ppm). Is there a significant difference in mean daily nitrogen dioxide levels? or in mean day-time levels? Test at a 5% significance level.

		Days			
		1	2	3	4
Time of Day	1	1.7	1.1	0.6	0.9
	2	1.1	1.1	0.7	0.4
	3	0.9	0.9	0.2	0.5
	4	0.8	0.1	0.1	0.1

16. Two alloy constituents of a certain steel are being studied for possible effect on tensile strength. Alloy A is varied over 3 levels and alloy B over 6 levels. Analyze at a 1% significance level the following coded responses

		A	
	1	2	3
1	40	35	25
2	35	37	32
B 3	41	33	21
4	33	37	20
5	30	40	22
6	38	43	30

17. The normal saturation level is approximately 5 parts of oxygen per million parts of water at the bottom of a large lake. It is known that thermal stratification decreases the oxygen level. Verify this by analysis of the following oxygen level responses for three different thermal stratification level periods. Test at a 5% significance level.

	Thermal	
1	2	3
3.1	2.2	1.0
5.2	4.1	2.1
4.2	3.6	1.3
5.3	2.2	2.1

18. A Latin square results from the addition of a third variable C_h to the analysis-of-variance model. Describe the conditions for the model

$$X_{ijh} = \mu + T_j + B_i + C_h + e_{ijh}$$

and explain the additional constraint on randomization procedure.

19. Three production factors T_j, B_i, and C_h possibly influence production time X_{ijh}. A Latin square experimental design is utilized and the following production times (min) are obtained as responses. Perform an analysis of variance at a 1% significance level.

	C_1	C_2	C_3	C_4
B_1	(T_1) 22.5	(T_3) 27.6	(T_2) 25.8	(T_4) 27.1
B_2	(T_2) 25.6	(T_4) 25.0	(T_1) 27.3	(T_3) 25.9
B_3	(T_3) 27.1	(T_1) 27.5	(T_4) 24.5	(T_2) 25.9
B_4	(T_4) 27.5	(T_2) 26.0	(T_3) 27.5	(T_1) 24.5

20. The effects of transistors on the output of a receiver are being studied. The transistors are produced by 5 different companies. Transistors T_j, receivers B_i, and test sets C_h affect the output responses. Perform an analysis of variance at a 5% significance level.

	C_1	C_2	C_3	C_4	C_5
B_1	(T_1) 65	(T_2) 34	(T_3) 67	(T_4) 52	(T_5) 51
B_2	(T_2) 41	(T_1) 54	(T_5) 29	(T_3) 29	(T_4) 32
B_3	(T_3) 44	(T_4) 29	(T_1) 80	(T_5) 36	(T_2) 23
B_4	(T_4) 39	(T_5) 35	(T_2) 45	(T_1) 60	(T_3) 40
B_5	(T_5) 46	(T_3) 62	(T_4) 28	(T_2) 30	(T_1) 71

21. Construct a Latin square design to investigate three treatments. How many degrees of freedom are assigned to the variables T_j, B_i, and C_h, and to error e_{ijh}?

22. A Graeco Latin square results from the addition of a third variable D_k to the experimental model. A nonrandomized layout and then a randomized layout for this model follows. Describe the randomization procedure and identify the *d.f.* associated with each term of the model, which is

$$X_{ijhk} = \mu + T_j + B_i + C_h + D_k + e_{ijhk}$$

	C_1	C_2	C_3	C_4	C_5
B_1	T_1D_1	T_2D_3	T_3D_5	T_4D_2	T_5D_4
B_2	T_2D_2	T_3D_4	T_4D_1	T_5D_3	T_1D_5
B_3	T_3D_3	T_4D_5	T_5D_2	T_1D_4	T_2D_1
B_4	T_4D_4	T_5D_1	T_1D_3	T_2D_5	T_3D_2
B_5	T_5D_5	T_1D_2	T_2D_4	T_3D_1	T_4D_3

	C_1	C_2	C_3	C_4	C_5
B_1	T_4D_4	T_2D_5	T_1D_3	T_5D_1	T_3D_2
B_2	T_3D_3	T_1D_4	T_5D_2	T_4D_5	T_2D_1
B_3	T_5D_5	T_3D_1	T_2D_4	T_1D_2	T_4D_3
B_4	T_1D_1	T_4D_2	T_3D_5	T_2D_3	T_5D_4
B_5	T_2D_2	T_5D_3	T_4D_1	T_3D_4	T_1D_5

23. Another variable D_k is added to the conditions of Problem 20 and the response values obtained are coded and recorded as follows. Perform an analysis of variance at a 5% significance level for this Graeco Latin square experimental design.

	C_1	C_2	C_3	C_4	C_5
B_1	T_4D_4 7	T_2D_5 7	T_1D_3 6	T_5D_1 6	T_3D_2 6
B_2	T_3D_3 7	T_1D_4 7	T_5D_2 8	T_4D_5 12	T_2D_1 7
B_3	T_5D_5 12	T_3D_1 6	T_2D_4 10	T_1D_2 7	T_4D_3 10
B_4	T_1D_1 6	T_4D_2 6	T_3D_5 13	T_2D_3 8	T_5D_4 13
B_5	T_2D_2 7	T_5D_3 11	T_4D_1 10	T_3D_4 12	T_1D_5 12

24. A simple factorial design experiment is described by the model

$$X_{ijk} = \mu + A_i + B_j + AB_{ij} + e_{k(ij)}$$

Explain what is being measured by each term of this model. Describe the randomization procedure.

25. Explain what is meant by *interaction effect.*

26. Two production variables A_i and B_j affect the time (min) of a processing operation. A factorial design experiment is performed for 3 levels of A_i and 2 levels of B_j with 3 observations for each AB combination. Complete an analysis of variance at a 5% significance level.

		A_i 1	2	3
B_j	1	2.3	3.6	4.5
		2.8	3.2	4.2
		2.5	3.5	5.1
	2	2.2	4.3	4.1
		1.8	4.4	4.3
		2.4	3.5	3.9

27. Analyze the following observations as a factorial design experiment. Test at a 5% significance level.

		A_i				
		1	2	3	4	5
B_j	1	43	12	30	21	23
		32	3	18	12	18
		36	9	23	20	29
		35	5	27	27	15
	2	43	17	47	40	28
		55	22	32	34	33
		48	19	40	38	30
		59	28	29	43	25
	3	72	39	63	57	67
		67	50	70	61	62
		65	40	59	54	55
		78	46	67	65	53

PROCESSING OF STATISTICAL DATA

All of the statistical computations for the various techniques of Chaps. 1–11 can be efficiently carried out on modern desk calculators or the various computers available to engineers. For almost all of the statistical tests, the calculations involve $\sum x$, $\sum x^2$, $\sum y$, $\sum y^2$, $\sum xy$, *and some algebraic manipulation of these values. By entering a series of* x- *and* y-*values into the keyboard of a calculator, all of these sums can be obtained at the same time. Also, most modern calculators possess automatic square root operations and thus values for the standard deviation, correlation coefficient, and other statistics can be easily generated from a single keyboard entry of the number for which the root is desired.*

12.1 Desk Calculator Computation

The input of data to a desk calculator is by means of number entries into a keyboard. However, in programmable desk calculators the data input may be either via punched entries into the keyboard, or magnetic cards, or other peripheral devices. An example of a statistical programmable printing calculator is given in Fig. 12.1. All of the statistical operations of Chaps. 1–11,

12

including three-variable regression analysis, can be handled by means of standard programs available from the calculator manufacturer's library. Or, one may prepare special programs on a magnetic card device and enter this into the calculator by depressing one switch.

The programmable calculator shown in Fig. 12.1 is a system of data processing consisting of the basic elements: input, processing, and output (Fig. 12.2). The *input* operation is one of putting data into the system. This operation is provided either by: (1) the keyboard, (2) data recorded on a magnetic card, in which case the calculator reads the data directly from the card, or (3) punch cards and mark-sense cards. Also, using accessory equipment, data can be entered into the calculator directly from test or measuring equipment in the form of electrical pulses.

The *processing* element of the system refers to each and every movement of the data within the system and every arithmetic and mathematical change or operation applied to the data. Processing implies that a planned set of actions is applied to the data; this sequence is called a *program*, and programming is the action of designing, writing, and testing a program. One advantage of a programmable calculator is that there is no machine language to master. Programming is done directly in arithmetic and mathematical language and

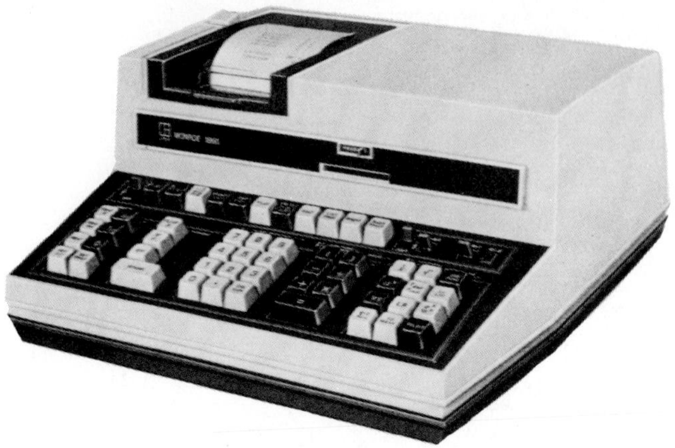

Figure 12.1. Statistical programmable calculator. (*Courtesy of Monroe, The Calculator Company, Division of Litton Industries, Inc.*)

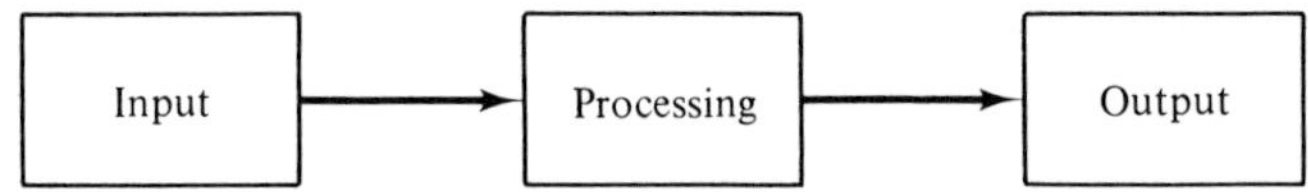

Figure 12.2. Basic data-processing cycle.

entered into the calculator in the same sequence as in the mathematical equation.

When data has been processed to its final form, it is ready for *output*—a term which refers to sending the processed data outside of the system for inspection by the operator or programmer. The output device on the programmable calculator is a printing mechanism that prints results on a continuous strip of paper. It is also possible to obtain output data on magnetic cards or on display light-emitting numerical readouts.

A more complete description of the data processing system of input, processing, and output elements is given by Fig. 12.3. The processing element is detailed to indicate the control and storage functions. The *control* function refers to calculator operations being controlled by either keyboard manipulations or program instructions. The *storage* function is the operation of data and instructions being stored into and accessed from the calculator's storage registers. The data storage consists of 10 scratch-pad registers and 64 main storage registers, which are expandable to 522. For complex problems, there are 512 basic memory steps that are expandable to 4096.

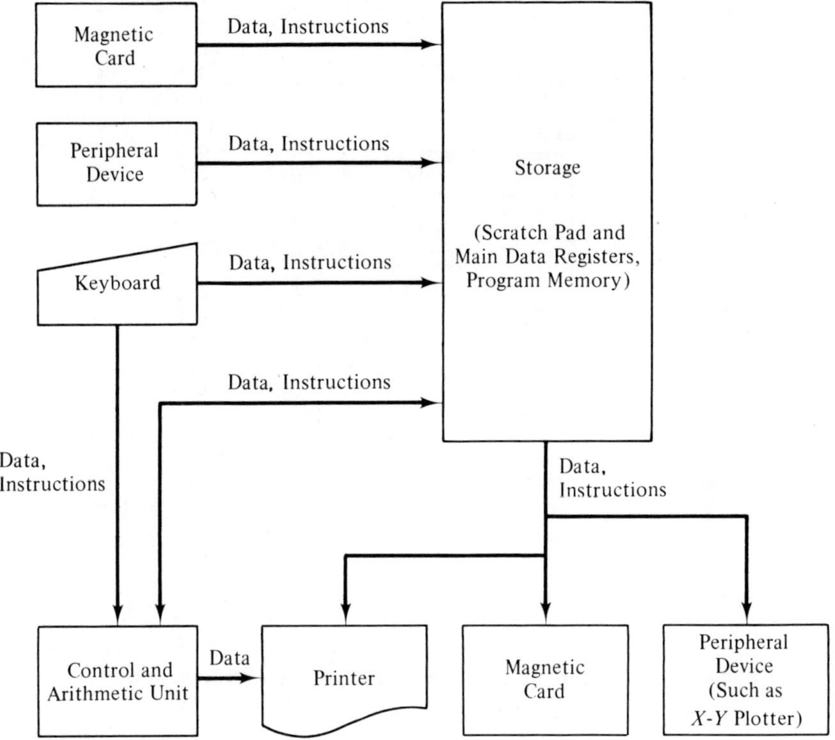

Figure 12.3. Data processing system. (*From Programming Reference Manual, Monroe, the Calculator Company, Division of Litton Industries, Inc.*)

EXAMPLE 12.1.1

The data of Problem 1.14 is repeated here as follows:

x_j	0.4	0.6	0.8	1.0	1.2	1.4	1.6	1.8	2.0	
f_j	1	3	5	6	4	2	2	1	1	25

The keyboard operation and readout strip of the output solution are shown in Fig. 12.4. The mean is 1.072, the variance is 0.1563, and the standard deviation is 0.3953.

EXAMPLE 12.1.2

Given binomial conditions are $n = 20$ and $p = 0.05$. The problem requirement is to compute the respective probabilities of $X = 0, 1, 2, \ldots$.

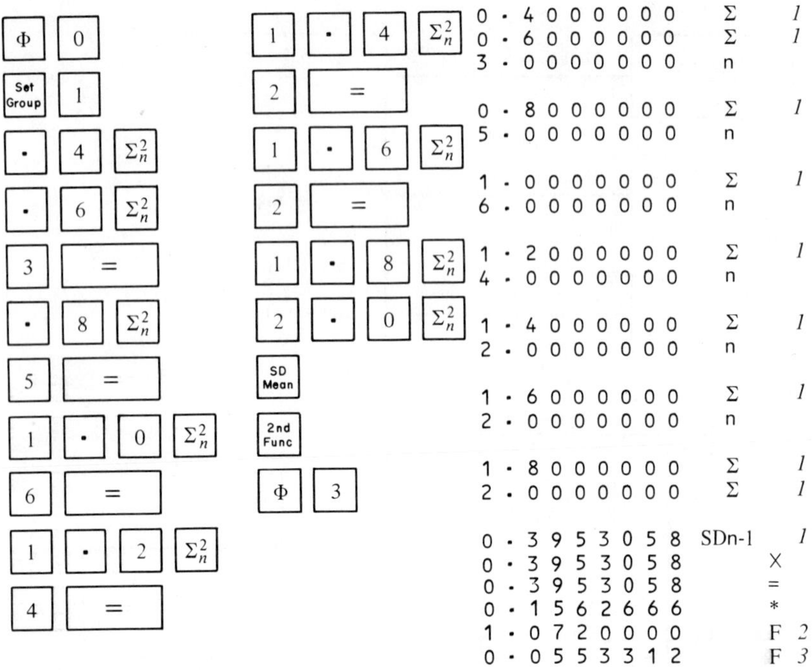

Figure 12.4. Example 12.1.1—keyboard input and readout strip for computation of mean, variance, and standard deviation.

A program based on Eq. (4.25) is prepared. Keyboard entries are $X = 10$, $n = 20$, and $p = 0.05$. The $X = 10$ entry identifies the last computation to be performed.

The output solution shown in Fig. 12.5 is obtained in 20 seconds. The program can be used again for other binomial computations without any changes; only the three keyboard entries vary. In fact, this program was used to construct the Appendix A table.

EXAMPLE 12.1.3

The multiple linear regression, Example 10.9.1, is repeated here. The data observations are:

x_{1i}	0	1	2	3	4	5
x_{2i}	0	2	2	3	3	4
y_i	2.0	5.0	5.6	7.4	8.1	9.0

The input program and readout strip of the output solution are given in Fig. 12.6. Readout strip symbols X, Y, and Z correspond to x_{1i}, x_{2i}, and y_i respectively. The regression line coefficients are $a_0 = 2.1465$, $a_1 = 0.4931$, and $a_2 = 1.2017$, with r^2 being equal to 0.9863.

```
. . . . . . . . . . . . . . . .

                10.00
                20.00
                 0.05

. . . . . . . . . . . . . . . .

                 0.                    X

0.35848592                        A

. . . . . . . . . . . . . . . .

                 1.                    X

0.37735360                        A

. . . . . . . . . . . . . . . .

                 2.                    X

0.18867680                        A

. . . . . . . . . . . . . . . .

                 3.                    X

0.05958215                        A

. . . . . . . . . . . . . . . .

                 4.                    X

0.01332759                        A

. . . . . . . . . . . . . . . .

                 5.                    X

0.00224465                        A

. . . . . . . . . . . . . . . .

                 6.                    X

0.00029535                        A

. . . . . . . . . . . . . . . .

                 7.                    X

0.00003109                        A

. . . . . . . . . . . . . . . .

                 8.                    X

0.00000266                        A

. . . . . . . . . . . . . . . .

                 9.                    X

0.00000019                        A

. . . . . . . . . . . . . . . .

                10.                    X

0.00000001                        A

. . . . . . . . . . . . . . . .
```

Figure 12.5. Example 12.1.2—readout strip of the output solution for computation of the respective probabilities of $X = 0, 1, 2, \ldots$.

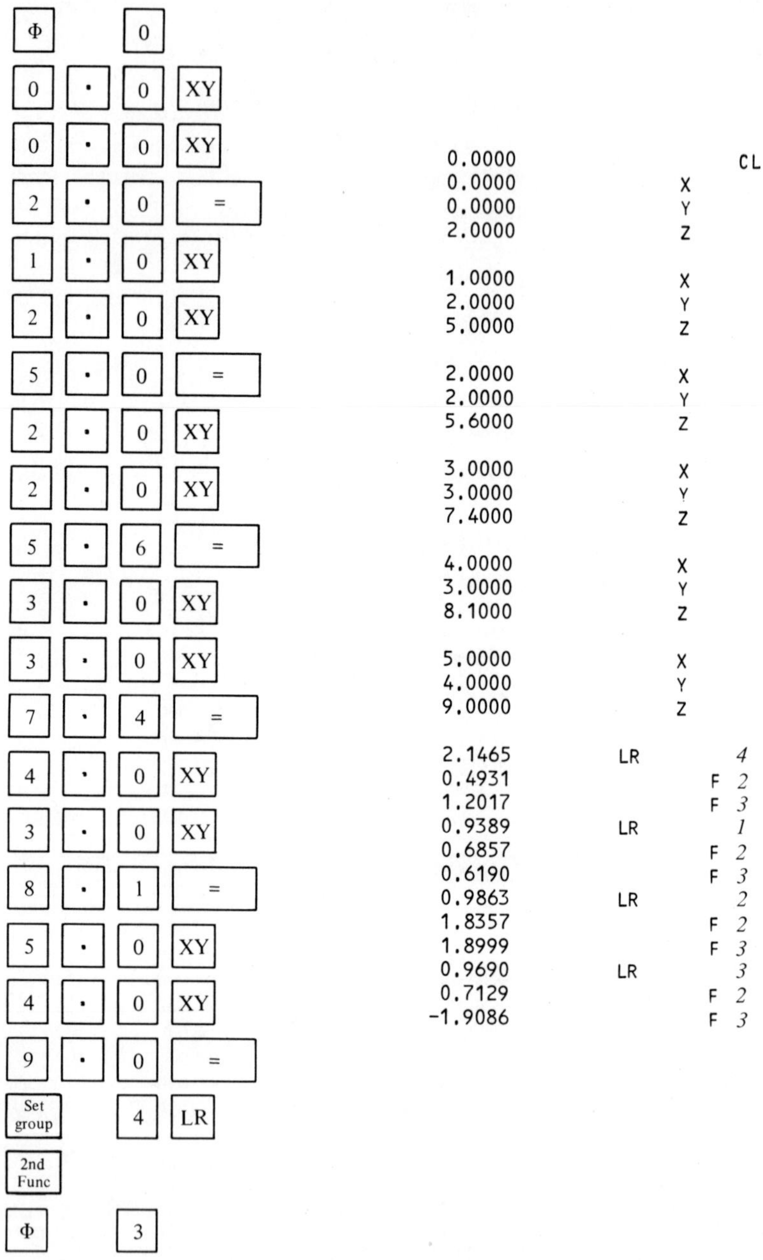

Figure 12.6. Example 12.1.3—input program and readout strip of the output solution of a multiple linear regression problem.

12.2 Digital Computer Calculation

Two examples are presented here to indicate the advantages of computer processing of statistical data. The first is a multiple linear regression analysis using a stepwise regression procedure that selects independent variables one at a time for inclusion in the regression model. The second example illustrates a simulation of an assembly distribution (see Sec. 5.9). Neither example can be efficiently handled on a desk calculator.

EXAMPLE 12.2.1

A metal cutting operation study* involves a relationship between tangential force Y (lb) and surface linear speed X_1 (in./min), feed X_2 (in./rev.), and depth of cut X_3 (in.). A first attempt to describe the relationship is by means of a multiple linear regression model:

$$\mu_{Y|X_1|X_2|X_3} = A_0 + A_1X_1 + A_2X_2 + A_3X_3$$

The input data is given as follows:

y	x_1	x_2	x_3
94.0	269	0.0040	0.060
136.0	269	0.0075	0.060
120.0	430	0.0075	0.060
68.0	430	0.0040	0.060
83.5	269	0.0040	0.060
120.0	269	0.0075	0.060
115.0	430	0.0075	0.060
68.0	430	0.0040	0.060
63.7	355	0.0075	0.030
63.7	269	0.0075	0.030
41.8	269	0.0040	0.030
54.5	355	0.0075	0.030
104.5	269	0.0150	0.030
104.5	430	0.0150	0.030
34.5	430	0.0040	0.030
57.5	355	0.0075	0.030
54.5	355	0.0075	0.030
57.5	355	0.0075	0.030
57.5	355	0.0075	0.030
53.4	355	0.0075	0.030

*This example is by courtesy of Mr. Jong-Ping Hsu, Purdue University, W. Lafayette, Ind.

```
SELECTION NO.      1- 0

SAMPLE SIZE        6
NO. OF VARIABLES  3      NO. OF VARIABLES DELETED  0   (FOR VARIABLES DELETED. SEE BELOW)
DEPENDENT VARIABLE IS NOW NO.  3

COEFFICIENT OF DETERMINATION     .9885
MULTIPLE CORR. COEFFICIENT       .9942

SUM OF SQUARES ATTRIBUTABLE TO REGRESSION          31.95557
SUM OF SQUARES OF DEVIATION FROM REGRESSION          .37276

VARIANCE OF ESTIMATE          .12425
STD. ERROR OF ESTIMATE        .35250

INTERCEPT (A VALUE)       2.14655
```

ANALYSIS OF VARIANCE FOR THE MULTIPLE LINEAR REGRESSION

SOURCE OF VARIATION	D.F.	SUM OF SQUARES	MEAN SQUARES	F VALUE
DUE TO REGRESSION............	2	31.95557	15.97779	128.5909
DEVIATION ABOUT REGRESSION...	3	.37276	.12425	
TOTAL...	5	32.32833		

VARIABLE NO.	MEAN	STD. DEVIATION	REG. COEFF.	STD.ERROR OF REG.COE.	COMPUTED T VALUE	PARTIAL CORR. COF.	SUM OF SQ. ADDED	PROP. VAR. CUM.
1	2.50000	1.87083	.49310	.24492	2.01335	.75808	30.36014	.93912
2	2.33333	1.36626	1.20172	.33537	3.58332	.90034	1.59543	.04935
3	6.18333	2.54277						

```
COMP. CHECK ON FINAL COEFF.          1.20172
```

Figure 12.7. Example 12.2.1—regression printout.

A BMDO3R program (Biomedical Computer Program/University of California) is used, and the output solution from this program is given by Fig. 12.7. The printout solution gives: (1) sums and sums of squares, (2) cross products of deviations, (3) regression coefficients a_0, a_1, a_2, and a_3, (4) standard errors of the regression coefficients s_{a_0}, s_{a_1}, s_{a_2}, and s_{a_3}, (4) computed t-values for making confidence interval estimates of A_0, A_1, A_2, and A_3, (5) sums of squares, mean squares, and F_{data} values for testing the significance of regression, and (6) correlation coefficients. A correlation coefficient is given at each stage of the inclusion of the independent variables into the regression model.

EXAMPLE 12.2.2

Probable assembly variation Y is being studied, where $Y = X_1 + X_2 + X_3 + X_4$ with the X_j being uniformly distributed over the respective intervals (0.501, 0.505), (0.496, 0.500), (0.504, 0.508), and (0.490, 0.494) in.

By the methods of Sect. 5.9, the standard deviation of Y is $\sigma_Y = 0.0028$ in. and for $\lambda = 0.0456$ the expected Y-variation is 0.0112 in. However, if the X_j distributions are nonnormal with nonhomogeneous variances, the methods of Sec. 5.9 are not applicable. A method of estimating the expected Y-variation is simulation. To illustrate this method, four sets of 20 punched cards each are prepared to repre-

```
                 PROGRAM MAIN(INPUT,OUTPUT,TAPE5=INPUT,TAPE6=OUTPUT)
000002           DIMENSION A(100),B(100),C(100),D(100),Y(100)
000002           REAL LCL
000002           X=0.
000003           U=0.
000004           DO 10 I=1.100
000005           READ (5,1) A(I)
000012           READ (5,1) B(I)
000020           READ (5,1) C(I)
000026           READ (5,1) D(I)
000034           Y(I)=A(I)+B(I)+C(I)+D(I)
000041         1 FORMAT (F5.3)
000042           X=X+Y(I)
000044           Z=Y(I)**2
000044           U=U+Z
000046        10 WRITE(6,3) I,Y(I)
000060         3 FORMAT (10X,I3,7X,F5.3)
000060           WRITE (6,5)
000064         5 FORMAT (1H1)
000064           CALL SSORT (Y,100,0,1)
000067           XMEAN=X/100.
000071           STDDEV=SQRT((U-100.*(XMEAN**2))/100.)
000077           UCL = XMEAN+3.*STDDEV
000101           LCL = XMEAN-3.*STDDEV
000103           WRITE (6,4) XMEAN,STDDEV, UCL, LCL
000116         4 FORMAT(//1X,≠MEAN = ≠F5.3//1X,≠STANDARD DEVIATION =≠F6.4,//1X,
                2,≠UPPER LIMIT = ≠F6.4,//1X,≠LOWER LIMIT = ≠F6.4)
000116           STOP
000120           END

PROGRAM LENGTH INCLUDING I/O BUFFERS
003247

UNUSED COMPILER SPACE
003500
```

Figure 12.8. Example 12.2.2—program for simulating the Y distribution.

sent the X-distributions:

x_1	f_1	x_2	f_2	x_3	f_3	x_4	f_4
0.501	20	0.496	20	0.504	20	0.490	20
0.502	20	0.497	20	0.505	20	0.491	20
0.503	20	0.498	20	0.506	20	0.492	20
0.504	20	0.499	20	0.507	20	0.493	20
0.505	20	0.500	20	0.508	20	0.494	20

The cards are randomly selected and a computer simulation of the Y-distribution is obtained. The program is shown in Fig. 12.8 and the printout solution is given by Fig. 12.9. The standard deviation of the simulated Y-distribution is $\sigma_Y = 0.003$

```
ORDERED DATA POINTS

2.00700E+00    2.00500E+00    2.00500E+00    2.00400E+00    2.00400E+00
2.00400E+00    2.00400E+00    2.00300E+00    2.00300E+00    2.00300E+00
2.00300E+00    2.00300E+00    2.00300E+00    2.00200E+00    2.00200E+00
2.00200E+00    2.00200E+00    2.00200E+00    2.00200E+00    2.00200E+00
2.00200E+00    2.00200E+00    2.00200E+00    2.00100E+00    2.00100E+00
2.00100E+00    2.00100E+00    2.00100E+00    2.00100E+00    2.00100E+00
2.00100E+00    2.00100E+00    2.00000E+00    2.00000E+00    2.00000E+00
2.00000E+00    2.00000E+00    2.00000E+00    2.00000E+00    2.00000E+00
2.00000E+00    2.00000E+00    2.00000E+00    2.00000E+00    1.99900E+00
1.99900E+00    1.99900E+00    1.99900E+00    1.99900E+00    1.99900E+00
1.99900E+00    1.99900E+00    1.99900E+00    1.99900E+00    1.99800E+00
1.99800E+00    1.99800E+00    1.99800E+00    1.99800E+00    1.99800E+00
1.99800E+00    1.99800E+00    1.99700E+00    1.99700E+00    1.99700E+00
1.99700E+00    1.99700E+00    1.99700E+00    1.99700E+00    1.99700E+00
1.99700E+00    1.99700E+00    1.99700E+00    1.99700E+00    1.99700E+00
1.99700E+00    1.99700E+00    1.99700E+00    1.99700E+00    1.99600E+00
1.99600E+00    1.99600E+00    1.99600E+00    1.99600E+00    1.99600E+00
1.99600E+00    1.99600E+00    1.99600E+00    1.99600E+00    1.99500E+00
1.99500E+00    1.99500E+00    1.99500E+00    1.99500E+00    1.99500E+00
1.99500E+00    1.99400E+00    1.99400E+00    1.99200E+00    1.99200E+00

MEAN = 1.999

STANDARD DEVIATION = .0030

UPPER LIMIT = 2.0080

LOWER LIMIT = 1.9900
```

Figure 12.9. Example 12.2.2—printout of the simulated Y distribution.

in. and the expected Y-variation is 0.012 in. Also, by counting Y-values that are outside of assembly specifications on the printout Y-distribution a simulated value for the risk value λ can be obtained.

REFERENCES

DIXON, W. J. (1967), *BMD Biomedical Computer Programs*, University of California Press, Berkeley, Calif.

Operating Instructions/Model 1860 Statistical Programmable Printing Calculator (1972), Monroe, The Calculator Company, Division of Litton Industries, Orange, N. J.

Programming Reference Manual/Model 1860 Statistical Programmable Printing Calculator (1972), Monroe, The Calculator Company, Division of Litton Industries, Orange, N. J.

STERLING, T. D. AND POLLACK, S. V. (1968), *Introduction to Statistical Data Processing*, Prentice-Hall, Inc., Englewood Cliffs, N. J.

APPENDICES

APPENDIX A

CUMULATIVE BINOMIAL DISTRIBUTION

n	X	0.01	0.05	0.10	0.15	p 0.20	0.25	0.30	0.40	0.50
2	0	.9801	.9025	.8100	.7225	.6400	.5625	.4900	.3600	.2500
	1	.9999	.9975	.9900	.9775	.9600	.9375	.9100	.8400	.7500
3	0	.9703	.8574	.7290	.6141	.5120	.4219	.3430	.2160	.1250
	1	.9997	.9928	.9720	.9392	.8960	.8438	.7840	.6480	.5000
	2		.9999	.9990	.9966	.9920	.9844	.9730	.9360	.8750
4	0	.9606	.8145	.6561	.5220	.4096	.3164	.2401	.1296	.0625
	1	.9994	.9860	.9477	.8905	.8192	.7383	.6517	.4752	.3125
	2		.9995	.9963	.9880	.9728	.9492	.9163	.8208	.6875
	3			.9999	.9995	.9984	.9961	.9919	.9744	.9375
5	0	.9510	.7738	.5905	.4437	.3277	.2373	.1681	.0778	.0312
	1	.9990	.9774	.9185	.8352	.7373	.6328	.5282	.3370	.1875
	2		.9988	.9914	.9734	.9421	.8965	.8369	.6826	.5000
	3			.9995	.9978	.9933	.9844	.9692	.9130	.8125
	4				.9999	.9997	.9990	.9976	.9898	.9688
6	0	.9415	.7351	.5314	.3771	.2621	.1780	.1176	.0467	.0156
	1	.9985	.9672	.8857	.7765	.6554	.5339	.4202	.2333	.1094
	2		.9978	.9842	.9527	.9011	.8306	.7443	.5443	.3438
	3		.9999	.9987	.9941	.9830	.9624	.9295	.8208	.6562
	4			.9999	.9996	.9984	.9954	.9891	.9590	.8906
	5					.9999	.9998	.9993	.9959	.9844
7	0	.9321	.6983	.4783	.3206	.2097	.1335	.0824	.0280	.0078
	1	.9980	.9556	.8503	.7166	.5767	.4449	.3294	.1586	.0625
	2		.9962	.9743	.9262	.8520	.7564	.6471	.4199	.2266
	3		.9998	.9973	.9879	.9667	.9294	.8740	.7102	.5000
	4			.9998	.9988	.9953	.9871	.9712	.9037	.7734
	5				.9999	.9996	.9987	.9962	.9812	.9375
	6						.9999	.9998	.9984	.9922
8	0	.9227	.6634	.4305	.2725	.1678	.1001	.0576	.0168	.0039
	1	.9973	.9428	.8131	.6572	.5033	.3671	.2553	.1064	.0352
	2	.9999	.9942	.9619	.8948	.7969	.6785	.5518	.3154	.1445
	3		.9996	.9950	.9786	.9437	.8862	.8059	.5941	.3633
	4			.9996	.9971	.9896	.9727	.9420	.8263	.6367
	5				.9998	.9988	.9958	.9887	.9502	.8555
	6					.9999	.9996	.9987	.9915	.9648
	7							.9999	.9993	.9961
9	0	.9135	.6302	.3874	.2316	.1342	.0751	.0404	.0101	.0020
	1	.9966	.9288	.7748	.5995	.4362	.3003	.1960	.0705	.0195
	2	.9999	.9916	.9470	.8591	.7382	.6007	.4628	.2318	.0898
	3		.9994	.9917	.9661	.9144	.8343	.7297	.4826	.2539
	4			.9991	.9944	.9804	.9511	.9012	.7334	.5000
	5			.9999	.9994	.9969	.9900	.9747	.9006	.7461
	6					.9997	.9987	.9957	.9750	.9102
	7						.9999	.9996	.9962	.9805
	8								.9997	.9980

Blank spaces above a first column entry are read 0.0, and blank spaces below the last column entry are read 1.0.

CUMULATIVE BINOMIAL DISTRIBUTION (Continued)

		p								
n	*X*	0.01	0.05	0.10	0.15	0.20	0.25	0.30	0.40	0.50
10	0	.9044	.5987	.3487	.1969	.1074	.0563	.0282	.0060	.0010
	1	.9957	.9139	.7361	.5443	.3758	.2440	.1493	.0464	.0107
	2	.9999	.9885	.9298	.8202	.6778	.5256	.3828	.1673	.0547
	3		.9990	.9872	.9500	.8791	.7759	.6496	.3823	.1719
	4		.9999	.9984	.9901	.9672	.9219	.8497	.6331	.3770
	5			.9999	.9986	.9936	.9803	.9527	.8338	.6230
	6				.9999	.9991	.9965	.9894	.9452	.8281
	7					.9999	.9996	.9984	.9877	.9453
	8							.9999	.9983	.9893
	9								.9999	.9990
15	0	.8600	.4633	.2059	.0874	.0352	.0134	.0047	.0005	
	1	.9904	.8290	.5490	.3186	.1671	.0802	.0353	.0052	.0005
	2	.9996	.9638	.8159	.6042	.3980	.2361	.1268	.0271	.0037
	3		.9945	.9444	.8227	.6482	.4613	.2969	.0905	.0176
	4		.9994	.9873	.9383	.8358	.6865	.5155	.2173	.0592
	5		.9999	.9978	.9832	.9389	.8516	.7216	.4032	.1509
	6			.9997	.9964	.9819	.9434	.8689	.6098	.3036
	7				.9996	.9958	.9827	.9500	.7869	.5000
	8				.9999	.9992	.9958	.9849	.9050	.6964
	9					.9999	.9992	.9963	.9662	.8491
	10						.9999	.9993	.9907	.9408
	11							.9999	.9981	.9824
	12								.9997	.9963
	13									.9995
20	0	.8179	.3585	.1216	.0388	.0115	.0032	.0008		
	1	.9831	.7358	.3917	.1756	.0692	.0243	.0076	.0005	
	2	.9990	.9245	.6769	.4049	.2061	.0913	.0355	.0036	.0002
	3	.9999	.9841	.8670	.6477	.4114	.2252	.1071	.0160	.0013
	4		.9974	.9568	.8298	.6296	.4148	.2375	.0510	.0059
	5		.9997	.9887	.9327	.8042	.6172	.4164	.1256	.0207
	6			.9976	.9781	.9133	.7858	.6080	.2500	.0577
	7			.9996	.9941	.9679	.8982	.7723	.4159	.1316
	8			.9999	.9987	.9900	.9591	.8867	.5956	.2517
	9				.9998	.9974	.9861	.9520	.7553	.4119
	10					.9994	.9961	.9829	.8725	.5881
	11					.9999	.9991	.9949	.9435	.7483
	12						.9998	.9987	.9790	.8684
	13							.9997	.9935	.9423
	14								.9984	.9793
	15								.9997	.9941
	16									.9987
										.9998
25	0	.7778	.2774	.0718	.0172	.0038	.0007	.0001		
	1	.9742	.6424	.2712	.0931	.0274	.0070	.0016		
	2	.9980	.8729	.5371	.2537	.0982	.0321	.0090	.0004	
	3	.9999	.9659	.7636	.4711	.2340	.0962	.0332	.0024	.0001
	4		.9928	.9020	.6821	.4207	.2137	.0905	.0095	.0005
	5		.9988	.9666	.8385	.6167	.3783	.1935	.0294	.0020
	6		.9998	.9905	.9305	.7800	.5611	.3406	.0736	.0073
	7			.9977	.9745	.8909	.7265	.5118	.1535	.0216
	8			.9995	.9920	.9532	.8506	.6769	.2735	.0539
	9			.9999	.9978	.9827	.9287	.8106	.4246	.1148
	10				.9995	.9944	.9703	.9022	.5858	.2122

CUMULATIVE BINOMIAL DISTRIBUTION (Continued)

		p								
n	*X*	0.01	0.05	0.10	0.15	0.20	0.25	0.30	0.40	0.50
25	11				.9999	.9985	.9803	.9557	.7323	.3450
	12					.9996	.9966	.9825	.8462	.5000
	13					.9999	.9991	.9940	.9222	.6550
	14						.9998	.9982	.9656	.7878
	15							.9995	.9868	.8852
	16							.9999	.9957	.9461
	17								.9988	.9784
	18								.9997	.9927
	19								.9999	.9980
	20									.9995
	21									.9999
50	0	.6050	.0769	.0051	.0003					
	1	.9106	.2794	.0338	.0029	.0002				
	2	.9862	.5405	.1117	.0142	.0013	.0001			
	3	.9984	.7604	.2503	.0460	.0057	.0005			
	4	.9998	.8964	.4312	.1121	.0185	.0021	.0002		
	5		.9622	.6161	.2193	.0480	.0070	.0007		
	6		.9882	.7702	.3613	.1034	.0194	.0025		
	7		.9968	.8778	.5187	.1904	.0453	.0073	.0001	
	8		.9992	.9421	.6681	.3073	.0916	.0182	.0002	
	9		.9998	.9755	.7911	.4437	.1637	.0402	.0008	
	10			.9906	.8801	.5836	.2622	.0788	.0022	
	11			.9968	.9372	.7107	.3816	.1390	.0057	
	12			.9990	.9699	.8139	.5110	.2229	.0132	.0001
	13			.9997	.9868	.8894	.6370	.3279	.0280	.0005
	14			.9999	.9947	.9393	.7481	.4468	.0540	.0013
	15				.9980	.9692	.8369	.5692	.0955	.0033
	16				.9993	.9856	.9017	.6839	.1561	.0077
	17				.9998	.9937	.9449	.7822	.2369	.0164
	18					.9975	.9713	.8594	.3356	.0324
	19					.9991	.9861	.9152	.4465	.0595
	20					.9997	.9937	.9522	.5610	.1013
	21					.9999	.9974	.9749	.6701	.1611
	22						.9990	.9877	.7660	.2399
	23						.9996	.9944	.8438	.3359
	24						.9999	.9976	.9022	.4439
	25							.9991	.9427	.5561
	26							.9997	.9686	.6641
	27							.9999	.9840	.7601
	28								.9924	.8389
	29								.9966	.8987
	30								.9986	.9405
	31								.9995	.9675
	32								.9998	.9836
	33								.9999	.9923
	34									.9967
	35									.9987
	36									.9995
	37									.9998
	38									.9999

CUMULATIVE BINOMIAL DISTRIBUTION (Continued)

n	*X*	0.01	0.05	0.10	0.15	*p* 0.20	0.25	0.30	0.40	0.50
100	0	.3660	.0059							
	1	.7358	.0371	.0003						
	2	.9206	.1183	.0019	.0001					
	3	.9816	.2578	.0078	.0002					
	4	.9966	.4360	.0237	.0004					
	5	.9995	.6160	.0576	.0016					
	6	.9999	.7660	.1172	.0048	.0001				
	7		.8720	.2060	.0125	.0003				
	8		.9369	.3209	.0283	.0009				
	9		.9718	.4513	.0567	.0023				
	10		.9885	.5832	.1024	.0057	.0001			
	11		.9957	.7030	.1684	.0126	.0004			
	12		.9985	.8018	.2547	.0253	.0010			
	13		.9995	.8761	.3578	.0469	.0025	.0001		
	14		.9999	.9274	.4709	.0804	.0054	.0002		
	15			.9601	.5854	.1285	.0111	.0004		
	16			.9794	.6926	.1923	.0211	.0010		
	17			.9900	.7862	.2712	.0376	.0022		
	18			.9954	.8623	.3621	.0630	.0045		
	19			.9980	.9203	.4602	.0995	.0089		
	20			.9992	.9617	.5595	.1488	.0165		
	21			.9999	.9896	.6540	.2114	.0288		
	22				.9999	.7389	.2864	.0479	.0001	
	23					.8109	.3702	.0755	.0002	
	24					.8686	.4617	.1136	.0006	
	25					.9125	.5535	.1631	.0012	
	26					.9442	.6417	.2244	.0024	
	27					.9658	.7224	.2964	.0046	
	28					.9800	.7925	.3768	.0084	
	29					.9887	.8505	.4623	.0148	
	30					.9939	.8962	.5491	.0248	
	31					.9969	.9306	.6331	.0398	.0001
	32					.9984	.9554	.7107	.0615	.0002
	33					.9993	.9724	.7793	.0912	.0004
	34					.9997	.9836	.8371	.1303	.0009
	35					.9998	.9906	.8839	.1795	.0018
	36					.9999	.9948	.9201	.2386	.0033
	37						.9972	.9469	.3068	.0060
	38						.9986	.9660	.3822	.0105
	39						.9993	.9790	.4621	.0176
	40						.9997	.9875	.5433	.0284
	41						.9998	.9928	.6225	.0443
	42						.9999	.9960	.6967	.0666
	43							.9979	.7635	.0967
	44							.9989	.8211	.1356
	45							.9995	.8689	.1841
	46							.9997	.9070	.2421
	47							.9999	.9362	.3087
	48								.9577	.3822
	49								.9729	.4602
	50								.9832	.5398
	51								.9900	.6178
	52								.9942	.6913
	53								.9968	.7979
	54								.9983	.8159
	55								.9991	.8644

CUMULATIVE BINOMIAL DISTRIBUTION (Continued)

n	X	0.01	0.05	0.10	0.15	p 0.20	0.25	0.30	0.40	0.50
100	56								.9996	.9033
	57								.9998	.9334
	58								.9999	.9557
	59									.9716
	60									.9824
	61									.9895
	62									.9940
	63									.9967
	64									.9982
	65									.9991
	66									.9996
	67									.9998
	68									.9999

APPENDIX B

CUMULATIVE POISSON DISTRIBUTION

X	μ: 0.01	0.05	0.10	0.15	0.20	0.25	0.30	0.35
0	.990	.951	.905	.861	.819	.779	.741	.705
1	.999	.998	.995	.990	.982	.974	.963	.951
2		.999	.999	.999	.999	.998	.996	.994
3							.999	.999

X	0.40	0.45	0.50	0.55	0.60	0.65	0.70	0.75
0	.670	.638	.607	.577	.549	.522	.497	.472
1	.938	.925	.910	.894	.878	.861	.844	.827
2	.992	.989	.986	.982	.977	.972	.966	.959
3	.999	.999	.998	.998	.997	.996	.994	.993
4			.999	.999	.999	.999	.999	.999

X	0.80	0.85	0.90	0.95	1.0	1.1	1.2	1.3
0	.449	.427	.407	.387	.368	.333	.301	.273
1	.809	.791	.772	.754	.736	.699	.663	.627
2	.953	.945	.937	.929	.920	.900	.879	.857
3	.991	.989	.987	.984	.981	.974	.966	.957
4	.999	.998	.998	.997	.996	.995	.992	.989
5					.999	.999	.998	.998

X	1.4	1.5	1.6	1.7	1.8	1.9	2.0	2.1
0	.247	.223	.202	.183	.165	.150	.135	.111
1	.592	.558	.525	.493	.463	.434	.406	.355
2	.833	.809	.783	.757	.731	.704	.677	.623
3	.946	.934	.921	.907	.891	.875	.857	.819
4	.986	.981	.976	.970	.964	.956	.947	.928
5	.997	.996	.994	.992	.990	.987	.983	.975
6	.999	.999	.999	.998	.997	.997	.995	.993
7					.999	.999	.999	.998

X	2.2	2.3	2.4	2.5	2.6	2.7	2.8	2.9
0	.110	.100	.090	.082	.074	.067	.060	.055
1	.354	.330	.308	.287	.267	.248	.231	.214
2	.622	.596	.569	.543	.518	.493	.469	.445
3	.819	.799	.778	.757	.736	.714	.691	.669
4	.927	.916	.904	.891	.877	.862	.847	.831
5	.975	.970	.964	.957	.950	.943	.934	.925
6	.992	.990	.988	.985	.982	.979	.975	.971
7	.998	.997	.996	.995	.994	.993	.991	.990
8	.999	.999	.999	.998	.998	.998	.997	.996
9				.999	.999	.999	.999	.999

Blank spaces above a first column entry are read 0.0, and blank spaces below the last column entry are read 1.0.

CUMULATIVE POISSON DISTRIBUTION (Continued)

X	3.0	3.2	3.4	3.6	3.8	4.0	4.2	4.4
0	.049	.041	.033	.027	.022	.018	.015	.012
1	.199	.171	.147	.126	.107	.091	.078	.066
2	.423	.380	.340	.303	.269	.238	.210	.185
3	.647	.603	.558	.515	.473	.433	.395	.359
4	.815	.781	.744	.706	.668	.628	.590	.551
5	.916	.895	.871	.844	.816	.785	.753	.720
6	.966	.955	.942	.927	.909	.889	.867	.844
7	.988	.983	.977	.969	.960	.948	.936	.921
8	.996	.994	.992	.988	.984	.978	.972	.964
9	.998	.998	.997	.996	.994	.991	.989	.985
10	.999	.999	.999	.999	.998	.997	.996	.994
11					.999	.999	.999	.998
12								.999

X	4.6	4.8	5.0	5.2	5.4	5.6	5.8	6.0
0	.010	.008	.007	.006	.005	.004	.003	.002
1	.056	.048	.040	.034	.029	.024	.021	.017
2	.163	.143	.124	.109	.095	.082	.072	.061
3	.326	.294	.265	.238	.213	.191	.170	.151
4	.513	.476	.440	.406	.373	.342	.313	.285
5	.686	.651	.615	.581	.546	.512	.478	.445
6	.818	.791	.762	.732	.702	.670	.638	.606
7	.905	.887	.866	.845	.822	.797	.771	.743
8	.955	.944	.931	.918	.903	.886	.867	.847
9	.980	.975	.968	.960	.951	.941	.929	.916
10	.992	.990	.986	.982	.977	.972	.965	.957
11	.997	.996	.994	.993	.990	.988	.984	.979
12	.999	.999	.997	.997	.996	.995	.993	.991
13			.999	.999	.999	.998	.997	.996
14						.999	.999	.998
15								.999

X	6.2	6.4	6.6	6.8	7.0	7.2	7.4	7.6
0	.002	.002	.001	.001	.001	.001	.001	.001
1	.015	.012	.010	.009	.007	.006	.005	.004
2	.054	.046	.040	.034	.030	.025	.022	.019
3	.134	.119	.105	.093	.082	.072	.063	.055
4	.259	.235	.213	.192	.173	.156	.140	.125
5	.414	.384	.355	.327	.301	.276	.253	.231
6	.574	.542	.511	.480	.450	.420	.392	.365
7	.716	.687	.658	.628	.599	.569	.539	.510
8	.826	.803	.780	.755	.729	.703	.676	.648
9	.902	.886	.869	.850	.830	.810	.788	.765
10	.949	.939	.927	.915	.901	.887	.871	.854
11	.975	.969	.963	.955	.947	.937	.926	.915
12	.989	.986	.982	.978	.973	.967	.961	.954
13	.995	.994	.992	.990	.987	.984	.980	.976
14	.998	.997	.997	.996	.994	.993	.991	.989
15	.999	.999	.999	.998	.998	.997	.996	.995
16				.999	.999	.999	.998	.998
17							.999	.999

CUMULATIVE POISSON DISTRIBUTION (Continued)

X	μ 7.8	8.0	8.5	9.0	9.5	10.0	11.0	12.0
0								
1	.004	.003	.002	.001	.001			
2	.016	.014	.009	.006	.004	.003	.001	.001
3	.048	.042	.030	.021	.015	.010	.005	.002
4	.112	.100	.074	.055	.040	.029	.015	.008
5	.210	.191	.150	.116	.089	.067	.038	.020
6	.338	.313	.256	.207	.165	.130	.079	.046
7	.481	.453	.386	.324	.269	.220	.143	.090
8	.620	.593	.523	.456	.392	.333	.232	.155
9	.741	.717	.653	.587	.522	.458	.341	.242
10	.835	.816	.763	.706	.645	.583	.460	.347
11	.902	.888	.849	.803	.752	.697	.579	.462
12	.945	.936	.909	.876	.836	.792	.689	.576
13	.971	.966	.949	.926	.898	.864	.781	.682
14	.986	.983	.973	.959	.940	.917	.854	.772
15	.993	.992	.986	.978	.967	.951	.907	.844
16	.997	.996	.993	.989	.982	.973	.944	.899
17	.999	.998	.997	.995	.991	.986	.968	.937
18		.999	.999	.998	.996	.993	.982	.963
19				.999	.998	.997	.991	.979
20					.999	.998	.995	.988
21						.999	.998	.994
22							.999	.997
23								.999

CUMULATIVE POISSON DISTRIBUTION (Continued)

	μ							
X	13.0	14.0	15.0	16.0	17.0	18.0	19.0	20.0
0								
1								
2								
3	.001							
4	.004	.002	.001					
5	.011	.006	.003	.001	.001			
6	.026	.014	.008	.004	.002	.001	.001	
7	.054	.032	.018	.010	.005	.003	.002	.001
8	.100	.062	.037	.022	.013	.007	.004	.002
9	.166	.109	.070	.043	.026	.015	.009	.005
10	.252	.176	.118	.077	.049	.030	.018	.011
11	.353	.260	.185	.127	.085	.055	.035	.021
12	.463	.358	.268	.193	.135	.092	.061	.039
13	.573	.464	.363	.275	.201	.143	.098	.066
14	.675	.570	.466	.368	.281	.208	.150	.105
15	.764	.669	.568	.467	.371	.287	.215	.157
16	.835	.756	.664	.566	.468	.375	.292	.221
17	.890	.827	.749	.659	.564	.469	.378	.297
18	.930	.883	.819	.742	.655	.562	.469	.381
19	.957	.923	.875	.812	.736	.651	.561	.470
20	.975	.952	.917	.868	.805	.731	.647	.559
21	.986	.971	.947	.911	.861	.799	.725	.644
22	.992	.983	.967	.942	.905	.855	.793	.721
23	.996	.991	.981	.963	.937	.899	.849	.787
24	.998	.995	.989	.978	.959	.932	.893	.843
25	.999	.997	.994	.987	.975	.955	.927	.888
26		.999	.997	.993	.985	.972	.951	.922
27			.998	.996	.991	.983	.969	.948
28			.999	.998	.995	.990	.980	.966
29				.999	.997	.994	.988	.978
30					.999	.997	.993	.987
31						.998	.996	.992
32						.999	.998	.995
33							.999	.997
34								.999

APPENDIX C

THE HYPERGEOMETRIC DISTRIBUTION

N	Np	Nq	X	F(X)	f(X)
10	1	1	0	0.900000	0.900000
10	1	1	1	1.000000	0.100000
10	2	1	0	0.800000	0.800000
10	2	1	1	1.000000	0.200000
10	2	2	0	0.622222	0.622222
10	2	2	1	0.977778	0.355556
10	2	2	2	1.000000	0.022222
10	3	1	0	0.700000	0.700000
10	3	2	1	1.000000	0.300000
10	3	2	0	0.466667	0.466667
10	3	2	1	0.933333	0.466667
10	3	2	2	1.000000	0.066667
10	3	3	0	0.291667	0.291667
10	3	3	1	0.816667	0.525000
10	3	3	2	0.991667	0.175000
10	3	3	3	1.000000	0.008333
10	4	1	0	0.600000	0.600000
10	4	1	1	1.000000	0.400000
10	4	2	0	0.333333	0.333333
10	4	2	1	0.866667	0.533333
10	4	2	2	1.000000	0.133333
10	4	3	0	0.166667	0.166667
10	4	3	1	0.666667	0.500000
10	4	3	2	0.966667	0.300000
10	4	3	3	1.000000	0.033333
10	4	4	0	0.071429	0.071429
10	4	4	1	0.452381	0.380952
10	4	4	2	0.880952	0.428571
10	4	4	3	0.995238	0.114286
10	4	4	4	1.000000	0.004762
10	5	1	0	0.500000	0.500000
10	5	1	1	1.000000	0.500000
10	5	2	0	0.222222	0.222222
10	5	2	1	0.777778	0.555556
10	5	2	2	1.000000	0.222222
10	6	6	3	0.452381	0.380952
10	6	6	4	0.880952	0.428571
10	6	6	5	0.995238	0.114286
10	6	6	6	1.000000	0.004762
10	7	1	0	0.300000	0.300000
10	7	1	1	1.000000	0.700000
10	7	2	0	0.066667	0.066667
10	7	2	1	0.533333	0.466667
10	7	2	2	1.000000	0.466667
10	7	3	0	0.008333	0.008333
10	5	3	0	0.083333	0.083333
10	5	3	1	0.500000	0.416667
10	5	3	2	0.916667	0.416667
10	5	3	3	1.000000	0.083333
10	5	4	0	0.023810	0.023810
10	5	4	1	0.261905	0.238095
10	5	4	2	0.738095	0.476190
10	5	4	3	0.976190	0.238095
10	5	4	4	1.000000	0.023810
10	5	5	0	0.003968	0.003968
10	5	5	1	0.103175	0.099206
10	5	5	2	0.500000	0.396825
10	5	5	3	0.896825	0.396825
10	5	5	4	0.996032	0.099206
10	5	5	5	1.000000	0.003968
10	6	1	0	0.400000	0.400000
10	6	1	1	1.000000	0.600000
10	6	2	0	0.133333	0.133333
10	6	2	1	0.666667	0.533333
10	6	2	2	1.000000	0.333333
10	6	3	0	0.033333	0.033333
10	6	3	1	0.333333	0.300000
10	6	3	2	0.833333	0.500000
10	6	3	3	1.000000	0.166667
10	6	4	0	0.004762	0.004762
10	6	4	1	0.119048	0.114286
10	6	4	2	0.547619	0.428571
10	6	4	3	0.928571	0.380952
10	6	4	4	1.000000	0.071429
10	6	5	1	0.023810	0.023810
10	6	5	2	0.261905	0.238095
10	6	5	3	0.738095	0.476190
10	6	5	4	0.976190	0.238095
10	6	5	5	1.000000	0.023810
10	6	6	2	0.071429	0.071429
10	8	3	2	0.533333	0.466667
10	8	3	3	1.000000	0.466667
10	8	4	2	0.133333	0.133333
10	8	4	3	0.666667	0.533333
10	8	4	4	1.000000	0.333333
10	8	5	3	0.222222	0.222222
10	8	5	4	0.777778	0.555556
10	8	5	5	1.000000	0.222222
10	8	6	4	0.333333	0.333333
10	8	6	5	0.866667	0.533333

*Extracted with permission from Gerald J. Lieberman and Donald B. Owen, *Tables of the Hypergeometric Probability Distribution*, Stanford University Press, Stanford, Calif., 1961.

THE HYPERGEOMETRIC DISTRIBUTION (Continued)

N	Np	Nq	X	$F(X)$	$f(X)$	N	Np	Nq	X	$F(X)$	$f(X)$
10	7	3	1	0.183333	0.175000	10	8	6	6	1.000000	0.133333
10	7	3	2	0.708333	0.525000	10	8	7	5	0.466667	0.466667
10	7	3	3	1.000000	0.291667	10	8	7	6	0.933333	0.466667
10	7	4	1	0.033333	0.033333	10	8	7	7	1.000000	0.066667
10	7	4	2	0.333333	0.300000	10	8	8	6	0.622222	0.622222
10	7	4	3	0.833333	0.500000	10	8	8	7	0.977778	0.355556
10	7	4	4	1.000000	0.166667	10	8	8	8	1.000000	0.022222
10	7	5	2	0.083333	0.083333	10	9	1	0	0.100000	0.100000
10	7	5	3	0.500000	0.416667	10	9	1	1	1.000000	0.900000
10	7	5	4	0.916667	0.416667	10	9	2	1	0.200000	0.200000
10	7	5	5	1.000000	0.083333	10	9	2	2	1.000000	0.800000
10	7	6	3	0.166667	0.166667	10	9	3	2	0.300000	0.300000
10	7	6	4	0.666667	0.500000	10	9	3	3	1.000000	0.700000
10	7	6	5	0.966667	0.300000	10	9	4	3	0.400000	0.400000
10	7	6	6	1.000000	0.033333	10	9	4	4	1.000000	0.600000
10	7	7	4	0.291667	0.291667	10	9	5	4	0.500000	0.500000
10	7	7	5	0.816667	0.525000	10	9	5	5	1.000000	0.500000
10	7	7	6	0.991667	0.175000	10	9	6	5	0.600000	0.600000
10	7	7	7	1.000000	0.008333	10	9	6	6	1.000000	0.400000
10	8	1	0	0.200000	0.200000	10	9	7	6	0.700000	0.700000
10	8	1	1	1.000000	0.800000	10	9	7	7	1.000000	0.300000
10	8	2	0	0.022222	0.022222	10	9	8	7	0.800000	0.800000
10	8	2	1	0.377778	0.355556	10	9	8	8	1.000000	0.200000
10	8	2	2	1.000000	0.622222	10	9	9	8	0.900000	0.900000
10	8	3	1	0.066667	0.066667	10	9	9	9	1.000000	0.100000

APPENDIX D

TAIL AREAS—THE NORMAL DISTRIBUTION*

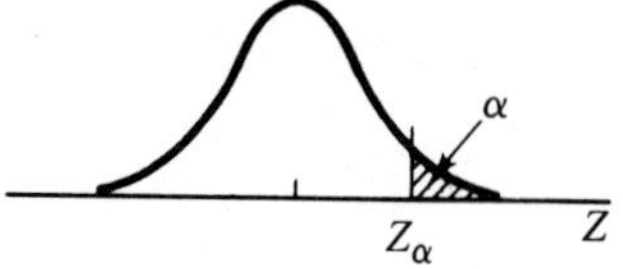

Z_α	.00	.01	.02	.03	.04	.05	.06	.07	.08	.09
0.0	.5000	.4960	.4920	.4880	.4840	.4801	.4761	.4721	.4681	.4641
0.1	.4602	.4562	.4522	.4483	.4443	.4404	.4364	.4325	.4286	.4247
0.2	.4207	.4168	.4129	.4090	.4052	.4013	.3974	.3936	.3897	.3859
0.3	.3821	.3783	.3745	.3707	.3669	.3632	.3594	.3557	.3520	.3483
0.4	.3446	.3409	.3372	.3336	.3300	.3264	.3228	.3192	.3156	.3121
0.5	.3085	.3050	.3015	.2981	.2946	.2912	.2877	.2843	.2810	.2776
0.6	.2743	.2709	.2676	.2643	.2611	.2578	.2546	.2514	.2483	.2451
0.7	.2420	.2389	.2358	.2327	.2296	.2266	.2236	.2206	.2177	.2148
0.8	.2119	.2090	.2061	.2033	.2005	.1977	.1949	.1922	.1894	.1867
0.9	.1841	.1814	.1788	.1762	.1736	.1711	.1685	.1660	.1635	.1611
1.0	.1587	.1562	.1539	.1515	.1492	.1469	.1446	.1423	.1401	.1379
1.1	.1357	.1335	.1314	.1292	.1271	.1251	.1230	.1210	.1190	.1170
1.2	.1151	.1131	.1112	.1093	.1075	.1056	.1038	.1020	.1003	.0985
1.3	.0968	.0951	.0934	.0918	.0901	.0885	.0869	.0853	.0838	.0823
1.4	.0808	.0793	.0778	.0764	.0749	.0735	.0721	.0708	.0694	.0681
1.5	.0668	.0655	.0643	.0630	.0618	.0606	.0594	.0582	.0571	.0559
1.6	.0548	.0537	.0526	.0516	.0505	.0495	.0485	.0475	.0465	.0455
1.7	.0446	.0436	.0427	.0418	.0409	.0401	.0392	.0384	.0375	.0367
1.8	.0359	.0351	.0344	.0336	.0329	.0322	.0314	.0307	.0301	.0294
1.9	.0287	.0281	.0274	.0268	.0262	.0256	.0250	.0244	.0239	.0233
2.0	.0228	.0222	.0217	.0212	.0207	.0202	.0197	.0192	.0188	.0183
2.1	.0179	.0174	.0170	.0166	.0162	.0158	.0154	.0150	.0146	.0143
2.2	.0139	.0136	.0132	.0129	.0125	.0122	.0119	.0116	.0113	.0110
2.3	.0107	.0104	.0102	.00990	.00964	.00939	.00914	.00889	.00866	.00842
2.4	.00820	.00798	.00776	.00755	.00734	.00714	.00695	.00676	.00657	.00639
2.5	.00621	.00604	.00587	.00570	.00554	.00539	.00523	.00508	.00494	.00480
2.6	.00466	.00453	.00440	.00427	.00415	.00402	.00391	.00379	.00368	.00357
2.7	.00347	.00336	.00326	.00317	.00307	.00298	.00289	.00280	.00272	.00264
2.8	.00256	.00248	.00240	.00233	.00226	.00219	.00212	.00205	.00199	.00193
2.9	.00187	.00181	.00175	.00169	.00164	.00159	.00154	.00149	.00144	.00139

Z_α	.0	.1	.2	.3	.4	.5	.6	.7	.8	.9
3	.00135	$.0^{3}968$	$.0^{3}687$	$.0^{3}483$	$.0^{3}337$	$.0^{3}233$	$.0^{3}159$	$.0^{3}108$	$.0^{4}723$	$.0^{4}481$
4	$.0^{4}317$	$.0^{4}207$	$.0^{4}133$	$.0^{5}854$	$,0^{5}541$	$.0^{5}340$	$.0^{5}211$	$.0^{5}130$	$.0^{6}793$	$.0^{6}479$
5	$.0^{6}287$	$.0^{6}170$	$.0^{7}996$	$.0^{7}579$	$.0^{7}333$	$.0^{7}190$	$.0^{7}107$	$.0^{8}599$	$.0^{8}332$	$.0^{8}182$
6	$.0^{9}987$	$.0^{9}530$	$.0^{9}282$	$.0^{9}149$	$.0^{10}777$	$.0^{10}402$	$.0^{10}206$	$.0^{10}104$	$.0^{11}523$	$.0^{11}260$

*Reprinted by permission from Frederick E. Croxton, *Elementary Statistics with Applications in Medicine*, Prentice-Hall, Inc., Englewood Cliffs, N. J., 1953, p. 323.

APPENDIX E

THE CHI-SQUARE DISTRIBUTION*

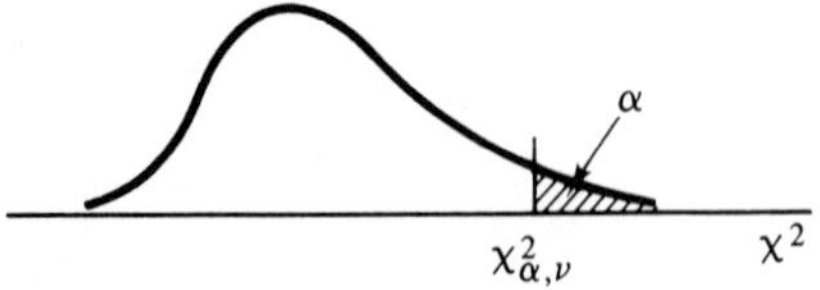

ν \ α	.995	.99	.98	.975	.95	.90	.80	.75	.70	.50
1	$.0^4393$	$.0^3157$	$.0^3628$	$.0^3982$	.00393	.0158	.0642	.102	.148	.455
2	.0100	.0201	.0404	.0506	.103	.211	.446	.575	.713	1.386
3	.0717	.115	.185	.216	.352	.584	1.005	1.213	1.424	2.366
4	.207	.297	.429	.484	.711	1.064	1.649	1.923	2.195	3.357
5	.412	.554	.752	.831	1.145	1.610	2.343	2.675	3.000	4.351
6	.676	.872	1.134	1.237	1.635	2.204	3.070	3.455	3.828	5.348
7	.989	1.239	1.564	1.690	2.167	2.833	3.822	4.255	4.671	6.346
8	1.344	1.646	2.032	2.180	2.733	3.490	4.594	5.071	5.527	7.344
9	1.735	2.088	2.532	2.700	3.325	4.168	5.380	5.899	6.393	8.343
10	2.156	2.558	3.059	3.247	3.940	4.865	6.179	6.737	7.267	9.342
11	2.603	3.053	3.609	3.816	4.575	5.578	6.989	7.584	8.148	10.341
12	3.074	3.571	4.178	4.404	5.226	6.304	7.807	8.438	9.034	11.340
13	3.565	4.107	4.765	5.009	5.892	7.042	8.634	9.299	9.926	12.340
14	4.075	4.660	5.368	5.629	6.571	7.790	9.467	10.165	10.821	13.339
15	4.601	5.229	5.985	6.262	7.261	8.547	10.307	11.036	11.721	14.339
16	5.142	5.812	6.614	6.908	7.962	9.312	11.152	11.912	12.624	15.338
17	5.697	6.408	7.255	7.564	8.672	10.085	12.002	12.792	13.531	16.338
18	6.265	7.015	7.906	8.231	9.390	10.865	12.857	13.675	14.440	17.338
19	6.844	7.633	8.567	8.907	10.117	11.651	13.716	14.562	15.352	18.338
20	7.434	8.260	9.237	9.591	10.851	12.443	14.578	15.452	16.266	19.337
21	8.034	8.897	9.915	10.283	11.591	13.240	15.445	16.344	17.182	20.337
22	8.643	9.542	10.600	10.982	12.338	14.041	16.314	17.240	18.101	21.337
23	9.260	10.196	11.293	11.688	13.091	14.848	17.187	18.137	19.021	22.337
24	9.886	10.856	11.992	12.401	13.848	15.659	18.062	19.037	19.943	23.337
25	10.520	11.524	12.697	13.120	14.611	16.473	18.940	19.939	20.867	24.337
26	11.160	12.198	13.409	13.844	15.379	17.292	19.820	20.843	21.792	25.336
27	11.808	12.879	14.125	14.573	16.151	18.114	20.703	21.749	22.719	26.336
28	12.461	13.565	14.847	15.308	16.928	18.939	21.588	22.657	23.647	27.336
29	13.121	14.256	15.574	16.047	17.708	19.768	22.475	23.567	24.577	28.336
30	13.787	14.953	16.306	16.791	18.493	20.599	23.364	24.478	25.508	29.336

*For values of $\nu > 30$, approximate values for χ^2 may be obtained from the expression $\nu[1 \pm 2/9\nu \pm x/\sigma(\sqrt{2/9\nu})]^3$, where x/σ is the normal deviate cutting off the tails of a normal distribution. If x/σ is taken at the 0.02 level, so that 0.01 of the normal distribution is in each tail, the expression yields χ^2 at the 0.99 and 0.01 points. For very large values of ν, it is sufficiently accurate to compute $\sqrt{2\chi^2}$, the distribution of which is approximately normal around a mean of $\sqrt{2\nu - 1}$ and with a standard deviation of 1.

This table is taken by consent from R. A. Fisher and F. Yates, *Statistical Tables for Biological, Agricultural, and Medical Research*, Oliver and Boyd, Edinburgh, and from Catherine M. Thomson, "Table of Percentage Points of the χ^2 Distribution," *Biometrika*, Vol. 32, Part II, October 1941, pp. 187–191. Reproduced in Croxton *Elementary Statistics with Applications in Medicine*, Appendix VI, pp. 328–329.

THE CHI-SQUARE DISTRIBUTION (Continued)

.30	.25	.20	.10	.05	.025	.02	.01	.005	.001	α / ν
1.074	1.323	1.642	2.706	3.841	5.024	5.412	6.635	7.879	10.827	1
2.408	2.773	3.219	4.605	5.991	7.378	7.824	9.210	10.597	13.815	2
3.665	4.108	4.642	6.251	7.815	9.348	9.837	11.345	12.838	16.268	3
4.878	5.385	5.989	7.779	9.488	11.143	11.668	13.277	14.860	18.465	4
6.064	6.626	7.289	9.236	11.070	12.832	13.388	15.086	16.750	20.517	5
7.231	7.841	8.558	10.645	12.592	14.449	15.033	16.812	18.548	22.457	6
8.383	9.037	9.803	12.017	14.067	16.013	16.622	18.475	20.278	24.322	7
9.524	10.219	11.030	13.362	15.507	17.535	18.168	20.090	21.955	26.125	8
10.656	11.389	12.242	14.684	16.919	19.023	19.679	21.666	23.589	27.877	9
11.781	12.549	13.442	15.987	18.307	20.483	21.161	23.209	25.188	29.588	10
12.899	13.701	14.631	17.275	19.675	21.920	22.618	24.725	26.757	31.264	11
14.011	14.845	15.812	18.549	21.026	23.337	24.054	26.217	28.300	32.909	12
15.119	15.984	16.985	19.812	22.362	24.736	25.472	27.688	29.819	34.528	13
16.222	17.117	18.151	21.064	23.685	26.119	26.873	29.141	31.319	36.123	14
17.322	18.245	19.311	22.307	24.996	27.488	28.259	30.578	32.801	37.697	15
18.418	19.369	20.465	23.542	26.296	28.845	29.633	32.000	34.267	39.252	16
19.511	20.489	21.615	24.769	27.587	30.191	30.995	33.409	35.718	40.790	17
20.601	21.605	22.760	25.989	28.869	31.526	32.346	34.805	37.156	42.312	18
21.689	22.718	23.900	27.204	30.144	32.852	33.687	36.191	38.582	43.820	19
22.775	23.828	25.038	28.412	31.410	34.170	35.020	37.566	39.997	45.315	20
23.858	24.935	26.171	29.615	32.671	35.479	36.343	38.932	41.401	46.797	21
24.939	26.039	27.301	30.813	33.924	36.781	37.659	40.289	42.796	48.268	22
26.018	27.141	28.429	32.007	35.172	38.076	38.968	41.638	44.181	49.728	23
27.096	28.241	29.553	33.196	36.415	39.364	40.270	42.980	45.558	51.179	24
28.172	29.339	30.675	34.382	37.652	40.646	41.566	44.314	46.928	52.620	25
29.246	30.434	31.795	35.563	38.885	41.923	42.856	45.642	48.290	54.052	26
30.319	31.528	32.912	36.741	40.113	43.194	44.140	46.963	49.645	55.476	27
31.391	32.620	34.027	37.916	41.337	44.461	45.419	48.278	50.993	56.893	28
32.461	33.711	35.139	39.087	42.557	45.722	46.693	49.588	52.336	58.302	29
33.530	34.800	36.250	40.256	43.773	46.979	47.962	50.892	53.672	59.703	30

APPENDIX F

THE t-DISTRIBUTION*

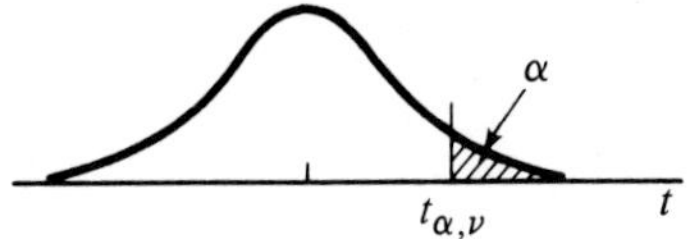

ν \ α	.40	.30	.20	.10	.050	.025	.010	.005	.001	0.005
1	.325	.727	1.376	3.078	6.314	12.71	31.82	63.66	318.3	636.6
2	.289	.617	1.061	1.886	2.920	4.303	6.965	9.925	22.33	31.60
3	.277	.584	.978	1.638	2.353	3.182	4.541	5.841	10.22	12.94
4	.271	.569	.941	1.533	2.132	2.776	3.747	4.604	7.173	8.610
5	.267	.559	.920	1.476	2.015	2.571	3.365	4.032	5.893	6.859
6	.265	.553	.906	1.440	1.943	2.447	3.143	3.707	5.208	5.959
7	.263	.549	.896	1.415	1.895	2.365	2.998	3.499	4.785	5.405
8	.262	.546	.889	1.397	1.860	2.306	2.896	3.355	4.501	5.041
9	.261	.543	.883	1.383	1.833	2.262	2.821	3.250	4.297	4.781
10	.260	.542	.879	1.372	1.812	2.228	2.764	3.169	4.144	4.587
11	.260	.540	.876	1.363	1.796	2.201	2.718	3.106	4.025	4.437
12	.259	.539	.873	1.356	1.782	2.179	2.681	3.055	3.930	4.318
13	.259	.538	.870	1.350	1.771	2.160	2.650	3.012	3.852	4.221
14	.258	.537	.868	1.345	1.761	2.145	2.624	2.977	3.787	4.140
15	.258	.536	.866	1.341	1.753	2.131	2.602	2.947	3.733	4.073
16	.258	.535	.865	1.337	1.746	2.120	2.583	2.921	3.686	4.015
17	.257	.534	.863	1.333	1.740	2.110	2.567	2.898	3.646	3.965
18	.257	.534	.862	1.330	1.734	2.101	2.552	2.878	3.611	3.922
19	.257	.533	.861	1.328	1.729	2.093	2.539	2.861	3.579	3.883
20	.257	.533	.860	1.325	1.725	2.086	2.528	2.845	3.552	3.850
21	.257	.532	.859	1.323	1.721	2.080	2.518	2.831	3.527	3.819
22	.256	.532	.858	1.321	1.717	2.074	2.508	2.819	3.505	3.792
23	.256	.532	.858	1.319	1.714	2.069	2.500	2.807	3.485	3.767
24	.256	.531	.857	1.318	1.711	2.064	2.492	2.797	3.467	3.745
25	.256	.531	.856	1.316	1.708	2.060	2.485	2.787	3.450	3.725
26	.256	.531	.856	1.315	1.706	2.056	2.479	2.779	3.435	3.707
27	.256	.531	.855	1.314	1.703	2.052	2.473	2.771	3.421	3.690
28	.256	.530	.855	1.313	1.701	2.048	2.467	2.763	3.408	3.674
29	.256	.530	.854	1.311	1.699	2.045	2.462	2.756	3.396	3.659
30	.256	.530	.854	1.310	1.697	2.042	2.457	2.750	3.385	3.646
40	.255	.529	.851	1.303	1.684	2.021	2.423	2.704	3.307	3.551
50	.255	.528	.849	1.298	1.676	2.009	2.403	2.678	3.262	3.495
60	.254	.527	.848	1.296	1.671	2.000	2.390	2.660	3.232	3.460
80	.254	.527	.846	1.292	1.664	1.990	2.374	2.639	3.195	3.415
100	.254	.526	.845	1.290	1.660	1.984	2.365	2.626	3.174	3.389
200	.254	.525	.843	1.286	1.653	1.972	2.345	2.601	3.131	3.339
500	.253	.525	.842	1.283	1.648	1.965	2.334	2.586	3.106	3.310
∞	.253	.524	.842	1.282	1.645	1.960	2.326	2.576	3.090	3.291

*This table is reproduced from *Statistical Tables and Formulas* by A. Hald, published by John Wiley & Sons. Inc., New York, the greater part of which has been reproduced from Table III of R. A. Fisher and F. Yates, *Statistical Tables*, Oliver & Boyd, Edinburgh, by permission of the authors and publishers.

APPENDIX G

THE F-DISTRIBUTION*

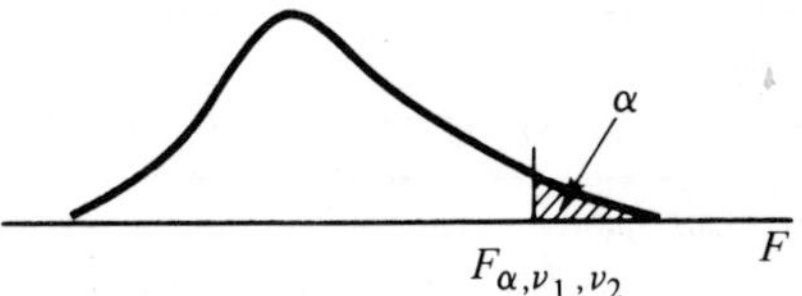

Table of $F_{0.10;\,\nu_1,\nu_2}$

Degrees of freedom for the denominator (ν_2)	Degrees of freedom for the numerator (ν_1)																	
	1	2	3	4	5	6	7	8	9	10	15	20	30	50	100	200	500	∞
1	39.9	49.5	53.6	55.8	57.2	58.2	58.9	59.4	59.9	60.2	61.2	61.7	62.3	62.7	63.0	63.2	63.3	63.3
2	8.53	9.00	9.16	9.24	9.29	9.33	9.35	9.37	9.38	9.39	9.42	9.44	9.46	9.47	9.48	9.49	9.49	9.49
3	5.54	5.46	5.39	5.34	5.31	5.28	5.27	5.25	5.24	5.23	5.20	5.18	5.17	5.15	5.14	5.14	5.14	5.13
4	4.54	4.32	4.19	4.11	4.05	4.01	3.98	3.95	3.94	3.92	3.87	3.84	3.82	3.80	3.78	3.77	3.76	3.76
5	4.06	3.78	3.62	3.52	3.45	3.40	3.37	3.34	3.32	3.30	3.24	3.21	3.17	3.15	3.13	3.12	3.11	3.10
6	3.78	3.46	3.29	3.18	3.11	3.05	3.01	2.98	2.96	2.94	2.87	2.84	2.80	2.77	2.75	2.73	2.73	2.72
7	3.59	3.26	3.07	2.96	2.88	2.83	2.78	2.75	2.72	2.70	2.63	2.59	2.56	2.52	2.50	2.48	2.48	2.47
8	3.46	3.11	2.92	2.81	2.73	2.67	2.62	2.59	2.56	2.54	2.46	2.42	2.38	2.35	2.32	2.31	2.30	2.29
9	3.36	3.01	2.81	2.69	2.61	2.55	2.51	2.47	2.44	2.42	2.34	2.30	2.25	2.22	2.19	2.17	2.17	2.16
10	3.28	2.92	2.73	2.61	2.52	2.46	2.41	2.38	2.35	2.32	2.24	2.20	2.16	2.12	2.09	2.07	2.06	2.06
11	3.23	2.86	2.66	2.54	2.45	2.39	2.34	2.30	2.27	2.25	2.17	2.12	2.08	2.04	2.00	1.99	1.98	1.97
12	3.18	2.81	2.61	2.48	2.39	2.38	2.28	2.24	2.21	2.19	2.10	2.06	2.01	1.97	1.94	1.92	1.91	1.90
13	3.14	2.76	2.56	2.43	2.35	2.28	2.23	2.20	2.16	2.14	2.05	2.01	1.96	1.92	1.88	1.86	1.85	1.85
14	3.10	2.73	2.52	2.39	2.31	2.24	2.19	2.15	2.12	2.10	2.01	1.96	1.91	1.87	1.83	1.82	1.80	1.80
15	3.07	2.70	2.49	2.36	2.27	2.21	2.16	2.12	2.09	2.06	1.97	1.92	1.87	1.83	1.79	1.77	1.76	1.76
16	3.05	2.67	2.46	2.33	2.24	2.18	2.13	2.09	2.06	2.03	1.94	1.89	1.84	1.79	1.76	1.74	1.73	1.72
17	3.03	2.64	2.44	2.31	2.22	2.15	2.10	2.06	2.03	2.00	1.91	1.86	1.81	1.76	1.73	1.71	1.69	1.69
18	3.01	2.62	2.42	2.29	2.20	2.13	2.08	2.04	2.00	1.98	1.89	1.84	1.78	1.74	1.70	1.68	1.67	1.66
19	2.99	2.61	2.40	2.27	2.18	2.11	2.06	2.02	1.98	1.96	1.86	1.81	1.76	1.71	1.67	1.65	1.64	1.63
20	2.97	2.59	2.38	2.25	2.16	2.09	2.04	2.00	1.96	1.94	1.84	1.79	1.74	1.69	1.65	1.63	1.62	1.61
22	2.95	2.56	2.35	2.22	2.13	2.06	2.01	1.97	1.93	1.90	1.81	1.76	1.70	1.65	1.61	1.59	1.58	1.57
24	2.93	2.54	2.33	2.19	2.10	2.04	1.98	1.94	1.91	1.88	1.78	1.73	1.67	1.62	1.58	1.56	1.54	1.53
26	2.91	2.52	2.31	2.17	2.08	2.01	1.96	1.92	1.88	1.86	1.76	1.71	1.65	1.59	1.55	1.53	1.51	1.50
28	2.89	2.50	2.29	2.16	2.06	2.00	1.94	1.90	1.87	1.84	1.74	1.69	1.63	1.57	1.53	1.50	1.49	1.48
30	2.88	2.49	2.28	2.14	2.05	1.98	1.93	1.88	1.85	1.82	1.72	1.67	1.61	1.55	1.51	1.48	1.47	1.46
40	2.84	2.44	2.23	2.09	2.00	1.93	1.87	1.83	1.79	1.76	1.66	1.61	1.54	1.48	1.43	1.41	1.39	1.38
50	2.81	2.41	2.20	2.06	1.97	1.90	1.84	1.80	1.76	1.73	1.63	1.57	1.50	1.44	1.39	1.36	1.34	1.33
60	2.79	2.39	2.18	2.04	1.95	1.87	1.82	1.77	1.74	1.71	1.60	1.54	1.48	1.41	1.36	1.33	1.31	1.29
80	2.77	2.37	2.15	2.02	1.92	1.85	1.79	1.75	1.71	1.68	1.57	1.51	1.44	1.38	1.32	1.28	1.26	1.24
100	2.76	2.36	2.14	2.00	1.91	1.83	1.78	1.73	1.70	1.66	1.56	1.49	1.42	1.35	1.29	1.26	1.23	1.21
200	2.73	2.33	2.11	1.97	1.88	1.80	1.75	1.70	1.66	1.63	1.52	1.46	1.38	1.31	1.24	1.20	1.17	1.14
500	2.72	2.31	2.10	1.96	1.86	1.79	1.73	1.68	1.64	1.61	1.50	1.44	1.36	1.28	1.21	1.16	1.12	1.09
∞	2.71	2.30	2.08	1.94	1.85	1.77	1.72	1.67	1.63	1.60	1.49	1.42	1.34	1.26	1.18	1.13	1.08	1.00

Example: $P\{F_{.10;\,8,\,20} < 2.00\} = 90\%$.

$F_{.90;\,\nu_1,\,\nu_2} = 1/F_{.10;\,\nu_2,\,\nu_1}$. Example: $F_{.90;\,8,\,20} = 1/F_{.10;\,20,\,8} = 1/2.42 = 0.413$.

Approximate formula for ν_1 and ν_2 larger than 30: $\log_{10} F_{.10;\,\nu_1,\,\nu_2} \simeq \dfrac{1.1131}{\sqrt{h - 0.77}} - 0.527\left(\dfrac{1}{\nu_1} - \dfrac{1}{\nu_2}\right)$, where $\dfrac{1}{h} = \dfrac{1}{2}\left(\dfrac{1}{\nu_1} + \dfrac{1}{\nu_2}\right)$.

*This table is abridged from A. Hald, *Statistical Tables and Formulas*, John Wiley and Sons, Inc., New York, a major part of which has been abridged from "Tables of Percentage Points of the Inverted Beta (F) Distribution," computed by M. Merrington and C. M. Thompson, *Biometrika*, Vol. 33, 1943, pp. 73–88, by permission of the proprietors, or reproduced from Table V of R. A. Fisher and F. Yates, *Statistical Tables*, Oliver and Boyd, Edinburgh, by permission of the authors and the publishers.

APPENDIX G THE F-DISTRIBUTION (Continued)

Table of $F_{0.05;\nu_1,\nu_2}$

Degrees of freedom for the numerator (ν_1)

Multiply the numbers of the first row ($\nu_2 = 1$) by 10.

Degrees of freedom for the denominator (ν_2)	1	2	3	4	5	6	7	8	9	10	11	12	13	14	15	16	17	18
1	16.1	20.0	21.6	22.5	23.0	23.4	23.7	23.9	24.1	24.2	24.3	24.4	24.5	24.5	24.6	24.6	24.7	24.7
2	18.5	19.0	19.2	19.2	19.3	19.3	19.4	19.4	19.4	19.4	19.4	19.4	19.4	19.4	19.4	19.4	19.4	19.4
3	10.1	9.55	9.28	9.12	9.01	8.94	8.89	8.85	8.81	8.79	8.76	8.74	8.73	8.71	8.70	8.69	8.68	8.67
4	7.71	6.94	6.59	6.39	6.26	6.16	6.09	6.04	6.00	5.96	5.94	5.91	5.89	5.87	5.86	5.84	5.83	5.82
5	6.61	5.79	5.41	5.19	5.05	4.95	4.88	4.82	4.77	4.74	4.70	4.68	4.66	4.64	4.62	4.60	4.59	4.58
6	5.99	5.14	4.76	4.53	4.39	4.28	4.21	4.15	4.10	4.06	4.03	4.00	3.98	3.96	3.94	3.92	3.91	3.90
7	5.59	4.74	4.35	4.12	3.97	3.87	3.79	3.73	3.68	3.64	3.60	3.57	3.55	3.53	3.51	3.49	3.48	3.47
8	5.32	4.46	4.07	3.84	3.69	3.58	3.50	3.44	3.39	3.35	3.31	3.28	3.26	3.24	3.22	3.20	3.19	3.17
9	5.12	4.26	3.86	3.63	3.48	3.37	3.29	3.23	3.18	3.14	3.10	3.07	3.05	3.03	3.01	2.99	2.97	2.96
10	4.96	4.10	3.71	3.48	3.33	3.22	3.14	3.07	3.02	2.98	2.94	2.91	2.89	2.86	2.85	2.83	2.81	2.80
11	4.84	3.98	3.59	3.36	3.20	3.09	3.01	2.95	2.90	2.85	2.82	2.79	2.76	2.74	2.72	2.70	2.69	2.67
12	4.75	3.89	3.49	3.26	3.11	3.00	2.91	2.85	2.80	2.75	2.72	2.69	2.66	2.64	2.62	2.60	2.58	2.57
13	4.67	3.81	3.41	3.18	3.03	2.92	2.83	2.77	2.71	2.67	2.63	2.60	2.58	2.55	2.53	2.51	2.50	2.48
14	4.60	3.74	3.34	3.11	2.96	2.85	2.76	2.70	2.65	2.60	2.57	2.53	2.51	2.48	2.46	2.44	2.43	2.41
15	4.54	3.68	3.29	3.06	2.90	2.79	2.71	2.64	2.59	2.54	2.51	2.48	2.45	2.42	2.40	2.38	2.37	2.35
16	4.49	3.63	3.24	3.01	2.85	2.74	2.66	2.59	2.54	2.49	2.46	2.42	2.40	2.37	2.35	2.33	2.32	2.30
17	4.45	3.59	3.20	2.96	2.81	2.70	2.61	2.55	2.49	2.45	2.41	2.38	2.35	2.33	2.31	2.29	2.27	2.26
18	4.41	3.55	3.16	2.93	2.77	2.66	2.58	2.51	2.46	2.41	2.37	2.34	2.31	2.29	2.27	2.25	2.23	2.22
19	4.38	3.52	3.13	2.90	2.74	2.63	2.54	2.48	2.42	2.38	2.34	2.31	2.28	2.26	2.23	2.21	2.20	2.18
20	4.35	3.49	3.10	2.87	2.71	2.60	2.51	2.45	2.39	2.35	2.31	2.28	2.25	2.22	2.20	2.18	2.17	2.15
21	4.32	3.47	3.07	2.84	2.68	2.57	2.49	2.42	2.37	2.32	2.28	2.25	2.22	2.20	2.18	2.16	2.14	2.12
22	4.30	3.44	3.05	2.82	2.66	2.55	2.46	2.40	2.34	2.30	2.26	2.23	2.20	2.17	2.15	2.13	2.11	2.10
23	4.28	3.42	3.03	2.80	2.64	2.53	2.44	2.37	2.32	2.27	2.23	2.20	2.18	2.15	2.13	2.11	2.09	2.07
24	4.26	3.40	3.01	2.78	2.62	2.51	2.42	2.36	2.30	2.25	2.21	2.18	2.15	2.13	2.11	2.09	2.07	2.05
25	4.24	3.39	2.99	2.76	2.60	2.49	2.40	2.34	2.28	2.24	2.20	2.16	2.14	2.11	2.09	2.07	2.05	2.04
26	4.23	3.37	2.98	2.74	2.59	2.47	2.39	2.32	2.27	2.22	2.18	2.15	2.12	2.09	2.07	2.05	2.03	2.02
27	4.21	3.35	2.96	2.73	2.57	2.46	2.37	2.31	2.25	2.20	2.17	2.13	2.10	2.08	2.06	2.04	2.02	2.00
28	4.20	3.34	2.95	2.71	2.56	2.45	2.36	2.29	2.24	2.19	2.15	2.12	2.09	2.06	2.04	2.02	2.00	1.99
29	4.18	3.33	2.93	2.70	2.55	2.43	2.35	2.28	2.22	2.18	2.14	2.10	2.08	2.05	2.03	2.01	1.99	1.97
30	4.17	3.32	2.92	2.69	2.53	2.42	2.33	2.27	2.21	2.16	2.13	2.09	2.06	2.04	2.01	1.99	1.98	1.96
32	4.15	3.29	2.90	2.67	2.51	2.40	2.31	2.24	2.19	2.14	2.10	2.07	2.04	2.01	1.99	1.97	1.95	1.94
34	4.13	3.28	2.88	2.65	2.49	2.38	2.29	2.23	2.17	2.12	2.08	2.05	2.02	1.99	1.97	1.95	1.93	1.92
36	4.11	3.26	2.87	2.63	2.48	2.36	2.28	2.21	2.15	2.11	2.07	2.03	2.00	1.98	1.95	1.93	1.92	1.90
38	4.10	3.24	2.85	2.62	2.46	2.35	2.26	2.19	2.14	2.09	2.05	2.02	1.99	1.96	1.94	1.92	1.90	1.88
40	4.08	3.23	2.84	2.61	2.45	2.34	2.25	2.18	2.12	2.08	2.04	2.00	1.97	1.95	1.92	1.90	1.89	1.87
42	4.07	3.22	2.83	2.59	2.44	2.32	2.24	2.17	2.11	2.06	2.03	1.99	1.96	1.93	1.91	1.89	1.87	1.86
44	4.06	3.21	2.82	2.58	2.43	2.31	2.23	2.16	2.10	2.05	2.01	1.98	1.95	1.92	1.90	1.88	1.86	1.84
46	4.05	3.20	2.81	2.57	2.42	2.30	2.22	2.15	2.09	2.04	2.00	1.97	1.94	1.91	1.89	1.87	1.85	1.83
48	4.04	3.19	2.80	2.57	2.41	2.29	2.21	2.14	2.08	2.03	1.99	1.96	1.93	1.90	1.88	1.86	1.84	1.82
50	4.03	3.18	2.79	2.56	2.40	2.29	2.20	2.13	2.07	2.03	1.99	1.95	1.92	1.89	1.87	1.85	1.83	1.81
55	4.02	3.16	2.77	2.54	2.38	2.27	2.18	2.11	2.06	2.01	1.97	1.93	1.90	1.88	1.85	1.83	1.81	1.79
60	4.00	3.15	2.76	2.53	2.37	2.25	2.17	2.10	2.04	1.99	1.95	1.92	1.89	1.86	1.84	1.82	1.80	1.78
65	3.99	3.14	2.75	2.51	2.36	2.24	2.15	2.08	2.03	1.98	1.94	1.90	1.87	1.85	1.82	1.80	1.78	1.76
70	3.98	3.13	2.74	2.50	2.35	2.23	2.14	2.07	2.02	1.97	1.93	1.89	1.86	1.84	1.81	1.79	1.77	1.75
80	3.96	3.11	2.72	2.49	2.33	2.21	2.13	2.06	2.00	1.95	1.91	1.88	1.84	1.82	1.79	1.77	1.75	1.73
90	3.95	3.10	2.71	2.47	2.32	2.20	2.11	2.04	1.99	1.94	1.90	1.86	1.83	1.80	1.78	1.76	1.74	1.72
100	3.94	3.09	2.70	2.46	2.31	2.19	2.10	2.03	1.97	1.93	1.89	1.85	1.82	1.79	1.77	1.75	1.73	1.71
125	3.92	3.07	2.68	2.44	2.29	2.17	2.08	2.01	1.96	1.91	1.87	1.83	1.80	1.77	1.75	1.72	1.70	1.69
150	3.90	3.06	2.66	2.43	2.27	2.16	2.07	2.00	1.94	1.89	1.85	1.82	1.79	1.76	1.73	1.71	1.69	1.67
200	3.89	3.04	2.65	2.42	2.26	2.14	2.06	1.98	1.93	1.88	1.84	1.80	1.77	1.74	1.72	1.69	1.67	1.66
300	3.87	3.03	2.63	2.40	2.24	2.13	2.04	1.97	1.91	1.86	1.82	1.78	1.75	1.72	1.70	1.68	1.66	1.64
500	3.86	3.01	2.62	2.39	2.23	2.12	2.03	1.96	1.90	1.85	1.81	1.77	1.74	1.71	1.69	1.66	1.64	1.62
1000	3.85	3.00	2.61	2.38	2.22	2.11	2.02	1.95	1.89	1.84	1.80	1.76	1.73	1.70	1.68	1.65	1.63	1.61
∞	3.84	3.00	2.60	2.37	2.21	2.10	2.01	1.94	1.88	1.83	1.79	1.75	1.72	1.69	1.67	1.64	1.62	1.60

Example: $P\{F_{.05;8,20} < 2.45\} = 95\%$.

$F_{.95;\nu_1,\nu_2} = 1/F_{.05;\nu_2,\nu_1}$. Example: $F_{.95;8,20} = 1/F_{.05;20,8} = 1/3.15 = 0.317$.

Table of $F_{0.05;\,\nu_1,\nu_2}$

Degrees of freedom for the numerator (ν_1)																	Degrees of freedom for the denominator (ν_2)
19	20	22	24	26	28	30	35	40	45	50	60	80	100	200	500	∞	
24.8	24.8	24.9	24.9	24.9	25.0	25.0	25.1	25.1	25.1	25.2	25.2	25.2	25.3	25.4	25.4	25.4	1
19.4	19.4	19.5	19.5	19.5	19.5	19.5	19.5	19.5	19.5	19.5	19.5	19.5	19.5	19.5	19.5	19.5	2
8.67	8.66	8.65	8.64	8.63	8.62	8.62	8.60	8.59	8.59	8.58	8.57	8.56	8.55	8.54	8.53	8.53	3
5.81	5.80	5.79	5.77	5.76	5.75	5.75	5.73	5.72	5.71	5.70	5.69	5.67	5.66	5.65	5.64	5.63	4
4.57	4.56	4.54	4.53	4.52	4.50	4.50	4.48	4.46	4.45	4.44	4.43	4.41	4.41	4.39	4.37	4.37	5
3.88	3.87	3.86	3.84	3.83	3.82	3.81	3.79	3.77	3.76	3.75	3.74	3.72	3.71	3.69	3.68	3.67	6
3.46	3.44	3.43	3.41	3.40	3.39	3.38	3.36	3.34	3.33	3.32	3.30	3.29	3.27	3.25	3.24	3.23	7
3.16	3.15	3.13	3.12	3.10	3.09	3.08	3.06	3.04	3.03	3.02	3.01	2.99	2.97	2.95	2.94	2.93	8
2.95	2.94	2.92	2.90	2.89	2.87	2.86	2.84	2.83	2.81	2.80	2.79	2.77	2.76	2.73	2.72	2.71	9
2.78	2.77	2.75	2.74	2.72	2.71	2.70	2.68	2.66	2.65	2.64	2.62	2.60	2.59	2.56	2.55	2.54	10
2.66	2.65	2.63	2.61	2.59	2.58	2.57	2.55	2.53	2.52	2.51	2.49	2.47	2.46	2.43	2.42	2.40	11
2.56	2.54	2.52	2.51	2.49	2.48	2.47	2.44	2.43	2.41	2.40	2.38	2.36	2.35	2.32	2.31	2.30	12
2.47	2.46	2.44	2.42	2.41	2.39	2.38	2.36	2.34	2.33	2.31	2.30	2.27	2.26	2.23	2.22	2.21	13
2.40	2.39	2.37	2.35	2.33	2.32	2.31	2.28	2.27	2.25	2.24	2.22	2.20	2.19	2.16	2.14	2.13	14
2.34	2.33	2.31	2.29	2.27	2.26	2.25	2.22	2.20	2.19	2.18	2.16	2.14	2.12	2.10	2.08	2.07	15
2.29	2.28	2.25	2.24	2.22	2.21	2.19	2.17	2.15	2.14	2.12	2.11	2.08	2.07	2.04	2.02	2.01	16
2.24	2.23	2.21	2.19	2.17	2.16	2.15	2.12	2.10	2.09	2.08	2.06	2.03	2.02	1.99	1.97	1.96	17
2.20	2.19	2.17	2.15	2.13	2.12	2.11	2.08	2.06	2.05	2.04	2.02	1.99	1.98	1.95	1.93	1.92	18
2.17	2.16	2.13	2.11	2.10	2.08	2.07	2.05	2.03	2.01	2.00	1.98	1.96	1.94	1.91	1.89	1.88	19
2.14	2.12	2.10	2.08	2.07	2.05	2.04	2.01	1.99	1.98	1.97	1.95	1.92	1.91	1.88	1.86	1.84	20
2.11	2.10	2.07	2.05	2.04	2.02	2.01	1.98	1.96	1.95	1.94	1.92	1.89	1.88	1.84	1.82	1.81	21
2.08	2.07	2.05	2.03	2.01	2.00	1.98	1.96	1.94	1.92	1.91	1.89	1.86	1.85	1.82	1.80	1.78	22
2.06	2.05	2.02	2.00	1.99	1.97	1.96	1.93	1.91	1.90	1.88	1.86	1.84	1.82	1.79	1.77	1.76	23
2.04	2.03	2.00	1.98	1.97	1.95	1.94	1.91	1.89	1.88	1.86	1.84	1.82	1.80	1.77	1.75	1.73	24
2.02	2.01	1.98	1.96	1.95	1.93	1.92	1.89	1.87	1.86	1.84	1.82	1.80	1.78	1.75	1.73	1.71	25
2.00	1.99	1.97	1.95	1.93	1.91	1.90	1.87	1.85	1.84	1.82	1.80	1.78	1.76	1.73	1.71	1.69	26
1.99	1.97	1.95	1.93	1.91	1.90	1.88	1.86	1.84	1.82	1.81	1.79	1.76	1.74	1.71	1.69	1.67	27
1.97	1.96	1.93	1.91	1.90	1.88	1.87	1.84	1.82	1.80	1.79	1.77	1.74	1.73	1.69	1.67	1.65	28
1.96	1.94	1.92	1.90	1.88	1.87	1.85	1.83	1.81	1.79	1.77	1.75	1.73	1.71	1.67	1.65	1.64	29
1.95	1.93	1.91	1.89	1.87	1.85	1.84	1.81	1.79	1.77	1.76	1.74	1.71	1.70	1.66	1.64	1.62	30
1.92	1.91	1.88	1.86	1.85	1.83	1.82	1.79	1.77	1.75	1.74	1.71	1.69	1.67	1.63	1.61	1.59	32
1.90	1.89	1.86	1.84	1.82	1.80	1.80	1.77	1.75	1.73	1.71	1.69	1.66	1.65	1.61	1.59	1.57	34
1.88	1.87	1.85	1.82	1.81	1.79	1.78	1.75	1.73	1.71	1.69	1.67	1.64	1.62	1.59	1.56	1.55	36
1.87	1.85	1.83	1.81	1.79	1.77	1.76	1.73	1.71	1.69	1.68	1.65	1.62	1.61	1.57	1.54	1.53	38
1.85	1.84	1.81	1.79	1.77	1.76	1.74	1.72	1.69	1.67	1.66	1.64	1.61	1.59	1.55	1.53	1.51	40
1.84	1.83	1.80	1.78	1.76	1.74	1.73	1.70	1.68	1.66	1.65	1.62	1.59	1.57	1.53	1.51	1.49	42
1.83	1.81	1.79	1.77	1.75	1.73	1.72	1.69	1.67	1.65	1.63	1.61	1.58	1.56	1.52	1.49	1.48	44
1.82	1.80	1.78	1.76	1.74	1.72	1.71	1.68	1.65	1.64	1.62	1.60	1.57	1.55	1.51	1.48	1.46	46
1.81	1.79	1.77	1.75	1.73	1.71	1.70	1.67	1.64	1.62	1.61	1.59	1.56	1.54	1.49	1.47	1.45	48
1.80	1.78	1.76	1.74	1.72	1.70	1.69	1.66	1.63	1.61	1.60	1.58	1.54	1.52	1.48	1.46	1.44	50
1.78	1.76	1.74	1.72	1.70	1.68	1.67	1.64	1.61	1.59	1.58	1.55	1.52	1.50	1.46	1.43	1.41	55
1.76	1.75	1.72	1.70	1.68	1.66	1.65	1.62	1.59	1.57	1.56	1.53	1.50	1.48	1.44	1.41	1.39	60
1.75	1.73	1.71	1.69	1.67	1.65	1.63	1.60	1.58	1.56	1.54	1.52	1.49	1.46	1.42	1.39	1.37	65
1.74	1.72	1.70	1.67	1.65	1.64	1.62	1.59	1.57	1.55	1.53	1.50	1.47	1.45	1.40	1.37	1.35	70
1.72	1.70	1.68	1.65	1.63	1.62	1.60	1.57	1.54	1.52	1.51	1.48	1.45	1.43	1.38	1.35	1.32	80
1.70	1.69	1.66	1.64	1.62	1.60	1.59	1.55	1.53	1.51	1.49	1.46	1.43	1.41	1.36	1.32	1.30	90
1.69	1.68	1.65	1.63	1.61	1.59	1.57	1.54	1.52	1.49	1.48	1.45	1.41	1.39	1.34	1.31	1.28	100
1.67	1.65	1.63	1.60	1.58	1.57	1.55	1.52	1.49	1.47	1.45	1.42	1.39	1.36	1.31	1.27	1.25	125
1.66	1.64	1.61	1.59	1.57	1.55	1.53	1.50	1.48	1.45	1.44	1.41	1.37	1.34	1.29	1.25	1.22	150
1.64	1.62	1.60	1.57	1.55	1.53	1.52	1.48	1.46	1.43	1.41	1.39	1.35	1.32	1.26	1.22	1.19	200
1.62	1.61	1.58	1.55	1.53	1.51	1.50	1.46	1.43	1.41	1.39	1.36	1.32	1.30	1.23	1.19	1.15	300
1.61	1.59	1.56	1.54	1.52	1.50	1.48	1.45	1.42	1.40	1.38	1.34	1.30	1.28	1.21	1.16	1.11	500
1.60	1.58	1.55	1.53	1.51	1.49	1.47	1.44	1.41	1.38	1.36	1.33	1.29	1.26	1.19	1.13	1.08	1000
1.59	1.57	1.54	1.52	1.50	1.48	1.46	1.42	1.39	1.37	1.35	1.32	1.27	1.24	1.17	1.11	1.00	∞

Approximate formula for ν_1 and ν_2 larger than 30: $\log_{10} F_{.05;\,\nu_1,\nu_2} \simeq \dfrac{1.4287}{\sqrt{h} - 0.95} - 0.681\left(\dfrac{1}{\nu_1} - \dfrac{1}{\nu_2}\right)$, where $\dfrac{1}{h} = \dfrac{1}{2}\left(\dfrac{1}{\nu_1} + \dfrac{1}{\nu_2}\right)$.

APPENDIX G THE *F*-DISTRIBUTION (Continued)

Table of $F_{0.025;\,\nu_1,\,\nu_2}$

Degrees of freedom for the numerator (ν_1); rows: Degrees of freedom for the denominator (ν_2)

Multiply the numbers of the first row ($\nu_2 = 1$) by 10.

ν_2	1	2	3	4	5	6	7	8	9	10	11	12	13	14	15	16	17	18
1	64.8	80.0	86.4	90.0	92.2	93.7	94.8	95.7	96.3	96.9	97.3	97.7	98.0	98.3	98.5	98.7	98.9	99.0
2	38.5	39.0	39.2	39.2	39.3	39.3	39.4	39.4	39.4	39.4	39.4	39.4	39.4	39.4	39.4	39.4	39.4	39.4
3	17.4	16.0	15.4	15.1	14.9	14.7	14.6	14.5	14.5	14.4	14.4	14.3	14.3	14.3	14.3	14.2	14.2	14.2
4	12.2	10.6	9.98	9.60	9.36	9.20	9.07	8.98	8.90	8.84	8.79	8.75	8.72	8.69	8.66	8.64	8.62	8.60
5	10.0	8.43	7.76	7.39	7.15	6.98	6.85	6.76	6.68	6.62	6.57	6.52	6.49	6.46	6.43	6.41	6.39	6.37
6	8.81	7.26	6.60	6.23	5.99	5.82	5.70	5.60	5.52	5.46	5.41	5.37	5.33	5.30	5.27	5.25	5.23	5.21
7	8.07	6.54	5.89	5.52	5.29	5.12	4.99	4.90	4.82	4.76	4.71	4.67	4.63	4.60	4.57	4.54	4.52	4.50
8	7.57	6.06	5.42	5.05	4.82	4.65	4.53	4.43	4.36	4.30	4.24	4.20	4.16	4.13	4.10	4.08	4.05	4.03
9	7.21	5.71	5.08	4.72	4.48	4.32	4.20	4.10	4.03	3.96	3.91	3.87	3.83	3.80	3.77	3.74	3.72	3.70
10	6.94	5.46	4.83	4.47	4.24	4.07	3.95	3.85	3.78	3.72	3.66	3.62	3.58	3.55	3.52	3.50	3.47	3.45
11	6.72	5.26	4.63	4.28	4.04	3.88	3.76	3.66	3.59	3.53	3.47	3.43	3.39	3.36	3.33	3.30	3.28	3.26
12	6.55	5.10	4.47	4.12	3.89	3.73	3.61	3.51	3.44	3.37	3.32	3.28	3.24	3.21	3.18	3.15	3.13	3.11
13	6.41	4.97	4.35	4.00	3.77	3.60	3.48	3.39	3.31	3.25	3.20	3.15	3.12	3.08	3.05	3.03	3.00	2.98
14	6.30	4.86	4.24	3.89	3.66	3.50	3.38	3.29	3.21	3.15	3.09	3.05	3.01	2.98	2.95	2.92	2.90	2.88
15	6.20	4.76	4.15	3.80	3.58	3.41	3.29	3.20	3.12	3.06	3.01	2.96	2.92	2.89	2.86	2.84	2.81	2.79
16	6.12	4.69	4.08	3.73	3.50	3.34	3.22	3.12	3.05	2.99	2.93	2.89	2.85	2.82	2.79	2.76	2.74	2.72
17	6.04	4.62	4.01	3.66	3.44	3.28	3.16	3.06	2.98	2.92	2.87	2.82	2.79	2.75	2.72	2.70	2.67	2.65
18	5.98	4.56	3.95	3.61	3.38	3.22	3.10	3.01	2.93	2.87	2.81	2.77	2.73	2.70	2.67	2.64	2.62	2.60
19	5.92	4.51	3.90	3.56	3.33	3.17	3.05	2.96	2.88	2.82	2.76	2.72	2.68	2.65	2.62	2.59	2.57	2.55
20	5.87	4.46	3.86	3.51	3.29	3.13	3.01	2.91	2.84	2.77	2.72	2.68	2.64	2.60	2.57	2.55	2.52	2.50
21	5.83	4.42	3.82	3.48	3.25	3.09	2.97	2.87	2.80	2.73	2.68	2.64	2.60	2.56	2.53	2.51	2.48	2.46
22	5.79	4.38	3.78	3.44	3.22	3.05	2.93	2.84	2.76	2.70	2.65	2.60	2.56	2.53	2.50	2.47	2.45	2.43
23	5.75	4.35	3.75	3.41	3.18	3.02	2.90	2.81	2.73	2.67	2.62	2.57	2.53	2.50	2.47	2.44	2.42	2.39
24	5.72	4.32	3.72	3.38	3.15	2.99	2.87	2.78	2.70	2.64	2.59	2.54	2.50	2.47	2.44	2.41	2.39	2.36
25	5.69	4.29	3.69	3.35	3.13	2.97	2.85	2.75	2.68	2.61	2.56	2.51	2.48	2.44	2.41	2.38	2.36	2.34
26	5.66	4.27	3.67	3.33	3.10	2.94	2.82	2.73	2.65	2.59	2.54	2.49	2.45	2.42	2.39	2.36	2.34	2.31
27	5.63	4.24	3.65	3.31	3.08	2.92	2.80	2.71	2.63	2.57	2.51	2.47	2.43	2.39	2.36	2.34	2.31	2.29
28	5.61	4.22	3.63	3.29	3.06	2.90	2.78	2.69	2.61	2.55	2.49	2.45	2.41	2.37	2.34	2.32	2.29	2.27
29	5.59	4.20	3.61	3.27	3.04	2.88	2.76	2.67	2.59	2.53	2.48	2.43	2.39	2.36	2.32	2.30	2.27	2.25
30	5.57	4.18	3.59	3.25	3.03	2.87	2.75	2.65	2.57	2.51	2.46	2.41	2.37	2.34	2.31	2.28	2.26	2.23
32	5.53	4.15	3.56	3.22	3.00	2.84	2.72	2.62	2.54	2.48	2.43	2.38	2.34	2.31	2.28	2.25	2.22	2.20
34	5.50	4.12	3.53	3.19	2.97	2.81	2.69	2.59	2.52	2.45	2.40	2.35	2.31	2.28	2.25	2.22	2.19	2.17
36	5.47	4.09	3.51	3.17	2.94	2.79	2.66	2.57	2.49	2.43	2.37	2.33	2.29	2.25	2.22	2.20	2.17	2.15
38	5.45	4.07	3.48	3.15	2.92	2.76	2.64	2.55	2.47	2.41	2.35	2.31	2.27	2.23	2.20	2.17	2.15	2.13
40	5.42	4.05	3.46	3.13	2.90	2.74	2.62	2.53	2.45	2.39	2.33	2.29	2.25	2.21	2.18	2.15	2.13	2.11
42	5.40	4.03	3.45	3.11	2.89	2.73	2.61	2.51	2.44	2.37	2.32	2.27	2.23	2.20	2.16	2.14	2.11	2.09
44	5.39	4.02	3.43	3.09	2.87	2.71	2.59	2.50	2.42	2.36	2.30	2.26	2.21	2.18	2.15	2.12	2.10	2.07
46	5.37	4.00	3.42	3.08	2.86	2.70	2.58	2.48	2.41	2.34	2.29	2.24	2.20	2.17	2.13	2.11	2.08	2.06
48	5.35	3.99	3.40	3.07	2.84	2.69	2.57	2.47	2.39	2.33	2.27	2.23	2.19	2.15	2.12	2.09	2.07	2.05
50	5.34	3.98	3.39	3.06	2.83	2.67	2.55	2.46	2.38	2.32	2.26	2.22	2.18	2.14	2.11	2.08	2.06	2.03
55	5.31	3.95	3.36	3.03	2.81	2.65	2.53	2.43	2.36	2.29	2.24	2.19	2.15	2.11	2.08	2.05	2.03	2.01
60	5.29	3.93	3.34	3.01	2.79	2.63	2.51	2.41	2.33	2.27	2.22	2.17	2.13	2.09	2.06	2.03	2.01	1.98
65	5.27	3.91	3.32	2.99	2.77	2.61	2.49	2.39	2.32	2.25	2.20	2.15	2.11	2.07	2.04	2.01	1.99	1.97
70	5.25	3.89	3.31	2.98	2.75	2.60	2.48	2.38	2.30	2.24	2.18	2.14	2.10	2.06	2.03	2.00	1.97	1.95
80	5.22	3.86	3.28	2.95	2.73	2.57	2.45	2.36	2.28	2.21	2.16	2.11	2.07	2.03	2.00	1.97	1.95	1.93
90	5.20	3.84	3.27	2.93	2.71	2.55	2.43	2.34	2.26	2.19	2.14	2.09	2.05	2.02	1.98	1.95	1.93	1.91
100	5.18	3.83	3.25	2.92	2.70	2.54	2.42	2.32	2.24	2.18	2.12	2.08	2.04	2.00	1.97	1.94	1.91	1.89
125	5.15	3.80	3.22	2.89	2.67	2.51	2.39	2.30	2.22	2.15	2.10	2.05	2.01	1.97	1.94	1.91	1.89	1.86
150	5.13	3.78	3.20	2.87	2.65	2.49	2.37	2.28	2.20	2.13	2.08	2.03	1.99	1.95	1.92	1.89	1.87	1.84
200	5.10	3.76	3.18	2.85	2.63	2.47	2.35	2.26	2.18	2.11	2.06	2.01	1.97	1.93	1.90	1.87	1.84	1.82
300	5.08	3.74	3.16	2.83	2.61	2.45	2.33	2.23	2.16	2.09	2.04	1.99	1.95	1.91	1.88	1.85	1.82	1.80
500	5.05	3.72	3.14	2.81	2.59	2.43	2.31	2.22	2.14	2.07	2.02	1.97	1.93	1.89	1.86	1.83	1.80	1.78
1000	5.04	3.70	3.13	2.80	2.58	2.42	2.30	2.20	2.13	2.06	2.01	1.96	1.92	1.88	1.85	1.82	1.79	1.77
∞	5.02	3.69	3.12	2.79	2.57	2.41	2.29	2.19	2.11	2.05	1.99	1.94	1.90	1.87	1.83	1.80	1.78	1.75

Example: $P\{F_{.025;\,8,\,20} < 2.91\} = 97.5\%$.

$F_{.975;\,\nu_1,\,\nu_2} = 1/F_{.025;\,\nu_2,\,\nu_1}$. Example: $F_{.975;\,8,\,20} = 1/F_{.025;\,8,\,20} = 1/4.00 = 0.250$.

Table of $F_{0.025;\,\nu_1,\nu_2}$

Degrees of freedom for the numerator (ν_1)																	
19	20	22	24	26	28	30	35	40	45	50	60	80	100	200	500	∞	Degrees of freedom for the denominators (ν_2)
99.2	99.3	99.5	99.7	99.9	100.0	100.1	100.4	100.6	100.7	100.8	101.0	101.2	101.3	101.6	101.7	101.8	1
39.4	39.4	39.5	39.5	39.5	39.5	39.5	39.5	39.5	39.5	39.5	39.5	39.5	39.5	39.5	39.5	39.5	2
14.2	14.2	14.1	14.1	14.1	14.1	14.1	14.1	14.0	14.0	14.0	14.0	14.0	14.0	13.9	13.9	13.9	3
8.58	8.56	8.53	8.51	8.49	8.48	8.46	8.44	8.41	8.39	8.38	8.36	8.33	8.32	8.29	8.27	8.26	4
6.35	6.33	6.30	6.28	6.26	6.24	6.23	6.20	6.18	6.16	6.14	6.12	6.10	6.08	6.05	6.03	6.02	5
5.19	5.17	5.14	5.12	5.10	5.08	5.07	5.04	5.01	4.99	4.98	4.96	4.93	4.92	4.88	4.86	4.85	6
4.48	4.47	4.44	4.42	4.39	4.38	4.36	4.33	4.31	4.29	4.28	4.25	4.23	4.21	4.18	4.16	4.14	7
4.02	4.00	3.97	3.95	3.93	3.91	3.89	3.86	3.84	3.82	3.81	3.78	3.76	3.74	3.70	3.68	3.67	8
3.68	3.67	3.64	3.61	3.59	3.58	3.56	3.53	3.51	3.49	3.47	3.45	3.42	3.40	3.37	3.35	3.33	9
3.44	3.42	3.39	3.37	3.34	3.33	3.31	3.28	3.26	3.24	3.22	3.20	3.17	3.15	3.12	3.09	3.08	10
3.24	3.23	3.20	3.17	3.15	3.13	3.12	3.09	3.06	3.04	3.03	3.00	2.97	2.96	2.92	2.90	2.88	11
3.09	3.07	3.04	3.02	3.00	2.98	2.96	2.93	2.91	2.89	2.87	2.85	2.82	2.80	2.76	2.74	2.72	12
2.96	2.95	2.92	2.89	2.87	2.85	2.84	2.80	2.78	2.76	2.74	2.72	2.69	2.67	2.63	2.61	2.60	13
2.86	2.84	2.81	2.79	2.77	2.75	2.73	2.70	2.67	2.65	2.64	2.61	2.58	2.56	2.53	2.50	2.49	14
2.77	2.76	2.73	2.70	2.68	2.66	2.64	2.61	2.58	2.56	2.55	2.52	2.49	2.47	2.44	2.41	2.40	15
2.70	2.68	2.65	2.63	2.60	2.58	2.57	2.53	2.51	2.49	2.47	2.45	2.42	2.40	2.36	2.33	2.32	16
2.63	2.62	2.59	2.56	2.54	2.52	2.50	2.47	2.44	2.42	2.41	2.38	2.35	2.33	2.29	2.26	2.25	17
2.58	2.56	2.53	2.50	2.48	2.46	2.44	2.41	2.38	2.36	2.35	2.32	2.29	2.27	2.23	2.20	2.19	18
2.53	2.51	2.48	2.45	2.43	2.41	2.39	2.36	2.33	2.31	2.30	2.27	2.24	2.22	2.18	2.15	2.13	19
2.48	2.46	2.43	2.41	2.39	2.37	2.35	2.31	2.29	2.27	2.25	2.22	2.19	2.17	2.13	2.10	2.09	20
2.44	2.42	2.39	2.37	2.34	2.33	2.31	2.27	2.25	2.23	2.21	2.18	2.15	2.13	2.09	2.06	2.04	21
2.41	2.39	2.36	2.33	2.31	2.29	2.27	2.24	2.21	2.19	2.17	2.14	2.11	2.09	2.05	2.02	2.00	22
2.37	2.36	2.33	2.30	2.28	2.26	2.24	2.20	2.18	2.15	2.14	2.11	2.08	2.06	2.01	1.99	1.97	23
2.35	2.33	2.30	2.27	2.25	2.23	2.21	2.17	2.15	2.12	2.11	2.08	2.05	2.02	1.98	1.95	1.94	24
2.32	2.30	2.27	2.24	2.22	2.20	2.18	2.15	2.12	2.10	2.08	2.05	2.02	2.00	1.95	1.92	1.91	25
2.29	2.28	2.24	2.22	2.19	2.17	2.16	2.12	2.09	2.07	2.05	2.03	1.99	1.97	1.92	1.90	1.88	26
2.27	2.25	2.22	2.19	2.17	2.15	2.13	2.10	2.07	2.05	2.03	2.00	1.97	1.94	1.90	1.87	1.85	27
2.25	2.23	2.20	2.17	2.15	2.13	2.11	2.08	2.05	2.03	2.01	1.98	1.94	1.92	1.88	1.85	1.83	28
2.23	2.21	2.18	2.15	2.13	2.11	2.09	2.06	2.03	2.01	1.99	1.96	1.92	1.90	1.86	1.83	1.81	29
2.21	2.20	2.16	2.14	2.11	2.09	2.07	2.04	2.01	1.99	1.97	1.94	1.90	1.88	1.84	1.81	1.79	30
2.18	2.16	2.13	2.10	2.08	2.06	2.04	2.00	1.98	1.95	1.93	1.91	1.87	1.85	1.80	1.77	1.75	32
2.15	2.13	2.10	2.07	2.05	2.03	2.01	1.97	1.95	1.92	1.90	1.88	1.84	1.82	1.77	1.74	1.72	34
2.13	2.11	2.08	2.05	2.03	2.00	1.99	1.95	1.92	1.90	1.88	1.85	1.81	1.79	1.74	1.71	1.69	36
2.11	2.09	2.05	2.03	2.00	1.98	1.96	1.93	1.90	1.87	1.85	1.82	1.79	1.76	1.71	1.68	1.66	38
2.09	2.07	2.03	2.01	1.98	1.96	1.94	1.90	1.88	1.85	1.83	1.80	1.76	1.74	1.69	1.66	1.64	40
2.07	2.05	2.02	1.99	1.96	1.94	1.92	1.89	1.86	1.83	1.81	1.78	1.74	1.72	1.67	1.64	1.62	42
2.05	2.03	2.00	1.97	1.95	1.93	1.91	1.87	1.84	1.82	1.80	1.77	1.73	1.70	1.65	1.62	1.60	44
2.04	2.02	1.99	1.96	1.93	1.91	1.89	1.85	1.82	1.80	1.78	1.75	1.71	1.69	1.63	1.60	1.58	46
2.02	2.01	1.97	1.94	1.92	1.90	1.88	1.84	1.81	1.79	1.77	1.73	1.69	1.67	1.62	1.58	1.56	48
2.01	1.99	1.96	1.93	1.91	1.88	1.87	1.83	1.80	1.77	1.75	1.72	1.68	1.66	1.60	1.57	1.55	50
1.99	1.97	1.93	1.90	1.88	1.86	1.84	1.80	1.77	1.74	1.72	1.69	1.65	1.62	1.57	1.54	1.51	55
1.96	1.94	1.91	1.88	1.86	1.83	1.82	1.78	1.74	1.72	1.70	1.67	1.62	1.60	1.54	1.51	1.48	60
1.95	1.93	1.89	1.86	1.84	1.82	1.80	1.76	1.72	1.70	1.68	1.65	1.60	1.58	1.52	1.48	1.46	65
1.93	1.91	1.88	1.85	1.82	1.80	1.78	1.74	1.71	1.68	1.66	1.63	1.58	1.56	1.50	1.46	1.44	70
1.90	1.88	1.85	1.82	1.79	1.77	1.75	1.71	1.68	1.65	1.63	1.60	1.55	1.53	1.47	1.43	1.40	80
1.88	1.86	1.83	1.80	1.77	1.75	1.73	1.69	1.66	1.63	1.61	1.58	1.53	1.50	1.44	1.40	1.37	90
1.87	1.85	1.81	1.78	1.76	1.74	1.71	1.67	1.64	1.61	1.59	1.56	1.51	1.48	1.42	1.38	1.35	100
1.84	1.82	1.79	1.75	1.73	1.71	1.68	1.64	1.61	1.58	1.56	1.52	1.48	1.45	1.38	1.34	1.30	125
1.82	1.80	1.77	1.74	1.71	1.69	1.67	1.62	1.59	1.56	1.54	1.50	1.45	1.42	1.35	1.31	1.27	150
1.80	1.78	1.74	1.71	1.68	1.66	1.64	1.60	1.56	1.53	1.51	1.47	1.42	1.39	1.32	1.27	1.23	200
1.77	1.75	1.72	1.69	1.66	1.64	1.62	1.57	1.54	1.51	1.48	1.45	1.39	1.36	1.28	1.23	1.18	300
1.76	1.74	1.70	1.67	1.64	1.62	1.60	1.55	1.51	1.49	1.46	1.42	1.37	1.34	1.25	1.19	1.14	500
1.74	1.72	1.69	1.65	1.63	1.60	1.58	1.54	1.50	1.47	1.44	1.41	1.35	1.32	1.23	1.16	1.09	1000
1.73	1.71	1.67	1.64	1.61	1.59	1.57	1.52	1.48	1.45	1.43	1.39	1.33	1.30	1.21	1.13	1.00	∞

Approximate formula for ν_1 and ν_2 larger than 30: $\log_{10} F_{.025;\,\nu_1,\nu_2} \simeq \dfrac{1.7023}{\sqrt{h} - 1.14} - 0.846\left(\dfrac{1}{\nu_1} - \dfrac{1}{\nu_2}\right)$, where $\dfrac{1}{h} = \dfrac{1}{2}\left(\dfrac{1}{\nu_1} + \dfrac{1}{\nu_2}\right)$.

APPENDIX G THE F-DISTRIBUTION (Continued)

Table of $F_{0.01;\,\nu_1,\,\nu_2}$

Degrees of freedom for the numerator (ν_1); rows: Degrees of freedom for the denominator (ν_2)

Multiply the numbers of the first row (ν_2 = 1) by 100.

ν_2	1	2	3	4	5	6	7	8	9	10	11	12	13	14	15	16	17	18
1	*40.5*	*50.0*	*54.0*	*56.3*	*57.6*	*58.6*	*59.3*	*59.8*	*60.2*	*60.6*	*60.8*	*61.1*	*61.3*	*61.4*	*61.6*	*61.7*	*61.8*	*61.9*
2	98.5	99.0	99.2	99.2	99.3	99.3	99.4	99.4	99.4	99.4	99.4	99.4	99.4	99.4	99.4	99.4	99.4	99.4
3	34.1	30.8	29.5	28.7	28.2	27.9	27.7	27.5	27.3	27.2	27.1	27.1	27.0	26.9	26.9	26.8	26.8	26.8
4	21.2	18.0	16.7	16.0	15.5	15.2	15.0	14.8	14.7	14.5	14.4	14.4	14.3	14.2	14.2	14.2	14.1	14.1
5	16.3	13.3	12.1	11.4	11.0	10.7	10.5	10.3	10.2	10.1	9.96	9.89	9.82	9.77	9.72	9.68	9.64	9.61
6	13.7	10.9	9.78	9.15	8.75	8.47	8.26	8.10	7.98	7.87	7.79	7.72	7.66	7.60	7.56	7.52	7.48	7.45
7	12.2	9.55	8.45	7.85	7.46	7.19	6.99	6.84	6.72	6.62	6.54	6.47	6.41	6.36	6.31	6.27	6.24	6.21
8	11.3	8.65	7.59	7.01	6.63	6.37	6.18	6.03	5.91	5.81	5.73	5.67	5.61	5.56	5.52	5.48	5.44	5.41
9	10.6	8.02	6.99	6.42	6.06	5.80	5.61	5.47	5.35	5.26	5.18	5.11	5.05	5.00	4.96	4.92	4.89	4.86
10	10.0	7.56	6.55	5.99	5.64	5.39	5.20	5.06	4.94	4.85	4.77	4.71	4.65	4.60	4.56	4.52	4.49	4.46
11	9.65	7.21	6.22	5.67	5.32	5.07	4.89	4.74	4.63	4.54	4.46	4.40	4.34	4.29	4.25	4.21	4.18	4.15
12	9.33	6.93	5.95	5.41	5.06	4.82	4.64	4.50	4.39	4.30	4.22	4.16	4.10	4.05	4.01	3.97	3.94	3.91
13	9.07	6.70	5.74	5.21	4.86	4.62	4.44	4.30	4.19	4.10	4.02	3.96	3.91	3.86	3.82	3.78	3.75	3.72
14	8.86	6.51	5.56	5.04	4.70	4.46	4.28	4.14	4.03	3.94	3.86	3.80	3.75	3.70	3.66	3.62	3.59	3.56
15	8.68	6.36	5.42	4.89	4.56	4.32	4.14	4.00	3.89	3.80	3.73	3.67	3.61	3.56	3.52	3.49	3.45	3.42
16	8.53	6.23	5.29	4.77	4.44	4.20	4.03	3.89	3.78	3.69	3.62	3.55	3.50	3.45	3.41	3.37	3.34	3.31
17	8.40	6.11	5.18	4.67	4.34	4.10	3.93	3.79	3.68	3.59	3.52	3.46	3.40	3.35	3.31	3.27	3.24	3.21
18	8.29	6.01	5.09	4.58	4.25	4.01	3.84	3.71	3.60	3.51	3.43	3.37	3.32	3.27	3.23	3.19	3.16	3.13
19	8.18	5.93	5.01	4.50	4.17	3.94	3.77	3.63	3.52	3.43	3.36	3.30	3.24	3.19	3.15	3.12	3.08	3.05
20	8.10	5.85	4.94	4.43	4.10	3.87	3.70	3.56	3.46	3.37	3.29	3.23	3.18	3.13	3.09	3.05	3.02	2.99
21	8.02	5.78	4.87	4.37	4.04	3.81	3.64	3.51	3.40	3.31	3.24	3.17	3.12	3.07	3.03	2.99	2.96	2.93
22	7.95	5.72	4.82	4.31	3.99	3.76	3.59	3.45	3.35	3.26	3.18	3.12	3.07	3.02	2.98	2.94	2.91	2.88
23	7.88	5.66	4.76	4.26	3.94	3.71	3.54	3.41	3.30	3.21	3.14	3.07	3.02	2.97	2.93	2.89	2.86	2.83
24	7.82	5.61	4.72	4.22	3.90	3.67	3.50	3.36	3.26	3.17	3.09	3.03	2.98	2.93	2.89	2.85	2.82	2.79
25	7.77	5.57	4.68	4.18	3.86	3.63	3.46	3.32	3.22	3.13	3.06	2.99	2.94	2.89	2.85	2.81	2.78	2.75
26	7.72	5.53	4.64	4.14	3.82	3.59	3.42	3.29	3.18	3.09	3.02	2.96	2.90	2.86	2.82	2.78	2.74	2.72
27	7.68	5.49	4.60	4.11	3.78	3.56	3.39	3.26	3.15	3.06	2.99	2.93	2.87	2.82	2.78	2.75	2.71	2.68
28	7.64	5.45	4.57	4.07	3.75	3.53	3.36	3.23	3.12	3.03	2.96	2.90	2.84	2.79	2.75	2.72	2.68	2.65
29	7.60	5.42	4.54	4.04	3.73	3.50	3.33	3.20	3.09	3.00	2.93	2.87	2.81	2.77	2.73	2.69	2.66	2.63
30	7.56	5.39	4.51	4.02	3.70	3.47	3.30	3.17	3.07	2.98	2.91	2.84	2.79	2.74	2.70	2.66	2.63	2.60
32	7.50	5.34	4.46	3.97	3.65	3.43	3.26	3.13	3.02	2.93	2.86	2.80	2.74	2.70	2.66	2.62	2.58	2.55
34	7.44	5.29	4.42	3.93	3.61	3.39	3.22	3.09	2.98	2.89	2.82	2.76	2.70	2.66	2.62	2.58	2.55	2.51
36	7.40	5.25	4.38	3.89	3.57	3.35	3.18	3.05	2.95	2.86	2.79	2.72	2.67	2.62	2.58	2.54	2.51	2.48
38	7.35	5.21	4.34	3.86	3.54	3.32	3.15	3.02	2.92	2.83	2.75	2.69	2.64	2.59	2.55	2.51	2.48	2.45
40	7.31	5.18	4.31	3.83	3.51	3.29	3.12	2.99	2.89	2.80	2.73	2.66	2.61	2.56	2.52	2.48	2.45	2.42
42	7.28	5.15	4.29	3.80	3.49	3.27	3.10	2.97	2.86	2.78	2.70	2.64	2.59	2.54	2.50	2.46	2.43	2.40
44	7.25	5.12	4.26	3.78	3.47	3.24	3.08	2.95	2.84	2.75	2.68	2.62	2.56	2.52	2.47	2.44	2.40	2.37
46	7.22	5.10	4.24	3.76	3.44	3.22	3.06	2.93	2.82	2.73	2.66	2.60	2.54	2.50	2.45	2.42	2.38	2.35
48	7.19	5.08	4.22	3.74	3.43	3.20	3.04	2.91	2.80	2.72	2.64	2.58	2.53	2.48	2.44	2.40	2.37	2.33
50	7.17	5.06	4.20	3.72	3.41	3.19	3.02	2.89	2.79	2.70	2.63	2.56	2.51	2.46	2.42	2.38	2.35	2.32
55	7.12	5.01	4.16	3.68	3.37	3.15	2.98	2.85	2.75	2.66	2.59	2.53	2.47	2.42	2.38	2.34	2.31	2.28
60	7.08	4.98	4.13	3.65	3.34	3.12	2.95	2.82	2.72	2.63	2.56	2.50	2.44	2.39	2.35	2.31	2.28	2.25
65	7.04	4.95	4.10	3.62	3.31	3.09	2.93	2.80	2.69	2.61	2.53	2.47	2.42	2.37	2.33	2.29	2.26	2.23
70	7.01	4.92	4.08	3.60	3.29	3.07	2.91	2.78	2.67	2.59	2.51	2.45	2.40	2.35	2.31	2.27	2.23	2.20
80	6.96	4.88	4.04	3.56	3.26	3.04	2.87	2.74	2.64	2.55	2.48	2.42	2.36	2.31	2.27	2.23	2.20	2.17
90	6.93	4.85	4.01	3.54	3.23	3.01	2.84	2.72	2.61	2.52	2.45	2.39	2.33	2.29	2.24	2.21	2.17	2.14
100	6.90	4.82	3.98	3.51	3.21	2.99	2.82	2.69	2.59	2.50	2.43	2.37	2.31	2.26	2.22	2.19	2.15	2.12
125	6.84	4.78	3.94	3.47	3.17	2.95	2.79	2.66	2.55	2.47	2.39	2.33	2.28	2.23	2.19	2.15	2.11	2.08
150	6.81	4.75	3.92	3.45	3.14	2.92	2.76	2.63	2.53	2.44	2.37	2.31	2.25	2.20	2.16	2.12	2.09	2.06
200	6.76	4.71	3.88	3.41	3.11	2.89	2.73	2.60	2.50	2.41	2.34	2.27	2.22	2.17	2.13	2.09	2.06	2.02
300	6.72	4.68	3.85	3.38	3.08	2.86	2.70	2.57	2.47	2.38	2.31	2.24	2.19	2.14	2.10	2.06	2.03	1.99
500	6.69	4.65	3.82	3.36	3.05	2.84	2.68	2.55	2.44	2.36	2.28	2.22	2.17	2.12	2.07	2.04	2.00	1.97
1000	6.66	4.63	3.80	3.34	3.04	2.82	2.66	2.53	2.43	2.34	2.27	2.20	2.15	2.10	2.06	2.02	1.98	1.95
∞	6.63	4.61	3.78	3.32	3.02	2.80	2.64	2.51	2.41	2.32	2.25	2.18	2.13	2.08	2.04	2.00	1.97	1.93

Example: $P\{F_{.01;\,8,\,20} < 3.56\} = 99\%$.

$F_{.99;\,\nu_1,\,\nu_2} = 1/F_{.01;\,\nu_2,\,\nu_1}$. Example: $F_{.99;\,8,\,20} = 1/F_{.01;\,20,\,8} = 1/5.36 = 0.187$.

Table of $F_{0.01;\nu_1,\nu_2}$

Degrees of freedom for the numerator (ν_1)

Multiply the numbers of the first row ($\nu_2 = 1$) by 10.

19	20	22	24	26	28	30	35	40	45	50	60	80	100	200	500	∞	Degrees of freedom for the denominator (ν_2)
62.0	*62.1*	*62.2*	*62.3*	*62.4*	*62.5*	*62.6*	*62.8*	*62.9*	*63.0*	*63.0*	*63.1*	*63.3*	*63.3*	*63.5*	*63.6*	*63.7*	1
99.4	99.4	99.5	99.5	99.5	99.5	99.5	99.5	99.5	99.5	99.5	99.5	99.5	99.5	99.5	99.5	99.5	2
26.7	26.7	26.6	26.6	26.6	26.5	26.5	26.5	26.4	26.4	26.4	26.3	26.3	26.2	26.2	26.1	26.1	3
14.0	14.0	14.0	13.9	13.9	13.9	13.8	13.8	13.7	13.7	13.7	13.7	13.6	13.6	13.5	13.5	13.5	4
9.58	9.55	9.51	9.47	9.43	9.40	9.38	9.33	9.29	9.26	9.24	9.20	9.16	9.13	9.08	9.04	9.02	5
7.42	7.40	7.35	7.31	7.28	7.25	7.23	7.18	7.14	7.11	7.09	7.06	7.01	6.99	6.93	6.90	6.88	6
6.18	6.16	6.11	6.07	6.04	6.02	5.99	5.94	5.91	5.88	5.86	5.82	5.78	5.75	5.70	5.67	5.65	7
5.38	5.36	5.32	5.28	5.25	5.22	5.20	5.15	5.12	5.09	5.07	5.03	4.99	4.96	4.91	4.88	4.86	8
4.83	4.81	4.77	4.73	4.70	4.67	4.65	4.60	4.57	4.54	4.52	4.48	4.44	4.42	4.36	4.33	4.31	9
4.43	4.41	4.36	4.33	4.30	4.27	4.25	4.20	4.17	4.14	4.12	4.08	4.04	4.01	3.96	3.93	3.91	10
4.12	4.10	4.06	4.02	3.99	3.96	3.94	3.89	3.86	3.83	3.81	3.78	3.73	3.71	3.66	3.62	3.60	11
3.88	3.86	3.82	3.78	3.75	3.72	3.70	3.65	3.62	3.59	3.57	3.54	3.49	3.47	3.41	3.38	3.36	12
3.69	3.66	3.62	3.59	3.56	3.53	3.51	3.46	3.43	3.40	3.38	3.34	3.30	3.27	3.22	3.19	3.17	13
3.53	3.51	3.46	3.43	3.40	3.37	3.35	3.30	3.27	3.24	3.22	3.18	3.14	3.11	3.06	3.03	3.00	14
3.40	3.37	3.33	3.29	3.26	3.24	3.21	3.17	3.13	3.10	3.08	3.05	3.00	2.98	2.92	2.89	2.87	15
3.28	3.26	3.22	3.18	3.15	3.12	3.10	3.05	3.02	2.99	2.97	2.93	2.89	2.86	2.81	2.78	2.75	16
3.18	3.16	3.12	3.08	3.05	3.03	3.00	2.96	2.92	2.89	2.87	2.83	2.79	2.76	2.71	2.68	2.65	17
3.10	3.08	3.03	3.00	2.97	2.94	2.92	2.87	2.84	2.81	2.78	2.75	2.70	2.68	2.62	2.59	2.57	18
3.03	3.00	2.96	2.92	2.89	2.87	2.84	2.80	2.76	2.73	2.71	2.67	2.63	2.60	2.55	2.51	2.49	19
2.96	2.94	2.90	2.86	2.83	2.80	2.78	2.73	2.69	2.67	2.64	2.61	2.56	2.54	2.48	2.44	2.42	20
2.90	2.88	2.84	2.80	2.77	2.74	2.72	2.67	2.64	2.61	2.58	2.55	2.50	2.48	2.42	2.38	2.36	21
2.85	2.83	2.78	2.75	2.72	2.69	2.67	2.62	2.58	2.55	2.53	2.50	2.45	2.42	2.36	2.33	2.31	22
2.80	2.78	2.74	2.70	2.67	2.64	2.62	2.57	2.54	2.51	2.48	2.45	2.40	2.37	2.32	2.28	2.26	23
2.76	2.74	2.70	2.66	2.63	2.60	2.58	2.53	2.49	2.46	2.44	2.40	2.36	2.33	2.27	2.24	2.21	24
2.72	2.70	2.66	2.62	2.59	2.56	2.54	2.49	2.45	2.42	2.40	2.36	2.32	2.29	2.23	2.19	2.17	25
2.69	2.66	2.62	2.58	2.55	2.53	2.50	2.45	2.42	2.39	2.36	2.33	2.28	2.25	2.19	2.16	2.13	26
2.66	2.63	2.59	2.55	2.52	2.49	2.47	2.42	2.38	2.35	2.33	2.29	2.25	2.22	2.16	2.12	2.10	27
2.63	2.60	2.56	2.52	2.49	2.46	2.44	2.39	2.35	2.32	2.30	2.26	2.22	2.19	2.13	2.09	2.06	28
2.60	2.57	2.53	2.49	2.46	2.44	2.41	2.36	2.33	2.30	2.27	2.23	2.19	2.16	2.10	2.06	2.03	29
2.57	2.55	2.51	2.47	2.44	2.41	2.39	2.34	2.30	2.27	2.25	2.21	2.16	2.13	2.07	2.03	2.01	30
2.53	2.50	2.46	2.42	2.39	2.36	2.34	2.29	2.25	2.22	2.20	2.16	2.11	2.08	2.02	1.98	1.96	32
2.49	2.46	2.42	2.38	2.35	2.32	2.30	2.25	2.21	2.18	2.16	2.12	2.07	2.04	1.98	1.94	1.91	34
2.45	2.43	2.38	2.35	2.32	2.29	2.26	2.21	2.17	2.14	2.12	2.08	2.03	2.00	1.94	1.90	1.87	36
2.42	2.40	2.35	2.32	2.28	2.26	2.23	2.18	2.14	2.11	2.09	2.05	2.00	1.97	1.90	1.86	1.84	38
2.39	2.37	2.33	2.29	2.26	2.23	2.20	2.15	2.11	2.08	2.06	2.02	1.97	1.94	1.87	1.83	1.80	40
2.37	2.34	2.30	2.26	2.23	2.20	2.18	2.13	2.09	2.06	2.03	1.99	1.94	1.91	1.85	1.80	1.78	42
2.35	2.32	2.28	2.24	2.21	2.18	2.15	2.10	2.06	2.03	2.01	1.97	1.92	1.89	1.82	1.78	1.75	44
2.33	2.30	2.26	2.22	2.19	2.16	2.13	2.08	2.04	2.01	1.99	1.95	1.90	1.86	1.80	1.75	1.73	46
2.31	2.28	2.24	2.20	2.17	2.14	2.12	2.06	2.02	1.99	1.97	1.93	1.88	1.84	1.78	1.73	1.70	48
2.29	2.27	2.22	2.18	2.15	2.12	2.10	2.05	2.01	1.97	1.95	1.91	1.86	1.82	1.76	1.71	1.68	50
2.25	2.23	2.18	2.15	2.11	2.08	2.06	2.01	1.97	1.93	1.91	1.87	1.81	1.78	1.71	1.67	1.64	55
2.22	2.20	2.15	2.12	2.08	2.05	2.03	1.98	1.94	1.90	1.88	1.84	1.78	1.75	1.68	1.63	1.60	60
2.20	2.17	2.13	2.09	2.06	2.03	2.00	1.95	1.91	1.88	1.85	1.81	1.75	1.72	1.65	1.60	1.57	65
2.18	2.15	2.11	2.07	2.03	2.01	1.98	1.93	1.89	1.85	1.83	1.78	1.73	1.70	1.62	1.57	1.54	70
2.14	2.12	2.07	2.03	2.00	1.97	1.94	1.89	1.85	1.81	1.79	1.75	1.69	1.66	1.58	1.53	1.49	80
2.11	2.09	2.04	2.00	1.97	1.94	1.92	1.86	1.82	1.79	1.76	1.72	1.66	1.62	1.54	1.49	1.46	90
2.09	2.07	2.02	1.98	1.94	1.92	1.89	1.84	1.80	1.76	1.73	1.69	1.63	1.60	1.52	1.47	1.43	100
2.05	2.03	1.98	1.94	1.91	1.88	1.85	1.80	1.76	1.72	1.69	1.65	1.59	1.55	1.47	1.41	1.37	125
2.03	2.00	1.96	1.92	1.88	1.85	1.83	1.77	1.73	1.69	1.66	1.62	1.56	1.52	1.43	1.38	1.33	150
2.00	1.97	1.93	1.89	1.85	1.82	1.79	1.74	1.69	1.66	1.63	1.58	1.52	1.48	1.39	1.33	1.28	200
1.97	1.94	1.89	1.85	1.82	1.79	1.76	1.71	1.66	1.62	1.59	1.55	1.48	1.44	1.35	1.28	1.22	300
1.94	1.92	1.87	1.83	1.79	1.76	1.74	1.68	1.63	1.60	1.56	1.52	1.45	1.41	1.31	1.23	1.16	500
1.92	1.90	1.85	1.81	1.77	1.74	1.72	1.66	1.61	1.57	1.54	1.50	1.43	1.38	1.28	1.19	1.11	1000
1.90	1.88	1.83	1.79	1.76	1.72	1.70	1.64	1.59	1.55	1.52	1.47	1.40	1.36	1.25	1.15	1.00	∞

Approximate formula for ν_1 and ν_2 larger than 30:

$$\log_{10} F_{.01;\nu_1,\nu_2} \simeq \frac{2.0206}{\sqrt{h} - 1.40} - 1.073\left(\frac{1}{\nu_1} - \frac{1}{\nu_2}\right), \text{ where } \frac{1}{h} = \frac{1}{2}\left(\frac{1}{\nu_1} + \frac{1}{\nu_2}\right).$$

APPENDIX G THE *F*-DISTRIBUTION (Continued)

Table of $F_{0.005;\, \nu_1, \nu_2}$

Degrees of freedom for the numerator (ν_1); Degrees of freedom for the denominator (ν_2)

Multiply the numbers of the first row ($\nu_2 = 1$) by 1,000.

ν_2 \ ν_1	1	2	3	4	5	6	7	8	9	10	11	12	13	14	15	16	17	18
1	*16.2*	*20.0*	*21.6*	*22.5*	*23.1*	*23.4*	*23.7*	*23.9*	*24.1*	*24.2*	*24.3*	*24.4*	*24.5*	*24.6*	*24.6*	*24.7*	*24.7*	*24.8*
2	198	199	199	199	199	199	199	199	199	199	199	199	199	199	199	199	199	199
3	55.6	49.8	47.5	46.2	45.4	44.8	44.4	44.1	43.9	43.7	43.5	43.4	43.3	43.2	43.1	43.0	42.9	42.9
4	31.3	26.3	24.3	23.2	22.5	22.0	21.6	21.4	21.1	21.0	20.8	20.7	20.6	20.5	20.4	20.4	20.3	20.3
5	22.8	18.3	16.5	15.6	14.9	14.5	14.2	14.0	13.8	13.6	13.5	13.4	13.3	13.2	13.1	13.1	13.0	13.0
6	18.6	14.5	12.9	12.0	11.5	11.1	10.8	10.6	10.4	10.2	10.1	10.0	9.95	9.88	9.81	9.76	9.71	9.66
7	16.2	12.4	10.9	10.0	9.52	9.16	8.89	8.68	8.51	8.38	8.27	8.18	8.10	8.03	7.97	7.93	7.87	7.83
8	14.7	11.0	9.60	8.81	8.30	7.95	7.69	7.50	7.34	7.21	7.10	7.01	6.94	6.87	6.81	6.76	6.72	6.68
9	13.6	10.1	8.72	7.96	7.47	7.13	6.88	6.69	6.54	6.42	6.31	6.23	6.15	6.09	6.03	5.98	5.94	5.90
10	12.8	9.43	8.08	7.34	6.87	6.54	6.30	6.12	5.97	5.85	5.75	5.66	5.59	5.53	5.47	5.42	5.38	5.34
11	12.2	8.91	7.60	6.88	6.42	6.10	5.86	5.68	5.54	5.42	5.32	5.24	5.16	5.10	5.05	5.00	4.96	4.92
12	11.8	8.51	7.23	6.52	6.07	5.76	5.52	5.35	5.20	5.09	4.99	4.91	4.84	4.77	4.72	4.67	4.63	4.59
13	11.4	8.19	6.93	6.23	5.79	5.48	5.25	5.08	4.94	4.82	4.72	4.64	4.57	4.51	4.46	4.41	4.37	4.33
14	11.1	7.92	6.68	6.00	5.56	5.26	5.03	4.86	4.72	4.60	4.51	4.43	4.36	4.30	4.25	4.20	4.16	4.12
15	10.8	7.70	6.48	5.80	5.37	5.07	4.85	4.67	4.54	4.42	4.33	4.25	4.18	4.12	4.07	4.02	3.98	3.95
16	10.6	7.51	6.30	5.64	5.21	4.91	4.69	4.52	4.38	4.27	4.18	4.10	4.03	3.97	3.92	3.87	3.83	3.80
17	10.4	7.35	6.16	5.50	5.07	4.78	4.56	4.39	4.25	4.14	4.05	3.97	3.90	3.84	3.79	3.75	3.71	3.67
18	10.2	7.21	6.03	5.37	4.96	4.66	4.44	4.28	4.14	4.03	3.94	3.86	3.79	3.73	3.68	3.64	3.60	3.56
19	10.1	7.09	5.92	5.27	4.85	4.56	4.34	4.18	4.04	3.93	3.84	3.76	3.70	3.64	3.59	3.54	3.50	3.46
20	9.94	6.99	5.82	5.17	4.76	4.47	4.26	4.09	3.96	3.85	3.76	3.68	3.61	3.55	3.50	3.46	3.42	3.38
21	9.83	6.89	5.73	5.09	4.68	4.39	4.18	4.01	3.88	3.77	3.68	3.60	3.54	3.48	3.43	3.38	3.34	3.31
22	9.73	6.81	5.65	5.02	4.61	4.32	4.11	3.94	3.81	3.70	3.61	3.54	3.47	3.41	3.36	3.31	3.27	3.24
23	9.63	6.73	5.58	4.95	4.54	4.26	4.05	3.88	3.75	3.64	3.55	3.47	3.41	3.35	3.30	3.25	3.21	3.18
24	9.55	6.66	5.52	4.89	4.49	4.20	3.99	3.83	3.69	3.59	3.50	3.42	3.35	3.30	3.25	3.20	3.16	3.12
25	9.48	6.60	5.46	4.84	4.43	4.15	3.94	3.78	3.64	3.54	3.45	3.37	3.30	3.25	3.20	3.15	3.11	3.08
26	9.41	6.54	5.41	4.79	4.38	4.10	3.89	3.73	3.60	3.49	3.40	3.33	3.26	3.20	3.15	3.11	3.07	3.03
27	9.34	6.49	5.36	4.74	4.34	4.06	3.85	3.69	3.56	3.45	3.36	3.28	3.22	3.16	3.11	3.07	3.03	2.99
28	9.28	6.44	5.32	4.70	4.30	4.02	3.81	3.65	3.52	3.41	3.32	3.25	3.18	3.12	3.07	3.03	2.99	2.95
29	9.23	6.40	5.28	4.66	4.26	3.98	3.77	3.61	3.48	3.38	3.29	3.21	3.15	3.09	3.04	2.99	2.95	2.92
30	9.18	6.35	5.24	4.62	4.23	3.95	3.74	3.58	3.45	3.34	3.25	3.18	3.11	3.06	3.01	2.96	2.92	2.89
32	9.09	6.28	5.17	4.56	4.17	3.89	3.68	3.52	3.39	3.29	3.20	3.12	3.06	3.00	2.95	2.90	2.86	2.83
34	9.01	6.22	5.11	4.50	4.11	3.84	3.63	3.47	3.34	3.24	3.15	3.07	3.01	2.95	2.90	2.85	2.81	2.78
36	8.94	6.16	5.06	4.46	4.06	3.79	3.58	3.42	3.30	3.19	3.10	3.03	2.96	2.90	2.85	2.81	2.77	2.73
38	8.88	6.11	5.02	4.41	4.02	3.75	3.54	3.39	3.25	3.15	3.06	2.99	2.92	2.87	2.82	2.77	2.73	2.70
40	8.83	6.07	4.98	4.37	3.99	3.71	3.51	3.35	3.22	3.12	3.03	2.95	2.89	2.83	2.78	2.74	2.70	2.66
42	8.78	6.03	4.94	4.34	3.95	3.68	3.48	3.32	3.19	3.09	3.00	2.92	2.86	2.80	2.75	2.71	2.67	2.63
44	8.74	5.99	4.91	4.31	3.92	3.65	3.45	3.29	3.16	3.06	2.97	2.89	2.83	2.77	2.72	2.68	2.64	2.60
46	8.70	5.96	4.88	4.28	3.90	3.62	3.42	3.26	3.14	3.03	2.94	2.87	2.80	2.75	2.70	2.65	2.61	2.58
48	8.66	5.93	4.85	4.25	3.87	3.60	3.40	3.24	3.11	3.01	2.92	2.85	2.78	2.72	2.67	2.63	2.59	2.55
50	8.63	5.90	4.83	4.23	3.85	3.58	3.38	3.22	3.09	2.99	2.90	2.82	2.76	2.70	2.65	2.61	2.57	2.53
55	8.55	5.84	4.77	4.18	3.80	3.53	3.33	3.17	3.05	2.94	2.85	2.78	2.71	2.66	2.61	2.56	2.52	2.49
60	8.49	5.80	4.73	4.14	3.76	3.49	3.29	3.13	3.01	2.90	2.82	2.74	2.68	2.62	2.57	2.53	2.49	2.45
65	8.44	5.75	4.68	4.11	3.73	3.46	3.26	3.10	2.98	2.87	2.79	2.71	2.65	2.59	2.54	2.49	2.45	2.42
70	8.40	5.72	4.65	4.08	3.70	3.43	3.23	3.08	2.95	2.85	2.76	2.68	2.62	2.56	2.51	2.47	2.43	2.39
80	8.33	5.67	4.61	4.03	3.65	3.39	3.19	3.03	2.91	2.80	2.72	2.64	2.58	2.52	2.47	2.43	2.39	2.35
90	8.28	5.62	4.57	3.99	3.62	3.35	3.15	3.00	2.87	2.77	2.68	2.61	2.54	2.49	2.44	2.39	2.35	2.32
100	8.24	5.59	4.54	3.96	3.59	3.33	3.13	2.97	2.85	2.74	2.66	2.58	2.52	2.46	2.41	2.37	2.33	2.29
125	8.17	5.53	4.49	3.91	3.54	3.28	3.08	2.93	2.80	2.70	2.61	2.54	2.47	2.42	2.37	2.32	2.28	2.24
150	8.12	5.49	4.45	3.88	3.51	3.25	3.05	2.89	2.77	2.67	2.58	2.51	2.44	2.38	2.33	2.29	2.25	2.21
200	8.06	5.44	4.41	3.84	3.47	3.21	3.01	2.85	2.73	2.63	2.54	2.47	2.40	2.35	2.30	2.25	2.21	2.18
300	8.00	5.39	4.37	3.80	3.43	3.17	2.97	2.81	2.69	2.59	2.51	2.43	2.37	2.31	2.26	2.21	2.17	2.14
500	7.95	5.36	4.33	3.76	3.40	3.14	2.94	2.79	2.66	2.56	2.48	2.40	2.34	2.28	2.23	2.19	2.14	2.11
1000	7.92	5.33	4.31	3.74	3.37	3.11	2.92	2.77	2.64	2.54	2.45	2.38	2.32	2.26	2.21	2.16	2.12	2.09
∞	7.88	5.30	4.28	3.72	3.35	3.09	2.90	2.74	2.62	2.52	2.43	2.36	2.29	2.24	2.19	2.14	2.10	2.06

Example: $P\{F_{.005;\,8,20} < 4.09\} = 99.5\%$.

$F_{.995;\,\nu_1,\nu_2} = 1/F_{.005;\,\nu_2,\nu_1}$. Example: $F_{.995;\,8,20} = 1/F_{.005;\,20,8} = 1/6.61 = 0.151$.

Table of $F_{0.005;\, \nu_1, \nu_2}$

Degrees of freedom for the numerator (ν_1)

Multiply the numbers of the first row ($\nu_2 = 1$) by 100.

19	20	22	24	26	28	30	35	40	45	50	60	80	100	200	500	∞	Degrees of freedom for the denominator (ν_2)
24.8	*24.8*	*24.9*	*24.9*	*25.0*	*25.0*	*25.0*	*25.1*	*25.1*	*25.2*	*25.2*	*25.3*	*25.3*	*25.3*	*25.4*	*25.4*	*25.5*	1
19.9	19.9	19.9	19.9	19.9	19.9	19.9	19.9	19.9	19.9	19.9	19.9	19.9	19.9	19.9	20.0	20.0	2
42.8	42.8	42.7	42.6	42.6	42.5	42.5	42.4	42.3	42.3	42.2	42.1	42.1	42.0	41.9	41.9	41.8	3
20.2	20.2	20.1	20.0	20.0	19.9	19.9	19.8	19.8	19.7	19.7	19.6	19.5	19.5	19.4	19.4	19.3	4
12.9	12.9	12.8	12.8	12.7	12.7	12.7	12.6	12.5	12.5	12.5	12.4	12.3	12.3	12.2	12.2	12.1	5
9.62	9.59	9.53	9.47	9.43	9.39	9.36	9.29	9.24	9.20	9.17	9.12	9.06	9.03	8.95	8.91	8.88	6
7.79	7.75	7.69	7.64	7.60	7.57	7.53	7.47	7.42	7.38	7.35	7.31	7.25	7.22	7.15	7.10	7.08	7
6.64	6.61	6.55	6.50	6.46	6.43	6.40	6.33	6.29	6.25	6.22	6.18	6.12	6.09	6.02	5.98	5.95	8
5.86	5.83	5.78	5.73	5.69	5.65	5.62	5.56	5.52	5.48	5.45	5.41	5.36	5.32	5.26	5.21	5.19	9
5.30	5.27	5.22	5.17	5.13	5.10	5.07	5.01	4.97	4.93	4.90	4.86	4.80	4.77	4.71	4.67	4.64	10
4.89	4.86	4.80	4.76	4.72	4.68	4.65	4.60	4.55	4.52	4.49	4.44	4.39	4.36	4.29	4.25	4.23	11
4.56	4.53	4.48	4.43	4.39	4.36	4.33	4.27	4.23	4.19	4.17	4.12	4.07	4.04	3.97	3.93	3.90	12
4.30	4.27	4.22	4.17	4.13	4.10	4.07	4.01	3.97	3.94	3.91	3.87	3.81	3.78	3.71	3.67	3.65	13
4.09	4.06	4.01	3.96	3.92	3.89	3.86	3.80	3.76	3.73	3.70	3.66	3.60	3.57	3.50	3.46	3.44	14
3.91	3.88	3.83	3.79	3.75	3.72	3.69	3.63	3.58	3.55	3.52	3.48	3.43	3.39	3.33	3.29	3.26	15
3.76	3.73	3.68	3.64	3.60	3.57	3.54	3.48	3.44	3.40	3.37	3.33	3.28	3.25	3.18	3.14	3.11	16
3.64	3.61	3.56	3.51	3.47	3.44	3.41	3.35	3.31	3.28	3.25	3.21	3.15	3.12	3.05	3.01	2.98	17
3.53	3.50	3.45	3.40	3.36	3.33	3.30	3.25	3.20	3.17	3.14	3.10	3.04	3.01	2.94	2.90	2.87	18
3.43	3.40	3.35	3.31	3.27	3.24	3.21	3.15	3.11	3.07	3.04	3.00	2.95	2.91	2.85	2.80	2.78	19
3.35	3.32	3.27	3.22	3.18	3.15	3.12	3.07	3.02	2.99	2.96	2.92	2.86	2.83	2.76	2.72	2.69	20
3.27	3.24	3.19	3.15	3.11	3.08	3.05	2.99	2.95	2.91	2.88	2.84	2.78	2.75	2.68	2.64	2.61	21
3.20	3.18	3.12	3.08	3.04	3.01	2.98	2.92	2.88	2.84	2.82	2.77	2.72	2.69	2.62	2.57	2.55	22
3.15	3.12	3.06	3.02	2.98	2.95	2.92	2.86	2.82	2.78	2.76	2.71	2.66	2.62	2.56	2.51	2.48	23
3.09	3.06	3.01	2.97	2.93	2.90	2.87	2.81	2.77	2.73	2.70	2.66	2.60	2.57	2.50	2.46	2.43	24
3.04	3.01	2.96	2.92	2.88	2.85	2.82	2.76	2.72	2.68	2.65	2.61	2.55	2.52	2.45	2.41	2.38	25
3.00	2.97	2.92	2.87	2.83	2.80	2.77	2.72	2.67	2.64	2.61	2.56	2.51	2.47	2.40	2.36	2.33	26
2.96	2.93	2.88	2.83	2.79	2.76	2.73	2.67	2.63	2.59	2.57	2.52	2.47	2.43	2.36	2.32	2.29	27
2.92	2.89	2.84	2.79	2.76	2.72	2.69	2.64	2.59	2.56	2.53	2.48	2.43	2.39	2.32	2.28	2.25	28
2.88	2.86	2.80	2.76	2.72	2.69	2.66	2.60	2.56	2.52	2.49	2.45	2.39	2.36	2.28	2.24	2.21	29
2.85	2.82	2.77	2.73	2.69	2.66	2.63	2.57	2.52	2.49	2.46	2.42	2.36	2.32	2.25	2.21	2.18	30
2.80	2.77	2.71	2.67	2.63	2.60	2.57	2.51	2.47	2.43	2.40	2.36	2.30	2.26	2.19	2.15	2.11	32
2.75	2.72	2.66	2.62	2.58	2.55	2.52	2.46	2.42	2.38	2.35	2.30	2.25	2.21	2.14	2.09	2.06	34
2.70	2.67	2.62	2.58	2.54	2.50	2.48	2.42	2.37	2.33	2.30	2.26	2.20	2.17	2.09	2.04	2.01	36
2.66	2.63	2.58	2.54	2.50	2.47	2.44	2.38	2.33	2.29	2.27	2.22	2.16	2.12	2.05	2.00	1.97	38
2.63	2.60	2.55	2.50	2.46	2.43	2.40	2.34	2.30	2.26	2.23	2.18	2.12	2.09	2.01	1.96	1.93	40
2.60	2.57	2.52	2.47	2.43	2.40	2.37	2.31	2.26	2.23	2.20	2.15	2.09	2.06	1.98	1.93	1.90	42
2.57	2.54	2.49	2.44	2.40	2.37	2.34	2.28	2.24	2.20	2.17	2.12	2.06	2.03	1.95	1.90	1.87	44
2.54	2.51	2.46	2.42	2.38	2.34	2.32	2.26	2.21	2.17	2.14	2.10	2.04	2.00	1.92	1.87	1.84	46
2.52	2.49	2.44	2.39	2.36	2.32	2.29	2.23	2.19	2.15	2.12	2.07	2.01	1.97	1.89	1.84	1.81	48
2.50	2.47	2.42	2.37	2.33	2.30	2.27	2.21	2.16	2.13	2.10	2.05	1.99	1.95	1.87	1.82	1.79	50
2.45	2.42	2.37	2.33	2.29	2.26	2.23	2.16	2.12	2.08	2.05	2.00	1.94	1.90	1.82	1.77	1.73	55
2.42	2.39	2.33	2.29	2.25	2.22	2.19	2.13	2.08	2.04	2.01	1.96	1.90	1.86	1.78	1.73	1.69	60
2.39	2.36	2.30	2.26	2.22	2.19	2.16	2.09	2.05	2.01	1.98	1.93	1.87	1.83	1.74	1.69	1.65	65
2.36	2.33	2.28	2.23	2.19	2.16	2.13	2.07	2.02	1.98	1.95	1.90	1.84	1.80	1.71	1.66	1.62	70
2.32	2.29	2.23	2.19	2.15	2.11	2.08	2.02	1.97	1.93	1.90	1.85	1.79	1.75	1.66	1.60	1.56	80
2.28	2.25	2.20	2.15	2.12	2.08	2.05	1.99	1.94	1.90	1.87	1.82	1.75	1.71	1.62	1.56	1.52	90
2.26	2.23	2.17	2.13	2.09	2.05	2.02	1.96	1.91	1.87	1.84	1.79	1.72	1.68	1.59	1.53	1.49	100
2.21	2.18	2.13	2.08	2.04	2.01	1.98	1.91	1.86	1.82	1.79	1.74	1.67	1.63	1.53	1.47	1.42	125
2.18	2.15	2.10	2.05	2.01	1.98	1.94	1.88	1.83	1.79	1.76	1.70	1.63	1.59	1.49	1.42	1.37	150
2.14	2.11	2.06	2.01	1.97	1.94	1.91	1.84	1.79	1.75	1.71	1.66	1.59	1.54	1.44	1.37	1.31	200
2.10	2.07	2.02	1.97	1.93	1.90	1.87	1.80	1.75	1.71	1.67	1.61	1.54	1.50	1.39	1.31	1.25	300
2.07	2.04	1.99	1.94	1.90	1.87	1.84	1.77	1.72	1.67	1.64	1.58	1.51	1.46	1.35	1.26	1.18	500
2.05	2.02	1.97	1.92	1.88	1.84	1.81	1.75	1.69	1.65	1.61	1.56	1.48	1.43	1.31	1.22	1.13	1000
2.03	2.00	1.95	1.90	1.86	1.82	1.79	1.72	1.67	1.63	1.59	1.53	1.45	1.40	1.28	1.17	1.00	∞

Approximate formula for ν_1 and ν_2 larger than 30: $\log_{10} F_{.005;\, \nu_1, \nu_2} \simeq \dfrac{2.2373}{\sqrt{h} - 1.61} - 1.250\left(\dfrac{1}{\nu_1} - \dfrac{1}{\nu_2}\right)$ where $\dfrac{1}{h} = \dfrac{1}{2}\left(\dfrac{1}{\nu_1} + \dfrac{1}{\nu_2}\right)$.

APPENDIX G THE F-DISTRIBUTION (Continued)

Table of $F_{0.001;\nu_1,\nu_2}$

Degrees of freedom for the numerator (ν_1); rows: Degrees of freedom for the denominator (ν_2)

Multiply the numbers of the first row ($\nu_1 = 1$) by 10,000.

ν_2	1	2	3	4	5	6	7	8	9	10	15	20	30	50	100	200	500	∞
1	*40.5*	*50.0*	*54.0*	*56.2*	*57.6*	*58.6*	*59.3*	*59.8*	*60.2*	*60.6*	*61.6*	*62.1*	*62.6*	*63.0*	*63.3*	*63.5*	*63.6*	*63.7*
2	998	999	999	999	999	999	999	999	999	999	999	999	999	999	999	999	999	999
3	168	148	141	137	135	133	132	131	130	129	127	126	125	125	124	124	124	124
4	74.1	61.2	59.2	53.4	51.7	50.5	49.7	49.0	48.5	48.0	46.8	46.1	45.4	44.9	44.5	44.3	44.1	44.0
5	47.0	36.6	33.2	31.1	29.8	28.8	28.2	27.6	27.2	26.9	25.9	25.4	24.9	24.4	24.1	23.9	23.8	23.8
6	35.5	27.0	23.7	21.9	20.8	20.0	19.5	19.0	18.7	18.4	17.6	17.1	16.7	16.3	16.0	15.9	15.8	15.8
7	29.2	21.7	18.8	17.2	16.2	15.5	15.0	14.6	14.3	14.1	13.3	12.9	12.5	12.2	11.9	11.8	11.7	11.7
8	25.4	18.5	15.8	14.4	13.5	12.9	12.4	12.0	11.8	11.5	10.8	10.5	10.1	9.80	9.57	9.46	9.39	9.34
9	22.9	16.4	13.9	12.6	11.7	11.1	10.7	10.4	10.1	9.89	9.24	8.90	8.55	8.26	8.04	7.93	7.86	7.81
10	21.0	14.9	12.6	11.3	10.5	9.92	9.52	9.20	8.96	8.75	8.13	7.80	7.47	7.19	6.98	6.87	6.81	6.76
11	19.7	13.8	11.6	10.4	9.58	9.05	8.66	8.35	8.12	7.92	7.32	7.01	6.68	6.41	6.21	6.10	6.04	6.00
12	18.6	13.0	10.8	9.63	8.89	8.38	8.00	7.71	7.48	7.29	6.71	6.40	6.09	5.83	5.63	5.52	5.46	5.42
13	17.8	12.3	10.2	9.07	8.35	7.86	7.49	7.21	6.98	6.80	6.23	5.93	5.62	5.37	5.17	5.07	5.01	4.97
14	17.1	11.8	9.73	8.62	7.92	7.43	7.08	6.80	6.58	6.40	5.85	5.56	5.25	5.00	4.80	4.70	4.64	4.60
15	16.6	11.3	9.34	8.25	7.57	7.09	6.74	6.47	6.26	6.08	5.53	5.25	4.95	4.70	4.51	4.41	4.35	4.31
16	16.1	11.0	9.00	7.94	7.27	6.81	6.46	6.19	5.98	5.81	5.27	4.99	4.70	4.45	4.26	4.16	4.10	4.06
17	15.7	10.7	8.73	7.68	7.02	6.56	6.22	5.96	5.75	5.58	5.05	4.78	4.48	4.24	4.05	3.95	3.89	3.85
18	15.4	10.4	8.49	7.46	6.81	6.35	6.02	5.76	5.56	5.39	4.87	4.59	4.30	4.06	3.87	3.77	3.71	3.67
19	15.1	10.2	8.28	7.26	6.61	6.18	5.84	5.59	5.39	5.22	4.70	4.43	4.14	3.90	3.71	3.61	3.55	3.51
20	14.8	9.95	8.10	7.10	6.46	6.02	5.69	5.44	5.24	5.08	4.56	4.29	4.01	3.77	3.58	3.48	3.42	3.38
22	14.4	9.61	7.80	6.81	6.19	5.76	5.44	5.19	4.99	4.83	4.32	4.06	3.77	3.53	3.34	3.25	3.19	3.15
24	14.0	9.34	7.55	6.59	5.98	5.55	5.23	4.99	4.80	4.64	4.14	3.87	3.59	3.35	3.16	3.07	3.01	2.97
26	13.7	9.12	7.36	6.41	5.80	5.38	5.07	4.83	4.64	4.48	3.99	3.72	3.45	3.20	3.01	2.92	2.86	2.82
28	13.5	8.93	7.19	6.25	5.66	5.24	4.93	4.69	4.50	4.35	3.86	3.60	3.32	3.08	2.89	2.79	2.73	2.70
30	13.3	8.77	7.05	6.12	5.53	5.12	4.82	4.58	4.39	4.24	3.75	3.49	3.22	2.98	2.79	2.69	2.63	2.59
40	12.6	8.25	6.60	5.70	5.13	4.73	4.43	4.21	4.02	3.87	3.40	3.15	2.87	2.64	2.44	2.34	2.28	2.23
50	12.2	7.95	6.34	5.46	4.90	4.51	4.22	4.00	3.82	3.67	3.20	2.95	2.68	2.44	2.24	2.14	2.07	2.03
60	12.0	7.76	6.17	5.31	4.76	4.37	4.09	3.87	3.69	3.54	3.08	2.83	2.56	2.31	2.11	2.01	1.93	1.89
80	11.7	7.54	5.97	5.13	4.58	4.21	3.92	3.70	3.53	3.39	2.93	2.68	2.40	2.16	1.95	1.84	1.77	1.72
100	11.5	7.41	5.85	5.01	4.48	4.11	3.83	3.61	3.44	3.30	2.84	2.59	2.32	2.07	1.87	1.75	1.68	1.62
200	11.2	7.15	5.64	4.81	4.29	3.92	3.65	3.43	3.26	3.12	2.67	2.42	2.15	1.90	1.68	1.55	1.46	1.39
500	11.0	7.01	5.51	4.69	4.18	3.82	3.54	3.33	3.16	3.02	2.58	2.33	2.05	1.80	1.57	1.43	1.32	1.23
∞	10.8	6.91	5.42	4.62	4.10	3.74	3.47	3.27	3.10	2.96	2.51	2.27	1.99	1.73	1.49	1.34	1.21	1.00

Example: $P\{F_{.001;8,20} < 5.44\} = 99.9\%$.

$F_{.999;\nu_1,\nu_2} = 1/F_{.001;\nu_2,\nu_1}$. Example $F_{.999;8,20} = 1/F_{.001;20,8} = 1/10.5 = 0.095$.

Approximate formula for ν_1 and ν_2 larger than 30: $\log_{10} F_{.001;\nu_1,\nu_2} \simeq \dfrac{2.6841}{\sqrt{h - 2.09}} - 1.672\left(\dfrac{1}{\nu_1} - \dfrac{1}{\nu_2}\right)$, where $\dfrac{1}{h} = \dfrac{1}{2}\left(\dfrac{1}{\nu_1} + \dfrac{1}{\nu_2}\right)$.

APPENDIX H

QUALITY CONTROL CHART COEFFICIENTS*

TABLE B2.—FACTORS FOR COMPUTING CONTROL CHARTLINES.

Number of Observations in Sample, n	Chart for Averages			Chart for Standard Deviations						Chart for Ranges						
	Factors for Control Limits			Factors for Central Line		Factors for Control Limits				Factors for Central Line		Factors for Control Limits				
	A	A_1	A_2	c_2	$1/c_2$	B_1	B_2	B_3	B_4	d_2	$1/d_2$	d_2	D_1	D_2	D_3	D_4
2......	2.121	3.760	1.880	0.5642	1.7725	0	1.843	0	3.267	1.128	0.8865	0.853	0	3.686	0	3.267
3......	1.732	2.394	1.023	0.7236	1.3820	0	1.858	0	2.568	1.693	0.5907	0.888	0	4.358	0	2.575
4......	1.500	1.880	0.729	0.7979	1.2533	0	1.808	0	2.266	2.059	0.4857	0.880	0	4.698	0	2.282
5......	1.342	1.596	0.577	0.8407	1.1894	0	1.765	0	2.089	2.326	0.4299	0.864	0	4.918	0	2.115
6......	1.225	1.410	0.483	0.8686	1.1512	0.026	1.711	0.030	1.970	2.534	0.3946	0.848	0	5.078	0	2.004
7......	1.134	1.277	0.419	0.8882	1.1259	0.105	1.672	0.118	1.882	2.704	0.3698	0.833	0.205	5.203	0.076	1.924
8......	1.061	1.175	0.373	0.9027	1.1078	0.167	1.638	0.185	1.815	2.847	0.3512	0.820	0.387	5.307	0.136	1.864
9......	1.000	1.094	0.337	0.9139	1.0942	0.219	1.609	0.239	1.761	2.970	0.3367	0.808	0.546	5.394	0.184	1.816
10......	0.949	1.028	0.308	0.9227	1.0837	0.262	1.584	0.284	1.716	3.078	0.3249	0.797	0.687	5.469	0.223	1.777
11......	0.905	0.973	0.285	0.9300	1.0753	0.299	1.561	0.321	1.679	3.173	0.3152	0.787	0.812	5.534	0.256	1.744
12......	0.866	0.925	0.266	0.9359	1.0684	0.331	1.541	0.354	1.646	3.258	0.3069	0.778	0.924	5.592	0.284	1.716
13......	0.832	0.884	0.249	0.9410	1.0627	0.359	1.523	0.382	1.618	3.336	0.2998	0.770	1.026	5.646	0.308	1.692
14......	0.802	0.848	0.235	0.9453	1.0579	0.384	1.507	0.406	1.594	3.407	0.2935	0.762	1.121	5.693	0.329	1.671
15......	0.775	0.816	0.223	0.9490	1.0537	0.406	1.492	0.428	1.572	3.472	0.2880	0.755	1.207	5.737	0.348	1.652
16......	0.750	0.788	0.212	0.9523	1.0501	0.427	1.478	0.448	1.552	3.532	0.2831	0.749	1.285	5.779	0.364	1.636
17......	0.728	0.762	0.203	0.9551	1.0470	0.445	1.465	0.466	1.534	3.588	0.2787	0.743	1.359	5.817	0.379	1.621
18......	0.707	0.738	0.194	0.9576	1.0442	0.461	1.454	0.482	1.518	3.640	0.2747	0.738	1.426	5.854	0.392	1.608
19......	0.688	0.717	0.187	0.9599	1.0418	0.477	1.443	0.497	1.503	3.689	0.2711	0.733	1.490	5.888	0.404	1.596
20......	0.671	0.697	0.180	0.9619	1.0396	0.491	1.433	0.510	1.490	3.735	0.2677	0.729	1.548	5.922	0.414	1.586
21......	0.655	0.679	0.173	0.9638	1.0376	0.504	1.424	0.523	1.477	3.778	0.2647	0.724	1.606	5.950	0.425	1.575
22......	0.640	0.662	0.167	0.9655	1.0358	0.516	1.415	0.534	1.466	3.819	0.2618	0.720	1.659	5.979	0.434	1.566
23......	0.626	0.647	0.162	0.9670	1.0342	0.527	1.407	0.545	1.455	3.858	0.2592	0.716	1.710	6.006	0.443	1.557
24......	0.612	0.632	0.157	0.9684	1.0327	0.538	1.399	0.555	1.445	3.895	0.2567	0.712	1.759	6.031	0.452	1.548
25......	0.600	0.619	0.135	0.9696	1.0313	0.548	1.392	0.565	1.435	3.931	0.2544	0.709	1.804	6.058	0.459	1.541
Over 25......	$\frac{3}{\sqrt{n}}$	$\frac{3}{\sqrt{n}}$	....	....	....	*	**	*	**	....	....	....	....	....	....	....

$*1 - \frac{3}{\sqrt{2n}}$ $\qquad$ $**1 + \frac{3}{\sqrt{2n}}$

*Reproduced by permission from *ASTM Manual on Quality Control of Materials*, American Society for Testing Materials, Philadelphia, Pa., 1951.

APPENDIX I

RANDOM NUMBERS*

	00 04	05 09	10 14	15 19	20 24	25 29	30 34	35 39	40 44	45 49
00	39591	66082	48626	95780	55228	87189	75717	97042	19696	48613
01	46304	97377	43462	21739	14566	72533	60171	29024	77581	72760
02	99547	60779	22734	23678	44895	89767	18249	41702	35850	40543
03	06743	63537	24553	77225	94743	79448	12753	95986	78088	48019
04	69568	65496	49033	88577	98606	92156	08846	54912	12691	13170
05	68198	69571	34349	73141	42640	44721	30462	35075	33475	47407
06	27974	12609	77428	64441	49008	60489	66780	55499	80842	57706
07	50552	20688	02769	63037	15494	71784	70559	58158	53437	46216
08	74687	02033	98290	62635	88877	28599	63682	35566	03271	05651
09	49303	76629	71897	30990	62923	36686	96167	11492	90333	84501
10	89734	39183	52026	14997	15140	18250	62831	51236	61236	09179
11	74042	40747	02617	11346	01884	82066	55913	72422	13971	64209
12	84706	31375	67053	73367	95349	31074	36908	42782	89690	48002
13	83664	21365	28882	48926	45435	60577	85270	02777	06878	27561
14	47813	74854	73388	11385	99108	97878	32858	17473	07682	20166
15	00371	56525	38880	53702	09517	47281	15995	98350	25233	79718
16	81182	48434	27431	55806	25389	40774	72978	16835	65066	28732
17	75242	35904	73077	24537	81354	48902	03478	42867	04552	66034
18	96239	80246	07000	09555	55051	49596	44629	88225	28195	44598
19	82988	17440	85311	03360	38176	51462	86070	03924	84413	92363
20	77599	29143	89088	57593	60036	17297	30923	36224	46327	96266
21	61433	33118	53488	82981	44709	63655	64388	00498	14135	57514
22	76008	15045	45440	84062	52363	18079	33726	44301	86246	99727
23	26494	76598	85834	10844	56300	02244	72118	96510	98388	80161
24	46570	88558	77533	33359	07830	84752	53260	45655	36881	98535
25	73995	41532	87933	79930	14310	64833	49020	70067	99726	97007
26	93901	38276	75544	19679	82899	11365	22896	42118	77165	08734
27	41925	28215	40866	93501	45446	27913	21708	01788	81404	15119
28	80720	02782	24326	41328	10357	86883	80086	77138	57072	12100
29	92596	39416	50362	04423	04561	58179	54188	44978	14322	97056
30	39693	58559	45839	47278	38548	38885	19875	26829	86711	57005
31	86923	37863	14340	30929	04079	65274	03030	15106	09362	82972
32	99700	79237	18177	58879	56221	65644	33331	87502	32961	40996
33	60248	21953	52321	16984	03252	90433	97304	50181	71026	01946
34	29136	71987	03992	67025	31070	78348	47823	11033	13037	47732
35	57471	42913	85212	42319	92901	97727	04775	94396	38154	25238
36	57324	93847	03269	56096	95028	14039	76128	63747	27301	65529
37	56768	71694	63361	80836	30841	71875	40944	54827	01887	54822
38	70400	81534	02148	41441	26582	27481	84262	14084	42409	62950
39	05454	88418	48646	99565	36635	85496	18894	77271	26894	00889
40	80934	56136	47063	96311	19067	59790	08752	68050	85685	83076
41	06919	46237	50676	11238	75637	43086	95323	52867	06891	32089
42	00152	23997	41751	74756	50975	75365	70158	67663	51431	46375
43	88505	74625	71783	82511	13661	63178	39291	76796	74736	10980
44	64514	80967	33545	09582	86329	58152	05931	35961	70069	12142
45	25280	53007	99651	96366	49378	80971	10419	12981	70572	11575
46	71292	63716	93210	59312	39493	24252	54849	29754	41497	79228
47	49734	50498	08974	05904	68172	02864	10994	22482	12912	17920
48	43075	09754	71880	92614	99928	94424	86353	87549	94499	11459
49	15116	16643	03981	06566	14050	33671	03814	48856	41267	76252

*Reproduced from George W. Snedecor's *Everyday Statistics*, published by Wm. C. Brown Company, Dubuque, Iowa, 1950, by permission of the author and publishers.

APPENDIX I RANDOM NUMBERS (Continued)

	50 54	**55 59**	**60 64**	**65 69**	**70 74**	**75 79**	**80 84**	**85 89**	**90 94**	**95 99**
00	25178	77518	41773	39926	09843	29694	43801	69276	44707	23455
01	45803	95106	85816	33366	37383	76832	37024	06581	22587	24827
02	15532	30898	14922	13923	44987	45122	86515	55836	96165	19650
03	99068	35453	42152	12078	04913	06083	06645	93310	40016	85421
04	70983	88359	95583	79848	24101	67502	25692	42496	77732	19278
05	71181	48289	03153	18779	65702	03612	64608	84071	47588	09982
06	44052	59163	74033	86112	27731	46135	63092	59171	44816	12354
07	91555	87708	70964	43346	56811	08725	75139	77674	82467	41899
08	54307	12188	58089	73745	35569	97352	77301	37684	36823	69218
09	63631	23919	06785	13891	89918	76211	09362	34292	17640	65907
10	46832	30801	98898	28954	97793	20825	36775	71974	15574	09184
11	05944	82632	39310	74857	61725	50569	81937	16820	85446	51168
12	28199	90116	59501	49025	73005	84954	11587	97691	90415	34685
13	08391	05600	00624	95068	33776	44985	01505	76911	45539	32181
14	29634	13021	96568	15124	55092	44043	31073	92371	51288	33378
15	61509	18842	79201	46451	68594	98120	68110	91062	42095	61839
16	87888	23033	69837	65661	15130	44649	42515	83861	50721	36110
17	94585	15218	74838	61809	92293	85490	46934	08531	70107	65707
18	82033	93915	34898	79913	70013	27573	39256	35167	35070	47095
19	79131	10022	82199	78976	22702	37936	10445	96846	84927	69745
20	79344	39236	41333	11473	15049	47930	99029	97150	82275	55149
21	15384	44585	18773	89733	40779	59664	83328	25162	58758	17761
22	38802	90957	32910	97485	10358	88588	95310	22252	19143	69011
23	85874	18400	28151	29541	63706	43197	65726	94117	22169	91806
24	26200	72680	12364	46010	92208	59103	60417	45389	56122	85353
25	13772	75282	81418	42188	66529	47981	92548	10079	68179	40915
26	91876	07434	96946	98382	97374	34444	17992	42811	01579	48741
27	31721	21713	83632	40605	24227	53219	05482	86768	53239	24812
28	92570	53242	98133	84706	78048	29645	79336	66091	05793	25922
29	02880	29307	73734	66448	64739	74645	29562	13999	17492	49891
30	80982	14684	31038	85302	98349	57313	86371	33938	10768	60837
31	38000	43364	94825	32413	46781	09685	69058	56644	85531	55173
32	14218	94289	79484	61868	40034	22546	68726	14736	80844	13466
33	74358	21940	40280	22233	09123	49375	55094	46113	54046	51771
34	39049	14986	94000	26649	13037	34609	45186	89515	63214	66886
35	48727	06309	91486	67316	84576	11100	37580	49629	83224	46321
36	22719	29784	40682	96715	40745	57458	70048	48306	50270	87424
37	33980	36769	51977	03689	79071	20279	64787	48877	44063	93733
38	23885	66721	16542	12648	65986	43104	45583	75729	35118	58742
39	85190	44068	78477	69133	58983	96504	44232	74809	25266	73872
40	33453	36333	45814	78128	55914	89829	43251	41634	48488	49153
41	98236	11489	97240	01678	30779	75214	80039	68895	95271	19654
42	21295	53563	43609	48439	87427	88065	09892	58524	43815	31340
43	28335	79849	69842	71669	38770	54445	48736	03242	83181	85403
44	95449	35273	62581	85522	35813	34475	97514	72839	10387	31649
45	88167	03878	89405	55461	73248	48620	31732	47317	06252	54652
46	86131	62596	98785	02360	54271	26242	93735	20752	17146	18315
47	71134	90264	30126	08586	97497	61678	81940	00907	39096	02082
48	02664	53438	76839	52290	77999	05799	93744	16634	84924	31344
49	90664	96876	16663	25608	67140	84619	67167	13192	81774	58619

APPENDIX I RANDOM NUMBERS (Continued)

	00 04	05 09	10 14	15 19	20 24	25 29	30 34	35 39	40 44	45 49
50	03873	86558	72524	02542	73184	37905	05882	15596	73646	50798
51	08761	47547	02216	48086	56490	89959	69975	04500	23779	76697
52	61270	98773	40298	26077	80396	08166	35723	61933	13985	19102
53	73758	15578	95748	02967	35122	36539	72822	68241	34803	42457
54	17132	32196	60523	00544	73700	70122	27962	85597	36011	79971
55	26175	29794	44838	84414	82748	22246	70694	57953	39780	17791
56	06004	04516	06210	03536	84451	30767	37928	26928	07396	64611
57	34687	73753	36327	73704	61564	99434	90938	03967	97420	19913
58	27865	08255	57859	04746	79700	68823	16002	58115	07580	12675
59	89423	51114	90820	26786	77404	05795	49036	34686	98767	32284
60	99039	80312	69745	87636	10058	84834	89485	08775	19041	61375
61	02852	54339	45496	20587	85921	06763	68873	35367	42627	54973
62	10850	42788	94737	74549	74296	13053	46816	32141	02533	25648
63	38391	18507	33151	69434	80103	02603	61110	89395	67621	67025
64	48181	95478	62739	90148	00156	09338	44558	53271	87549	45974
65	23098	23720	76508	69083	56584	00423	21634	35990	09234	95116
66	25104	82019	21120	06165	44324	77577	15774	44091	69687	67576
67	22205	40198	86884	28103	57306	54915	03426	66700	45993	36668
68	64975	05064	29617	40622	20330	18518	45312	57921	23188	82361
69	58710	75278	47730	26093	16436	38868	76861	85914	14162	21984
70	12140	72905	26022	07675	16362	34504	47740	39923	04081	03162
71	73226	39840	47958	97249	14146	34543	76162	74158	59739	67447
72	12320	86217	66162	70941	58940	58006	80731	66680	02183	94678
73	41364	64156	23000	23188	64945	33815	32884	76955	56574	61666
74	97881	80867	70117	72041	03554	29087	19767	71838	80545	61402
75	88295	87271	82812	97588	09960	06312	03050	77332	25977	18385
76	95321	89836	78230	46037	72483	87533	74571	88859	26908	55626
77	24337	14264	30185	36753	22343	81737	62926	76494	93536	75502
78	00718	66303	75009	91431	64245	61863	16738	23127	89435	45109
79	38093	10328	96998	91386	34967	40407	48380	09115	59367	49596
80	87661	31701	29974	56777	66751	35181	63887	95094	20056	84990
81	87142	91818	51857	85061	17890	39057	44506	00969	32942	54794
82	60634	27142	21199	50137	04685	70252	91453	75952	66753	50664
83	73356	64431	05068	56334	34487	78253	67684	69916	63885	88491
84	29889	11378	65915	66776	95034	81557	98035	16815	68432	63020
85	48257	36438	48479	72173	31418	14035	84239	02032	40409	11715
86	38425	29462	79880	45713	90049	01136	72426	25077	64361	94284
87	48226	31868	38629	12135	28346	17552	03293	42618	44151	78438
88	80189	30031	15435	76730	58565	29817	36775	64007	47912	16754
89	33208	33475	95219	29832	74569	50667	90569	66717	46958	04820
90	19750	48564	49690	43352	53884	80125	47795	99701	06800	22794
91	62820	23174	71124	36040	34873	95650	79059	23894	58534	78296
92	95737	34362	81520	79481	26442	37826	76886	01850	83713	94272
93	64642	62961	37566	41064	69372	84369	92823	91391	61056	44495
94	77636	60163	14915	50744	95611	99346	38741	04407	72940	87936
95	43633	52102	93561	31010	11299	52661	79014	17910	88492	60753
96	93686	41960	61280	96529	52924	87371	34855	67125	40279	10186
97	23775	33402	28647	42314	51213	29116	26243	40243	32137	25177
98	91325	64698	58868	63107	08993	96000	66854	11567	80604	72299
99	58129	44367	31924	73586	24422	92799	28963	36444	01315	10226

APPENDIX I RANDOM NUMBERS (Continued)

	50 54	55 59	60 64	65 69	70 74	75 79	80 84	85 89	90 94	95 99
50	37686	78520	31209	83677	99115	94024	09286	58927	24078	16770
51	58108	29344	11825	51955	50618	99753	02200	50503	32466	50055
52	71545	42326	66429	93607	55276	85482	24449	41764	19884	46443
53	93303	90557	79166	90097	01627	96690	77434	06402	05379	59549
54	36731	37929	13079	83036	31525	35811	59131	65257	03731	86703
55	49781	31581	80391	84608	23390	30433	08240	85136	80060	43651
56	65995	94208	68785	04370	44192	91852	01129	28739	08705	54538
57	19663	09309	02836	10223	90814	92786	96747	46014	54765	76001
58	88479	24307	63812	47615	17220	27942	11785	49933	03923	35432
59	95407	95006	95421	20811	76761	47475	58865	06204	36543	81002
60	22789	87011	61926	97996	10604	80855	48714	52754	98279	96467
61	96783	18403	36729	18760	30810	73087	94565	68682	15792	60020
62	68933	05665	12264	23954	01583	75411	04460	83939	66528	22576
63	68794	13000	20066	98963	93483	51165	63358	12343	13877	37580
64	40537	31604	60323	51235	65546	85117	15647	09617	73520	48525
65	41249	42504	91773	81579	02882	74657	73765	10932	74607	83825
66	08813	84525	30329	33144	76884	89996	07834	67266	96820	15128
67	46609	30917	29996	10848	39555	09233	58988	82131	69232	76762
68	68543	69424	92072	57937	05563	80727	67053	35431	00881	56541
69	09926	84249	30089	08843	24998	27105	18397	79071	40738	73876
70	30515	76316	49597	37000	98604	05857	51729	19006	15239	27129
71	21611	26346	04877	71584	55724	39616	64648	36811	60915	34108
72	47410	83767	56454	96768	27001	83712	01245	27256	57991	75758
73	18572	31214	41015	64110	61807	72472	78059	69701	78681	17356
74	28078	02819	02459	33308	96540	15817	78694	81476	87856	99737
75	56644	50430	34562	75842	67724	02918	55603	55195	88219	39676
76	27331	48055	18928	47763	61966	64507	06559	81329	29481	03660
77	32080	21524	32929	07739	00836	39497	94476	27433	96857	52987
78	27027	69762	65362	90214	89572	52054	43067	73017	87664	03293
79	56471	68839	09969	45853	72627	71793	49920	64544	71874	74053
80	22689	19799	18870	49272	74783	38777	76176	40961	18089	32499
81	71263	82247	66684	90239	67686	48963	30842	59354	33551	87966
82	64084	57386	89278	27187	52142	96305	87393	80164	95518	82742
83	23121	10194	09911	37062	43446	09107	47156	70179	00858	92326
84	78906	48080	76745	65814	51167	87755	66884	12718	14951	47937
85	87257	26005	21544	37223	53288	72056	96396	67099	49416	91891
86	39529	98126	33694	29025	94308	24426	63072	51444	04718	49891
87	89632	11606	87159	89408	06295	31055	15530	46432	49871	37982
88	23708	98919	14407	53722	58779	92849	94176	24870	56688	25405
89	51445	46758	42024	27940	64237	10086	95601	53923	85209	79385
90	23849	65272	24743	39960	27313	99925	29743	87270	05773	21797
91	78613	15441	34568	57398	25872	61792	94599	60944	90908	38948
92	90694	27996	94181	87428	41135	29461	72716	68056	67871	72459
93	86772	86829	36403	40087	67456	21071	39039	91037	45280	00066
94	24527	40701	56894	56894	00789	97573	09303	41704	05772	95372
95	31596	70876	46807	06741	29352	23829	52465	00336	24155	61871
96	31613	99249	17260	05242	19535	52702	64761	66694	06150	13820
97	02911	09514	50864	80622	20017	59019	43450	75942	08567	40547
98	02484	74068	04671	19646	41951	05111	34013	57443	87481	48994
99	69259	75535	73007	15236	01572	44870	53280	25132	70276	87334

INDEX

G

H

I

K

L

R

S

T